*S*pringer **M**onographs in **M**athematics

Springer
*New York
Berlin
Heidelberg
Barcelona
Hong Kong
London
Milan
Paris
Singapore
Tokyo*

Vladimir V. Andrievskii   Hans-Peter Blatt

Discrepancy of Signed Measures and Polynomial Approximation

 Springer

Vladimir V. Andrievskii
Department of Mathematics and Computer Science
Kent State University
Kent, OH 44242
USA
andriyev@mcs.kent.edu

Hans-Peter Blatt
Mathematisch-Geographische Fakultät
Katholische Universität Eichstätt
D-85072 Eichstätt
Germany
hans.blatt@ku-eichstaett.de

Mathematics Subject Classification (2000): 26C10, 31A15, 30E10, 42CO5, 4

Library of Congress Cataloging-in-Publication Data
Andrievskii, V. V., 1953–
  Discrepancy of signed measures and polynomial approximation / Vladimir V.
Andrievskii, Hans-Peter Blatt.
    p. cm.— (Springer monographs in mathematics)
  Includes bibliographical references and index.
  ISBN 0-387-98652-9 (alk. paper)
    1. Approximation theory. 2. Orthogonal polynomials. I. Blatt, Hans-Peter. II. Title. III.
Series.
  QA221 .A3567 2001
  511.4—dc21   2001032837

Printed on acid-free paper.

© 2002 Springer-Verlag New York, Inc.
All rights reserved. This work may not be translated or copied in whole or in part without the written permission of the publisher (Springer-Verlag New York, Inc., 175 Fifth Avenue, New York, NY 10010, USA), except for brief excerpts in connection with reviews or scholarly analysis. Use in connection with any form of information storage and retrieval, electronic adaptation, computer software, or by similar or dissimilar methodology now known or hereafter developed is forbidden.
The use of general descriptive names, trade names, trademarks, etc., in this publication, even if the former are not especially identified, is not to be taken as a sign that such names, as understood by the Trade Marks and Merchandise Marks Act, may accordingly be used freely by anyone.

Production managed by Steven Pisano; manufacturing supervised by Erica Bresler.
Photocomposed pages prepared from the authors' LaTeX files.
Printed and bound by Sheridan Books, Ann Arbor, MI
Printed in the United States of America.

9 8 7 6 5 4 3 2 1

ISBN 0-387-98652-9 Springer-Verlag New York Berlin Heidelberg   SPIN 10696722

To Elena
    Olga
    Sophie
    Sarah
    Simon
    Markus

Preface

In many situations in approximation theory the distribution of points in a given set is of interest. For example, the suitable choice of interpolation points is essential to obtain satisfactory estimates for the convergence of interpolating polynomials. Zeros of orthogonal polynomials are the nodes for Gauss quadrature formulas. Alternation points of the error curve characterize the best approximating polynomials.

In classical complex analysis an interesting feature is the location of zeros of approximants to an analytic function. In 1918 R. Jentzsch [91] showed that every point of the circle of convergence of a power series is a limit point of zeros of its partial sums. This theorem of Jentzsch was sharpened by Szegő [170] in 1923. He proved that for a power series with finite radius of convergence there is an infinite sequence of partial sums, the zeros of which are "equidistributed" with respect to the angular measure.

In 1929 Bernstein [27] stated the following theorem. Let f be a positive continuous function on $[-1, 1]$; if almost all zeros of the polynomials of best approximation to f (in a weighted L^2-norm) are outside of an open ellipse $\mathcal{E}$ with foci at -1 and 1, then f has a continuous extension that is analytic in $\mathcal{E}$.

Results extremely useful for the study of "equidistribution" were published by Erdős and Turán [55, 57], who established estimates for the distribution of zeros of a monic polynomial the uniform norm of which is known on the interval $[-1, 1]$ or on the unit disk.

From another point of view, the theorem of Jentzsch tells us that every boundary point of the domain of uniform convergence of the partial sums of a power series is a limit point of zeros of partial sums. This, of course, is not true for general sequences of polynomials. Ostrowski [136] and Szegő [169] have given supplementary sufficient conditions to ensure that the boundary points of the domain of uniform convergence for a sequence of polynomials are limit points of their zeros.

Walsh [181] considered, among other things, interpolation points on the boundary of a set E where the function f has to be approximated by polynomials. He discovered that the distribution of the interpolation points has to be related to the equilibrium distribution of E to obtain sufficiently good approximations.

In 1963 Kadec [92] proved the surprising result that the distribution of alternation points in Chebyshev polynomial approximation on a real interval is related to its equilibrium measure.

Each of the results just mentioned has been generalized in different directions and has become a source of many fruitful constructions in function theory. The crucial key to these generalizations can be found in a potential-theoretic interpretation of the classical results above. There is a huge literature dealing with this field. We cite here only the main results directly related to the subject under consideration. The topics of this book reflect the authors' personal mathematical interests and point of view on key ideas and results concerning Erdős–Turán and Jentzsch–Szegő type theorems and their applications.

The subjects and goals of the book are apparent from the table of contents. To our regret, it was not possible to touch questions related to the distribution of zeros, poles, and extreme points in complex rational approximation; generalizations of Ostrowski gaps in power series; Tsuji points; Leja points; zeros of polynomials of best approximation to finitely and infinitely differentiable functions; or to consider zeros and values of Faber polynomials.

At the end of the main chapters we have included a section entitled "Historical Comments," which contains citations for many of the theorems. It would be presumptuous to trace a complete history of the ideas presented in this book. Nevertheless, in general, the absence of a reference implies originality on our part, to our best knowledge.

Our main tool is the application of some basic notions and facts from potential theory and conformal invariants, such as the module of a family of curves. Moreover, some methods and results from the theory of quasiconformal mappings in the complex plane will play an important role, too.

Mario Götz and Jörg Hüsing have carefully studied the manuscript and have contributed various corrections and improvements to the presentation. We would like to extend to them our deepest gratitude. Further, we want to thank Wolfgang Schmidt for drawing the figures.

To D. Gaier, R. Grothmann, H.N. Mhaskar, Ch. Pommerenke, E.B. Saff, H. Stahl, and V. Totik we extend our sincere appreciation for their encouragement and helpful comments.

Thanks go to Katharina Surovcik for typing part of the manuscript, as well as to Markus and Simon for their great support to tie the figures into the text and to put together the final manuscript.

Last but not least we want to thank Ina Lindemann and the editorial staff at Springer-Verlag for their continuous support and unending patience.

Eichstätt, Kent *V. V. Andrievskii, H.-P. Blatt*

Contents

Preface   vii

1 Auxiliary Facts   1
  1.1 Basic Potential-Theoretical Concepts 1
      1.1.1 Sets and Curves 1
      1.1.2 Harmonic and Subharmonic Functions,
             Maximum Principle 2
      1.1.3 Green's Formula 6
      1.1.4 Harmonic Measure 8
      1.1.5 Energy, Capacity, Equilibrium Measure 10
      1.1.6 Maximum Principle 14
      1.1.7 Weak* Topology, Helly's Theorem 14
      1.1.8 Green's Function 14
      1.1.9 Fekete Points, Transfinite Diameter,
             Chebyshev Constant 17
  1.2 Conformal and Quasiconformal Mappings 19
      1.2.1 Basic Principles of Complex Analysis 19
      1.2.2 Equilibrium Measure and Conformal Mappings . . . 21
      1.2.3 Distortion Theorems 22
      1.2.4 Module of a Family of Curves and Arcs 24
      1.2.5 Quasiconformal Curves and Arcs 27
      1.2.6 Dini-Smooth Curves and Arcs 32
  1.3 Faber Polynomials and Grunsky Coefficients 37
  1.4 Monic Polynomials 40
  1.5 Linear Approximation 44
      1.5.1 Polynomial Approximation 45
      1.5.2 Averaging and Smoothing 46
  1.6 Ordering Symbols 48
  1.7 Historical Comments 48

2 Zero Distribution of Polynomials 49
- 2.1 Jentzsch–Szegő Type Theorems 50
- 2.2 Erdős–Turán Type Theorems for Quasiconformal Curves and Arcs 56
  - 2.2.1 Polynomial Bounds on Quasiconformal Curves . . . 57
  - 2.2.2 Polynomial Bounds on Quasiconformal Arcs 68
- 2.3 Sharpness and Extensions of Erdős–Turán Type Theorems . 72
- 2.4 Erdős–Turán Type Theorems on Compact Sets with Smooth Curves and Arcs 78
- 2.5 Erdős–Turán Type Theorems on a System of Intervals . . . 86
- 2.6 Historical Comments 92

3 Discrepancy Theorems via Two–Sided Bounds for Potentials 95
- 3.1 Estimates for Quasiconformal Curves and Arcs 96
- 3.2 Local Estimates for Intervals 108
- 3.3 Local Estimates for Dini–Smooth Arcs and Curves 114
- 3.4 Historical Comments 128

4 Discrepancy Theorems via One-Sided Bounds for Potentials 129
- 4.1 Outer Bounds for Potentials 130
- 4.2 Inner and Outer Bounds for Potentials 150
- 4.3 Inner Bounds for Potentials of Signed Measures on Analytic Jordan Curves 152
- 4.4 Another Approach for Dini-Smooth Arcs 161
- 4.5 Historical Comments 172

5 Discrepancy Theorems via Energy Integrals 173
- 5.1 Modulus of a Doubly Split Plane 174
- 5.2 Estimates of Mass Distributions from Their Energies 178
- 5.3 Historical Comments 186

6 Applications of Jentzsch–Szegő and Erdős–Turán Type Theorems 187
- 6.1 Polynomials of Best Uniform Approximation 188
- 6.2 Polynomials of Near-Best Approximation 193
- 6.3 Polynomials of Maximal Convergence 202
- 6.4 a-Values of Orthogonal Polynomials on Quasidisks 208
- 6.5 Zeros of Bieberbach Polynomials and Their Derivatives . . . 215
- 6.6 Historical Comments 223

7 Applications of Discrepancy Theorems 225
- 7.1 Distribution of Fekete Points 226
- 7.2 Fekete Points for Domains with Analytic Boundary 229

	7.3 Extreme Points in Uniform Polynomial Approximation	241
	7.4 Alternation Points in Chebyshev Approximation	248
	7.5 Sign Changes in L^p-Approximation	254
	7.5.1 The Case $1 < p < \infty$	256
	7.5.2 The Case $p = 1$	266
	7.6 Extremal Polynomials on the Real Line	274
	7.7 Zeros of Orthogonal Polynomials	297
	7.7.1 Orthogonal Polynomials on Real Intervals	297
	7.7.2 Orthogonal Polynomials on the Unit Circle	312
	7.8 Historical Comments	317
8	**Special Topics**	**321**
	8.1 Polynomial Approximation to a Piecewise Analytic Function on Touching Domains	321
	8.2 Uniform Convergence of Bieberbach Polynomials in Domains with Quasiconformal Boundary	336
	8.3 Historical Comments	340
A	**Conformally Invariant Characteristics of Curve Families**	**341**
	A.1 Module and Extremal Length of a Curve Family	341
	A.2 Reduced Module	357
B	**Basics in the Theory of Quasiconformal Mappings**	**361**
	B.1 Quasiconformal Mappings	361
	B.2 Quasiconformal Curves and Arcs	369
C	**Constructive Theory of Functions of a Complex Variable**	**381**
	C.1 Jackson Type Kernels	383
	C.2 Polynomial Kernels Approximating the Cauchy Kernel	386
	C.3 Inverse Theorems	393
	C.4 Polynomial Approximation in Domains with Smooth Boundary	403
D	**Miscellaneous Topics**	**407**
	D.1 The Regularized Distance	407
	D.2 Green's Function for a System of Intervals	409
	Bibliography	**415**
	Notation	**427**
	Index	**435**

1
Auxiliary Facts

The aim of this preliminary chapter is to recall some of the results and general principles of potential theory, geometric function theory, the theory of quasiconformal mappings, and approximation theory. In many cases the statements of theorems and lemmas are supported by proofs if they are not published in monographs or by exact reference to where they can be found. For detailed proofs of all other facts and further discussion we recommend standard introductory texts such as Tsuji [178], Saff and Totik [156] (potential theory), Goluzin [75], Pommerenke [146, 147], Duren [52] (geometric function theory), Ahlfors [3, 4], Lehto and Virtanen [112] (theory of quasiconformal mappings in the plane), Walsh [181], Smirnov and Lebedev [165], Gaier [62], Andrievskii, Belyi, and Dzjadyk [17] (approximation theory in the complex plane).

1.1 Basic Potential-Theoretical Concepts

1.1.1 Sets and Curves

In this book we shall use $\mathbb{C}$ for the complex plane and $\overline{\mathbb{C}} = \mathbb{C} \cup \{\infty\}$ for the extended complex plane, which is also referred to as the *Riemann sphere*.

We identify $\overline{\mathbb{C}}$ by the stereographic projection with a sphere in the 3-dimensional space $\mathbb{R}^3$. The Euclidean metric on this sphere, called *chordal metric*, induces the topology on $\overline{\mathbb{C}}$ and is equivalent on bounded sets in $\mathbb{C}$ to the standard Euclidean metric.

Moreover, we use

$$D(a, r) = \{z \in \mathbb{C} : |z - a| < r\} \quad (a \in \mathbb{C}, r > 0),$$
$$D(r) = D(0, r) = \{z \in \mathbb{C} : |z| < r\} \quad (r > 0),$$
$$\mathbb{D} = D(0, 1) = \{z \in \mathbb{C} : |z| < 1\} \quad \text{(unit disk)},$$
$$\mathbb{T} = \{z \in \mathbb{C} : |z| = 1\} = \partial \mathbb{D} \quad \text{(unit circle)},$$
$$\Delta = \{z \in \overline{\mathbb{C}} : |z| > 1\} \quad \text{(exterior of the unit circle)}.$$

A *domain* is a nonempty open connected set in $\overline{\mathbb{C}}$. A domain is said to be *simply connected* if its complement is connected. An (open) *neighborhood* of a set $E \in \mathbb{C}$ is an open set that contains E. The *closure* $\overline{E}$ of a set $E \in \mathbb{C}$ is the smallest closed set containing E. The *interior* E° of a set E is the largest open set contained in E. It may well happen that E° is empty. ∂E denotes the *boundary* of E.

For subsets $A, B \subset \mathbb{C}$ we define the *distance* $d(A, B)$ between A and B by

$$d(A, B) = \text{dist}(A, B) := \inf_{z \in A, \zeta \in B} |z - \zeta|.$$

A *Jordan curve* is a set that is homeomorphic to a circle, and a *Jordan arc* is the homeomorphic image of a *line segment*, i.e., a bounded connected subset of the real line $\mathbb{R}$. A Jordan arc is called *open* or *closed* depending on whether it is homeomorphic to an open or a closed segment. Often we will omit the word "Jordan" and write briefly *curve*, *arc*, *open* or *closed arc*.

A closed arc always has two *endpoints*, while an open arc γ has endpoints only when its closure $\overline{\gamma}$ is a closed arc. A closed arc has a connected complement (with respect to $\overline{\mathbb{C}}$), while the Jordan curve theorem (see [133, p. 115]) asserts that every curve L divides the plane into two regions, the *interior* int L and the *exterior* ext L of the curve L. The interior or exterior of a Jordan curve is called a *Jordan domain*.

A *crosscut* γ of a domain $G \subset \overline{\mathbb{C}}$ is an open Jordan arc in G such that its closure $\overline{\gamma} = \gamma \cup \{a, b\}$ with $a, b \in \partial G$; we allow that $a = b$. The points a and b are called endpoints of the crosscut γ.

A curve or arc is *smooth* if it has a continuously varying tangent (for details, see [147, p. 43]).

1.1.2 Harmonic and Subharmonic Functions, Maximum Principle

We denote by

$$\Delta := \frac{\partial^2}{\partial x^2} + \frac{\partial^2}{\partial y^2}, \quad z = x + iy,$$

the *Laplacian operator* (or briefly *Laplacian*) in $\mathbb{R}^2$. There is no confusion with the same notation for the exterior of the unit disk, since the Laplacian operator is always used in connection with a function.

Let the real-valued function $u(z) = u(x,y)$ be defined in a domain $G \subset \overline{\mathbb{C}}$. Suppose its partial derivatives up to the second order are continuous in $G \setminus \{\infty\}$, satisfy *Laplace's equation*

$$\Delta u(z) = 0, \quad z \in G \setminus \{\infty\}, \tag{1.1}$$

and if $\infty \in G$, suppose u is bounded in a punctured neighborhood of ∞. Then u is said to be *harmonic* in G. We note: If u is harmonic and bounded in some punctured disk with center z_0 then u can be defined at z_0 so that u is harmonic at z_0 (cf. Ahlfors [5]).

If $f = u + iv$ is analytic in G, then both $u = \operatorname{Re} f$ and $v = \operatorname{Im} f$, are harmonic in G. This well-known fact follows immediately from the *Cauchy–Riemann equations*

$$\frac{\partial u}{\partial x} = \frac{\partial v}{\partial y}, \quad \frac{\partial v}{\partial x} = -\frac{\partial u}{\partial y}. \tag{1.2}$$

In the converse direction we have the following result.

Theorem 1.1. *A function u is harmonic in a simply connected domain G if and only if there exists a (single-valued) function f analytic in G such that $u = \operatorname{Re} f$ in G.*

For a proof, see [156, p. 6].

If $f = u + iv$ is analytic in G, then v is called the *harmonic conjugate* of u, and f is an *analytic completion* of u. If u is given, then the functions v and f (if they exist) are unique up to an additive constant.

Harmonic functions satisfy the so-called *mean-value property*: If u is harmonic in a neighborhood of the closed disk $\overline{D(z,r)}$, then

$$u(z) = \frac{1}{2\pi} \int_0^{2\pi} u\left(z + re^{i\theta}\right) d\theta. \tag{1.3}$$

Using polar coordinates we obtain from (1.3)

$$u(z) = \frac{1}{\pi r^2} \int_{D(z,r)} u \, dm. \tag{1.4}$$

One of the important consequences of the mean-value property is the *maximum principle* for harmonic functions:

Theorem 1.2. *If u is harmonic in a domain $G \subset \overline{\mathbb{C}}$ and attains its maximum or minimum value in G, then u is constant.*

For a proof, see [5, p. 166].

An important tool in the theory of harmonic functions is Poisson's integral. Let $L^1(\mathbb{T})$ denote the collection of functions that are integrable in Lebesgue's sense. Then, for $g \in L^1(\mathbb{T})$ *Poisson's integral* is defined by

$$u(z) = u(re^{it}) = \frac{1}{2\pi} \int_0^{2\pi} P(r, t-\theta) g(\zeta) d\theta, \quad \zeta = e^{i\theta} \in \mathbb{T}, \quad (1.5)$$

where

$$P(r, \theta) = \frac{1-r^2}{1 - 2r\cos\theta + r^2}, \quad 0 < r < 1, \quad \theta \in [0, 2\pi],$$

is the *Poisson kernel*. The function $u(z)$ in (1.5) is harmonic for $z \in \mathbb{D}$. This can be seen easily, since Poisson's kernel can be written as

$$P(r, t-\theta) = \frac{1-|z|^2}{|\zeta-z|^2} = \operatorname{Re} \frac{\zeta+z}{\zeta-z}.$$

Hence, (1.5) leads to an explicit expression for any function $f(z)$, analytic in $\mathbb{D}$, if $u = \operatorname{Re} f$ has a continuous extension to $\overline{\mathbb{D}}$, namely

$$f(z) = \frac{1}{2\pi} \int_0^{2\pi} \frac{\zeta+z}{\zeta-z} u(\zeta) d\theta + i \operatorname{Im} f(0)$$

$$= \frac{1}{2\pi i} \int_{\mathbb{T}} \frac{\zeta+z}{\zeta-z} u(\zeta) \frac{d\zeta}{\zeta} + i \operatorname{Im} f(0), \quad z \in \mathbb{D}, \quad (1.6)$$

which is known as *Schwarz's formula*. Differentiating (1.6) we obtain

$$f'(z) = \frac{1}{\pi i} \int_{\mathbb{T}} \frac{u(\zeta)}{(\zeta-z)^2} d\zeta, \quad z \in \mathbb{D}, \quad (1.7)$$

where $\mathbb{T}$ is positively oriented.

The question of boundary values was settled by H.A. Schwarz for continuous g. Precisely, in (1.5)

$$\lim_{z \to e^{i\theta_0}} u(z) = g(e^{i\theta_0})$$

provided that $g(e^{i\theta})$ is continuous at θ_0.

We will often use the following fact. Let $f = u + iv$ be analytic in the domain G. Then by the Cauchy–Riemann equations (1.2)

$$|\operatorname{grad} u(z)| = \left(\left|\frac{\partial u(z)}{\partial x}\right|^2 + \left|\frac{\partial u(z)}{\partial y}\right|^2 \right)^{\frac{1}{2}} = |f'(z)|, \quad z = x + iy \in G.$$

$$(1.8)$$

Suppose g is a given real-valued continuous function on $\mathbb{T}$. By the same symbol we denote the function that is continuous on $\overline{\mathbb{C}}$, harmonic in $\mathbb{D} \cup \Delta$, and coincides with g on $\mathbb{T}$. It is obvious that the continuity properties of the function g in a neighborhood of $\mathbb{T}$ depend on the values of g on $\mathbb{T}$.

The following result is stated for the unit disk. For the exterior Δ of $\mathbb{T}$, the corresponding result is quite the same, and we do not dwell on it.

Lemma 1.3. *Let* $g \in L^1(\mathbb{T})$, $e^{i\theta_0} \in \mathbb{T}$ *fixed,* $0 < \delta \leq \varepsilon < 1$, $M > 0$, *and let* g *satisfy the inequalities*

$$\left|g\left(e^{i\theta}\right) - g\left(e^{i\theta_0}\right)\right| \leq M(|\theta - \theta_0| + \delta), \quad |\theta - \theta_0| \leq \varepsilon,$$

and

$$\left|g\left(e^{i\theta}\right)\right| \leq M.$$

Then for $0 < 1 - r < \delta$

$$\left|g\left(re^{i\theta_0}\right) - g\left(e^{i\theta_0}\right)\right| \leq cM\delta \log \frac{1}{\delta}$$

with some constant $c = c(\varepsilon) > 0$.

Proof. We may assume that $\theta_0 = 0$, i.e., $e^{i\theta_0} = 1$. By Poisson's integral formula,

$$|g(r) - g(1)| \leq \frac{1 - r^2}{2\pi} \int_{-\pi}^{\pi} \frac{|g(e^{i\theta}) - g(1)|}{1 - 2r\cos\theta + r^2} d\theta.$$

Note that

$$1 - 2r\cos\theta + r^2 = (1-r)^2 + 4r\sin^2\frac{\theta}{2}.$$

We subdivide the integral

$$\frac{1 - r^2}{2\pi} \int_{-\pi}^{\pi} \frac{|g(e^{i\theta}) - g(1)|}{1 - 2r\cos\theta + r^2} d\theta = \int_{|\theta|\leq\delta} + \int_{\delta\leq|\theta|\leq\varepsilon} + \int_{\varepsilon\leq|\theta|\leq\pi} = I_1 + I_2 + I_3$$

into three integrals and obtain

$$I_1 \leq 4M\delta^2 \frac{1-r^2}{2\pi} \int_{-\pi}^{\pi} \frac{1}{1-2r\cos\theta + r^2} d\theta = 4M\delta^2,$$

$$I_2 \leq \frac{\pi M}{2(1-\varepsilon)} \delta \int_{\delta}^{\varepsilon} \frac{\theta + \delta}{\theta^2} d\theta \leq \frac{\pi M}{2(1-\varepsilon)} \delta \left(\log\frac{1}{\delta} + 1\right),$$

$$I_3 \leq \frac{\pi M}{1-\varepsilon} \delta \int_{\varepsilon}^{\pi} \frac{1}{\theta^2} d\theta \leq \frac{\pi M}{\varepsilon(1-\varepsilon)} \delta.$$

Summarizing, we have

$$|g(r) - g(1)| \le c(\varepsilon)\,\delta\,\log\frac{1}{\delta}.$$

□

Let $K \subset \overline{\mathbb{C}}$ and let u be an extended (i.e., values $\pm\infty$ are possible) real-valued function on K. Then u is said to be *upper semicontinuous* if $-\infty \le u < +\infty$ on K and if for every $z \in K$ and $a > u(z)$ there exists a neighborhood V of z in K such that for every $\zeta \in V$ we have $u(\zeta) < a$.

If $-u$ is upper semi-continuous, then u is said to be *lower semi-continuous*.

Let $G \subset \mathbb{C}$ be a domain. An extended real-valued function u on G is called *subharmonic* on G if it satisfies the following three conditions:

(i) u is upper semi-continuous in G.

(ii) $u(z) \le \dfrac{1}{2\pi}\displaystyle\int_0^{2\pi} u(z + re^{i\theta})\,d\theta$ for all $z \in G$ and $r > 0$ such that $\overline{D(z,r)} \subset G$.

(iii) None of the integrals in (ii) is equal to $-\infty$.

A function u is called *superharmonic* in G if $-u$ is subharmonic in G.

Note that the integrals in (ii) always exist and are not equal to $+\infty$, since (i) and (ii) imply that u is bounded on every compact set in G. Hence (ii) says that the integrands belong to $L^1(\mathbb{T})$.

A rich supply of subharmonic functions is provided by the following example. If f is analytic in a domain $G \subset \mathbb{C}$ and $\alpha > 0$, then $|f(z)|^\alpha$ is subharmonic; furthermore, $\log|f(z)|$ is subharmonic in G, provided that f is not identically zero (see [156, p. 20]).

We observe that a function u is harmonic in G if and only if it is subharmonic and superharmonic in G.

1.1.3 Green's Formula

Let $G \subset \mathbb{C}$ be a bounded domain whose boundary consists of a finite number of smooth curves, and let u and v be twice continuously differentiable functions defined in a neighborhood of $\overline{G}$. Then

$$\int_G (v\Delta u - u\Delta v)\,dm = \int_{\partial G}\left(v\,\frac{\partial u}{\partial \mathbf{n}} - u\,\frac{\partial v}{\partial \mathbf{n}}\right)|dz| \qquad (1.9)$$

is called *Green's formula*. In (1.9) $\partial/\partial \mathbf{n} = \partial/\partial \mathbf{n}_z$ denotes differentiation in the direction of the outward normal of ∂G at z, $|dz|$ indicates integration with respect to the arc length on ∂G, dm means integration with respect to the two-dimensional Lebesgue measure (area), and Δ is the Laplacian operator.

The same name, Green's formula, has the following integral representation of a function u (the assumptions on u and G are the same as in (1.9)):

$$u(z_0) = \frac{1}{2\pi} \int_{\partial G} \left(u \frac{\partial}{\partial \mathbf{n}} \log|z - z_0| - \frac{\partial u}{\partial \mathbf{n}} \log|z - z_0| \right) |dz|$$

$$+ \frac{1}{2\pi} \int_G \Delta u \log|z - z_0| \, dm, \quad z_0 \in G. \tag{1.10}$$

If u is harmonic in G, then (1.10) reduces to

$$u(z_0) = \frac{1}{2\pi} \int_{\partial G} \left(u \frac{\partial}{\partial \mathbf{n}} \log|z - z_0| - \frac{\partial u}{\partial \mathbf{n}} \log|z - z_0| \right) |dz|.$$

In order to verify (1.10) we take $D(z_0, \rho)$ and ρ so small that $\partial G \cap D(z_0, \rho) = \emptyset$. Applying (1.9) to the region bounded by $\partial G \cup D(z_0, \rho)$ and to the functions $u(z)$ and $v(z) = \log|z - z_0|$ we obtain formula (1.10) after passage to the limit as $\rho \to 0$.

If u is a twice continuously differentiable function with compact support in $\mathbb{C}$, then by (1.10) we have for $z_0 \in \mathbb{C}$ and $z = x + iy \in \mathbb{C}$,

$$u(z_0) = \frac{1}{2\pi} \int_{\mathbb{C}} \Delta u \log|z - z_0| dm$$

$$= -\frac{1}{2\pi} \int_{\mathbb{C}} \left(\frac{\partial u}{\partial x} \frac{\partial}{\partial x} \log|z - z_0| + \frac{\partial u}{\partial y} \frac{\partial}{\partial y} \log|z - z_0| \right) dm. \tag{1.11}$$

Note two consequences of Green's formula for harmonic functions. Let $G \subset \overline{\mathbb{C}}$ be a domain bounded by a finite number of smooth curves, and let u be harmonic in a neighborhood of $\overline{G}$. If $\infty \notin G$, then taking $v \equiv 1$ in (1.9) we have

$$\int_{\partial G} \frac{\partial u}{\partial \mathbf{n}} |dz| = 0. \tag{1.12}$$

If $\infty \in G$, then for any $z_0 \notin \overline{G}$,

$$u(\infty) = \frac{1}{2\pi} \int_{\partial G} \left(\frac{\partial u}{\partial \mathbf{n}} \log|z - z_0| - u \frac{\partial}{\partial \mathbf{n}} \log|z - z_0| \right) |dz|. \tag{1.13}$$

Indeed, taking $R > 0$ so large that $\partial G \subset D(z_0, R)$ and applying (1.9) in $G \cap D(z_0, R)$ to the functions $u(z)$ and $v(z) = \log|z - z_0|$ we get

$$0 = \int_{\partial G} \left(u \frac{\partial}{\partial \mathbf{n}} \log|z - z_0| - \frac{\partial u}{\partial \mathbf{n}} \log|z - z_0| \right) |dz|$$

$$+ \int_0^{2\pi} u\left(z_0 + Re^{i\theta}\right) d\theta - R \log R \int_0^{2\pi} \frac{\partial u\left(z_0 + Re^{i\theta}\right)}{\partial R} d\theta. \quad (1.14)$$

Carrying out the substitution $z_0 + Re^{i\theta} \mapsto \xi^{-1}$ we see that by (1.12) the second integral on the right-hand side of (1.14) vanishes, and by the mean-value property the third integral tends to $2\pi u(\infty)$ as $R \to \infty$.

1.1.4 Harmonic Measure

The notion of harmonic measure proves extremely useful for some of our constructions. We need it in a relatively simple form that can be described as follows (for details, see [4, Chapter 3] or [132, Chapter III, Section 1]).

Let $G \subset \overline{\mathbb{C}}$ be a domain whose boundary $L := \partial G$ consists of a finite number of disjoint curves. Let $J \subset L$ consist of a finite number of arcs and curves; it is not important whether the endpoints of the arcs are included or not. There exists a unique bounded harmonic function $\omega(z) = \omega(z, G, J)$ in G such that $\omega(z) \to 1$ if z tends to an interior point of J and $\omega(z) \to 0$ as z tends to an interior point of $L \setminus J$. The number $\omega(z)$ is called the *harmonic measure* of J at the point z with respect to the domain G.

According to the generalized maximum principle (cf. Theorem 1.9 below)

$$0 \leq \omega(z, G, L) \leq 1, \quad z \in G. \quad (1.15)$$

The following assertion is also a simple consequence of the generalized maximum principle (see [132, p. 69] or [2, p. 39]).

Theorem 1.4. *Let $G_1, G_2 \subset \overline{\mathbb{C}}$ be domains whose boundaries consist of a finite number of disjoint curves. If*

$$G_1 \subset G_2, \quad J_1 \subset J_2 \subset \partial G_1 \cap \partial G_2,$$

where $J_j, j = 1, 2$, consist of a finite number of arcs and curves, then

$$\omega(z, G_1, J_1) \leq \omega(z, G_2, J_2), \quad z \in G_1.$$

If G coincides with the unit disk $\mathbb{D}$, then the harmonic measure of any subarc $J \subset \mathbb{T}$ of the unit circle $\mathbb{T}$ can be written in the form of the Poisson integral of the characteristic function of J,

$$\omega(z, \mathbb{D}, J) = \frac{1}{2\pi} \int_J \frac{1 - |z|^2}{|\zeta - z|^2} |d\zeta|, \quad z \in \mathbb{D}. \quad (1.16)$$

In particular,
$$\omega(0, \mathbb{D}, J) = \frac{1}{2\pi} \int_J |d\zeta| = \frac{1}{2\pi}|J|, \qquad (1.17)$$
where $|\gamma|$ denotes the length of a rectifiable arc or curve $\gamma \subset \mathbb{C}$.

We note that an analytic completion of $\omega(z, \mathbb{D}, J)$ can be written in the form (1.6) where g is the characteristic function of J. Therefore, (1.7) and (1.8) yield
$$|\operatorname{grad} \omega(z, \mathbb{D}, J)| \le \frac{1}{\pi} \int_J \frac{|d\zeta|}{|\zeta - z|^2}, \qquad z \in \mathbb{D}. \qquad (1.18)$$

The next lemma is an immediate consequence of Poisson's and Schwarz's representation formulas.

Lemma 1.5. *Let z_1 and z_2 be the endpoints of an arc $l \subset \mathbb{T}$. Then for any $z \in \mathbb{D}$ the inequalities*
$$\omega(z, \mathbb{D}, l) \le 8 \frac{1 - |z|}{d(z, l)}, \qquad (1.19)$$
$$|\operatorname{grad} \omega(z, \mathbb{D}, l)| \le 8 \left(\min_{j=1,2} |z - z_j| \right)^{-1} \qquad (1.20)$$
hold.

Proof. We begin with the proof of (1.19). Using rotation with respect to the origin we can reduce the situation to the case when
$$0 < z < 1, \quad z_j = e^{i\theta_j}, \quad 0 < \theta_1 < \theta_2 < 2\pi + \theta_1, \quad l = \left\{ e^{i\theta} : \theta_1 \le \theta \le \theta_2 \right\}.$$
Moreover, we can assume that $\theta_2 < 2\pi$, since otherwise, $d(z, l) = 1 - |z|$ and (1.19) is trivially valid. Set
$$l_1 := \{\zeta \in l : \operatorname{Im} \zeta \ge 0\}, \quad l_2 := l \setminus l_1.$$
We assume that $l_1 \ne \emptyset$. A simple geometric argument shows that for $\zeta = e^{i\theta} \in l_1$,
$$|\zeta - z| \ge \frac{1}{\pi}(\theta - \theta_1), \quad |\zeta - z| \ge |z - z_1|.$$
Therefore, by (1.16),
$$\omega(z, \mathbb{D}, l_1) = \frac{1 - |z|^2}{2\pi} \int_{l_1} \frac{|d\zeta|}{|\zeta - z|^2} \le \frac{1}{\pi}(1 - |z|) \int_{l_1} \frac{|d\zeta|}{|\zeta - z|^2}$$
$$\le 4\pi(1 - |z|) \int_{\theta_1}^{\pi} \frac{d\theta}{(\pi|z - z_1| + \theta - \theta_1)^2}$$
$$\le \frac{4(1 - |z|)}{|z - z_1|} \le \frac{4(1 - |z|)}{d(z, l)}.$$

Writing the same estimate for $\omega(z, \mathbb{D}, l_2)$ and taking their sum we get (1.19).

In order to prove (1.20) let us first assume that $d(z, l) > 1 - |z|$. Then applying (1.18) and reasoning as above we obtain

$$|\operatorname{grad} \omega(z, \mathbb{D}, l)| \leq \frac{1}{\pi} \int_l \frac{|d\zeta|}{|\zeta - z|^2} \leq \frac{8}{d(z, l)}.$$

Since $\omega(z, \mathbb{D}, l) + \omega(z, \mathbb{D}, \mathbb{T} \setminus l) \equiv 1$, we have

$$|\operatorname{grad} \omega(z, \mathbb{D}, l)| = |\operatorname{grad} \omega(z, \mathbb{D}, \mathbb{T} \setminus l)|$$

and

$$|\operatorname{grad} \omega(z, \mathbb{D}, l)| \leq \frac{8}{\max(d(z, l), d(z, \mathbb{T} \setminus l))} \leq \frac{8}{\min_{j=1,2} |z_j - z|}.$$

and (1.20) is proved. □

The harmonic measure is one of the most important conformal invariants. Namely, let $G \subset \overline{\mathbb{C}}$ be a Jordan domain and let φ map G conformally onto $\mathbb{D}$. By Carathéodory's theorem (cf. section 2.1) φ has a continuous injective (i.e., one-to-one) extension to $\overline{G}$. Since harmonic functions are invariant under conformal mappings, we have

$$\omega(z, G, J) = \omega(\varphi(z), \mathbb{D}, \varphi(J)). \tag{1.21}$$

1.1.5 Energy, Capacity, Equilibrium Measure

We are going to introduce some terminology from potential theory. Recall that the *support* of a positive measure σ, denoted by $\operatorname{supp}(\sigma)$, consists of all points z such that $\sigma(D(z, r)) > 0$ for every disk $D(z, r)$. Let $E \subset \mathbb{C}$ be compact and let $\mathcal{M}(E)$ denote the collection of all positive unit Borel measures supported on E. For $\sigma \in \mathcal{M}(E)$ the *energy* of σ is defined by

$$I[\sigma] := \int U^\sigma(z) \, d\sigma(z),$$

where

$$U^\sigma(z) := \int \log \frac{1}{|\zeta - z|} \, d\sigma(\zeta), \quad z \in \mathbb{C},$$

is the *logarithmic potential* of σ.

The function U^σ is lower semi-continuous. Moreover, U^σ is superharmonic in $\mathbb{C}$ and harmonic in $\Omega \setminus \{\infty\}$, where $\Omega := \overline{\mathbb{C}} \setminus E$ (see [156, p. 21]). If

$$V(E) := \inf_{\sigma \in \mathcal{M}(E)} I[\sigma],$$

1.1 Basic Potential-Theoretical Concepts

then the (logarithmic) *capacity* of E is defined by

$$\operatorname{cap} E := \exp\{-V(E)\}.$$

Hence, if $V(E) < \infty$, then $\operatorname{cap} E > 0$, and if $V(E) = \infty$, then $\operatorname{cap} E = 0$.
The capacity of any set $E \subset \mathbb{C}$ is defined as

$$\operatorname{cap} E := \sup\{\operatorname{cap} K : K \subset E, K \text{ compact}\}.$$

A property is said to hold *quasi-everywhere (q.e.)* on a set E if the set of exceptional points is of capacity zero.

If $\operatorname{cap} E > 0$, then there exists a unique measure $\mu_E \in \mathcal{M}(E)$ such that

$$I[\mu_E] = V(E).$$

The measure μ_E is called the *equilibrium measure (distribution)* of E.

The behavior of $U^{\mu_E}(z)$ is described by the *fundamental theorem of Frostman*:

Theorem 1.6. *Let E be compact and $\operatorname{cap} E > 0$. Then the following properties hold:*

(i) $U^{\mu_E}(z) \leq V(E)$ *for all* $z \in \overline{\mathbb{C}}$.

(ii) $U^{\mu_E}(z) = V(E)$ *quasi-everywhere on E.*

For a proof see [178, p. 60].

The following theorem provides comparisons between $U^{\mu_E}(z)$ and any $U^{\nu}(z)$, $\nu \in \mathcal{M}(E)$.

Theorem 1.7. *Let E be compact, $\operatorname{cap}(E) > 0$, and $\nu \in \mathcal{M}(E)$. Then*

$$\inf_{z \in E} U^{\nu}(z) \leq V(E) \leq \sup_{z \in E} U^{\nu}(z).$$

For a proof see [178, p. 62].

The combination of Theorem 1.7 with the following lemma provides useful methods to determine the equilibrium distribution for special cases.

Lemma 1.8. *Let $\operatorname{cap} E > 0$ and $\nu \in \mathcal{M}(\partial E)$ be such that*

$$U^{\nu}(z) \leq U^{\mu_E}(z), \quad z \in \Omega. \tag{1.22}$$

Then $\nu = \mu_E$.

Proof. Since

$$U^{\mu_E}(z) \leq V(E), \quad z \in \mathbb{C},$$

and since $U^{\nu}(z)$ is lower semi-continuous, we obtain from (1.22) that

$$U^{\nu}(z) \leq V(E), \quad z \in \partial E.$$

Integrating this inequality with respect to ν we get
$$I[\nu] \leq V(E).$$
Hence, by the uniqueness of the solution of the minimum energy problem, the assertion $\nu = \mu_E$ follows. □

In the main parts of the book the most general compact sets that we are going to consider are the admissible sets. Precisely, $E \subset \mathbb{C}$ is called *admissible* if E is compact, cap $E > 0$, and the complement $\Omega = \overline{\mathbb{C}} \setminus E$ is connected.

As simple applications of the preceding results, let us first consider the unit circle $E = \mathbb{T}$. Set $d\sigma = ds/2\pi$, where ds denotes the arc length on $\mathbb{T}$. Then for $|z| > 1$ the function $\log(1/|z-\zeta|)$ is harmonic for $|\zeta| \leq 1$, and the mean-value property yields
$$U^\sigma(z) = \log \frac{1}{|z|} \quad \text{for } |z| > 1.$$

For $|z| = 1$, the definition of the logarithmic potential immediately yields $U^\sigma(z) = 0$. For $|z| < 1$, the function
$$\log \frac{1}{|z-\zeta|} - \log \frac{1}{|z|}$$
is harmonic for $|\zeta| \geq 1$. Hence, again the mean-value property yields
$$U^\sigma(z) = 0 \quad \text{for } |z| < 1.$$

Summarizing,
$$U^\sigma(z) = \begin{cases} 0 & \text{for } |z| \leq 1, \\ \log \dfrac{1}{|z|} & \text{for } |z| \geq 1. \end{cases}$$

Hence, U^σ is constant on $\mathbb{T}$. By Theorem 1.7, cap $\mathbb{T} = 1$, and by Theorem 1.6 and Lemma 1.8, $d\mu_E = ds/2\pi$ and cap $\mathbb{T} = $ cap $\overline{\mathbb{D}}$.

From the definition, we have
$$\operatorname{cap} T(E) = r \operatorname{cap} E$$
for any linear transformation $T(z) = rz + a$, and therefore for $r > 0$ the measure $d\mu = ds/2\pi r$ is the equilibrium measure of $\overline{D(a,r)}$, where ds again denotes the arc length measure on the circle $|z - a| = r$ and
$$U^\mu(z) = \begin{cases} \log \dfrac{1}{r} & \text{for } |z - a| \leq r, \\ \log \dfrac{1}{|z-a|} & \text{for } |z - a| \geq r. \end{cases}$$

Moreover, cap $\overline{D(a,r)} = r$.

Let $E = [-1, 1]$. Then the Joukowski map $z = \Psi(\zeta) = \frac{1}{2}(\zeta + \zeta^{-1})$ maps the exterior of the unit circle conformally to $\overline{\mathbb{C}} \setminus [-1, 1]$, and Ψ can be continuously extended to the unit circle $\mathbb{T}$. Let $\widetilde{\sigma}$ be the measure supported on $[-1, 1]$ and defined by

$$d\widetilde{\sigma}(x) := \frac{1}{\pi} \frac{dx}{\sqrt{1-x^2}}, \qquad -1 \leq x \leq 1.$$

For $z = \Psi(\zeta) \in \mathbb{C} \setminus [-1, 1]$, its logarithmic potential can be written as

$$U^{\widetilde{\sigma}}(z) = \frac{1}{2\pi} \int_{-\pi}^{\pi} \log \frac{1}{|\Psi(t) - \Psi(\zeta)|} d\theta$$

$$= \frac{1}{2\pi} \int_{-\pi}^{\pi} \log \left| \frac{t - \zeta}{\Psi(t) - \Psi(\zeta)} \right| d\theta + \frac{1}{2\pi} \int_{-\pi}^{\pi} \log \frac{1}{|t - \zeta|} d\theta,$$

where $t = e^{i\theta}$. From the previous arguments for $E = \mathbb{T}$ we know that

$$\frac{1}{2\pi} \int_{-\pi}^{\pi} \log \frac{1}{|t - \zeta|} d\theta = \log \frac{1}{|\zeta|} = \log \frac{1}{|\Psi^{-1}(z)|}.$$

The function

$$h(w) = \log \left| \frac{w - \zeta}{\Psi(w) - \Psi(\zeta)} \right|$$

is harmonic in Δ and continuous on the boundary. Hence, the mean-value property yields

$$\frac{1}{2\pi} \int_{-\pi}^{\pi} \log \left| \frac{t - \zeta}{\Psi(t) - \Psi(\zeta)} \right| d\theta = h(\infty) = \log 2,$$

and therefore

$$U^{\widetilde{\sigma}}(z) = -\log |\Psi^{-1}(z)| - \log \frac{1}{2}, \quad z \in \Omega = \overline{\mathbb{C}} \setminus [-1, 1].$$

Since $\widetilde{\sigma}$ is supported on $[-1, 1]$ and $U^{\widetilde{\sigma}}$ is lower semi-continuous, we have

$$U^{\widetilde{\sigma}}(z) \leq -\log \frac{1}{2} \qquad \text{for } z \in [-1, 1],$$

and moreover,

$$\liminf_{z \to [-1,1],\, z \in \Omega} U^{\widetilde{\sigma}}(z) = -\log \frac{1}{2}.$$

Hence, Theorem 1.7 yields

$$\operatorname{cap}[-1,1] = \frac{1}{2},$$

and by Lemma 1.8,

$$\mu_{[-1,1]} = \widetilde{\sigma}.$$

1.1.6 Maximum Principle

We formulate the so-called *generalized maximum principle* for subharmonic functions, which will be used frequently throughout the book.

Theorem 1.9. *Let $G \subset \overline{\mathbb{C}}$ be a domain and u a subharmonic function on G that is bounded from above and for which*

$$\limsup_{\zeta \to z,\, \zeta \in G} u(z) \leq m$$

is satisfied for quasi-every $z \in \partial G$. Then

$$u(z) < m, \quad z \in G,$$

unless u is constant.

For a proof, see [156, pp. 39–40].

1.1.7 Weak* Topology, Helly's Theorem

We say that a sequence of Borel measures $\{\mu_n\}$ on a compact set E of $\overline{\mathbb{C}}$ converges to a measure μ as $n \to \infty$ *weakly* or *weak**, and denote this convergence by the symbol $\mu_n \xrightarrow{*} \mu$, if

$$\lim_{n \to \infty} \int f\, d\mu_n = \int f\, d\mu \qquad (1.23)$$

for every function f continuous in E.

The following particular case of *Helly's selection theorem* will frequently be used.

Theorem 1.10. *If E is compact in $\overline{\mathbb{C}}$, then from each sequence $\{\mu_n\} \subset \mathcal{M}(E)$ it is possible to select a weak* convergent subsequence.*

For a proof, see [156, p. 4].

1.1.8 Green's Function

Next, we are going to introduce one of the central notions of potential theory: the Green function. We restrict ourselves to the case of admissible sets E. Then the Green function $G(z) = G(z, \infty) = G_\Omega(z, \infty)$ of Ω with pole at ∞ is defined as the unique function with the following properties:

(i) $G(z)$ is nonnegative and harmonic in $\Omega \setminus \{\infty\}$.

(ii) $\lim_{z \to \infty} (G(z) - \log|z|) = -\log \operatorname{cap} E$.

(iii) $\lim_{z \to \zeta,\, z \in \Omega} G(z) = 0$ for quasi-every $\zeta \in \partial\Omega$.

It is a well-known fact (see [156, p. 53]) that $G(z)$ is connected with the logarithmic potential of μ_E by

$$U^{\mu_E}(z) = -G(z) - \log \operatorname{cap} E, \quad z \in \Omega. \tag{1.24}$$

The notion of the Green function $G(z, a)$ with pole at some finite point $a \in \Omega$ is similar: The Green function $G(z, a) = G_\Omega(z, a)$ of Ω with pole at a is defined as the unique function on Ω satisfying the following properties (see [156, p. 109]):

(i) $G(z, a)$ is nonnegative and harmonic in $\Omega \setminus \{a\}$.

(ii) $G(z, a) + \log |z - a|$ is bounded in a punctured neighborhood of a.

(iii) $\lim_{z \to \zeta,\, z \in \Omega} G(z, a) = 0$ for quasi-every $\zeta \in \partial \Omega$.

The Green function is a symmetric function in its arguments:

$$G(z, \zeta) = G(\zeta, z), \quad z, \zeta \in \Omega, \quad z \neq \zeta, \tag{1.25}$$

(see [156, p. 119] or [83, p. 255]).

We say that E has a *regular boundary* (with respect to Ω), or Ω is a *regular domain*, if $G(z)$ extends continuously to the boundary of E. Then $G(z) = 0$ for $z \in \partial\Omega$.

Each *continuum* E, i.e., each compact connected set in $\mathbb{C}$ containing more than one point, with connected complement has a regular boundary.

Indeed, by the Riemann mapping theorem (see Section 1.2) there exists a unique conformal mapping Φ of Ω onto Δ normalized by $\Phi = \infty$, $\Phi'(\infty) \geq 0$. It follows from the definition of Green's function that

$$G(z) = \log |\Phi(z)|, \quad z \in \Omega.$$

Since $|\Phi(z)| \to 1$ as $z \to \partial\Omega$, we conclude that Ω is regular.

In general, a celebrated theorem of Wiener (see [156, p. 54]) or [178, p. 104]) characterizes compact sets with regular boundary.

The regularity of Ω plays an important role in solving Dirichlet's problem in Ω with prescribed continuous boundary function on $\partial\Omega$. In fact, in this case the solution of the Dirichlet problem is continuous in Ω.

Let $E \subset \mathbb{C}$ be admissible. Then we define the *level lines of Green's function* $G(z)$ by

$$L_r := \{z \in \Omega : G(z) = \log(1 + r)\}, \quad r > 0. \tag{1.26}$$

Moreover, we set

$$E_r := E \cup \{z \in \Omega : G(z) < \log(1 + r)\}, \quad r > 0. \tag{1.27}$$

If E has a regular boundary, then E_r consists of finitely many simply connected components that are bounded by Jordan curves (which are subsets

16   1. Auxiliary Facts

of L_r). Then the solution of Dirichlet's problem in $E_r \setminus E$ to a continuous boundary function is again continuous at every boundary point of $E_r \setminus E$.

Let $A \subset \mathbb{C}$. Then for any complex-valued function f we define by

$$\|f\|_A := \sup_{z \in A} |f(z)|$$

the *supremum norm* of f on the subset A. For this norm we also use the term *Chebyshev norm* or *uniform norm*.

Let $\mathbb{N} = \{1, 2, \dots\}$ denote the *natural numbers*, $\mathbb{N}_0 = \mathbb{N} \cup \{0\}$.

We denote by $\mathbb{P}_n$, $n \in \mathbb{N}_0$, the class of all polynomials p_n of degree at most n. For $p_n \in \mathbb{P}_n$, the function $\log |p_n(z)|$ is subharmonic in $\mathbb{C}$. Thus, comparing it in Ω with $G(z)$ we obtain the following extremely useful conclusion, known as the *Bernstein–Walsh lemma*.

Lemma 1.11. *Let E be an admissible set. For each $p_n \in \mathbb{P}_n$,*

$$|p_n(z)| \leq \|p_n\|_E \exp(n\, G(z)), \quad z \in \Omega \setminus \{\infty\}.$$

Proof. We consider

$$h(z) := \log |p_n(z)| - nG(z).$$

Then h is subharmonic in Ω and

$$\lim_{z \to \zeta,\, \zeta \in \partial\Omega} h(z) \leq \log \|p_n\|_E \quad \text{for quasi-every } \zeta \in \partial\Omega.$$

Then the maximum principle (Theorem 1.6) yields the inequality of the lemma. □

We denote by $\mathbb{M}_n$, $n \in \mathbb{N}_0$, the set of all *monic polynomials* $p(z) = p_n(z) = z^n + \cdots$ of degree n. Then the Chebyshev norm $\|p_n\|_E$ can be estimated from below by the following lemma

Lemma 1.12. *Let E be compact in $\mathbb{C}$. Then for any $p_n \in \mathbb{M}_n$, $n \in \mathbb{N}_0$,*

$$\|p_n\|_E \geq (\operatorname{cap} E)^n.$$

Proof. If $\operatorname{cap} E = 0$, then the assertion is trivial. Otherwise, let Ω be the unbounded component of $\overline{\mathbb{C}} \setminus E$. Then $E \subset \overline{\mathbb{C}} \setminus \Omega$ and $\operatorname{cap} E \leq \operatorname{cap}(\overline{\mathbb{C}} \setminus \Omega)$. Moreover, $\|p_n\|_E = \|p_n\|_{\partial\Omega} = \|p_n\|_{\overline{\mathbb{C}} \setminus \Omega}$, and the set $\overline{\mathbb{C}} \setminus \Omega$ is admissible.

Because of Lemma 1.11, we have for $h(z) := \log |p_n(z)| - nG(z)$ and $z \in \Omega$,

$$\limsup_{z \to \zeta,\, \zeta \in \partial\Omega} h(z) = \limsup_{z \to \zeta,\, \zeta \in \partial\Omega} (\log |p_n(z)| - nG(z)) \leq \log \|p_n\|_E.$$

Since $h(z)$ is subharmonic in Ω, it follows from the maximum principle that $h(\infty) \leq \log \|p_n\|_E$, i.e.,

$$h(\infty) = n \log \operatorname{cap} E \leq \log \|p_n\|_E,$$

which is equivalent to the inequality of the lemma. □

1.1.9 Fekete Points, Transfinite Diameter, Chebyshev Constant

Now we are going to discuss the notions of Fekete points and Fekete polynomials. Let $E \subset \mathbb{C}$ be a compact set. We call

$$\Delta_n = \Delta_n(E) := \max_{z_1,\ldots,z_n \in E} \prod_{k=1}^{n} \prod_{\substack{l=1 \\ k \neq l}}^{n} |z_k - z_l|, \qquad n \geq 2, \qquad (1.28)$$

the n-th *discriminant* of E; points $z_k = z_{n,k}$, $k = 1,\ldots,n$, for which the maximum (1.28) is attained are called nth *Fekete points*, and $\{z_{n,k}\}_{k=1}^{n}$ is called nth *Fekete point set*.

We can write Δ_n with the Vandermonde determinant as

$$\Delta_n = \left| \det_{k=1,\ldots,n} \left(1, z_k, \ldots, z_k^{n-1}\right) \right|^2,$$

and we call the expression

$$q_n(z) := \prod_{k=1}^{n} (z - z_k)$$

an nth *Fekete polynomial*.

Let

$$M_n := \max_{z \in E} |q_n(z)|, \qquad M'_n := \min_{k=1,\ldots,n} |q'_n(z_k)|.$$

Lemma 1.13. *For* $n = 2, 3, \ldots,$

$$M_n \leq \left(\frac{\Delta_{n+1}}{\Delta_n}\right)^{1/2} \leq M'_{n+1} \leq \Delta_{n+1}^{1/(n+1)}.$$

For a proof, see [146, p. 332].

It follows from Lemma 1.13 that $\Delta_{n+1}^{n-1} \leq \Delta_n^{n+1}$. Hence $\Delta_n^{1/(n(n-1))}$ is decreasing to a limit $\Delta(E)$ as $n \to \infty$. $\Delta(E)$ is called the *transfinite diameter* of E.

For any compact set $E \subset \mathbb{C}$, we introduce the *Chebyshev numbers*

$$t_n(E) := \inf \{\|p\|_E : p \in \mathbb{M}_n\},$$

where $\mathbb{M}_n$ denotes the set of all monic polynomials of degree n. It is well known that in the definition of $t_n(E)$ the infimum is actually a minimum. Moreover, if E contains at least n points, then there is a unique polynomial $T_n \in \mathbb{M}_n$ such that $t_n(E) = \|T_n\|_E$. T_n is called *Chebyshev polynomial* of degree n associated with E.

It is obvious that

$$t_{n+m}(E) \leq t_n(E) t_m(E), \qquad n, m \in \mathbb{N}_0. \qquad (1.29)$$

18   1. Auxiliary Facts

Moreover, it turns out that

$$t(E) := \lim_{n \to \infty} t_n(E)^{1/n} \qquad (1.30)$$

exists. Indeed, this is trivial if $t_n(E) = 0$ for some $n \in \mathbb{N}$. Otherwise, let us define $a_n := \log t_n(E)$ and let $c := \liminf_{n \to \infty} a_n/n$, where $c = -\infty$ is possible. Let $\alpha > c$. We fix $m \in \mathbb{N}$ such that $a_m/m < \alpha$. Because of (1.29) we know that

$$a_{n+m} \leq a_n + a_m, \quad n, m \in \mathbb{N},$$

and consequently, for $n = qm + r$, $0 \leq r < m - 1$,

$$\frac{a_n}{n} \leq \frac{a_{qm} + a_r}{n} \leq \frac{qm}{qm+r} \frac{a_m}{m} + \frac{a_r}{n} < \alpha$$

for n sufficiently large. Hence,

$$c = \liminf_{n \to \infty} \frac{a_n}{n} = \limsup_{n \to \infty} \frac{a_n}{n} = \lim_{n \to \infty} \frac{a_n}{n}$$

and

$$t(E) = \lim_{n \to \infty} t_n(E)^{1/n} = \lim_{n \to \infty} \exp\left(\frac{a_n}{n}\right) = e^{-c}.$$

The number $t(E)$ is called the *Chebyshev constant* associated with E. It is important that the capacity, the transfinite diameter, and the Chebyshev constant are the same.

Lemma 1.14. *If E is compact in $\mathbb{C}$, then $\operatorname{cap} E = \Delta(E) = t(E)$.*

Proof. Because of Lemma 1.13 we know that

$$t(E) \leq \Delta(E).$$

Moreover, Lemma 1.12 yields

$$t(E) \geq \operatorname{cap} E.$$

It remains to prove that $\Delta(E) \leq \operatorname{cap} E$. If $\Delta(E) = 0$, then this inequality is trivial. For $\Delta(E) > 0$, let $z_{n,k}$, $k = 1, \ldots, n$ be a set of nth Fekete points. The measure ν_n that associates the mass $1/n$ to each Fekete point has infinite energy. But if we use the truncated kernel

$$\log^{(N)} \frac{1}{|x|} = \min\left(N, \log \frac{1}{|x|}\right),$$

then we may write

$$\iint \log^{(N)} \frac{1}{|\xi - z|} d\nu_n(\xi) d\nu_n(z) \leq \frac{1}{n^2} \sum_{i \neq k} \log \frac{1}{|z_{n,i} - z_{n,k}|} + \frac{N}{n}.$$

By Helly's theorem we may assume that $\nu_n \xrightarrow{*} \nu \in \mathcal{M}(E)$. Because of the continuity of $\log^{(N)}(1/|\xi - z|)$ we obtain

$$\iint \log^{(N)} \frac{1}{|\xi - z|} \, d\nu(\xi) d\nu(z) \leq -\log \Delta(E).$$

Let $N \to \infty$, then the theorem of Beppo Levi yields

$$\iint \log \frac{1}{|\xi - z|} \, d\nu(\xi) d\nu(z) = I[\nu] \leq -\log \Delta(E),$$

or

$$\operatorname{cap} E \geq \exp(-I[\nu]) \geq \Delta(E).$$

This finishes the proof. $\square$

Remark 1.15. Lemma 1.13 and Lemma 1.14 yield

$$\lim_{n \to \infty} \|q_n\|^{1/n} = \operatorname{cap} E, \tag{1.31}$$

Moreover, if E is admissible, then Lemma 1.11 induces

$$\lim_{n \to \infty} \frac{1}{n} \log |q_n(z)| = G(z) + \log \operatorname{cap} E, \quad z \in \Omega. \tag{1.32}$$

1.2 Conformal and Quasiconformal Mappings

1.2.1 Basic Principles of Complex Analysis

One of the basic principles of the theory of analytic functions is the argument principle or, in an especially useful form, Rouché's theorem.

Rouché's Theorem. *Let f and g be analytic inside and on a rectifiable Jordan curve L, with $|g(z)| < |f(z)|$ on L. Then f and $(f + g)$ have the same number of zeros, counted according to multiplicity, inside L.*

If a sequence $\{f_n\}$ of functions analytic in a domain G converges to a function f uniformly on each compact subset of G, then f is also analytic in G. Moreover, there is a close connection between the zeros of f and the zeros of the functions f_n.

Hurwitz's Theorem. *Let f_n be analytic in a domain G and suppose $\{f_n(z)\}$ converge to $f(z)$ as $n \to \infty$ uniformly on each compact subset of G. Then either $f(z) \equiv 0$ in G, or every zero of f is a limit point of a sequence of zeros of the functions f_n.*

The next objective is the behavior of analytic functions in a normal family. The remarkable fact that for such functions pointwise convergence actually implies uniform convergence on each compact subset is known as Vitali's theorem.

Vitali's Theorem. *Let the functions f_n be analytic and locally bounded in a domain G, that is, for each compact subset B of G,*

$$|f_n(z)| \leq M, \quad z \in B,$$

where the bound M depends only on B. Suppose that $\{f_n(z)\}$ converges at each point of some subset of G, having a cluster point in G. Then $\{f_n(z)\}$ converges uniformly on each compact subset of G.

One of the key facts in geometric function theory is the following statement.

Riemann Mapping Theorem. *Let G be a simply connected domain that is a proper subset of the complex plane. Let $\zeta \in G$ be a given point. Then there is a unique function φ that maps G conformally onto the unit disk with the normalization $\varphi(\zeta) = 0$, $\varphi'(\zeta) > 0$.*

For proofs of these theorems see [52].

If G is a Jordan domain, the Riemann mapping φ can be extended continuously to the boundary, and the extended function maps the boundary bijectively to the unit circle. This important result is due to Carathéodory.

Carathéodory Extension Theorem. *Let G be a Jordan domain, and let φ map G conformally onto $\mathbb{D}$. Then f can be extended to a homeomorphism of $\overline{G}$ onto the closed disk $\overline{\mathbb{D}}$.*

For a proof, see [75].

We remark that for a simply connected domain G with $\infty \in G$, a slight variation of the Riemann mapping theorem implies the existence of a unique conformal mapping Φ of G onto the exterior Δ of the unit disk $\mathbb{D}$ normalized by the conditions

$$\Phi(\infty) = \infty, \quad \Phi'(\infty) := \lim_{z \to \infty} \frac{\Phi(z)}{z} > 0. \tag{2.1}$$

Moreover, if $E = \overline{\mathbb{C}} \setminus G$ is admissible, the Green's function satisfies

$$G(z) = \log |\Phi(z)|,$$

and for the equilibrium distribution μ_E, we conclude from (1.24) that

$$\lim_{z \to \infty} (U^{\mu_E}(z) + G(z)) = \log \Phi'(\infty) = -\log \operatorname{cap} E,$$

or

$$\operatorname{cap} E = 1/\Phi'(\infty). \tag{2.2}$$

1.2.2 Equilibrium Measure and Conformal Mappings

Let $L \subset \mathbb{C}$ be a bounded Jordan curve or a closed Jordan arc and let Φ denote the Riemann mapping that maps the domain

$$\Omega := \begin{cases} \overline{\mathbb{C}} \setminus L, & \text{if } L \text{ is an arc,} \\ \text{ext } L, & \text{if } L \text{ is a curve,} \end{cases}$$

conformally onto Δ, where Φ is normalized by the conditions (2.1). Then $\text{cap } L = 1/\Phi'(\infty)$.

The equilibrium measure of L can be interpreted by the conformal mappings Φ and $\Psi := \Phi^{-1}$. Namely, if $E = \overline{\mathbb{D}}$, then $\mu_E = \mu_{\overline{\mathbb{D}}} = \mu_{\mathbb{T}}$ is the normalized arc length measure of $\mathbb{T}$ and

$$U^{\mu_{\mathbb{T}}}(z) = \log \frac{1}{|z|}, \quad |z| \geq 1.$$

If L is a curve, then Φ can be extended to a homeomorphism $\Phi : \overline{\Omega} \to \overline{\Delta}$, and for any subarc $J \subset L$,

$$\mu_L(J) = \frac{1}{2\pi}|\Phi(J)|. \tag{2.3}$$

Indeed, the function $\log |\Phi(z)|$, $z \in \Omega$, possesses all properties of the Green function. Therefore, by (1.24),

$$U^{\mu_L}(z) = -\log|\Phi(z)| - \log \text{cap } L.$$

Further, let $d\sigma := ds/(2\pi)$, where ds denotes the arc length measure on $\mathbb{T}$. Denote by $\tilde{\sigma}$ the measure supported on L given for any Borel set $U \subset L$ by

$$\tilde{\sigma}(U) := \sigma(\Phi(U)).$$

For $z \in \Omega \setminus \{\infty\}$, its potential can be written in the form

$$U^{\tilde{\sigma}}(z) = \int \log \frac{1}{|\Phi(\zeta) - \Phi(z)|} d\tilde{\sigma}(\zeta) + \int \log \left|\frac{\Phi(\zeta) - \Phi(z)}{\zeta - z}\right| d\tilde{\sigma}(\zeta)$$
$$=: I_1(z) + I_2(z).$$

Since

$$I_1(z) = U^{\sigma}(\Phi(z)) = -\log|\Phi(z)|$$

and since by the mean-value property of harmonic functions (cf. (1.3))

$$I_2(z) = \frac{1}{2\pi} \int_0^{2\pi} \log \left|\frac{e^{i\theta} - \Phi(z)}{\Psi(e^{i\theta}) - z}\right| d\theta = \log \Phi'(\infty),$$

we get for $z \in \Omega \setminus \{\infty\}$,
$$U^{\tilde{\sigma}}(z) = -\log|\Phi(z)| + \log \Phi'(\infty) = U^{\mu_L}(z).$$

Hence, Lemma 1.8 yields $\tilde{\sigma} = \mu_L$.

For the case of an arc L, by slightly modifying the reasoning above we obtain a similar interpretation of the equilibrium measure. Namely, if L is an arc, then Ψ can be extended continuously to a function $\Psi : \overline{\Delta} \to \overline{\mathbb{C}}$, and for any subarc $J \subset L$, there exist two preimages, i.e., arcs $J'_1, J'_2 \subset \mathbb{T}$ such that $\Psi(J'_1) = \Psi(J'_2) = J$ and $J'_1 \cap J'_2$ consists of at most two points. In this case

$$\mu_L(J) = \frac{1}{2\pi}(|J'_1| + |J'_2|). \tag{2.4}$$

1.2.3 Distortion Theorems

Let S denote the class of functions
$$f(z) = z + a_2 z^2 + \cdots, \quad |z| < 1,$$

that are analytic and univalent in the unit disk $\mathbb{D}$, normalized by the conditions $f(0) = 0$, $f'(0) = 1$. Then Koebe proved sharp upper and lower bounds for $|f'(z)|$ and $|f(z)|$ as f ranges over the class S. These results are known as *Koebe's distortion theorem*.

Theorem 2.1. *If $f \in S$, for $z \in \mathbb{D}$,*

$$\frac{1-|z|}{(1+|z|)^3} \leq |f'(z)| \leq \frac{1+|z|}{(1-|z|)^3}, \tag{2.5}$$

$$\frac{|z|}{(1+|z|)^2} \leq |f(z)| \leq \frac{|z|}{(1-|z|)^2}. \tag{2.6}$$

For a proof see [146, p. 21].

If $f \in S$ and $F = f(\mathbb{D})$, then (2.6) yields

$$d(0, \partial F) = \liminf_{z \to 1} |f(z)| \geq \lim_{z \to 1} \frac{|z|}{(1+|z|)^2} = \frac{1}{4},$$

which is called *Koebe's one-quarter theorem*.

Theorem 2.2. *If $f \in S$ and $F = f(\mathbb{D})$, then*

$$d(0, \partial F) \geq \frac{1}{4}.$$

Let $f : G_1 \to G_2$ be a conformal mapping between domains G_1 and G_2 in $\mathbb{C}$. Then we can obtain distortion estimates for f using the classical theorems above.

1.2 Conformal and Quasiconformal Mappings

Lemma 2.3. *Let f be a conformal mapping of a domain $G_1 \subset \mathbb{C}$ onto a domain $G_2 \subset \mathbb{C}$. Then, for each $z \in G_1$,*

$$\frac{1}{4} \frac{d(f(z), \partial G_2)}{d(z, \partial G_1)} \leq |f'(z)| \leq 4 \frac{d(f(z), \partial G_2)}{d(z, \partial G_1)}. \qquad (2.7)$$

Moreover, if $|\xi - z| \leq d(z, \partial G_1)/2$, then

$$\frac{1}{16} \frac{d(f(z), \partial G_2)}{d(z, \partial G_1)} |\xi - z| \leq |f(\xi) - f(z)| \leq 16 \frac{d(f(z), \partial G_2)}{d(z, \partial G_1)} |\xi - z|. \qquad (2.8)$$

Proof. Let $d_1 := d(z, \partial G_1)$ and $d_2 := d(f(z), \partial G_2)$ and consider the function

$$h(\zeta) := \frac{f(z + d_1 \zeta) - f(z)}{f'(z) d_1}, \quad \zeta \in \mathbb{D}.$$

Then h is analytic and univalent in $\mathbb{D}$, normalized by the conditions $h(0) = 0$ and $h'(0) = 1$. Then Koebe's one-quarter theorem implies that the range of h contains the disk $D\left(0, \frac{1}{4}\right)$; i.e.,

$$\liminf_{|\zeta| \to 1} |h(\zeta)| = \frac{d(f(z), \partial G_2)}{|f'(z)| d_1} \geq \frac{1}{4},$$

or

$$|f'(z)| \leq 4 \frac{d_2}{d_1},$$

which is the second inequality of (2.7). Replacing f by f^{-1} we obtain the first inequality. Moreover, Koebe's distortion theorem yields

$$\frac{|\zeta|}{4} \leq |h(\zeta)| \leq 4|\zeta|, \quad |\zeta| < \frac{1}{2}.$$

Inserting $\zeta = (\xi - z)/d_1$ and using (2.7) we immediately obtain (2.8). □

Let G be a simply connected domain in $\mathbb{C}$, $G \neq \mathbb{C}$, $L = \partial G$. Let $z_0 \in G$ be fixed and let us consider the conformal mapping $\varphi : G \to \mathbb{D}$ normalized by $\varphi(z_0) = 0$, $\varphi'(z_0) > 0$. For $0 < r < 1$ we denote by

$$L_r = \{z \in G : |\varphi(z)| = 1 - r\}, \qquad (2.9)$$

the level lines of the mapping φ.

Lemma 2.4. *Let $\varphi : G \to \mathbb{D}$ be the conformal mapping normalized by $\varphi(z_0) = 0$, $\varphi'(z_0) > 0$. Then*

$$d(L, L_r) \geq \frac{1}{32\, \varphi'(z_0)} r^2, \quad 0 \leq r < 1,$$

where $L = \partial G$ and L_r are the level lines of φ defined by (2.9).

24   1. Auxiliary Facts

Proof. Set $\psi = \varphi^{-1}$ and define
$$h(z) := (\psi(z) - z_0)/\psi'(0) = \varphi'(z_0)(\psi(z) - z_0).$$

Then $h \in S$, and Koebe's distortion result (2.5) yields for $|z| = 1 - r$ that $|h'(z)| \geq r/8$. Moreover, let $F = h(\mathbb{D})$. Then by (2.7),
$$d(h(z), \partial F) = \varphi'(z_0) d(\psi(z), L) \geq \frac{1}{4} |h'(z)| \, d(z, \mathbb{T}),$$

or
$$d(L, L_r) \geq \frac{r^2}{32 \varphi'(z_0)} \, .$$

$\square$

1.2.4 Module of a Family of Curves and Arcs

The main idea of the proofs of some special metric relations for the conformal mappings used in this book is to estimate in different ways the (equal) modules of some family Γ of curves and arcs γ and of the image of Γ under conformal mappings. For the sake of later reference, we briefly recall the definition of the module of a family of curves and arcs (for more information, see Appendix A or [3, 4, 112]). We often call a family of curves and arcs simply a *curve family*.

A Borel measurable function $\rho \geq 0$ on $\mathbb{C}$ is said to be *admissible* or *metric* if
$$A(\rho) := \int_{\mathbb{C}} \rho^2(z) \, dm(z) < \infty,$$

where $dm(z)$ is the 2-dimensional Lebesgue measure (area) on $\mathbb{C}$.

For a family Γ of curves and arcs in $\mathbb{C}$ we set
$$L_\rho(\Gamma) := \inf_{\gamma \in \Gamma} \int_\gamma \rho(z) \, |dz|$$

(if the latter integral is not defined for some $\gamma \in \Gamma$, then, by definition, we set it equal to infinity).

The quantity
$$m(\Gamma) := \inf_\rho \frac{A(\rho)}{(L_\rho(\Gamma))^2},$$

where the infimum is taken with respect to all admissible functions ρ, is called the *module* of Γ.

We recall some well-known facts concerning this notion. The most important property of a module is its *conformal invariance*. This means that for any conformal mapping f of some domain $G \subset \overline{\mathbb{C}}$ and any family Γ of curves and arcs $\gamma \subset G$,

$$m(f(\Gamma)) = m(\Gamma) \qquad (2.10)$$

(cf. Theorem A.1.1).

Further, let Γ' and Γ'' be two families such that every $\gamma' \in \Gamma'$ contains a $\gamma'' \in \Gamma''$. Then (see Theorem A.1.2)

$$m(\Gamma') \leq m(\Gamma'') \qquad (2.11)$$

(this is the so-called *comparison principle*).

In addition to the comparison principle there are two *composition laws* that express a relationship between the modules of three families (see Theorem A.1.4). Namely, let G' and G'' be disjoint domains, and let Γ' and Γ'' consist of arcs and curves in G' and G'', respectively. If every $\gamma \in \Gamma$ (a third family) contains a $\gamma' \in \Gamma'$ and a $\gamma'' \in \Gamma''$, then

$$\frac{1}{m(\Gamma)} \geq \frac{1}{m(\Gamma')} + \frac{1}{m(\Gamma'')}. \qquad (2.12)$$

Next, if every $\gamma' \in \Gamma'$ and every $\gamma'' \in \Gamma''$ contains a $\gamma \in \Gamma$, then

$$m(\Gamma) \geq m(\Gamma') + m(\Gamma''). \qquad (2.13)$$

It is useful to compare the last relation with the inequality

$$m(\Gamma' \cup \Gamma'') \leq m(\Gamma') + m(\Gamma''), \qquad (2.14)$$

which is valid for arbitrary Γ' and Γ'' (see Theorem A.1.3).

There is also an *integrated version of the composition laws*, which we formulate in the following way.

Let G be a Jordan domain, $z_0 \in G$ be fixed, $z \in \partial G$. We fix $\varepsilon = \varepsilon(z_0, G) > 0$ such that on each circle $\{\zeta : |\zeta - z| = r\}$ with $0 < r < \varepsilon$ there is at least one crosscut of G, which we denote by $\gamma_z(r)$, separating z and z_0, which means that $\gamma_z(r)$ intersects each Jordan arc joining in G the points z and z_0. If $\gamma_z(r)$ is not uniquely determined, we agree to choose it so that in dividing G by $\gamma_z(r)$ into two subdomains, the domain containing z_0 is as large as possible for given z and r (see Figure 1.1).

Then for any $0 < r_1 < r_2 < \varepsilon$ and $z \in \partial G$,

$$m(\{\gamma_z(r) : r_1 \leq r \leq r_2\}) = \int_{r_1}^{r_2} \frac{dr}{|\gamma_z(r)|}. \qquad (2.15)$$

Indeed, let

$$E := \bigcup_{r_1 \leq r \leq r_2} \gamma_z(r), \quad \Gamma_0 := \Gamma := \{\gamma_z(r) : r_1 \leq r \leq r_2\},$$

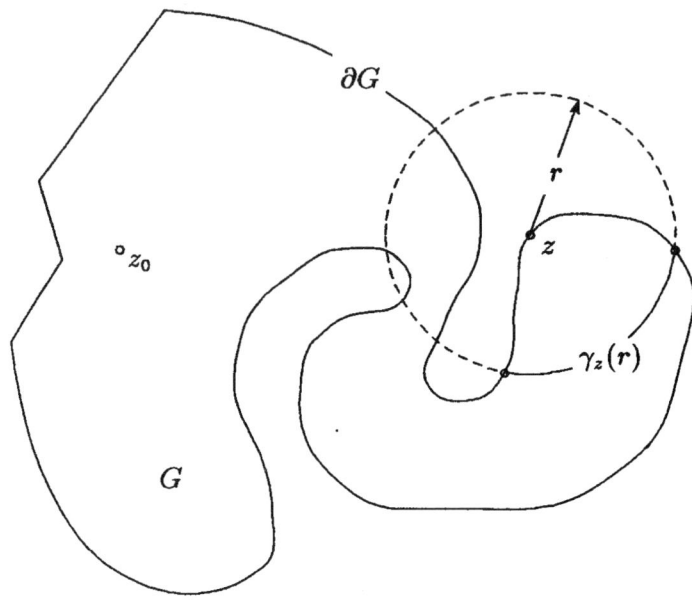

Fig. 1.1

$$\rho_0(\zeta) := \begin{cases} |\gamma_z(|\zeta|)|^{-1}, & \text{if } \zeta \in E, \\ 0, & \text{otherwise}. \end{cases}$$

By the Beurling criterion (see Theorem A.1.6),

$$m(\Gamma) = A(\rho_0) = \int_{r_1}^{r_2} \frac{dr}{|\gamma_z(r)|}.$$

and (2.15) is true.

There are some other particular standard cases in which a module can be calculated precisely, which are discussed in Appendix A. For example, the module of the family

$$\Gamma_0 = \Gamma_0(r_1, r_2, \theta_1, \theta_2) := \{\gamma_\theta := \{re^{i\theta} : r_1 < r < r_2\} : \theta_1 < \theta < \theta_2\},$$

where $0 < r_1 < r_2 < \infty$, $0 < \theta_2 - \theta_1 \leq 2\pi$, can be determined exactly, namely,

$$m(\Gamma_0) = \frac{\theta_2 - \theta_1}{\log(r_2/r_1)} \qquad (2.16)$$

(cf. (A.1.6)).

However, in the most important cases it is enough to have estimates for a module up to constants. One of these cases concerns the following connection between harmonic measure and the module of a curve family. Let G be a Jordan domain, $z \in G$, and let J be a subarc of ∂G. Denote by $\Gamma = \Gamma(J, z)$ the family of all crosscuts of G that separate in G the point z from J and by $\omega(z, G, J)$ the harmonic measure of J at the point z with respect to G. If $G = \mathbb{D}$, $z = 0$, then the harmonic measure of J coincides with $|J|/2\pi$. Since the harmonic measure and the module of a curve family are conformal invariants, we can rewrite inequality (A.1.12) in the form

$$c_1 \exp\{-\pi m(\Gamma)\} \leq \omega(z, G, J) \leq c_2 \exp\{-\pi m(\Gamma)\}, \qquad (2.17)$$

where $c_1 = 1/\pi$, $c_2 = 2e^{2\pi}/\pi$. This inequality holds for a Jordan domain G, any point $z \in G$, and every boundary subarc $J \subset \partial G$.

We complete this section with an estimate that characterizes the distortion properties of conformal mappings.

Lemma 2.5. *If E is an admissible continuum, $L := \partial E$, then for any $r > 0$,*

$$d(L, L_r) \geq \frac{r^2}{1+r} \operatorname{cap} E.$$

For the proof, see Corollary A.2.5.

We remark that the weaker inequality

$$d(L, L_r) \geq c(r^2 + r) \operatorname{cap} E$$

(with an absolute constant $c > 0$, independent of E) can be deduced directly from Lemma 2.4 (compare [161]).

1.2.5 Quasiconformal Curves and Arcs

Throughout this book L will be mainly a quasiconformal curve or arc. We recall that by definition, a *K-quasiconformal* ($K \geq 1$), or, briefly, *quasiconformal*, curve is the image of the unit circle under some K-quasiconformal mapping $F : \overline{\mathbb{C}} \to \overline{\mathbb{C}}$. Any subarc of a K-quasiconformal curve is called a K-quasiconformal, or, briefly, a quasiconformal, arc.

A domain $G \subset \mathbb{C}$ is called a *quasiconformal domain* or a *quasidisk* if it is bounded by a quasiconformal curve.

The basis of the theory of quasiconformal mappings is discussed in Appendix B. We cite these results that suffice for an understanding of the main part of this book (concerning the distribution of zeros of polynomials).

There exists a geometric characterization of quasiconformal curves and arcs. For example, for curves it is the well-known *Ahlfors's theorem* (see Theorem B.2.1 or [112, p. 100]), which can be formulated as follows: L is a quasiconformal curve if and only if there exists a constant $c > 0$, depending only on L, such that for $z_1, z_2 \in L$,

$$\min\{\text{diam } L', \text{diam } L''\} \leq c|z_1 - z_2|, \tag{2.18}$$

where L' and L'' denote the two arcs of which $L \setminus \{z_1, z_2\}$ consists and the diameter of a set $A \subset \mathbb{C}$ is defined by

$$\text{diam } A := \sup_{z, \zeta \in A} |z - \zeta|.$$

Moreover, the constant c in (2.18) and the coefficient K of quasiconformality of L are mutually dependent.

The corresponding result for arcs is a reformulation of Ahlfors's theorem.

Lemma 2.6. *Assume that L is a bounded Jordan arc. The following three conditions are equivalent.*

(i) *For any three points $z_1, z_2, z_3 \in L$ arranged on L in index-ascending order*

$$|z_1 - z_2| \leq c|z_1 - z_3| \tag{2.19}$$

with some constant $c \geq 1$.

(ii) *There is a K-quasiconformal mapping $F : \overline{\mathbb{C}} \to \overline{\mathbb{C}}$ such that $F(L) = [-1, 1]$, $F(\infty) = \infty$.*

(iii) *L is quasiconformal.*

Moreover, if K is given, then c depends only on K, and conversely, if c is given, then $K = K(c)$.

For the proof, see Appendix B, Section 2.2, or [17, p. 144].

Using these criteria, we can easily verify that convex curves, curves of bounded variation without cusps, and rectifiable Jordan curves that have locally the same order of arc length and chord length are quasiconformal. For example, smooth curves and piecewise smooth arcs without cusps are quasiconformal. On the other hand, it is of interest to know that a quasiconformal curve can be everywhere nonrectifiable [24, p. 42].

One of the most important properties of quasiconformal mappings, which by Väisälä's theorem (see [112, p. 171]) can be taken as their definition, is the *quasi-invariance of the module* of a family of arcs and curves. This means that for any K-quasiconformal mapping F of some domain $G \subset \mathbb{C}$ and any family Γ of arcs and curves $\gamma \subset G$,

$$K^{-1}m(F(\Gamma)) \leq m(\Gamma) \leq Km(F(\Gamma)). \tag{2.20}$$

The following result describing the change of the order of three points under quasiconformal automorphisms of the plane is useful in the study of metric properties of the conformal mappings Φ and Ψ.

Theorem 2.7. *Suppose that the function $w = F(\zeta)$ is a K-quasiconformal mapping of the plane onto itself, $F(\infty) = \infty$. Assume also that $\zeta_j \in \mathbb{C}$, $w_j := F(\zeta_j)$, $j = 1, 2, 3$. Then:*

(i) *The conditions $|\zeta_1 - \zeta_2| \leq c_1|\zeta_1 - \zeta_3|$ and $|w_1 - w_2| \leq c_2|w_1 - w_3|$ are equivalent; besides, the constants c_1 and c_2 are mutually dependent and depend on K but not on $\zeta_1, \zeta_2, \zeta_3$;*

(ii) *if $|\zeta_1 - \zeta_2| \leq c_1|\zeta_1 - \zeta_3|$, then*

$$c_3 \left|\frac{w_1 - w_3}{w_1 - w_2}\right|^{1/K} \leq \left|\frac{\zeta_1 - \zeta_3}{\zeta_1 - \zeta_2}\right| \leq c_4 \left|\frac{w_1 - w_3}{w_1 - w_2}\right|^K, \qquad (2.21)$$

where $c_j = c_j(c_1, K)$, $j = 3, 4$.

For a proof, see Appendix B, Section 1.6 or [17, pp. 98–99].

In the following, we shall often use the fact that a conformal mapping $\Phi : \Omega \to \Delta$ can be extended to a K^2-quasiconformal mapping $\Phi : \overline{\mathbb{C}} \to \overline{\mathbb{C}}$ [3, Chapter IV] if L is a K-quasiconformal curve. The inverse mapping $\Psi := \Phi^{-1}$ will then also be K^2-quasiconformal (in $\overline{\mathbb{C}}$).

In what follows we denote by $c, c_1, \ldots$ constants (each time different in general) that either are absolute or depend on parameters not essential for the arguments; otherwise, such a depending will be indicated.

Applying Theorem 2.7 to the function $F := \Phi$, it is easily verified that for any points $\zeta_j \in L$, $w_j := \Phi(\zeta_j)$, $j = 1, 2$, the double inequality

$$c_1|\zeta_1 - \zeta_2|^{K^2} \leq |w_1 - w_2| \leq c_2|\zeta_1 - \zeta_2|^{1/K^2}$$

holds with some constants $c_j = c_j(L)$, $j = 1, 2$.

Hence, for any arc $J \subset L$,

$$c_3(\operatorname{diam} J)^{K^2} \leq \mu_L(J) \leq c_4(\operatorname{diam} J)^{1/K^2}, \qquad (2.22)$$

where $c_j = c_j(L)$, $j = 3, 4$.

Next, we note that the same inequality is true for the case of a quasiconformal arc.

To be more precise, let L be a K-quasiconformal arc. Denote by z_1 and z_2 its endpoints, and for $j = 1, 2$, set

$$\begin{aligned} & w_j := \Phi(z_j), \\ & \Delta_1 := \{w : |w| > 1; \arg w_1 < \arg w < \arg w_2\}, \\ & \Delta_2 := \Delta \setminus \overline{\Delta}_1, \quad \Omega_j := \Psi(\Delta_j). \end{aligned} \qquad (2.23)$$

A routine category argument involving Ahlfors's geometric characterization of quasiconformal curves shows that $\partial \Delta_1$ and $\partial \Delta_2$ are quasiconformal curves.

Moreover, the following is true.

Lemma 2.8. *The boundary of Ω_1 (which is equal to the boundary of Ω_2) is a quasiconformal curve.*

Proof. Let $\Gamma_j := \{\zeta : \zeta \in \Omega, \arg \Phi(\zeta) = \arg w_j\}$, $j = 1, 2$. We assert the quasiconformality of the curve $L \cup \Gamma_1 \cup \Gamma_2$. For the proof it is enough to show that both $L \cup \Gamma_1$ and $L \cup \Gamma_2$ are quasiconformal. We verify, for instance, the quasiconformality of the arc $L \cup \Gamma_1$. Let F_1 be a quasiconformal mapping of the plane onto itself that carries L to the interval $[-1, 1]$ with the normalization $F_1(z_1) = 1$, $F_1(z_2) = -1$, $F_1(\infty) = \infty$ (the existence of a quasiconformal mapping satisfying this special normalization follows simply from the definition of quasiconformal arcs and Lemma 2.6), and let $F_2(\tau) = \frac{1}{2}(\tau + 1/\tau)$ be the Joukowski map. Then the function

$$F := F_2^{-1} \circ F_1 \circ \Psi$$

forms a quasiconformal mapping of the exterior of the unit disk onto itself. We extend this to a quasiconformal mapping of the plane onto itself by setting for w with $|w| < 1$,

$$F(w) := \overline{\left(F\left(\frac{1}{\overline{w}}\right)\right)^{-1}}.$$

For short, we write $K := \Phi(\Gamma_1)$, $M := F_1(\Gamma_1)$, $N := F_2^{-1}(M)$, $S := \{\tau : 1/\overline{\tau} \in N\}$, $\overline{N} := \{\tau : \overline{\tau} \in N\}$,

$$T_+ := \{\tau : |\tau| = 1, \operatorname{Im} \tau \geq 0\},$$
$$T_- := \{\tau : |\tau| = 1, \operatorname{Im} \tau \leq 0\}$$

(see Figure 1.2). As images of quasiconformal arcs under the quasiconformal mapping F, the arcs $N \cup S$, $S \cup T_\pm$, and $N \cup T_\pm$ are quasiconformal. Now let $\tau \in N$, $|\tau - 1| \leq 2$. We construct points τ_1 and τ_2 with the properties $|\tau_1| = 1$, $|\tau_1 - 1| = |\tau - 1|$, and $\tau_2 = \overline{\tau}_1$. By Theorem 2.7 we have

$$|\tau - 1| \leq c |\tau - \tau_j|, \quad j = 1, 2.$$

Consequently, the arc N forms a nonzero angle with the unit circumference. Therefore, the arc $\overline{N} \cup S$ is also quasiconformal, which, as is easily verified by using Theorem 2.7 and the explicit form of F_2, implies the quasiconformality of the arc $M \cup [-1, 1]$. Thus, we have also established the quasiconformality of $L \cup \Gamma_1$ as the image of $M \cup [-1, 1]$ under the quasiconformal mapping F_1^{-1}. □

By Lemma 2.8 the restriction Φ_j of the function Φ to the region Ω_j can be extended to a K_1-quasiconformal mapping $\Phi_j : \overline{\mathbb{C}} \to \overline{\mathbb{C}}$ with a suitable constant $K_1 = K_1(L) > 1$ (see [3, Chapter IV]).

1.2 Conformal and Quasiconformal Mappings   31

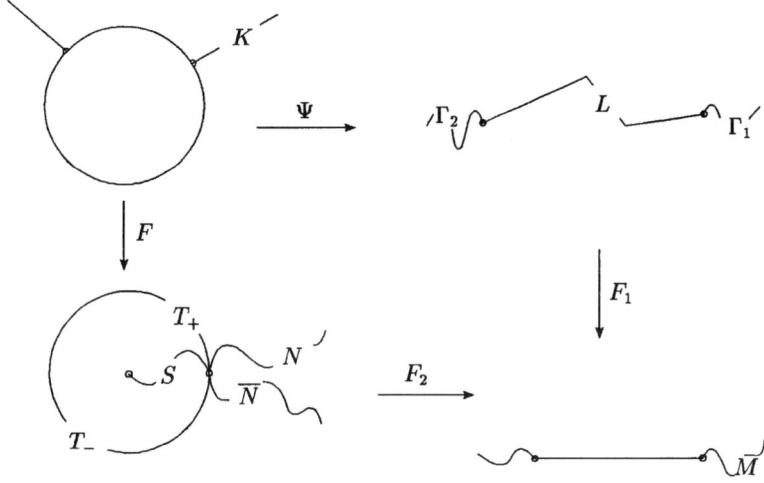

Fig. 1.2

Using Theorem 2.7 for the function $F := \Phi_j$, we find that for each pair of points ζ_1 and $\zeta_2 \in L$,

$$c_1|\zeta_1 - \zeta_2|^{K_1} \leq |\Phi_j(\zeta_1) - \Phi_j(\zeta_2)| \leq c_2|\zeta_1 - \zeta_2|^{1/K_1},$$

where $c_j = c_j(L)$, $j = 1, 2$.

Finally, recalling the properties of the equilibrium measure, we may conclude that for any subarc $J \subset L$, the inequalities

$$c_3(\text{diam } J)^{K_1} \leq \mu_L(J) \leq c_4(\text{diam } J)^{1/K_1} \qquad (2.24)$$

are satisfied with some $c_j = c_j(L)$, $j = 3, 4$.

It is well known that the harmonic measure is a conformal invariant. In the next lemma we claim that it is also a quasiconformal "quasiinvariant."

Lemma 2.9. *Let $G_j \subset \mathbb{C}$, $j = 1, 2$, be arbitrary Jordan domains, and let F be a K-quasiconformal ($K \geq 1$) mapping of G_1 onto G_2 extended continuously to the boundary ∂G_1. Then, for each arc $l \subset \partial G_1$ and each point $z \in G_1$,*

$$\omega(z, G_1, l) \leq 8\pi\,\omega(F(z), G_2, F(l))^{1/K}.$$

Proof. We remark that F can be extended continuously to ∂G_1 ([112, Theorem 8.2, p. 98]). According to the previous remark concerning the conformal invariance of the harmonic measure we have to consider only the case $G_1 = G_2 = \mathbb{D}$ and $z = F(z) = 0$. Note that by a Theorem of Mori [112, p.

66], for any $z_1, z_2 \in \overline{\mathbb{D}}$,

$$|z_1 - z_2| \leq 16\,|F(z_1) - F(z_2)|^{1/K}. \qquad (2.25)$$

Let z_1 and z_2 be the endpoints of the arc $l \subset \mathbb{T}$. The interesting case is

$$\omega(0, \mathbb{D}, F(l)) < (8\pi)^{-K}.$$

Then, by virtue of (2.25),

$$\omega(0, \mathbb{D}, l) < \frac{\pi}{2}|z_1 - z_2| \leq 8\pi|F(z_1) - F(z_2)|^{1/K} \leq 8\pi\omega(0, \mathbb{D}, F(l))^{1/K}.$$

$\square$

1.2.6 *Dini-Smooth Curves and Arcs*

A smooth Jordan curve L is called *Dini-smooth* if the angle $\beta(s)$ of the tangent, considered as a function of the arc length s, satisfies

$$|\beta(s_2) - \beta(s_1)| < h(s_2 - s_1), \quad s_1 < s_2,$$

where $h(x)$ is an increasing function for which

$$\int_0^1 \frac{h(x)}{x}\,dx < \infty. \qquad (2.26)$$

In the following we will use some obvious geometrical fact: If L is a Dini-smooth curve, then for any point $z(s) \in L$ the relation

$$L \cap \{\zeta : |\zeta - z(s)| \leq \varepsilon\} \subset \{\zeta = z(s) + re^{i\theta} : 0 \leq r \leq \varepsilon, |\theta - \beta(s)| \leq h(cr)$$
$$\text{or } |\theta + \pi - \beta(s)| \leq h(cr)\} \qquad (2.27)$$

holds with some constants $\varepsilon > 0$ and $c > 0$ independent of s (see Figure 1.3).

We call a Jordan arc Dini-smooth if it is a subarc of some Dini-smooth curve. Finally, a Jordan curve or arc is said to be *piecewise Dini-smooth* if it consists of a finite number of Dini-smooth arcs that form nonzero angles at their corners. If G is a domain bounded by a Dini-smooth Jordan curve, then each conformal mapping φ of G onto the unit disk $\mathbb{D}$ has a continuously differentiable extension to $\overline{\mathbb{D}}$ with $\varphi'(z) \neq 0$ (see [147, p. 48]). For domains bounded by a piecewise Dini-smooth curve or piecewise Dini-smooth arc this result has some consequences and generalizations, which we formulate as Lemmas 2.10–2.12. We will give only the proof of Lemma 2.10 in detail. The proofs of the other two lemmas are quite similar and need only some obvious technical modifications.

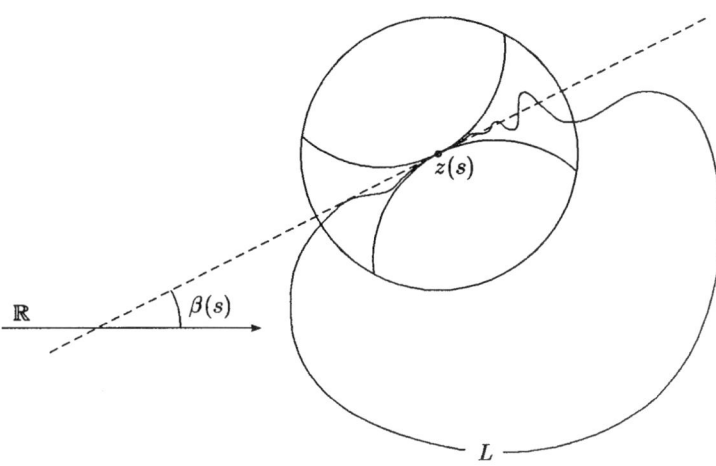

Fig. 1.3

Lemma 2.10. *Let L be a piecewise Dini-smooth curve, $\Omega = \overline{\mathbb{C}} \setminus L$, $z_0 \in L$ a corner with inner angle $\alpha\pi$, $0 < \alpha < 2$. Then there exists a constant $\varepsilon = \varepsilon(L, z_0) > 0$ such that for all points $z \in L$, $\zeta \in \overline{\Omega}$ with $|z - \zeta| \leq |z - z_0| \leq \varepsilon$ and their images $w := \Phi(z)$, $\tau := \Phi(\zeta)$, $w_0 := \Phi(z_0)$ the inequalities*

$$c_1 \left|\frac{w - w_0}{w - \tau}\right| \leq \left|\frac{z - z_0}{z - \zeta}\right| \leq c_2 \left|\frac{w - w_0}{w - \tau}\right|, \tag{2.28}$$

$$c_3 |w - w_0|^{2-\alpha} \leq |z - z_0| \leq c_4 |w - w_0|^{2-\alpha}, \tag{2.29}$$

hold with some constants $c_j = c_j(L, z_0, \varepsilon, \alpha) > 0$, $j = 1, \ldots, 4$.

Proof. First we note that there exists $\varepsilon > 0$ such that the intersection $D(z, x) \cap \Omega$ consists of one simple arc for all $0 < x \leq \varepsilon$ and all $|z - z_0| \leq \varepsilon$. We denote this subarc of $D(z, x)$ by $s_z(x)$. The set $Q_z(\delta, t)$, $0 < \delta < t < \varepsilon$, is the quadrilateral bounded by the arcs $s_z(\delta)$ and $s_z(t)$ and the subarcs of L joining their endpoints; $\Gamma_z(\delta, t)$ is the family of all arcs in $Q_z(\delta, t)$ that separate the sides $s_z(\delta)$ and $s_z(t)$ (see Figure 1.4); $m_z(\delta, t)$ is the module of $\Gamma_z(\delta, t)$.

By Theorem 2.7 for any triplet of points $\xi_1, \xi_2, \xi_3 \in \overline{\Omega}$ with $|\xi_1 - \xi_2| = |\xi_1 - \xi_3|$ we have

$$|\Phi(\xi_1) - \Phi(\xi_2)| \asymp |\Phi(\xi_1) - \Phi(\xi_3)|.$$

Here and in the following the notion $a \asymp b$ means that $a \leq c_1 b$ and $b \leq c_2 a$ with fixed constants c_1 and c_2 (cf. Section 1.6).

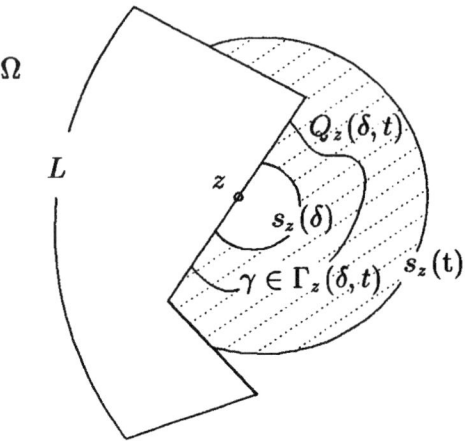

Fig. 1.4

Hence, according to Lemmas A.1.15 and A.1.16, for any triplet $\xi_1 \in L, \xi_2, \xi_3 \in \overline{\Omega}$ with $|\xi_1 - \xi_2| < |\xi_1 - \xi_3| < \varepsilon$ we obtain

$$\left|\frac{\Phi(\xi_1) - \Phi(\xi_3)}{\Phi(\xi_1) - \Phi(\xi_2)}\right| \asymp \exp(\pi m_{\xi_1}(|\xi_1 - \xi_2|, |\xi_1 - \xi_3|)).$$

Therefore, the inequalities (2.28) and (2.29) are equivalent to

$$-c_5 \leq m_z(\delta, t) - \frac{1}{\pi} \log \frac{t}{\delta} \leq c_6, \quad 0 < \delta < t = |z - z_0|,$$

$$-c_7 \leq m_{z_0}(\delta, \varepsilon) - \frac{1}{\pi(2-\alpha)} \log \frac{\varepsilon}{\delta} \leq c_8, \quad 0 < \delta < \varepsilon.$$

Each of these inequalities is an obvious consequence of the comparison principle for the modules of families of curves, the relation (2.27), and the following fact: Let $h(x)$, $0 < x < \varepsilon$, be a positive increasing function satisfying (2.26). For $j = 1, 2$ and $0 < \beta < 2$ as well as $0 < \delta < t \leq \varepsilon e^{-2\pi}$ consider the quadrilateral $Q^j = Q^j(\delta, t)$ where

$$Q^j(\delta, t) := \{z = re^{i\theta} : \delta < r < t,\ (-1)^j h(r) < \theta < \beta\pi + (-1)^{j+1} h(r)\}.$$

Let $m(\Gamma^j)$ be the module of the family Γ^j of all crosscuts of Q^j that separate the circular subarcs of ∂Q^j.

We assert that

$$\frac{1}{\beta\pi} \log \frac{t}{\delta} - c_9 \leq m(\Gamma^1) \leq m(\Gamma^2) \leq \frac{1}{\beta\pi} \log \frac{t}{\delta} + c_{10}. \quad (2.30)$$

1.2 Conformal and Quasiconformal Mappings   35

Indeed, the middle inequality is an immediate consequence of the comparison principle (2.11). The left-hand inequality follows from the integrated version of the composition laws for modules (2.15):

$$m(\Gamma^1) \geq \int_\delta^t \frac{dx}{x(\beta\pi + 2h(x))} \geq \int_\delta^t \frac{1}{\beta\pi x}\left(1 - 2\frac{h(x)}{\beta\pi x}\right) dx$$

$$\geq \frac{1}{\beta\pi}\int_\delta^t \frac{dx}{x} - \frac{2}{(\beta\pi)^2}\int_0^\varepsilon \frac{h(x)}{x} dx \geq \frac{1}{\beta\pi}\log\frac{t}{\delta} - c_9.$$

In order to establish the right-hand side of (2.30) we assume that $h(\varepsilon) < \beta\pi/4$ and consider the functions

$$v(x) := \min\{h(e^{2\pi}x), h(t)\}, \qquad \delta < x < t,$$

$$\rho(z) := \frac{1}{|z|(\beta\pi - 2v(|z|))}, \qquad z \in Q^2.$$

Note that for any $\gamma \in \Gamma^2$,

$$\int_\gamma \rho(z)\,|dz| \geq 1.$$

In fact, set

$$r_{\max} := \sup_{z \in \gamma} |z|, \qquad r_{\min} := \inf_{z \in \gamma} |z|.$$

If $r_{\max} \geq e^{2\pi} r_{\min}$, then there exists an arc $\gamma_1 \subset \gamma$ contained in the annulus $\{z : r_{\min} < |z| < r_{\max}\}$ whose endpoints connect the two boundary components of the annulus. Therefore,

$$\int_\gamma \rho(z)\,|dz| \geq \frac{1}{2\pi}\left|\int_{\gamma_1}\frac{dz}{z}\right| \geq \frac{1}{2\pi}\log\frac{r_{\max}}{r_{\min}} \geq 1.$$

If $r_{\max} < e^{2\pi} r_{\min}$, then $h(r_{\max}) < v(r_{\min})$ and

$$\int_\gamma \rho(z)\,|dz| \geq \frac{1}{\beta\pi - 2v(r_{\min})}\left|\int_\gamma \frac{dz}{z}\right| \geq \frac{\beta\pi - h(r_{\min}) - h(r_{\max})}{\beta\pi - 2v(r_{\min})} \geq 1.$$

The definition of the module yields

$$m(\Gamma^2) \le \int_{Q^2} \rho^2(z)\,dm(z) \le \int_\delta^t \frac{(\beta\pi - 2h(r))\,dr}{r(\beta\pi - 2v(r))^2}$$

$$\le \frac{1}{\beta\pi}\int_\delta^t \frac{dr}{r} + 4\int_\delta^t \frac{v(r)\,dr}{r(\beta\pi - 2v(r))^2}$$

$$\le \frac{1}{\beta\pi}\log\frac{t}{\delta} + \frac{16}{(\beta\pi)^2}\int_{e^{2\pi}\delta}^{e^{2\pi}t} \frac{h(r)}{r}\,dr \le \frac{1}{\beta\pi}\log\frac{t}{\delta} + c_{10}\,.$$

$\square$

Now let L be a piecewise Dini-smooth arc, i.e., L consists of a finite number of Dini-smooth arcs $l_1, \ldots, l_m$. Denote by z_j and z_{j+1} the endpoints of l_j (the endpoints of L are denoted by z_1 and z_{m+1}).

Lemma 2.11. *Let L be a piecewise Dini-smooth arc with corners at the points z_j, $j = 2, \ldots, m$, and angles $\alpha_{j,i}\pi$, $0 < \alpha_{j,i} < 2$, with respect to Ω_i, $i = 1, 2$, where Ω_i are the domains defined by (2.23). Then there exists a constant $\varepsilon = \varepsilon(L) > 0$ such that for all points $z \in l_{j-1} \cup l_j$, $\zeta \in \overline{\Omega}_i$ with*

$$|z - z_{j-1}| \ge \varepsilon, \quad |z - z_{j+1}| \ge \varepsilon,$$
$$|z - \zeta| \le |z - z_j|,$$

and their images $\tau_i := \Phi_i(\zeta)$, $w_{j,i} := \Phi_i(z_j)$, $w_i := \Phi_i(z)$ the inequalities

$$c_1 \left|\frac{w_i - w_{j,i}}{w_i - \tau_i}\right| \le \left|\frac{z - z_j}{z - \zeta}\right| \le c_2 \left|\frac{w_i - w_{j,i}}{w_i - \tau_i}\right|, \qquad (2.31)$$

$$c_3 |w_i - w_{j,i}|^{\alpha_{j,i}} \le |z - z_j| \le c_4 |w_i - w_{j,i}|^{\alpha_{j,i}} \qquad (2.32)$$

hold with some constants $c_k = c_k(L, \varepsilon, \alpha_{j,i}) > 0$, $k = 1, \ldots, 4$.

To complete the distortion properties of the conformal mapping Φ in the case of an arc L we describe its behavior in a neighborhood of the endpoints z_1 and z_{m+1} of L as follows.

Lemma 2.12. *Let L be a piecewise Dini-smooth arc as above and let z_j, $j = 1, m+1$, be its endpoints. Then there exists a constant $\varepsilon = \varepsilon(L) > 0$ such that for any $i = 1, 2$, and all points $z \in L$, $\zeta \in \overline{\Omega}_i$ with*

$$|z - \zeta| \le |z - z_j| \le \varepsilon$$

and their images $\tau_i := \Phi_i(\zeta)$, $w_i := \Phi_i(z)$, $w_{j,i} := \Phi_i(z_j)$, the inequalities (2.31) and (2.32) hold with $\alpha_{j,i} = 2$.

1.3 Faber Polynomials and Grunsky Coefficients

Let $E \subset \mathbb{C}$ be a continuum with connected complement $\Omega = \overline{\mathbb{C}} \setminus E$, and let $\Phi : \Omega \to \Delta$ be the Riemann mapping function normalized as in (2.1). Set $\Psi := \Phi^{-1}$ and denote by $\Phi_k(z)$, $k \in \mathbb{N}_0$, the *Faber polynomials* for E.

There are many different definitions for Faber polynomials. One that is broadly accepted is based on the Laurent expansion of the generating function

$$\frac{w\,\Psi'(w)}{\Psi(w)-z} = \sum_{k=0}^{\infty} \frac{\Phi_k(z)}{w^k}, \quad z \in E, \tag{3.1}$$

in the exterior of the unit disk.

Since Faber polynomials play a significant role in complex analysis, there are many articles and monographs devoted to their investigation (see, for example, [165, 50, 168]). Among the different generalizations of this notion we mention the so-called standard polynomials, which are the most suitable ones for the construction of the approximation of Cauchy kernels (for details, see [50]).

More precisely, we denote by *Faber standard generalized polynomials* for the continuum E those polynomials $\Pi_n(z)$ of order $n-1$, $n \in \mathbb{N}$, such that these polynomials are the coefficients of the Laurent expansion

$$\frac{1}{\Psi(w)-z} = \sum_{k=1}^{\infty} \frac{\Pi_k(z)}{w^k}, \quad z \in E, \tag{3.2}$$

of the function $(\Psi(w)-z)^{-1}$ with respect to the powers of w in a neighborhood of the point $w = \infty$.

We now introduce the Grunsky coefficients. Let cap $E = 1$, i.e.,

$$\Psi(w) = w + \sum_{n=0}^{\infty} c_n w^{-n}, \quad w \in \Delta. \tag{3.3}$$

Since $\Psi(w) \neq \Psi(\tau)$ for $w \neq \tau$ and since $\Psi'(w) \neq 0$, we can write

$$\log \frac{\Psi(w)-\Psi(\tau)}{w-\tau} = -\sum_{k=1}^{\infty} \sum_{l=1}^{\infty} a_{kl}\, w^{-k} \tau^{-l}, \quad w, \tau \in \Delta. \tag{3.4}$$

The numbers a_{kl}, $k, l \in \mathbb{N}$, are called the *Grunsky coefficients* of the function Ψ. It follows directly from the definition that

$$a_{kl} = a_{lk}, \quad k, l \in \mathbb{N}.$$

For fixed $z \in \mathbb{C}$, the function $\log[(\Psi(\tau)-z)/\tau]$ is analytic in a neighborhood of ∞ and vanishes at ∞ because of (3.3). Hence, it can be expanded in a series around ∞, which by virtue of (3.1) has the form

$$\log\frac{\Psi(\tau)-z}{\tau} = -\sum_{k=1}^{\infty}\frac{1}{k}\Phi_k(z)\tau^{-k}. \tag{3.5}$$

Setting $z = \Psi(w)$ in (3.5), we have

$$\log\frac{\Psi(\tau)-\Psi(w)}{\tau-w} = -\sum_{k=1}^{\infty}\frac{1}{k}\left[\Phi_k(\Psi(w))-w^k\right]\tau^{-k}.$$

Together with (3.4) this shows that

$$\Phi_k(\Psi(w)) = w^k + k\sum_{l=1}^{\infty}a_{kl}\,w^{-l}, \qquad w \in \Delta. \tag{3.6}$$

The following statement is a generalization of the classical area theorem and is known in geometrical function theory as the *Grunsky inequality*.

Theorem 3.1. *Let Ψ satisfy (3.3), and let c_k, $k \in \mathbb{N}$, be arbitrary complex numbers. Then*

$$\sum_{n=1}^{\infty}n\left|\sum_{k=1}^{\infty}a_{nk}\,c_k\right|^2 \leq \sum_{k=1}^{\infty}\frac{1}{k}|c_k|^2, \tag{3.7}$$

provided that the last series converges.

For a proof, see [146, pp. 61–62].

The next estimate for Faber polynomials is a direct consequence of the theorem above.

Theorem 3.2. *Let Ψ satisfy (3.3). Then*

$$\max_{z\in E}\sum_{k=1}^{n}\frac{1}{k}|\Phi_k(z)|^2 < 4\log n + 8\,, \quad n > 1.$$

Proof. Let $|w| =: r > 1$, $f_k(w) := \Phi_k(\Psi(w))$. We begin with the following elementary inequality:

$$\sum_{k=1}^{n}\frac{1}{k}|f_k(w)|^2 \leq 2\sum_{k=1}^{\infty}\frac{1}{k}\left|f_k(w)-w^k\right|^2 + 2\sum_{k=1}^{n}\frac{1}{k}r^{2k}. \tag{3.8}$$

Further, by (3.6) and Theorem 3.1 we have

$$\sum_{k=1}^{n}\frac{1}{k}\left|f_k(w)-w^k\right|^2 = \sum_{k=1}^{n}k\left|\sum_{l=1}^{\infty}a_{kl}w^{-l}\right|^2$$

$$\leq \sum_{k=1}^{\infty}\frac{1}{k}r^{-2k} = \log\frac{r^2}{r^2-1}. \tag{3.9}$$

Taking into account that

$$\sum_{k=1}^{n}\frac{1}{k}r^{2k}=\sum_{k=1}^{n}\frac{1}{k}+2\int_{1}^{r}\sum_{k=1}^{n}t^{2k-1}\,dt$$

$$<\log n+1+2n\int_{1}^{r}t^{2n-1}\,dt=\log n+r^{2n},$$

we obtain, according to (3.8) and (3.9), that

$$\sum_{k=1}^{n}\frac{1}{k}|f_k(w)|^2<2\log\frac{r^2}{r^2-1}+2\log n+2r^{2n}.$$

This means that for each

$$z\in L_{R-1}=\{z\in\Omega:|\Phi(z)|=R\}$$

with $R:=(1-1/n)^{-1/2}$,

$$\sum_{k=1}^{n}\frac{1}{k}|\Phi_k(z)|^2=\sum_{k=1}^{n}\frac{1}{k}|f_k(\Phi(z))|^2<4\log n+8.$$

Since L_{R-1} encloses E, the assertion of the theorem follows from the last estimate and the maximum principle for subharmonic functions. □

Now we are able to continue the chain of inequalities in Lemma 1.13 above in the following way.

Theorem 3.3. *Let E be a continuum. Then for $n=2,3,\ldots$,*

$$\Delta_n^{1/n}\leq\left(\frac{4}{e}\log n+4\right)n\,(\mathrm{cap}\,E)^{n-1}. \qquad (3.10)$$

Proof. Since $\Delta_n(cE)=c^{n(n-1)}\Delta_n(E)$, it is enough to prove (3.10) only for the case $\mathrm{cap}\,E=1$. Let $z_1,\ldots,z_n\in E$ be nth Fekete points, i.e.,

$$\Delta_n=\prod_{k=1}^{n}\prod_{\substack{l=1\\l\neq k}}^{n}|z_k-z_l|.$$

Since $\Phi_k(z)=z^k+\cdots$ are monic polynomials, we can write this equality by means of the Vandermonde determinant in the form

$$\Delta_n=\left|\det_{k=1,\ldots,n}(1,z_k,\ldots,z_k^{n-1})\right|^2$$

$$=\left|\det_{k=1,\ldots,n}(1,\Phi_1(z_k),\ldots,\Phi_{n-1}(z_k))\right|^2$$

$$=(n-1)!\left|\det_{k=1,\ldots,n}\left(1,\frac{1}{\sqrt{1}}\Phi_1(z_k),\ldots,\frac{1}{\sqrt{n-1}}\Phi_{n-1}(z_k)\right)\right|^2.$$

The well-known Hadamard inequality of linear algebra says that the absolute value of a determinant is not greater than the product of the Euclidean lengths of the row vectors. So we have

$$\Delta_n \leq (n-1)! \prod_{k=1}^{n} \left(1 + |\Phi_1(z_k)|^2 + \frac{1}{2}|\Phi_2(z_k)|^2 + \cdots + \frac{1}{n-1}|\Phi_{n-1}(z_k)|^2\right).$$

Hence, applying Theorem 3.2 and Stirling's formula we have

$$\Delta_n^{1/n} \leq ((n-1)!)^{1/n}(4\log(n-1) + 9) < ne^{-1}(4\log n + 4e),$$

and the theorem is proved. □

1.4 Monic Polynomials

Let $p \in \mathbb{P}_n \setminus \mathbb{P}_{n-1}$, $n \geq 1$. Then we associate with p the normalized counting measure for its zeros.

Definition 4.1. For any $B \subset \mathbb{C}$ the measure

$$\nu_p(B) := \frac{\text{number of zeros of } p \text{ in } B}{n},$$

where the zeros are counted with respect to their multiplicity, is called the *(normalized) zero counting measure of p*.

Moreover, in Chapter 5 we shall use the notion of *normalized counting measure $\nu(X)$* of a finite set X in $\mathbb{C}$; i.e., if m is the number of points of X, then $\nu(X)$ is the measure that associates the mass $1/m$ with each point $x \in X$.

One of the main ideas of our investigation below is to compare for $p \in \mathbb{M}_n$ the measure ν_p with the equilibrium measure μ_E of E. We will do this by using information about the difference of their logarithmic potentials

$$U^{\mu_E - \nu_p}(z) := U^{\mu_E}(z) - U^{\nu_p}(z),$$

where

$$U^{\nu_p}(z) = \int \log\frac{1}{|z-\zeta|}\,d\nu_p(\zeta) = -\frac{1}{n}\log|p(z)|, \quad z \in \mathbb{C}.$$

Lemma 4.2 (Upper bounds for $U^{\mu_E - \nu_p}$). *Let E be compact, $\operatorname{cap} E > 0$. If $p \in \mathbb{M}_n$, $n \geq 1$, and*

$$\|p\|_E \leq A_n(\operatorname{cap} E)^n, \tag{4.1}$$

where $A_n \geq 1$ is constant, then

$$U^{\mu_E - \nu_p}(z) \leq \frac{1}{n} \log A_n \quad \text{for } z \in \mathbb{C}. \tag{4.2}$$

Proof. Let Ω denote the unbounded component of $\overline{\mathbb{C}} \setminus E$. Then (4.1) yields

$$|p(z)| \leq A_n (\operatorname{cap} E)^n \quad \text{for } z \in \overline{\mathbb{C}} \setminus \Omega.$$

Taking the logarithm we obtain

$$\frac{1}{n} \log |p(z)| \leq \frac{\log A_n}{n} + \log \operatorname{cap} E \quad \text{for } z \in \mathbb{C} \setminus \Omega.$$

Since $U^{\mu_E}(z) \leq -\log \operatorname{cap} E$ for all $z \in \mathbb{C}$ (Theorem 1.6), we get

$$U^{\mu_E - \nu_p}(z) \leq \frac{\log A_n}{n} \quad \text{for } z \in \mathbb{C} \setminus \Omega. \tag{4.3}$$

Because of $U^{\mu_E - \nu_p}(\infty) = 0$ the function $U^{\mu_E - \nu_p}$ is subharmonic in Ω including ∞. By the maximum principle the inequality (4.3) is also true for all $z \in \Omega$, and (4.2) is proved. $\square$

The upper bound

$$\varepsilon = \varepsilon_p := \frac{1}{n} \log A_n$$

on the right-hand side of (4.2) plays an essential role in estimating the number of zeros of p_n in the exterior of the level line L_δ of Green's function G of $\Omega = \overline{\mathbb{C}} \setminus E$, where

$$L_\delta = \{z \in \Omega : G(z) = \log(1 + \delta)\} \tag{4.4}$$

and E is admissible.

Lemma 4.3. *Let E be admissible and let $p \in \mathbb{M}_n$ satisfy (4.1). Then*

$$\nu_p(\operatorname{ext} L_\delta) \leq \frac{\varepsilon}{\log(1 + \delta)}, \quad \delta > 0.$$

Proof. Fix $\delta > 0$, $\Omega = \mathbb{C} \setminus E$, and define

$$h(z) := U^{\mu_E - \nu_p}(z) + \frac{1}{n} \sum_{k=1}^{m} G(z, z_k),$$

where $z_1, \ldots, z_m$ are the zeros of p in $\operatorname{ext} L_\delta$, listed with respect to multiplicity, and $G(z, z_k)$ is Green's function for Ω with pole at z_k. Since h is subharmonic in Ω, bounded from above, and

$$\limsup_{z \to \xi,\, z \in \Omega} h(z) \leq \varepsilon$$

42   1. Auxiliary Facts

for quasi-every $\xi \in \partial\Omega$, the maximum principle (Theorem 1.9) yields

$$h(\infty) = \frac{1}{n}\sum_{k=1}^{m} G(\infty, z_k) = \frac{1}{n}\sum_{k=1}^{m} G(z_k) \leq \varepsilon,$$

where we have used the symmetry of Green's functions, namely

$$G(\infty, z_k) = G(z_k, \infty) = G(z_k).$$

Thus,

$$\frac{m}{n} = \nu_p(\text{ext } L_\delta) \leq \frac{\varepsilon}{\log(1+\delta)}.$$

□

Let E be a Jordan domain with $L = \partial E$. We fix $z_0 \in E$ and denote by $\omega = \varphi(z)$ the conformal mapping of E onto $\mathbb{D}$ with the normalization $\varphi(z_0) = 0$, $\varphi'(z_0) > 0$. Then we consider the level lines of φ, i.e.,

$$L_\delta^- = \{\zeta : |\varphi(\zeta)| = 1 - \delta\}, \quad 0 < \delta < 1. \tag{4.5}$$

Then, by same method, the number of zeros of p_n in the interior of the inner level line L_δ^- can be estimated.

Lemma 4.4. *Let L_δ^- be as in (4.5). Then*

$$\nu_p(\text{int } L_\delta^-) \leq \frac{\varepsilon - U^{\mu_E - \nu_p}(z_0)}{\log(1/(1-\delta))}, \quad 0 < \delta < 1.$$

Proof. Let $\delta > 0$ and define

$$h(z) := U^{\mu_E - \nu_p}(z) + \frac{1}{n}\sum_{k=1}^{m} \widetilde{G}(z, z_k),$$

where $\widetilde{G}(z, z_k)$ is Green's function for the Jordan domain E with pole at $z_k \in E$. Moreover, $z_1, \ldots, z_m$ are the zeros of p in int L_δ^-, listed again with multiplicity. The function h is subharmonic in E, and

$$\limsup_{z \to \xi, z \in G} h(z) \leq \varepsilon \quad \text{for } \xi \in \partial E.$$

Consequently, by the maximum principle

$$\varepsilon \geq h(z_0) = U^{\mu_E - \nu_p}(z_0) + \frac{1}{n}\sum_{k=1}^{m} \widetilde{G}(z_0, z_k)$$

$$= U^{\mu_E - \nu_p}(z_0) + \frac{1}{n}\sum_{k=1}^{m} \widetilde{G}(z_k, z_0)$$

$$= U^{\mu_E - \nu_p}(z_0) + \frac{m}{n}\log\frac{1}{1-\delta},$$

which proves the lemma. □

Let $p \in \mathbb{M}_n$ and let $Z(p)$ denote the set of all zeros of p. In the case that $Z(p) \subset E$ and all zeros of p are simple it is possible to obtain lower bounds for $U^{\mu_E - \nu_p}$.

Lemma 4.5 (Lower bounds for $U^{\mu_E-\nu_p}$). *Let $p \in \mathbb{M}_n$, $n \geq 2$, $0 < \delta < 1$ and let E be an admissible continuum with $Z(p) \subset E$. If there exists a constant $B_n \geq n$ such that*

$$|p'(z)| \geq \frac{1}{B_n} (\operatorname{cap} E)^n \quad \text{for } z \in Z(p), \tag{4.6}$$

then

$$U^{\mu_E-\nu_p}(z) \geq -\left(\frac{2 \log B_n}{n} + \frac{2 \log 1/\delta}{n} + \delta\right) - \frac{c_1}{n} \quad \text{for } z \in L_\delta, \tag{4.7}$$

where $c_1 = c_1(E)$ is a positive constant.
Moreover, if E is Jordan domain and $Z(p) \subset \partial E$, then

$$U^{\mu_E-\nu_p}(z) \geq -\left(\frac{2 \log B_n}{n} + \frac{2 \log 1/\delta}{n}\right) - \frac{c_2}{n} \quad \text{for } z \in L_\delta^-, \tag{4.8}$$

where $c_2 = c_2(E, z_0)$ is a positive constant.

Proof. Let $L := \partial E$. The Lagrange interpolation formula yields for $z \in L_\delta$ that

$$1 = \left|\sum_{k=1}^n \frac{p(z)}{p'(z_k)(z-z_k)}\right| \leq \frac{n B_n |p(z)|}{(\operatorname{cap} E)^n d(L, L_\delta)}.$$

Now, by Lemma 2.5 there exists a constant $c_1^* = c_1^*(E) > 0$ such that

$$d(L, L_\delta) \geq c_1^* \delta^2, \quad 0 < \delta < 1.$$

Then taking the logarithm of both sides of inequality (1.4) we obtain

$$U^{\mu_E-\nu_p}(z) + \frac{1}{n} \log \frac{nB_n}{c_1^* \delta^2} + \log(1+\delta) \geq 0 \quad \text{for } z \in L_\delta,$$

which proves (4.7).

If E is a Jordan domain, then again Lagrange interpolation yields

$$1 \leq \frac{n B_n |p(z)|}{(\operatorname{cap} E)^n d(L, L_\delta^-)}.$$

By Lemma 2.4 we know that $d(L, L_\delta^-) \geq c_2^* \delta^2$, where $c_2^* = c_2^*(E, z_0)$. Again, by taking the logarithm we get

$$U^{\mu_E-\nu_p}(z) + \frac{1}{n} \log \frac{nB_n}{c_2^* \delta^2} \geq 0 \quad \text{for } z \in L_\delta^-,$$

which proves (4.8). □

1.5 Linear Approximation

Let R be a normed linear space with norm $\|\cdot\|$, V a finite-dimensional subspace of R. Then for each $f \in R$ there exists an element $v^* \in V$ such that
$$\|f - v^*\| = \min_{v \in V} \|f - v\|.$$

If the best approximating element v^* is unique, we denote by $p : R \to V$ the operator that associates with f this best approximating element.

Theorem 5.1. *If each element $f \in R$ has a unique best approximating element in V, then the operator $p : R \to V$ is continuous.*

Proof. Let us assume that p is not continuous at the point $f_0 \in R$. Then there exist a sequence $f_n \to f_0$ and $\varepsilon > 0$ such that
$$\|p(f_0) - p(f_n)\| > \varepsilon.$$

Since
$$\|p(f_n)\| \le \|f_n\| + \|f_n - p(f_n)\| \le \|f_n\| + \|f_n - p(f_0)\|$$
$$\le \|f_n\| + \|f_0 - f_n\| + \|f_0 - p(f_0)\|,$$

the sequence $\{p(f_n)\}_{n=1}^\infty$ is bounded. By passing to a subsequence if necessary, we may assume that $p(f_n) \to v_0 \in V$ as $n \to \infty$ and $v_0 \ne p(f_0)$. Now,
$$\|f_0 - v_0\| \le \|f_0 - f_n\| + \|f_n - p(f_n)\| + \|p(f_n) - v_0\|$$
$$\le \|f_0 - f_n\| + \|f_n - p(f_0)\| + \|p(f_n) - v_0\|$$
$$\le 2\|f_0 - f_n\| + \|f_0 - p(f_0)\| + \|p(f_n) - v_0\|.$$

If we pass to the limit as $n \to \infty$, we obtain
$$\|f_0 - v_0\| \le \|f_0 - p(f_0)\|.$$

Hence, v_0 is a best approximating element, which contradicts the uniqueness of $p(f_0)$. $\square$

The following theorem is often used to construct functions with specific properties. Especially, it is used in the literature to calculate lower bounds for n-widths.

Theorem 5.2 (Krein). *Let X_j, $j = n, n+1$, be j-dimensional subspaces of a normed linear space. Then there exists an element $y \in X_{n+1}$ that has 0 as best approximant with respect to X_n.*

For a proof see [115, p. 132] or [140, p. 12].

1.5.1 Polynomial Approximation

Let $E \subset \mathbb{C}$ be compact. We denote by $C(E)$ the set of all continuous (complex-valued) functions on E endowed with the supremum norm $\|\cdot\|_E$. For $f \in C(E)$ and $n \in \mathbb{N}_0$, let

$$E_n(f, E) := \inf_{p \in \mathbb{P}_n} \|f - p\|_E.$$

We denote by $p_n^*(f)$ the *best uniform approximation* of $f \in C(E)$ with respect to $\mathbb{P}_n$ (*best uniform approximant*), i.e.,

$$\|f - p_n^*(f)\|_E = E_n(f, E).$$

It is a well-known fact that $p_n^*(f)$ exists and is unique. By Theorem 5.1 the operator $p_n^*(f)$ is continuous with respect to f.

Other examples for Theorem 5.1 can be found in L^p-approximation on an interval: Let τ be a positive, finite Borel measure on $I = [-1, 1]$ with infinite support. For $1 \leq p \leq \infty$, let L_τ^p be the class of all τ-measurable functions f on I such that $|f|^p$ is integrable with respect to τ. We denote by

$$\|f\|_{p,\tau} := \left(\int |f|^p \, d\tau \right)^{1/p}$$

the weighted L^p-norm on $L_\tau^p(I)$.

For $f \in L_\tau^p(I)$ and $n \in \mathbb{N}$, set $B_{n,p} = B_{n,p}(f)$ for a best L_τ^p-approximant of f on I with respect to $\mathbb{P}_n$, i.e.,

$$\|f - B_{n,p}\|_{p,\tau} = E_{n,p}(f) := \min_{P_n \in \mathbb{P}_n} \|f - P_n\|_{p,\tau}.$$

For $p > 1$ the normed linear space $L_\tau(I)$ is strictly convex. Therefore, the best approximant is unique. For $p = 1$ uniqueness of the best approximants does not hold in general. But for $f \in C(I)$ uniqueness holds due to a theorem of Jackson and Krein (cf. [45, Corollary 25]).

Let E be compact in $\mathbb{C}$ with connected complement and let $A(E)$ denote the subset of $C(E)$ consisting of all functions analytic in the interior of E. Mergelyan's theorem (see [165, pp. 115–117]) implies that

$$\lim_{n \to \infty} E_n(f, E) = 0, \quad f \in A(E). \tag{5.1}$$

The rate of decrease of $E_n(f, E)$ as $n \to \infty$ and the structural properties of $f \in A(E)$ are closely connected. We will need the following well-known result.

Theorem 5.3. *Suppose E is admissible and has a regular boundary. Let $R(f) \geq 1$ be the largest number of all $R \geq 1$ such that $f \in A(E)$ can be analytically continued to*

$$E_{R-1} := E \cup \{z \in \Omega : G(z) < \log R\}.$$

46   1. Auxiliary Facts

Then
$$\limsup_{n\to\infty} E_n(f,E)^{1/n} = \frac{1}{R(f)}.$$

For a proof, see [181, pp. 75–79].

Finally, we mention the classical inequality of Remez, which will be used in Chapter 5 for applications of discrepancy theorems.

Theorem 5.4 (Remez inequality). *The inequality*
$$\|p_n\|_I \leq T_n\left(\frac{2+s}{2-s}\right)$$
holds for every $p_n \in \mathbb{P}_n$ and $s \in (0,2)$ satisfying
$$\mathrm{mes}(\{x \in I : |p_n(x)| \leq 1\}) \geq 2 - s.$$

Here T_n is the Chebyshev polynomial of degree n for I, and $\mathrm{mes}(E)$ denotes the Lebesgue measure for sets $E \subset \mathbb{R}$.

For a proof, see [42, Theorem 5.1.1].

1.5.2 Averaging and Smoothing

Let $G \subset \mathbb{C}$ be a bounded domain, $\delta > 0$ an arbitrary fixed number. Set
$$G_\delta := \{z \in \mathbb{C} : \mathrm{dist}\,(z, \overline{G}) \leq \delta\}.$$

If f is a bounded real-valued function on G_δ, then there is a standard way to approximate f on G by convolution.

Let $K(z)$, $z \in \mathbb{C}$, be a nonnegative, continuous, and real-valued function that vanishes outside the unit disk, $K_\delta(z) := K(z/\delta)/\delta^2$. Hence, $K_\delta(z)$ vanishes outside $D(0,\delta)$. The convolution $f * K_\delta$ defined for $z \in G$ by

$$(f * K_\delta)(z) := \int_{D(0,\delta)} f(z-\zeta) K_\delta(\zeta)\, dm(\zeta)$$
$$= \int_{G_\delta} f(\zeta) K_\delta(z-\zeta)\, dm(\zeta) \quad (5.2)$$

is continuous in G. This follows immediately from
$$(f * K_\delta)(z) - (f * K_\delta)(z_0) = \int_{G_\delta} f(\zeta) \left(K_\delta(z-\zeta) - K_\delta(z_0-\zeta)\right) dm(\zeta).$$

If $K(z)$ is continuously differentiable, i.e., $K \in C^1(\mathbb{C})$, then the mean-value theorem implies that the expression
$$\frac{K_\delta(z+h) - K_\delta(z)}{h} - \frac{\partial}{\partial x} K_\delta(z),$$

where $h \neq 0$ is real, tends to zero as $h \to 0$ uniformly in $\mathbb{C}$. This fact implies for $z \in G$,

$$\lim_{h \to 0} \frac{(f * K_\delta)(z+h) - (f * K_\delta)(z)}{h} - \left(f * \frac{\partial}{\partial x} K_\delta\right)(z)$$

$$= \lim_{h \to 0} \int_{G_\delta} f(\zeta) \left(\frac{K_\delta(z+h-\zeta) - K_\delta(z-\zeta)}{h} - \frac{\partial}{\partial x} K_\delta(z-\zeta)\right) dm(\zeta) = 0.$$

Thus in G $f * K_\delta$ has the partial derivative

$$\frac{\partial}{\partial x}(f * K_\delta) = f * \frac{\partial}{\partial x} K_\delta,$$

which is continuous in G since $\partial K_\delta / \partial x$ is continuous. In the same way we can show that the partial derivative $\partial(f * K_\delta)/\partial y$ exists and equals $f * (\partial K_\delta / \partial y)$.

We say that a function g belongs to $C^\infty(G)$ if it is infinitely many times continuously differentiable in G. By repeating the above argument we deduce that in general, $f * K_\delta \in C^\infty(G)$ if $K \in C^\infty(\mathbb{C})$.

In addition to $K \in C^\infty(\mathbb{C})$, let us from now on assume that

$$K(z) = K(|z|) \geq 0, \qquad z \in \mathbb{C}, \tag{5.3}$$

$$K(z) = 0, \qquad |z| \geq 1, \tag{5.4}$$

$$\int_\mathbb{C} K(z) \, dm(z) = 1. \tag{5.5}$$

A function $K(z)$ with properties as above will be called an *averaging kernel*.

Note two of its remarkable properties. If $f(z)$ is harmonic in $D(z, \delta)$ for $z \in \mathbb{C}$, $\delta > 0$, then

$$(f * K_\delta)(z) = \int_0^\delta r K_\delta(r) \int_0^{2\pi} f(z - re^{i\theta}) \, d\theta \, dr$$

$$= f(z) \int_0^\delta r 2\pi K_\delta(r) \, dr = f(z) \int_\mathbb{C} K_\delta(\zeta) \, dm(\zeta) = f(z), \tag{5.6}$$

i.e., the averaging process preserves the values of a function harmonic in a neighborhood of the corresponding point.

Moreover, for $z \in G$ the inequality

$$|\Delta(f * K_\delta)(z)| = |\Delta((f - c) * K_\delta)(z)| = |((f - c) * \Delta K_\delta)(z)|$$

$$\leq \frac{c_1}{\delta^2} \sup_{z \in D(z,\delta)} |f(z) - c| \tag{5.7}$$

holds, where $c \in \mathbb{C}$ is an arbitrary constant and the constant $c_1 > 0$ depends only on $K(z)$.

48   1. Auxiliary Facts

1.6 Ordering Symbols

In many parts of the book we make use of some special ordering symbols for real-valued functions.

Let M be a set and let $f, g : M \longrightarrow \mathbb{R}$ be two real valued functions. Then we write
$$f \preceq g$$
if there exists a constant $C > 0$ such that $f(m) \leq Cg(m)$ for all $m \in M$.

In all cases where we use the symbol "$\preceq$" the domain M of the functions f and g is obvious from the actual situation, and we write
$$f(m) \preceq g(m)$$
without writing out explicitly the domain M.

Analogously,
$$f(m) \succeq g(m) \text{ if and only if } g(m) \preceq f(m)$$
and
$$f(m) \asymp g(m) \text{ if and only if } f(m) \preceq g(m) \text{ and } f(m) \succeq g(m).$$

1.7 Historical Comments

The first section contains the standard knowledge in potential theory. Note that μ_E in Lemma 1.8 can be replaced by any $\mu \in \mathcal{M}(\partial E)$ (see Carleson, [43]). However, the proof becomes much more complicated.

The second section is devoted to the theory of conformal mappings and their generalizations: quasiconformal mappings.

Apparently, Lemma 2.5 was first proved by Löwner [117]. The statement of Lemma 2.6, which is the analogue of the Ahlfors geometrical characterization of quasiconformal curves, is essentially due to Rickman [149].

Lemma 2.8 was proved in [10] and Lemma 2.5 in [14].

Investigating the distortion properties of the Riemann mapping function in the case of a domain with piecewise Dini-smooth boundary, we give the proofs of Lemmas 2.10–2.12 based on the appropriate general distortion theory constructed in [26, 25]. The other way to prove these statements is due to Warschawski (see [147]).

Theorems 3.2 and 3.3 were proved by Pommerenke [142].

Theorem 5.1 is well known; its proof follows Cheney [44].

2
Zero Distribution of Polynomials

The classical theorems of Jentzsch and Szegő concern the limiting behavior of the zeros of the partial sums of a power series. More precisely, if

$$s_n(z) = \sum_{k=0}^{n} a_k z^k, \quad n = 0, 1, 2, \ldots,$$

are the partial sums of a power series $f(z) = \sum_{k=0}^{\infty} a_k z^k$ having *finite* positive radius of convergence ρ, then Jentzsch [91] proved that each point of the circle of convergence $C_\rho := \{z : |z| = \rho\}$ is a limit point of zeros of polynomials $s_n(z)$, $n = 1, 2, \ldots$. Szegő [170] substantially improved this result by showing that there is a subsequence $\{n_k\}_{k=1}^{\infty}$ for which the zeros of the partial sums $s_{n_k}(z)$ are uniformly distributed in angle; that is, if $S(\alpha, \beta)$ is the sector

$$S(\alpha, \beta) = \{z \in \mathbb{C} : \alpha < \arg z < \beta\}, \quad \alpha < \beta < 2\pi + \alpha,$$

and $Z_n(A)$ denotes the number of zeros of s_n in the set A, then

$$\lim_{k \to \infty} \frac{Z_{n_k}(S(\alpha, \beta))}{n_k} = \frac{\beta - \alpha}{2\pi} \tag{0.1}$$

for all sectors $S(\alpha, \beta)$.

Erdős and Turán [57] obtained estimates about the speed of convergence of the zero distribution of s_n by considering the following situation: Let $p_n \in \mathbb{M}_n$, $p_n(z) = z^n + \cdots + a_0$, $a_0 \neq 0$, and let ν_n denote the normalized zero-counting measure of p_n. Then

$$\left|\nu_n(S(\alpha,\beta)) - \frac{\beta-\alpha}{2\pi}\right| \leq 16\sqrt{\frac{\log P(\rho)}{n}}, \qquad (0.2)$$

where
$$P(\rho) = \frac{1}{\sqrt{|a_0|\rho^n}}\|p_n\|_{C_\rho}.$$

If all zeros of $p_n \in \mathbb{M}_n$ are located in the interval $[-1,1]$, then Erdős and Turán proved in an earlier paper [55] that for any subinterval $[a,b] \subset [-1,1]$,

$$|\nu_n([a,b]) - \mu([a,b])| \leq \frac{8}{\log 3}\sqrt{\frac{\log \widetilde{P}}{n}}, \qquad (0.3)$$

where $\mu = \mu_{[-1,1]}$ is the equilibrium distribution of $[-1,1]$ and

$$\widetilde{P} = 2^n \max_{-1 \leq x \leq 1} |p_n(x)|.$$

Hence, the limiting distribution of the zeros of polynomials is determined by the supremum norm of p_n on circles or intervals, or in the case of partial sums, by approximating properties of these partial sums to the analytic function $f(z)$.

In this chapter we present a general and joint approach to theorems of Jentzsch–Szegő as well as of Erdős–Turán type. Basically, methods of potential theory are introduced, which furthermore, have the advantage of simplifying the previous proofs.

2.1 Jentzsch–Szegő Type Theorems

Throughout this section, E will be a compact subset of $\mathbb{C}$ such that $\Omega = \overline{\mathbb{C}} \setminus E$ is connected and regular, i.e., there is a Green's function $G(z) = G(z, \infty)$ on $\overline{\mathbb{C}} \setminus E$ with pole at ∞ with

$$\lim_{z \to \infty} (G(z) - \log|z|) = -\log \operatorname{cap} E, \qquad (1.1)$$

and $G(z) \to 0$ as $z \to \partial E$.

Furthermore, we suppose that $\{p_n\}_{n=1}^\infty$ is a sequence of complex polynomials, and $\nu_n := \nu_{p_n}$ is the zero-counting measure associated with p_n.

We begin with the following statement.

Theorem 1.1. *Let $k_n \in \mathbb{N}$ with $k_n \geq \deg p_n$ and assume that the following conditions hold:*

$$\limsup_{n \to \infty} \left(\frac{1}{k_n} \log \|p_n\|_E\right) \leq 0; \qquad (1.2)$$

$$\lim_{n \to \infty} \nu_n(M) = 0 \qquad (1.3)$$

for every compact set $M \subset E^\circ$, where E° is the open interior of E; there exists a compact set $S \subset \overline{\mathbb{C}} \setminus E$ with

$$\liminf_{n\to\infty} \left[\max_{z\in S}\left(\frac{1}{k_n}\log|p_n(z)| - G(z)\right)\right] \geq 0. \tag{1.4}$$

Then $\nu_n \xrightarrow{*} \mu_E$ as $n \to \infty$.

Note that the functions

$$h_n(z) := \frac{1}{k_n}\log|p_n(z)| - G(z) \tag{1.5}$$

are subharmonic in $\Omega = \overline{\mathbb{C}} \setminus E$ if we define

$$h_n(\infty) := \lim_{z\to\infty} h_n(z).$$

Therefore, the maximum of h_n in S, which is used in (1.4), exists.

Remark 1.2. By the maximum principle for subharmonic functions we conclude from (1.2) and (1.4) that really both left-hand sides are zero. Moreover, "lim inf" and "lim sup" can be replaced by "lim." But we prefer the above formulation in Theorem 1.1 because it is more appropriate for applications.

Remark 1.3. If (1.4) holds for some compact set $S \subset \Omega$, then (1.4) is true for any compact set $\widetilde{S} \subset \Omega$ with regular boundary and connected complement $\overline{\mathbb{C}} \setminus \widetilde{S}$.

This can be shown as follows: We may assume that $\widetilde{S} \cap S = \emptyset$. The set $\Omega \setminus \widetilde{S}$ is connected and regular for the Dirichlet problem. Hence, if the assertion is false for $\widetilde{S}$, there exists a harmonic function u on $\Omega \setminus \widetilde{S}$ such that

$$u(t) = 0, \qquad t \in \partial E,$$
$$u(t) = \liminf_{n\to\infty}\left(\max_{z\in\widetilde{S}} h_n(z)\right) < 0, \qquad t \in \partial\widetilde{S}.$$

Thus, by the maximum principle (see Theorem 1.1.9),

$$\liminf_{n\to\infty}\left(\max_{z\in S} h_n(z)\right) \leq \max_{z\in S} u(z) < 0,$$

which contradicts (1.4).

Remark 1.4. Let $\widetilde{k}_n$ denote the exact degree of p_n. Because of (1.2)–(1.4), we have

$$\lim_{n\to\infty} \widetilde{k}_n = \infty.$$

For otherwise, the polynomial sequence $\{p_n\}$ would have a limit point p, which is a polynomial of degree k. Then (1.3) implies that p has no zeros in E°. From Remark 1.2 and Remark 1.3 we know that

$$\max_{z \in \widetilde{S}} \left[\frac{1}{k} \log |p(z)| - G(z) \right] = 0$$

for any closed disk $\widetilde{S} \subset \Omega$. Hence, p has no zeros in $\overline{\Omega}$, which is a contradiction.

Remark 1.5. If (1.2) and (1.4) hold, then these inequalities are also valid if we replace k_n by the exact degree $\widetilde{k}_n$ of p_n.

Indeed, by the maximum principle for subharmonic functions (see Theorem 1.1.9)

$$\frac{1}{k_n} \log |p_n(z)| - G(z) = \frac{1}{k_n} (\log |p_n(z)| - \widetilde{k}_n G(z)) + \left(\frac{\widetilde{k}_n}{k_n} - 1 \right) G(z)$$

$$\leq \frac{1}{k_n} \log \|p_n\|_E + \left(\frac{\widetilde{k}_n}{k_n} - 1 \right) G(z)$$

for $z \in S \setminus \{\infty\}$. Thus, by (1.2) and (1.4),

$$0 \leq \liminf_{n \to \infty} \left(\frac{\widetilde{k}_n}{k_n} - 1 \right) \inf_{z \in S} G(z),$$

from which $\lim_{n \to \infty} \widetilde{k}_n / k_n = 1$ follows. Since for $z \in S \setminus \{\infty\}$,

$$\frac{1}{\widetilde{k}_n} \log |p_n(z)| - G(z) = \frac{k_n}{\widetilde{k}_n} \left(\frac{1}{k_n} \log |p_n(z)| - G(z) \right) + \left(\frac{k_n}{\widetilde{k}_n} - 1 \right) G(z),$$

we easily get (1.2) and (1.4) with k_n replaced by $\widetilde{k}_n$.

Lemma 1.6. *Under the conditions (1.2) and (1.4) we have for any compact set $V \subset \overline{\mathbb{C}} \setminus E$,*

$$\lim_{n \to \infty} \nu_n(V) = 0. \tag{1.6}$$

Proof. We may assume $V \cap S = \emptyset$. With (1.5) let us define

$$\widetilde{h}_n(z) := h_n(z) + \frac{1}{k_n} \sum_\nu G(z, z_{\nu n}),$$

where the $z_{\nu,n}$ are the zeros of p_n in V and $G(z,\xi)$ denotes the Green function of $\overline{\mathbb{C}}\setminus E$ with pole at ξ. Then $\widetilde{h}_n$ is subharmonic in $\overline{\mathbb{C}}\setminus E$. Thus, by the maximum principle (see Theorem 1.1.9)

$$\limsup_{n\to\infty}\left(\max_{z\in S}\widetilde{h}_n(z)\right)\leq 0,$$

where we have used (1.2). Since $\widetilde{h}_n \geq h_n$ on $\mathbb{C}\setminus E$, we get from (1.4) that

$$\lim_{n\to\infty}\left(\inf_{z\in S}(\widetilde{h}_n - h_n)(z)\right) = 0. \tag{1.7}$$

Set

$$\inf_{z\in S}\inf_{\xi\in V} G(z,\xi) =: \delta > 0.$$

Then from (1.7),

$$0 = \lim_{n\to\infty}\left(\inf_{z\in S}\frac{1}{k_n}\sum_{\nu} G(z, z_{\nu,n})\right) \geq \delta \limsup_{n\to\infty} \nu_n(V).$$

Thus

$$\lim_{n\to\infty} \nu_n(V) = 0,$$

and the lemma is proved. $\square$

Proof of Theorem 1.1. According to Remark 1.5 we may assume that $k_n =$ degree p_n.

Since $\{\nu_n\}$ is a sequence of unit measures on the Borel sets of $\overline{\mathbb{C}}$, by Helly's selection theorem every subsequence of $\{\nu_n\}$ possesses a weak limit point that is supported on ∂E by (1.3) and Lemma 1.6. Let ν be the weak limit of the subsequence $\{\nu_{n_l}\}$. It remains to show that $\nu = \mu_E$.

Fix $\varepsilon > 0$ and $\zeta \in \mathbb{C}\setminus E$. Then there exists $r_\zeta > 0$ such that the disk $D(\zeta, r_\zeta)$ is contained in $\mathbb{C}\setminus E$. Let $\mathcal{D} := \overline{D(\zeta, r_\zeta/2)}$. Then (1.4) holds with S replaced by $\mathcal{D}$. Since U^ν and U^{μ_E} are continuous in $\mathbb{C}\setminus E$, we can choose r_ζ so small that for all $z \in \mathcal{D}$,

$$|U^\nu(z) - U^\nu(\zeta)| \leq \varepsilon, \qquad |U^{\mu_E}(z) - U^{\mu_E}(\zeta)| \leq \varepsilon. \tag{1.8}$$

Let $R_0 > 0$ be such that

$$V := \{z \in \mathbb{C} : |z| \geq R_0\} \subset \Omega.$$

Decompose $p_n = \widetilde{p}_n q_n$, such that

$$q_n(z) = a_n \prod_{j=1}^{m_n}(z - z_{j,n}),$$

where the $z_{j,n}$, $j = 1, \ldots, m_n$, are the zeros of p_n in $V \cup D(\zeta, r_\zeta)$ and a_n is the highest coefficient of p_n. Because of Lemma 1.6, $m_n = o(k_n)$ as $n \to \infty$.

By the maximum principle, (1.1), (1.2), and (1.6),

$$\varepsilon > h_n(z) = \frac{1}{k_n} \log |\widetilde{p}_n(z)| - G(z) + \frac{1}{k_n} \log |q_n(z)|$$

$$\geq \log \operatorname{cap} E - \varepsilon + \frac{1}{k_n} \log |q_n(z)|,$$

for $|z| = R_\varepsilon > R_0$ and $n \geq N_\varepsilon$, where R_ε and N_ε are chosen large enough. By the maximum principle applied to $q_n(z)$,

$$\limsup_{n \to \infty} \left(\frac{1}{k_n} \log \|q_n\|_\mathcal{D} \right) \leq -\log \operatorname{cap} E.$$

Then there is an $N \in \mathbb{N}$, such that for $n \geq N$,

$$\max_{z \in \mathcal{D}} \frac{1}{k_n - m_n} \log |q_n(z)| \leq -\log \operatorname{cap} E + \varepsilon$$

and

$$\max_{z \in \mathcal{D}} \left(\frac{1}{k_n - m_n} \log |p_n(z)| - G(z) \right) \geq -\varepsilon,$$

where we have used (1.4) and $m_n = o(k_n)$ as $n \to \infty$.

Let $\widetilde{\nu}_n$ denote the zero counting measure associated with $\widetilde{p}_n$. Then by (1.6), ν is the weak limit of $\widetilde{\nu}_{n_l}$. Thus, for all $z \in \mathcal{D}$,

$$\lim_{l \to \infty} U^{\widetilde{\nu}_{n_l}}(z) = -\lim_{l \to \infty} \int \log |z - \xi| \, d\widetilde{\nu}_{n_l}(\xi)$$

$$= -\int \log |z - \xi| \, d\nu(\xi) = U^\nu(z), \qquad (1.9)$$

and the convergence in (1.9) is uniform in $\mathcal{D}$.

Thus, for $n \geq N$,

$$\inf_{z \in \mathcal{D}} U^{\widetilde{\nu}_n}(z) = \inf_{z \in \mathcal{D}} \frac{1}{k_n - m_n} \log \frac{1}{|\widetilde{p}_n(z)|}$$

$$\leq \max_{z \in \mathcal{D}} (-\log \operatorname{cap} E - G(z)) + 2\varepsilon$$

$$= \max_{z \in \mathcal{D}} U^{\mu_E}(z) + 2\varepsilon.$$

Since the functions $U^{\nu_{n_l}}$ converge uniformly in $\mathcal{D}$ to U^ν as $l \to \infty$, we obtain

$$\liminf_{l \to \infty} \inf_{z \in \mathcal{D}} U^{\nu_{n_l}}(z) = \inf_{z \in \mathcal{D}} U^\nu(z) \leq \max_{z \in \mathcal{D}} U^{\mu_E}(z) + 2\varepsilon,$$

and with (1.8),

$$U^\nu(\zeta) \leq U^{\mu_E}(\zeta) + 4\varepsilon.$$

Since $\varepsilon > 0$ is arbitrary, we have
$$U^\nu(\zeta) \le U^{\mu_E}(\zeta), \quad \zeta \in \mathbb{C}\setminus E.$$
Finally, by Lemma 1.1.8 we get $\nu = \mu_E$. □

The special case $S = \{\infty\}$ is very fruitful in applications.

Theorem 1.7. *Let $\Lambda \subset \mathbb{N}$ be an infinite subset of positive integers and let $\{p_n\}_{n \in \Lambda}$ be a sequence of monic polynomials of respective degrees precisely n satisfying*
$$\limsup_{\substack{n \to \infty \\ n \in \Lambda}} \|p_n\|_E^{1/n} \le \operatorname{cap} E$$
and
$$\lim_{\substack{n \to \infty \\ n \in \Lambda}} \nu_n(M) = 0$$
for every compact set $M \subset E^\circ$, where $\nu_n = \nu_{p_n}$ is the zero-counting measure associated with p_n. Then $\nu_n \xrightarrow{*} \mu_E$ as $n \to \infty$, $n \in \Lambda$.

Proof. Define
$$\widetilde{p}_n(z) := (\operatorname{cap} E)^{-n} p_n(z).$$
We get the assertion applying Theorem 1.1 to $\{\widetilde{p}_n\}$ with $S = \{\infty\}$. □

We are now able to prove the results of Jentzsch and Szegő mentioned in the introduction: Let $f(z) = \sum_{n=0}^\infty a_n z^n$ be analytic in the disk $D(0, \rho)$, $\rho > 0$, where ρ is the radius of convergence of the power series. Then
$$\frac{1}{\rho} = \limsup_{n \to \infty} |a_n|^{1/n},$$
and consequently, there exists a subsequence $\{n_k\}_{k=1}^\infty$ such that $a_{n_k} \ne 0$ and
$$\frac{1}{\rho} = \lim_{k \to \infty} |a_n|^{1/n_k}.$$
Moreover, the partial sums s_n converge locally uniformly to $f(z)$ in the interior of C_ρ. Hence, fix $\varepsilon > 0$ and $0 < r < \rho$. Then there exists $N \in \mathbb{N}$ such that
$$\|s_n\|_{C_r} \le \|f\|_{C_r} + \varepsilon, \quad n \ge N.$$
The Bernstein–Walsh lemma (Lemma 1.1.11) yields
$$\|s_n\|_{C_\rho} \le (\|f\|_{C_r} + \varepsilon) \left(\frac{\rho}{r}\right)^n,$$
since $G(z) = \log(|z|/r)$ is the Green function for $\Omega = \{z : |z| > r\}$. Therefore, the monic polynomials $p_n(z) = a_n^{-1} s_n(z)$, $n = n_1, n_2, \ldots$, satisfy
$$\limsup_{k \to \infty} \|p_{n_k}\|_{C_\rho}^{1/n_k} \le \operatorname{cap} C_\rho = \rho.$$

Moreover, Hurwitz's theorem yields for any compact $M \subset \operatorname{int} C_\rho$ that

$$\lim_{n\to\infty} \nu_n(M) = 0,$$

where ν_n is the zero counting measure of s_n, respectively p_n. Thereby, all conditions of Theorem 1.7 are satisfied, and $\nu_n \xrightarrow{*} \mu_{C_\rho}$ and

$$\lim_{k\to\infty} \nu_{n_k}(S(\alpha,\beta)) = \frac{\beta - \alpha}{2\pi}$$

for any sector $S(\alpha,\beta) = \{z \in \mathbb{C} : \alpha < \arg z < \beta\}$, $\alpha < \beta < 2\pi + \alpha$.

2.2 Erdős–Turán Type Theorems for Quasiconformal Curves and Arcs

Let $E = C_\rho$ be the circle with center 0 and radius ρ, and let μ denote the equilibrium measure of C_ρ. Moreover, let $p_n \in \mathbb{M}_n$ and $p_n(z) = z^n + \cdots + a_0$, $a_0 \neq 0$. Then the Erdős–Turán estimate (0.2) for the distribution of the zeros of p_n with respect to the sectors $S(\alpha,\beta)$ was stated in terms of

$$\frac{\log P(\rho)}{n} = \frac{1}{n} \log \frac{\|p_n\|_{C_\rho}}{\sqrt{|a_0|\rho^n}}.$$

Note that

$$\frac{1}{n}\log|a_0| - \log\rho = \frac{1}{n}\log|p_n(0)| - \log\operatorname{cap} C_\rho = U^{\mu-\nu_n}(0),$$

and therefore

$$\frac{\log P(\rho)}{n} = \frac{1}{n}\log\|p_n\|_{C_\rho} - \log\rho + \frac{1}{2}U^{\nu_n-\mu}(0).$$

If we define

$$\varepsilon := \sup_{z \in C_\rho} U^{\mu-\nu_n}(z) = \frac{1}{n}\log\|p_n\|_{C_\rho} - \log\rho$$

and

$$\delta := \varepsilon - U^{\mu-\nu_n}(0),$$

then the Erdős–Turán estimate can be reformulated as

$$|(\mu - \nu_n)(S(\alpha,\beta))| \leq c(\varepsilon + \delta)^{1/2},$$

where $c = 8\sqrt{2}$.

2.2 Erdős–Turán Type Theorems for Quasiconformal Curves and Arcs

On the other hand, we know by Lemma 1.4.3 and by Lemma 1.4.4 that

$$\nu_n(\mathbb{C} \setminus D((1+t)\rho)) \leq \frac{\varepsilon}{\log(1+t)}, \quad t > 0,$$

$$\nu_n(D((1-t)\rho)) \leq \frac{\delta}{\log(1/(1-t))}, \quad 0 < t < 1,$$

where $D(r) = \{z : |z| < r\}$, $r > 0$. Hence, for

$$A_t := \{z : (1-t)\rho \leq |z| \leq (1+t)\rho\}$$

we obtain

$$\nu_n(\mathbb{C} \setminus A_t) \leq \frac{1}{\log 2} \frac{\varepsilon + \delta}{t}, \quad 0 < t < 1.$$

Choosing $t := r_0 := (\varepsilon + \delta)^{1/2}$, with

$$|(\mu - \nu_n)(S(\alpha, \beta) \cap A_r)| \leq c_1 (\varepsilon + \delta)^{1/2}, \quad r \geq r_0,$$

we have obtained global estimates of order $(\varepsilon + \delta)^{1/2}$ for the difference of the two measures μ and ν_n in special neighborhoods of any subarcs of C_ρ.
These interpretations are motivations for the forthcoming theorems.

2.2.1 Polynomial Bounds on Quasiconformal Curves

Let $E := L \subset \mathbb{C}$ be a K-quasiconformal Jordan curve and let $p = p_n \in \mathbb{M}_n$ be a monic polynomial. We associate with p the normalized counting measure for its zeros ν_p given by Definition 1.4.1. Let μ_L be the equilibrium measure for L. Our basic results will be formulated in terms of the following quantities:

$$\varepsilon_{p,L} := \sup_{z \in \mathbb{C}} U^{\mu_L - \nu_p}(z) = \frac{1}{n} \log \|p\|_L - \log \operatorname{cap} L, \tag{2.1}$$

and

$$\delta_{p,L} := \varepsilon_{p,L} - U^{\mu_L - \nu_p}(z_0) = \frac{1}{n} \log \frac{\|p\|_L}{|p(z_0)|}, \tag{2.2}$$

where $z_0 \in G := \operatorname{int} L$ is an arbitrary, but fixed, point with $p(z_0) \neq 0$. That means that $\varepsilon_{p,L}$ is a one-sided bound for the logarithmic potential $U^{\mu_L - \nu_p}$. The term $\varepsilon_{p,L}$ plays an essential role in Lemma 1.4.3, and $\delta_{p,L}$ is the analogous parameter in Lemma 1.4.4 for the interior of L.

Let Φ denote again the Riemann mapping of the domain $\Omega = \operatorname{ext} L$ onto Δ normalized by $\Phi(\infty) = \infty$, $\Phi'(\infty) > 0$ and set $\Psi = \Phi^{-1}$. Denote by $w = \varphi(z)$ the conformal mapping of G onto $\mathbb{D}$ with the normalization $\varphi(z_0) = 0, \varphi'(z_0) > 0$. Set $\psi := \varphi^{-1}$.

58   2. Zero Distribution of Polynomials

The functions $\Phi, \Psi, \varphi, \psi$ can be naturally extended to quasiconformal homeomorphisms of $\overline{\mathbb{C}}$ with ∞ as the fixed point (cf. Theorem B.2.1), and we keep the previous notation for these extensions.

Further, for $\zeta \in \mathbb{C} \setminus \{z_0\}$ set

$$\zeta_L := \begin{cases} \Psi\left(\dfrac{\Phi(\zeta)}{|\Phi(\zeta)|}\right), & \text{if } \zeta \in \Omega \setminus \{\infty\}, \\ \psi\left(\dfrac{\varphi(\zeta)}{|\varphi(\zeta)|}\right), & \text{if } \zeta \in G \setminus \{z_0\}, \\ \zeta, & \text{if } \zeta \in L. \end{cases}$$

Moreover,

$$L_r^+ := \{\zeta : |\Phi(\zeta)| = 1 + r\} \quad r > 0,$$
$$L_r^- := \{\zeta : |\varphi(\zeta)| = 1 - r\}, \quad 0 < r < 1.$$

Let J be an arbitrary subarc of L. For $\sigma > 0$ and $0 < \tau < 1$ define

$$J_\sigma^+ := \{\zeta \in L_\sigma^+ : \zeta_L \in J\},$$
$$J_\tau^- := \{\zeta \in L_\tau^- : \zeta_L \in J\},$$
$$E_{\sigma,\tau} := (\text{ext } L_\tau^-) \cap (\text{int } L_\sigma^+),$$
$$A_{\sigma,\tau}(J) := \{\zeta \in E_{\sigma,\tau} : \zeta_L \in J\} \tag{2.3}$$

(see Figure 2.1). These characteristic sets $A_{\sigma,\tau}(J)$ will play an essential

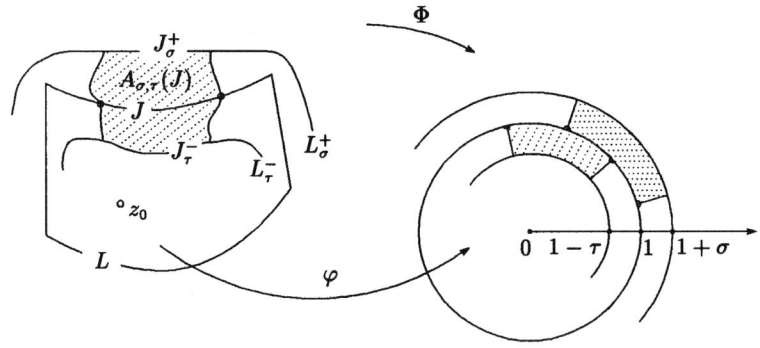

Fig. 2.1

role in the following estimates, since these are the sets where the main zeros near J live. We use for $A_{\sigma,\tau}(J)$ the notion of (σ, τ)-*neighborhood for* $J \subset L$ *with respect to the curve* L. Note that in general $A_{\sigma,\tau}(J)$ is not a

2.2 Erdős–Turán Type Theorems for Quasiconformal Curves and Arcs 59

neighborhood of J in $\mathbb{C}$ in the standard sense, only if J is an open subarc of L.

Consider the function $\Phi \circ \psi$ (given on the unit circle $\mathbb{T}$) and its modulus of continuity

$$\omega_{\Phi \circ \psi}(x) := \sup_{\substack{|w|=|t|=1, \\ |w-t| \leq x}} |(\Phi \circ \psi)(w) - (\Phi \circ \psi)(t)|, \quad x > 0.$$

The functions Φ and ψ satisfy a Hölder condition. The validity of this well-known fact follows, for example, from Theorem 1.2.7.

Hence,

$$\omega_{\Phi \circ \psi}(x) \leq C x^\alpha, \quad x > 0, \tag{2.4}$$

with some constants $C > 0$ and $0 < \alpha \leq 1$.

Theorem 2.1. *Let L be a K-quasiconformal curve, and let the function $\Phi \circ \psi$ satisfy condition (2.4). Let $z_0 \in G := \operatorname{int} L$ and $0 < c_0 < 1$ be fixed. Suppose that $p \in \mathbb{M}_n$, $n \in \mathbb{N}$, is a monic polynomial such that $(\varepsilon_{p,L} + \delta_{p,L}) < 1$, where $\varepsilon_{p,L}$ and $\delta_{p,L}$ are defined by (2.1) and (2.2). Then there exists a constant $c > 0$, depending only on K, α, C, z_0, c_0, such that*

$$|(\mu_L - \nu_p)(A_{\sigma,\tau}(J))| \leq c\, (\varepsilon_{p,L} + \delta_{p,L})^{\alpha/(1+\alpha)} \tag{2.5}$$

for all subarcs J of L and all σ and τ satisfying the properties $\sigma \geq \sigma_0$ and $1 > \tau \geq \tau_0$, where $\sigma_0 := c_0\, (\varepsilon_{p,L} + \delta_{p,L})^{\alpha/(1+\alpha)}$ and $\tau_0 := \sigma_0^{1/\alpha}$.

Before starting the proof of Theorem 2.1 we are going to outline its basic ideas: Given a subarc J of L, we construct a function g that is a good approximant of the characteristic function of $A_{\sigma,\tau}(J)$. Moreover, g is constructed in such a way that it has compact support and has partial derivatives of all orders in $\mathbb{C}$. Then Green's formula (1.1.10) yields

$$g(z) = \frac{1}{2\pi} \int_{\mathbb{C}} \Delta g(\xi) \log|z - \xi|\, dm(\xi), \quad z \in \mathbb{C}.$$

By integration we obtain

$$\int g\, (d\nu_p - d\mu_L) = \frac{1}{2\pi} \int_{\mathbb{C}} U^{\mu_L - \nu_p}(\xi) \Delta g(\xi)\, dm(\xi). \tag{2.6}$$

If g is an appropriate good approximation of the characteristic function of $A_{\sigma,\tau}(J)$, the left-hand side of (2.6) provides an approximation of

$$(\mu_L - \nu_p)(A_{\sigma,\tau})(J),$$

whereas the right-hand side of (2.6) can be estimated in terms of $(\varepsilon_{p,L} + \delta_{p,L})$.

First of all, we discuss the construction of the above-mentioned function g. Since we need such functions g for different subarcs of L, we use in the following construction an arbitrary subarc l of L instead of J.

Let l be an arbitrary subarc of the K-quasiconformal curve L, and let α, C be the constants in (2.4).

Denote by
$$\omega(z) := \omega(z, E_{\sigma,\tau}, l_\sigma^+ \cup l_\tau^-),$$
where $\sigma > 0$, $0 < \tau < 1$, the harmonic measure of $l_\sigma^+ \cup l_\tau^-$ at the point $z \in E_{\sigma,\tau}$ with respect to $E_{\sigma,\tau}$.

Lemma 2.2. *Let* $z \in A_{\sigma,\tau}(L \setminus l)$, $0 < \sigma < 1$, $\tau = \sigma^{1/\alpha}$. *There exist constants* $c_j = c_j(K, \alpha, C) > 0$, $j = 1, 2$, *such that*
$$\omega(z) \le c_1 \exp\left\{-c_2 \frac{d}{\sigma}\right\}, \qquad (2.7)$$
where $d := d(\Phi(z_L), \Phi(l))$ *is the distance between* $\Phi(z_L)$ *and* $\Phi(l)$.

Proof. The function $\omega(z)$ shows that we may suppose without loss of generality that $d \ge c_3 \sigma$, where the sufficiently large constant $c_3 > 1$ will be chosen later.

Let ζ_1 and ζ_2 be the endpoints of the arc l. Consider the quadrilateral $Q := A_{\sigma,\tau}(L \setminus l)$ whose sides are the arcs
$$\gamma_j := \{\zeta \in E_{\sigma,\tau} : \zeta_L = \zeta_j\}, \quad j = 1, 2,$$
$$\gamma_3 := (L \setminus l)_\tau^-, \, \gamma_4 := (L \setminus l)_\sigma^+.$$

An elementary argument involving the maximum principle for harmonic functions (see Theorem 1.1.2) shows that
$$\omega(z) \le \sum_{j=1}^{2} \omega(z, Q, \gamma_j), \qquad (2.8)$$
where $\sigma_0 := c_0 \left(\varepsilon_{p,L} + \delta_{p,L}\right)^{\alpha/(1+\alpha)}$. The arc
$$\gamma_5 := \{\zeta \in E_{\sigma,\tau} : \zeta_L = z_L\}$$
divides Q into two new quadrilaterals Q_1 and Q_2. For definiteness we assume that
$$\gamma_j \subset \partial Q_j, \quad j = 1, 2.$$
Denote by Γ_j, $j = 1, 2$, the family of all locally rectifiable arcs $\gamma \subset Q_j$ separating in Q_j the sides γ_j and γ_5, i.e., arcs with one endpoint on γ_3 and the other one on γ_4 (see Figure 2.2).

Applying (1.2.17) and the comparison principle for modules of families of arcs and curves (1.2.11), we find that
$$\omega(z, Q, \gamma_j) \preceq \exp\{-\pi \, m(\Gamma_j)\}, \quad j = 1, 2. \qquad (2.9)$$

2.2 Erdős–Turán Type Theorems for Quasiconformal Curves and Arcs 61

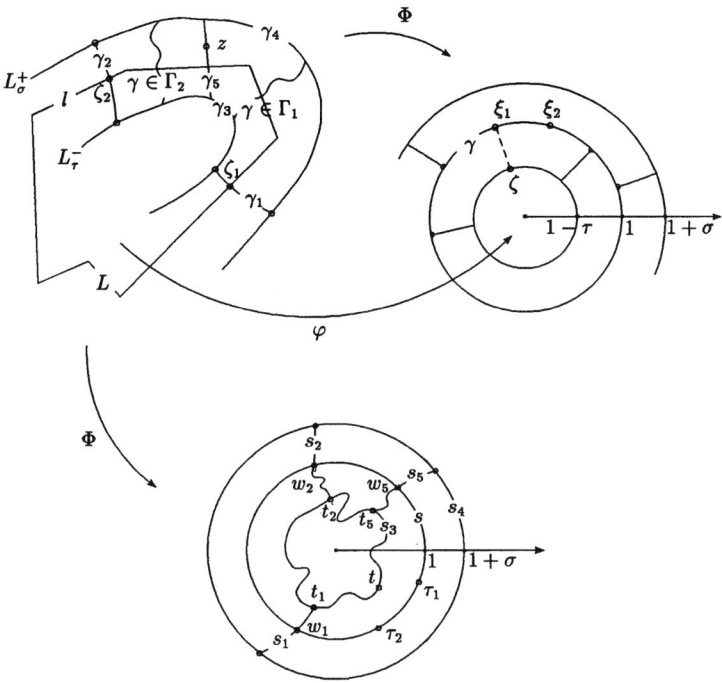

Fig. 2.2

Thus, the problem of a suitable estimation of the quantity $m(\Gamma_j)$ from below has to be our next objective.

We begin with the inequality

$$m(\Gamma_j) \succeq m(\Gamma'_j), \tag{2.10}$$

where $\Gamma'_j := \Phi(\Gamma_j)$, which is a direct consequence of relation (1.2.20).

Now set

$$s := \Phi(L \setminus l) = \{e^{i\theta} : \theta_1 < \theta < \theta_2\},$$
$$s_k := \Phi(\gamma_k), \quad k = 1, \ldots, 5,$$
$$t_j := \bar{s}_3 \cap \bar{s}_j, \ w_j := \bar{s} \cap \bar{s}_j, \quad j = 1, 2, 5.$$

Theorem 1.2.7 applied to the quasiconformal mapping $\Phi \circ \psi$ implies that for $j = 1, 2, 5$,

$$\operatorname{diam}(s_j \cap \mathbb{D}) \leq c_4 |w_j - t_j| \leq c_5 (1 - |t_j|).$$

Furthermore, for any $t \in s_3$,

$$1 - |t| \leq c_6 \sigma.$$

Indeed, consider the point $\zeta := (\varphi \circ \Psi)(t)$ and the arc $\gamma := (\varphi \circ \Psi)(s)$. Since by our assumption $|s| \geq \sigma$, we have
$$|\gamma| \geq c_7 \sigma^{1/\alpha}.$$
We select the points ξ_1 and $\xi_2 \in \gamma$ so that
$$\arg \xi_1 = \arg \zeta, \quad |\xi_1 - \xi_2| = \frac{1}{2 c_7} \sigma^{1/\alpha}.$$
It is important to recall that the mapping $\varphi \circ \Psi$ is quasiconformal, since this implies via Theorem 1.2.7 that for $\tau_j := (\Phi \circ \psi)(\xi_j)$, $j = 1, 2$, we have
$$1 - |t| \leq |t - \tau_1| \leq c_8 |\tau_1 - \tau_2| \leq c_9 |\xi_1 - \xi_2|^\alpha = c_6 \sigma.$$
Consider the family of arcs
$$\Gamma''_j := \{\gamma_\theta := \{re^{i\theta} : 1 - c_6 \sigma < r < 1 + \sigma\}, \theta_1 + c_{10} \sigma \leq \theta \leq \theta_2 - c_{10} \sigma\},$$
where $c_{10} := \frac{\pi}{2} c_5 c_6$.

It should be noted that Γ''_j will be defined correctly if we choose $c_3 > 2 c_{10}$.

Moreover, if $c_3 > 3 c_{10}$, then according to (1.2.16) and the comparison principle (1.2.11), we get
$$m(\Gamma'_j) \geq m(\Gamma''_j) = \frac{\theta_2 - \theta_1 - 2 c_{10} \sigma}{\log(1 + \sigma) - \log(1 - c_6 \sigma)} \geq c_{11} \frac{d}{\sigma}. \quad (2.11)$$
Now (2.7) follows from (2.8)–(2.11). $\square$

Corollary 2.3. *Writing the assertion of Lemma 2.2 for the arc $L \setminus l$ instead of l, we obtain the following inequality:*
$$1 - \omega(z) \leq c_1 \, exp\left\{-\frac{c_2}{\sigma} d(\Phi(z), \Phi(L \setminus l))\right\}, \quad z \in A_{\sigma, \sigma^{1/\alpha}}(l).$$

Now, for $w \in \mathbb{C} \setminus \mathbb{T}$, $0 < \sigma < 1$, and $\tau := \sigma^{1/\alpha}$ set
$$\widetilde{\omega}(w) := \begin{cases} \omega[\Psi(w)], & \text{if } 1 < |w| \leq 1 + \sigma, \\ \omega[\psi(w)], & \text{if } 1 - \tau \leq |w| < 1, \\ 0, & \text{if } |w| > 1 + \sigma \text{ or } |w| < 1 - \tau. \end{cases}$$

We average this function in $\mathbb{D}$ and Δ separately in the following way.

Let $K(z)$, $z \in \mathbb{C}$, be an arbitrary averaging kernel, i.e., $K(z) \in C^\infty(\mathbb{C})$ satisfies (1.5.3)–(1.5.5). Consider the function
$$\widetilde{g}(w) := \begin{cases} \dfrac{16}{\sigma^2} \displaystyle\int_{\mathbb{C}} \widetilde{\omega}(t) K\left(\dfrac{4(t-w)}{\sigma}\right) dm(t), & \text{if } 1 + \dfrac{1}{2}\sigma \leq |w| \leq 1 + \dfrac{3}{2}\sigma, \\ \dfrac{16}{\tau^2} \displaystyle\int_{\mathbb{C}} \widetilde{\omega}(t) K\left(\dfrac{4(t-w)}{\tau}\right) dm(t), & \text{if } 1 - \dfrac{3}{2}\tau \leq |w| \leq 1 - \dfrac{1}{2}\tau, \\ \widetilde{\omega}(w), & \text{elsewhere in } \mathbb{C} \setminus \mathbb{T}. \end{cases}$$

2.2 Erdős–Turán Type Theorems for Quasiconformal Curves and Arcs

According to the results of Section 1.5.2, the function $\widetilde{g}(w)$ has in $\mathbb{C} \setminus \mathbb{T}$ partial derivatives of all orders and satisfies the inequalities

$$0 \le \widetilde{g}(w) \le 1, \quad w \in \mathbb{C} \setminus \mathbb{T}, \tag{2.12}$$

$$|\Delta \widetilde{g}(w)| \preceq \begin{cases} \sigma^{-2}, & \text{if } |w| > 1, \\ \tau^{-2}, & \text{if } |w| < 1. \end{cases} \tag{2.13}$$

Therefore, the function

$$g(z) := \begin{cases} \widetilde{g}[\Phi(z)], & \text{if } z \in \operatorname{ext} L, \\ \widetilde{g}[\varphi(z)], & \text{if } z \in \operatorname{int} L, \\ \omega(z), & \text{if } z \in L, \end{cases}$$

has in $\mathbb{C}$ partial derivatives of all orders.

Moreover, we note that by Green's formula (1.1.9)

$$\int_{\mathbb{C}} \Delta g(z)\, dm(z) = 0. \tag{2.14}$$

Proof of Theorem 2.1. Let l be an arbitrary subarc of L. For abbreviation we use $\nu := \nu_p$, $\mu := \mu_L$, $\varepsilon := \varepsilon_{p,L}$, and $\delta := \delta_{p,L}$. Let $0 < \sigma < 1$ and $\tau = \sigma^{1/\alpha}$. We construct the function g as above and obtain according to (2.6)

$$\int g(d\nu - d\mu) = \frac{1}{2\pi} \int U^{\mu-\nu}(\xi) \Delta g(\xi)\, dm(\xi). \tag{2.15}$$

First, we establish the inequality

$$\left| \int g\, (d\nu - d\mu) \right| \preceq \frac{\varepsilon + \delta}{\tau}. \tag{2.16}$$

Indeed, setting

$$\widetilde{U}^{\mu-\nu}(w) := \begin{cases} U^{\mu-\nu}[\Psi(w)], & \text{if } |w| > 1, \\ U^{\mu-\nu}[\psi(w)], & \text{if } |w| < 1, \end{cases}$$

and using the representation of the function g by the Green formula (1.1.11)

$$g(z) = \frac{1}{2\pi} \iint_{\mathbb{C}} \Delta g(\zeta) \log|z - \zeta|\, dm(\zeta), \quad z \in \mathbb{C},$$

we obtain according to (2.13), (2.14), and (1.1.3)

$$\left| \int g\,(d\nu - d\mu) \right| = \frac{1}{2\pi} \left| \iint_{\mathbb{C}} U^{\mu-\nu}(\zeta)\Delta g(\zeta)\,dm(\zeta) \right|$$

$$= \frac{1}{2\pi} \left| \iint_{\mathbb{C}} (\varepsilon - U^{\mu-\nu}(\zeta))\Delta g(\zeta)\,dm(\zeta) \right|$$

$$\leq \frac{1}{2\pi} \iint_{\Delta \cup \mathbb{D}} (\varepsilon - \widetilde{U}^{\mu-\nu}(w))|\Delta \widetilde{g}(w)|\,dm(w)$$

$$\preceq \sigma^{-2} \int_{1+\sigma/2}^{1+2\sigma} r\,\frac{1}{2\pi} \int_0^{2\pi} (\varepsilon - \widetilde{U}^{\mu-\nu}(re^{i\theta}))\,d\theta dr$$

$$+ \tau^{-2} \int_{1-2\tau}^{1-\tau/2} r\,\frac{1}{2\pi} \int_0^{2\pi} (\varepsilon - \widetilde{U}^{\mu-\nu}(re^{i\theta}))\,d\theta dr$$

$$\preceq \frac{\varepsilon}{\sigma} + \frac{\varepsilon - \widetilde{U}^{\mu-\nu}(z_0)}{\tau} \preceq \frac{\varepsilon + \delta}{\tau}.$$

Next, we show that the left-hand side of (2.15) is a good approximation of $(\nu - \mu)(A_{\sigma,\tau})(l)$. Therefore, we set $\sigma_1 := \sigma/2, \tau_1 := \tau/2$. By Green's formula (1.1.10), applied to the domain E_{σ_1,τ_1}, we have for $z \in L$,

$$g(z) = \frac{1}{2\pi} \int_{L_{\sigma_1}^+ \cup L_{\tau_1}^-} \left(\omega(\zeta)\frac{\partial}{\partial \mathbf{n}} \log|\zeta - z| - \frac{\partial}{\partial \mathbf{n}}\omega(\zeta) \log|\zeta - z| \right) |d\zeta|,$$

where $\partial/\partial \mathbf{n}$ denotes the operator of differentiation with respect to the appropriate normal.

Integrating the last relation we get

$$\int g(z)\,d\mu(z) = -\frac{1}{2\pi} \int_{L_{\sigma_1}^+ \cup L_{\tau_1}^-} \left(\omega(\zeta)\frac{\partial}{\partial \mathbf{n}} U^\mu(\zeta) - \frac{\partial}{\partial \mathbf{n}}\omega(\zeta) U^\mu(\zeta) \right) |d\zeta|$$

$$= \frac{1}{2\pi} \int_{L_{\sigma_1}^+} \left(\omega(\zeta)\frac{\partial}{\partial \mathbf{n}} \log|\Phi(\zeta)| - \frac{\partial}{\partial \mathbf{n}}\omega(\zeta) \log|\Phi(\zeta)| \right) |d\zeta|.$$

(2.17)

2.2 Erdős–Turán Type Theorems for Quasiconformal Curves and Arcs

For $w \in \Delta$ set $\widetilde{U}^\mu(w) := U^\mu(\Psi(w))$. Next, we analyze the integrals on the right-hand side of (2.17). Note that

$$\frac{1}{2\pi} \int_{L_{\sigma_1}^+} \left(\omega(\zeta) \frac{\partial}{\partial \mathbf{n}} \log |\Phi(\zeta)| \right) |d\zeta| = \frac{1}{2\pi} \int_{|w|=1+\sigma_1} \frac{\widetilde{\omega}(w)}{|w|} |dw|$$

$$= \mu(l) + \frac{1}{2\pi(1+\sigma_1)} \int_{|w|=1+\sigma_1} (\widetilde{\omega}(w) - \chi(w)) |dw|,$$

where

$$\chi(w) := \begin{cases} 1, & \text{if } \dfrac{w}{1+\sigma_1} \in \Phi(l), \\ 0, & \text{elsewhere in } \mathbb{C}. \end{cases}$$

Since by Lemma 2.2 and Corollary 2.3

$$\int_{|w|=1+\sigma_1} (\widetilde{\omega}(w) - \chi(w)) |dw| \preceq \int_0^{2\pi} e^{-c_1 x/\sigma} \, dx \preceq \sigma,$$

we have

$$\left| \frac{1}{2\pi} \int_{L_{\sigma_1}^+} \omega(\zeta) \frac{\partial}{\partial \mathbf{n}} \log |\Phi(\zeta)| \, |d\zeta| - \mu(l) \right| \preceq \sigma. \qquad (2.18)$$

The same reasoning can be applied to the second integral. We need only add the following simple consequence of Schwarz's formula (cf. (1.1.7)): For w with $|w| = 1 + \sigma_1$,

$$|\operatorname{grad} \widetilde{\omega}(w)| \leq \frac{1}{\pi} \int_{|\tau-w|=\sigma_1} \frac{|\widetilde{\omega}(\tau) - \widetilde{\omega}(w)|}{|\tau-w|^2} |d\tau|$$

$$\preceq \frac{1}{\sigma} \exp \left\{ -c_2 \frac{d(w/(1+\sigma_1), \{w_1, w_2\})}{\sigma} \right\},$$

where w_1 and w_2 are the endpoints of $\Phi(l)$, and therefore

$$\left| \int_{|w|=1+\sigma_1} \frac{\partial}{\partial \mathbf{n}} \widetilde{\omega}(w) \, |dw| \right| \preceq \frac{1}{\sigma} \int_0^{2\pi} e^{-c_2 x/\sigma} \, dx \preceq 1.$$

Hence,

$$\left| \frac{1}{2\pi} \int_{L_{\sigma_1}^+} \frac{\partial}{\partial \mathbf{n}} \omega(\zeta) \log |\Phi(\zeta)| \, |d\zeta| \right|$$

$$= \frac{1}{2\pi} \log(1 + \sigma_1) \left| \int_{|w|=1+\sigma_1} \frac{\partial}{\partial \mathbf{n}} \widetilde{\omega}(w) \, |dw| \right| \preceq \sigma. \qquad (2.19)$$

Combining (2.17)–(2.19), we get

$$\left| \int g \, d\mu - \mu(l) \right| \preceq \sigma,$$

and by (2.16),

$$\left| \int g \, d\nu - \mu(l) \right| \preceq \left(\sigma + \frac{\varepsilon + \delta}{\tau} \right). \qquad (2.20)$$

By virtue of Lemma 1.4.3 and Lemma 1.4.4, we have

$$\nu(\operatorname{ext} L_t^+) \leq \frac{\varepsilon}{\log(1+t)}, \quad t > 0,$$

$$\nu(\operatorname{int} L_t^-) \leq \frac{\delta}{\log(1/(1-t))}, \quad 0 < t < 1.$$

Therefore, for $0 < \tau < \sigma \leq \frac{1}{2}$,

$$|(\mu - \nu)(E_{\sigma,\tau})| = \nu(\mathbb{C} \setminus E_{\sigma,\tau}) \preceq \frac{\varepsilon}{\sigma} + \frac{\delta}{\tau} \leq \frac{\varepsilon + \delta}{\tau}. \qquad (2.21)$$

To prove (2.5) we may assume that $(\varepsilon + \delta)$ is sufficiently small such that $\sigma_0 < \frac{1}{2}$. Moreover, we prove (2.5) first for the case $\sigma = \sigma_0$ and $\tau = \tau_0 = \sigma^{1/\alpha}$ and begin with the assertion

$$(\mu - \nu)(A_{\sigma_0, \tau_0}(J)) \preceq \sigma_0. \qquad (2.22)$$

For $t > 0$ set

$$\gamma := \Phi(J) = \{e^{i\theta} : \theta_1 \leq \theta \leq \theta_2\},$$
$$\gamma_t := \{e^{i\theta} : \theta_1 - t\sigma_0 \leq \theta \leq \theta_2 + t\sigma_0\},$$
$$J_t := \Psi(\gamma_t). \qquad (2.23)$$

We distinguish two cases:

$$\mu(J) \leq \sigma_0 \quad \text{and} \quad \mu(J) > \sigma_0. \qquad (2.24)$$

2.2 Erdős–Turán Type Theorems for Quasiconformal Curves and Arcs

If $\mu(J) \leq \sigma_0$, then (2.22) is trivial. But we need more information about $\nu(A_{\sigma_0,\tau_0}(J))$ in the following. For this reason, let $M > 0$, $l := J_M$, and $\sigma := 2\sigma_0$. Here g denotes the function constructed above for l. By Corollary 2.3 we can choose M so large that

$$g(z) = w(z) \geq \frac{1}{2}, \qquad z \in A_{\sigma_0,\tau_0}(J).$$

Therefore, by (2.20),

$$\frac{1}{2}\nu(A_{\sigma_0,\tau_0}(J)) \leq \int g\,d\nu \preceq \sigma_0. \tag{2.25}$$

Now let $\mu(J) > \sigma_0$. Then we set

$$a := \mu(J). \tag{2.26}$$

Without loss of generality we assume that

$$\gamma = \Phi(J) = \{e^{i\theta} : -a\pi \leq \theta \leq a\pi\}.$$

For $l := J$ and $\sigma := \sigma_0$ consider the function g from above. From (2.20) we obtain

$$\int g\,d\nu \geq a - c_1\sigma_0. \tag{2.27}$$

To estimate the left-hand term in (2.27) we construct the following subdivision of $\mathbb{T} \setminus \gamma$.

We again distinguish two cases.

If $1 - a \leq \sigma_0$, then

$$\Gamma_1 := \left\{e^{i\theta} : a\pi \leq \theta \leq \pi\right\}.$$

If $1 - a > \sigma_0$, then

$$\Gamma_{k+1} := \left\{e^{i\theta} : (a + k\sigma_0)\pi \leq \theta \leq (a + (k+1)\sigma_0)\pi\right\}$$
$$\left(k = 0, \ldots, k_0 := \left[\frac{1-a}{\sigma_0}\right] - 1\right),$$
$$\Gamma_{k_0+2} := \left\{e^{i\theta} : (a + (k_0+1)\sigma_0)\pi \leq \theta \leq \pi\right\},$$

where $[a]$ denotes the integer part of $a \geq 0$. In both cases let

$$\Gamma_{-k} := \left\{e^{i\theta} : e^{-i\theta} \in \Gamma_k\right\},$$
$$\Gamma'_{\pm k} := \Psi(\Gamma_{\pm k}).$$

Since by (2.27)

$$\nu(A_{\sigma_0,\tau_0}(\Gamma'_{\pm k})) \preceq \sigma_0,$$

we have owing to Lemma 2.2, Corollary 2.3, and (2.25),

$$\int g\, d\nu \leq \nu(\mathbb{C} \setminus E_{\sigma_0,\tau_0}) + \nu(A_{\sigma_0,\tau_0}(J))$$
$$+ c_2 \sum_{k=0}^{k_0+1} e^{-c_3 k} \left(\nu(A_{\sigma_0,\tau_0}(\Gamma'_{k+1})) + \nu(A_{\sigma_0,\tau_0}(\Gamma'_{-(k+1)})) \right)$$
$$\leq \nu(A_{\sigma_0,\tau_0}(J)) + c_4\, \sigma_0.$$

Comparing the last inequality with (2.27), we have for any subarc $J \subset L$,

$$(\mu - \nu)(A_{\sigma_0,\tau_0}(J)) \leq c_5\, \sigma_0. \tag{2.28}$$

To get the lower estimate of $(\mu - \nu)(A_{\sigma_0,\tau_0}(J))$ we use

$$(\nu - \mu)(A_{\sigma_0,\tau_0}(J)) = (\mu - \nu)(A_{\sigma_0,\tau_0}(L \setminus J)) - \nu(\mathbb{C} \setminus E_{\sigma_0,\tau_0})$$

and apply (2.21) and (2.28) to the arc $L \setminus J$. Hence we have proved (2.5) for $\sigma = \sigma_0$ and $\tau = \tau_0$.

Let $\sigma > \sigma_0$ and $1 > \tau > \tau_0$. Then (2.21) yields

$$|(\mu - \nu)(\mathbb{C} \setminus E_{\sigma,\tau})| \leq |(\mu - \nu)(\mathbb{C} \setminus E_{\sigma_0,\tau_0})| \leq \frac{\varepsilon + \delta}{\tau_0} = c_0^{-1/\alpha}\, (\varepsilon + \delta)^{\alpha/(1+\alpha)},$$

and (2.5) holds for all $\sigma \geq \sigma_0$ and $1 \geq \tau \geq \tau_0$. □

2.2.2 Polynomial Bounds on Quasiconformal Arcs

If L is a K-quasiconformal arc, an analogue of Theorem 2.1 holds as well. Denote by z_1 and z_2 the endpoints of L. Since the normalized Riemann mapping Φ of ext L onto Δ can be extended continuously to these points, we set for $r > 0$ and $j = 1, 2$

$$t_j := \Phi(z_j), \quad \Delta_1 := \{t : |t| > 1, \arg t_1 < \arg t < \arg t_2\},$$
$$\Delta_2 := \Delta \setminus \overline{\Delta}_1, \quad \Omega_j := \Psi(\Delta_j), \quad J_j := \overline{\Delta}_j \cap \mathbb{T},$$
$$L_r := \{\zeta \in \Omega : |\Phi(\zeta)| = 1 + r\}.$$

A standard argument (see (B.2.1)) shows that $\partial \Delta_1$ and $\partial \Delta_2$ are both quasiconformal curves. Moreover, the curve $\partial \Omega_1 = \partial \Omega_2$ is quasiconformal, too (cf. Lemma 1.2.8). Therefore, the restriction Φ_j, $j = 1, 2$, of the function Φ to the region Ω_j can be extended to a K_1-quasiconformal mapping of the extended complex plane $\overline{\mathbb{C}}$ onto itself for a suitable constant $K_1 = K_1(L) > 1$ (cf. Theorem B.2.1). Set $\Psi_j := \Phi_j^{-1}$, $j = 1, 2$.

Using Theorem 1.2.7, we see that this fact makes it possible to obtain for $x > 0$

$$\omega_{\Phi_1 \circ \Psi_2}(x) + \omega_{\Phi_2 \circ \Psi_1}(x) \leq C x^\alpha, \tag{2.29}$$

2.2 Erdős–Turán Type Theorems for Quasiconformal Curves and Arcs

where $C > 0$ and $0 < \alpha \leq 1$ are some constants depending only on L, and $\omega_{\Phi_1 \circ \Psi_2}$, and $\omega_{\Phi_2 \circ \Psi_1}$ denote the moduli of continuity of the functions $\Phi_1 \circ \Psi_2$ and $\Phi_2 \circ \Psi_1$ on J_2 and J_1, respectively.

For an arbitrary subarc J of L and $\sigma > 0$ set $E_\sigma := \text{int } L_\sigma$ and

$$A_\sigma(J) := \{\zeta \in E_\sigma : \zeta_L \in J\} \tag{2.30}$$

(see Figure 2.3), where we use the notation

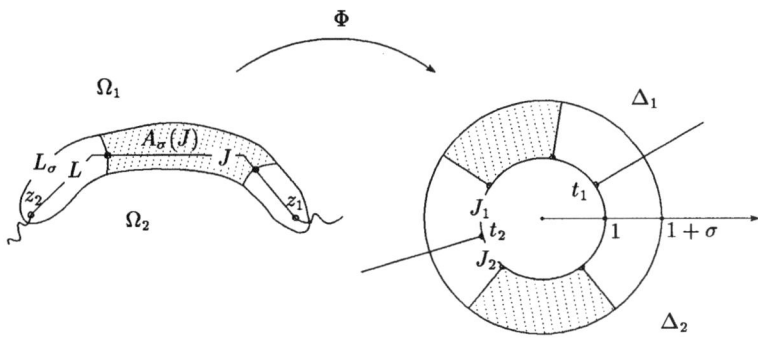

Fig. 2.3

$$\zeta_L := \Psi_j \left(\frac{\Phi_j(\zeta)}{|\Phi_j(\zeta)|} \right), \quad \zeta \in \overline{\Omega}_j .$$

Analogously to (2.3) the set $A_\sigma(J)$ is called σ-*neighborhood for J with respect to the arc L*.

Theorem 2.4. *Let L be a quasiconformal arc satisfying condition (2.29), and let $c_0 > 0$ be fixed. Suppose that $p \in \mathbb{M}_n$, $n \in \mathbb{N}$, is a monic polynomial such that $\varepsilon_{p,L} \leq 1$, where $\varepsilon_{p,L}$ is defined by (2.1). Then there exists a constant $c > 0$, depending only on L, α, C, and c_0, such that*

$$|(\mu_L - \nu_p)(A_\sigma(J))| \leq c \varepsilon_{p,L}^{\alpha/(1+\alpha)} \tag{2.31}$$

for all subarcs J of L and all σ with $\sigma \geq \sigma_0 := c_0 \varepsilon_{p,L}^{1/(1+\alpha)}$.

Notice that the proof of Theorem 2.4 follows the same ideas as the proof of Theorem 2.1. Therefore, we give only a sketch of the reasoning to show how the scheme has to be modified in order to obtain the result.

Let z_1 and z_2 be the endpoints of the quasiconformal arc L. Let l be a subarc of L. Consider for $\sigma > 0$ the arc

$$l_\sigma := \{\zeta \in L_\sigma : \zeta_L \in l\}$$

70   2. Zero Distribution of Polynomials

and the harmonic measure
$$\omega(z) := \omega(z, E_\sigma, l_\sigma)$$
of l_σ at z with respect to E_σ.

For $\zeta, \xi \in L$ we denote by $L(\zeta, \xi)$ the *subarc of L joining the points* ζ *and* ξ.

First we establish the analogue of Lemma 2.2. We use again $\mu := \mu_L$, $\nu := \nu_p$.

Lemma 2.5. *Let one of the endpoints of l coincide with z_1. Denote by $z_3 \in L$ the other endpoint of l. Then there exist constants $c_j = c_j(K, \alpha, C)$, $j = 1, 2$, such that for $z \in A_\sigma(L \setminus l)$ the relation*

$$\omega(z) \leq c_1 \exp\left\{-c_2 \frac{\mu(L(z_3, z_L))}{\sigma^\alpha}\right\} \tag{2.32}$$

holds, where $C > 0$, $0 < \alpha \leq 1$ are the constants from inequality (2.29).

Proof. Set $S := L(z_3, z_L)$. We may assume that $\mu(S) \geq c_3 \sigma^\alpha$, where the constant $c_3 > 1$ is large enough.

Divide the subarc S by points $\zeta_0 := z_L, \zeta_1, \ldots, \zeta_k := z_3$ into subarcs $S_i := L(\zeta_i, \zeta_{i+1})$, $i = 0, \ldots, k-1$, such that

$$\sigma/2 \leq \min_{j=1,2} \operatorname{diam} \Phi_j(S_i) \leq \sigma, \quad i = 1, \ldots, k-1$$

(see Figure 2.4). The possibility to perform this procedure for sufficiently large c_3 follows from (2.29).

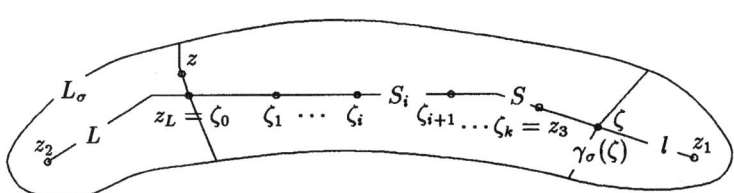

Fig. 2.4

Moreover, according to (2.29),
$$\max_{j=1,2} \operatorname{diam} \Phi_j(S_i) \preceq \sigma^\alpha.$$

Therefore,
$$k \succeq \mu(S)\,\sigma^{-\alpha}.$$

2.2 Erdős–Turán Type Theorems for Quasiconformal Curves and Arcs

We consider the arc
$$\gamma_\sigma(\zeta) := \{\xi \in E_\sigma : \xi_L = \zeta\}, \quad \zeta \in L.$$

Arguing exactly as in the proof of Lemma 2.2 we can show that
$$m(\{\gamma_\sigma(\zeta) : \zeta \in S_i\}) \succeq 1.$$

Recalling the composition law (1.2.13), we have
$$m(\{\gamma_\sigma(\zeta) : \zeta \in S\}) \geq \sum_{i=0}^{k-1} m(\{\gamma_\sigma(\zeta) : \zeta \in S_i\}) \succeq k \succeq \frac{\mu(S)}{\sigma^\alpha}.$$

Finally, the comparison principle (1.2.11) and relation (1.2.17) yield the desired estimate (2.32). □

Observe that the inequality
$$1 - \omega(z) \leq c_1 \exp\left\{-c_2 \frac{\mu(L(z_3, z_L))}{\sigma^\alpha}\right\}, \quad z \in A_\sigma(l), \tag{2.33}$$

is an immediate consequence of relation (2.32) written for $L \setminus l$.

If $l = L(\zeta_1, \zeta_2)$, $\zeta_1, \zeta_2 \in L$, is an arbitrary subarc of L with endpoints ζ_1 and ζ_2, we can express the function $\omega(z)$ as a difference of two functions of the same type as in Lemma 2.5 (corresponding to the arcs $L(z_1, \zeta_2)$ and $L(z_1, \zeta_1)$). Therefore, in this case we can also write the appropriate analogues of (2.32) and (2.33) describing the behavior of $\omega(z)$.

Now consider the following functions:
$$\widetilde{\omega}(w) := \begin{cases} \omega[\Psi(w)], & \text{if } 1 < |w| \leq 1 + \sigma, \\ 0, & \text{if } |w| > 1 + \sigma, \end{cases}$$

and
$$\widetilde{g}(w) := \begin{cases} \dfrac{16}{\sigma^2} \iint_{\mathbb{C}} \widetilde{\omega}(t) K\left(\dfrac{4(t-w)}{\sigma}\right) dm(t), & \text{if } 1 + \tfrac{1}{2}\sigma \leq |w| \leq 1 + \tfrac{3}{2}\sigma, \\ \widetilde{\omega}(w), & \text{elsewhere in } \Delta, \end{cases}$$

where $K(z)$ is some averaging kernel,
$$g(z) := \begin{cases} \widetilde{g}[\Phi(z)], & \text{if } z \in \Omega, \\ \omega(z), & \text{if } z \in L. \end{cases}$$

For the function g the analogue of inequality (2.16) takes the form
$$\left|\int g\,(d\nu - d\mu)\right| \preceq \frac{\varepsilon_{p,L}}{\sigma},$$

and the analogue of inequality (2.20) has the form

$$\left| \int g \, d\mu - \mu(l) \right| \preceq \sigma^\alpha.$$

The case of interest is as before where $\sigma = \sigma_0$ is small enough. To see that $(\nu - \mu)(A_{\sigma_0}(J))$ is appropriately bounded, we have only to repeat the last reasoning in the proof of Theorem 2.1.

2.3 Sharpness and Extensions of Erdős–Turán Type Theorems

It is far from the purposes of this book to investigate the connection between the geometry of L and the exponent α in (2.4) or in (2.29). More precise information about this subject can be derived, for example, from [113, 114]. We state only the following remark concerning piecewise smooth L.

It is simple to see from Lemmas 1.2.10–1.2.12 that for a Dini-smooth curve or arc we can set $\alpha = 1$ in Theorems 2.1 and 2.4. Hence, the Erdős–Turán estimates (0.2) and (0.3) are special cases of these theorems. Moreover, if a Jordan curve L consists of a finite number m of Dini-smooth arcs that form inner angles $\beta_j \pi$, $0 < \beta_j < 2$, $j = 1, \ldots, m$, with respect to int L (see Figure 2.5), then (2.4) is valid with

$$\alpha = \min\left\{ \min_{1 \leq j \leq m} \frac{\beta_j}{2 - \beta_j}, 1 \right\}.$$

Note that by this formula the value of α is the same (and equal to 1) for smooth L and piecewise smooth L with $\beta_j \geq 1$, $j = 1, \ldots, m$.

An analogous statement can be made in the case of a piecewise Dini-smooth arc L. Let $\beta_j \pi$, $0 < \beta_j < 2$, $j = 1, 2, \ldots, m$, be the angles of L formed by the Dini-smooth subarcs. Then with β_j the angle $2 - \beta_j$ occurs, too, and (2.29) holds with

$$\alpha = \min_{1 \leq j \leq m} \frac{\beta_j}{2 - \beta_j}.$$

In Note 1 of the paper [55] Erdős and Turán mentioned that their result for the interval (cf. (0.3)) is sharp up to the constant $8/\log 3$ in front. This can be seen easily by the following example.

Let

$$p(x) := c_n (x+1)^k P_{n-k}^{(0,2k)}(x),$$

2.3 Sharpness and Extensions of Erdős–Turán Type Theorems   73

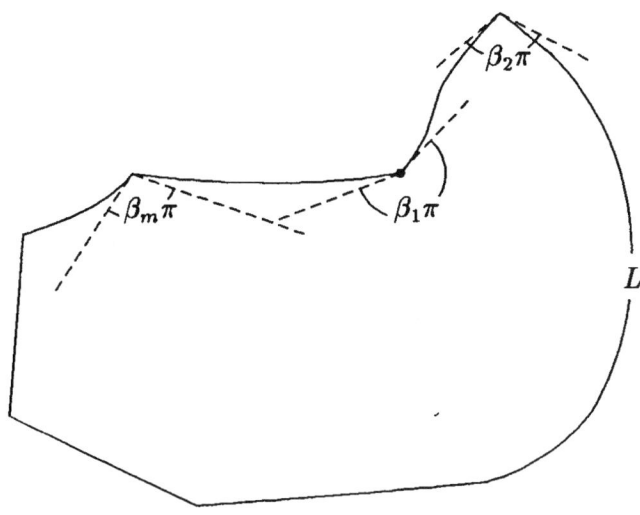

Fig. 2.5

where
$$k := \left[\sqrt{n \log n}\right] + 1,$$
$$c_n := 2^{n-k} \binom{2n}{n-k}^{-1},$$

and $P_m^{(\alpha,\beta)}(x)$ is the Jacobi polynomial of degree m with respect to the parameters α and β. Then $p \in \mathbb{M}_n$ and p has k zeros at -1. If μ denotes the equilibrium measure of $[-1.1]$, then

$$|(\mu - \nu_p)(J)| \geq \frac{k}{n} \geq \sqrt{\frac{\log n}{n}}$$

for a sufficiently small interval $J = [-1, -1 + \delta_n]$, $\delta_n > 0$.

On the other hand, because of the monotony of the weight function $w(x) = (x+1)^{2k}$ the maximum of the function

$$h(x) := \left|w(x)^{1/2} P_{n-k}^{(0,2k)}(x)\right| = \left|(x+1)^k P_{n-k}^{0,2k)}(x)\right|$$

is attained at the point 1 (cf. Szegő [173, Theorem 7.2]). Moreover, the normalization of the Jacobi polynomials imply that

$$P_m^{(\alpha,\beta)}(1) = \binom{m+\alpha}{n}$$

and therefore $P_{n-k}^{(0,2k)}(1) = 1$. Altogether we have

$$\|p\|_{[-1,1]} = 2^n \binom{2n}{n-k}^{-1}.$$

Now, Stirling's formula implies

$$\binom{2n}{n-k}^{-1} \leq c_0 \frac{n^\kappa}{2^{2n}}$$

with some positive constants c_0 and κ, independent of n. Hence

$$\|p\|_{[-1,1]} \leq c_0 \frac{n^\kappa}{2^n}$$

and by Theorem 2.4 (or (0.3)),

$$|(\mu - \nu_p)(J)| \leq c_1 \sqrt{\frac{\log n}{n}}$$

with some absolute constant $c_1 > 0$. This proves the sharpness of Theorem 2.4.

By Joukowski's map, this example can be transformed to an extremal example for the circle. Another example for the circle is given by Amoroso and Mignotte [7].

In the formulation of Theorems 2.1 and 2.4 we have used global estimates from above, i.e., estimates on $\mathbb{C}$ for the potential $U^{\mu_L - \nu_p}(z)$. However, in the proof of these results we needed estimates of $U^{\mu_L - \nu_p}(z)$ only on some subsets of $\mathbb{C}$, namely on

$$\text{ext } L_{\sigma_0/2}^+ \cup \text{int } L_{\sigma_0^{1/\alpha}/2}^-$$

in Theorem 2.1 and on

$$\text{ext } L_{\sigma_0/2}$$

in Theorem 2.4.

Moreover, if $U^{\mu_L - \nu_p}(z)$ is harmonic in $\mathbb{C} \setminus L$, which corresponds to the fact that all zeros of p belong to L, a lower bound of $U^{\mu_L - \nu_p}(z)$ can be used in the proof of the basic inequality (2.16) as well.

We display this approach writing the analogue of Theorem 2.4 only.

Theorem 3.1. *Let L be a quasiconformal arc satisfying condition (2.29). Suppose that p is a monic polynomial such that all of its zeros belong to L. Then there exists a constant $c > 0$, depending only on L, α, C, such that*

$$|(\mu_L - \nu_p)(J)| \leq c \left(\frac{\varepsilon_{p,L}^*(\sigma)}{\sigma} + \sigma^\alpha \right), \qquad \sigma > 0, \qquad (3.1)$$

for all subarcs J of L, where

$$\varepsilon_{p,L}^*(\sigma) := - \inf_{z \in L_\sigma} U^{\mu_L - \nu_p}(z).$$

2.3 Sharpness and Extensions of Erdős–Turán Type Theorems

We point out some possible applications of Theorem 3.1. If all zeros $z_1, z_2, \ldots, z_n$ of $p \in \mathbb{M}_n$, $n \geq 2$, are simple and for some $B_n \geq n$ satisfy

$$|p'(z_j)| \geq \frac{1}{B_n} (\operatorname{cap} L)^n, \qquad j = 1, \ldots, n, \tag{3.2}$$

then, by Lemma 1.4.5 we have

$$\varepsilon^*_{p,L}(\sigma) \leq \varepsilon^*_{p,L}\left(\frac{1}{n}\right) \leq c_1 \frac{\log B_n}{n}, \qquad \frac{1}{n} \leq \sigma < 1.$$

Thus, taking in Theorem 3.1

$$\sigma = \left(\frac{\log B_n}{n}\right)^{1/(1+\alpha)},$$

we get for any subarc J of L that

$$|(\mu_L - \nu_p)(J)| \leq c_2 \left(\frac{\log B_n}{n}\right)^{\alpha/(1+\alpha)}. \tag{3.3}$$

Concerning the sharpness of the last estimate, let $L = [-1, 1]$. Then we can choose $\alpha = 1$. For $2 \leq B_n < e^n$ let

$$\delta_n := \frac{\log B_n}{n}$$

and consider the monic polynomial

$$P_n(x) := \frac{(1 - \delta_n)^n}{2^{n-1}} \cos\left(n \arccos \frac{x}{1 - \delta_n}\right), \qquad |x| \leq 1 - \delta_n. \tag{3.4}$$

At the zeros $x_i \in E_n := [-1 + \delta_n, 1 - \delta_n]$, $i = 1, \ldots, n$, of P_n, we have

$$|P'_n(x_i)| = \frac{(1 - \delta_n)^{n-1}}{2^{n-1}} n \left(1 - \left(\frac{x_i}{1 - \delta_n}\right)^2\right)^{-1/2} \geq \frac{1}{B_n} \frac{1}{2^n}.$$

Since the polynomial P_n, considered on $[-1, 1]$, satisfies the condition (3.2), we obtain (3.3) with $\alpha = 1$. On the other hand,

$$|(\mu_L - \nu_p)([1 - \delta_n, 1])| = \mu_L([1 - \delta_n, 1]) \geq \frac{\sqrt{2}}{\pi} \sqrt{\delta_n}.$$

Finally, we note the following peculiarity of the example above. It is not difficult to see that

$$\varepsilon_{P_n, L} \succeq \sqrt{\delta_n}.$$

Hence, for P_n defined by (3.4), inequality (3.3) gives an essentially better estimate for the discrepancy between the measures μ and ν than that given by Theorem 2.4.

Next, we show that the requirement of quasiconformality of the curve or arc imposed in the theorems of this section is, in a certain sense, essential. To be more specific we describe its importance in Theorem 2.4.

If all zeros of a monic polynomial $p \in \mathbb{M}_n$ belong to a quasiconformal arc L, then, according to inequality (2.31), the discrepancy between the measures μ and ν can be estimated from above by some power of $\varepsilon_{p,L}$. We are going to show that the absence of quasiconformality of L leads, in general, to a violation of this fact.

Consider a function $f \in C^2([0,1])$ satisfying the conditions
$$f^{(j)}(x) > 0, \quad 0 < x < 1,$$
$$\lim_{x \to +0} f^{(j)}(x) = 0,$$
for $j = 0, 1, 2$. It is easily seen that the Jordan arc $L := \Gamma_1 \cup \Gamma_2$, where
$$\Gamma_1 := \{z = x + if(x) : 0 \leq x \leq 1\}, \quad \Gamma_2 := [0,1],$$
has a cusp at the origin (see Figure 2.6). According to Lemma 1.1.13 and

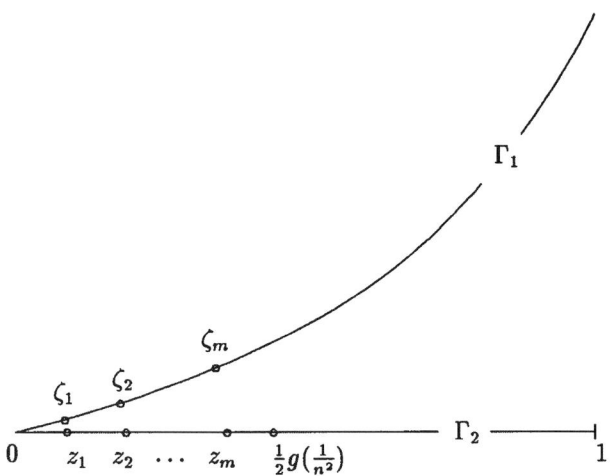

Fig. 2.6

Theorem 1.3.3 a Fekete polynomial q_n of degree n for the arc L satisfies
$$\|q_n\|_L \preceq n^2 (\operatorname{cap} L)^n. \tag{3.5}$$
Let $z_1, \ldots, z_m$, $m \leq n$, be the zeros of q_n that belong to the segment $[0, \frac{1}{2}g(\frac{1}{n^2})]$, where $g := f^{-1}$. We select points $\zeta_1, \ldots, \zeta_m \in \Gamma_1$ such that
$$|z_j - \zeta_j| = d(z_j, \Gamma_1), \quad j = 1, \ldots, m.$$

2.3 Sharpness and Extensions of Erdős–Turán Type Theorems

Let us consider the polynomial

$$p(z) = p_n(z) := q_n(z) \prod_{j=1}^{m} \frac{z - \zeta_j}{z - z_j}.$$

It is easily verified that

$$|p(z)| \leq |q_n(z)|, \quad z \in \Gamma_1. \tag{3.6}$$

Furthermore, let Φ_1 be the conformal mapping of $\overline{\mathbb{C}} \setminus \Gamma_1$ onto Δ satisfying $\Phi_1(\infty) = \infty$, $\Phi_1'(\infty) > 0$. By Lemma 1.2.5 we have for $z \in [0, g(\frac{1}{n^2})]$

$$\frac{1}{n^2} \geq d(z, \Gamma_1) \succeq (|\Phi_1(z)| - 1)^2.$$

Hence, the Bernstein–Walsh lemma (Lemma 1.1.11) yields for any $z \in [0, g(\frac{1}{n^2})]$ that

$$|p(z)| \preceq \|p\|_{\Gamma_1} \leq \|q_n\|_L. \tag{3.7}$$

By our assumption, in some neighborhood of the origin we have $f(x) \leq x^2$. Then, for sufficiently large n, we obtain

$$\frac{1}{n} \leq g\left(\frac{1}{n^2}\right),$$

and therefore

$$\left|\frac{p(z)}{q_n(z)}\right| \leq \left(1 + \frac{2/n^2}{g(1/n^2)}\right)^m \leq \left(1 + \frac{2}{n}\right)^n \leq e^2 \tag{3.8}$$

for all $z \in [g(\frac{1}{n^2}), 1]$. Comparing (3.6) – (3.8), we get

$$\|p\|_L \preceq \|q_n\|_L \preceq n^2 (\operatorname{cap} L)^n.$$

Hence, if n is large enough, we have

$$\varepsilon_{p,L} \preceq \frac{\log n}{n}.$$

On the other hand

$$\left|(\mu_L - \nu_p)\left(\left[0, \frac{1}{2} g\left(\frac{1}{n^2}\right)\right]\right)\right| = \mu_L\left(\left[0, \frac{1}{2} g\left(\frac{1}{n^2}\right)\right]\right) \succeq g\left(\frac{1}{n^2}\right).$$

It is obvious that $g(x)$ can be chosen such that it tends to zero arbitrarily slowly (as $x \to 0$). In particular, we may choose the function $f(x)$ such that it will be impossible to estimate the quantity $g(1/n^2)$ from above by some power of $\varepsilon_{p,L}$.

Moreover, we come to the conclusion that in general, the discrepancy between the measures μ and ν cannot be estimated from above by any universal function of $\varepsilon_{p,L}$ unless we impose restrictions on L. Therefore, if we want to establish inequalities like (2.5) and (2.31) for rather complicated curves or arcs, we have to take into account some information about their geometry.

2.4 Erdős–Turán Type Theorems on Compact Sets with Smooth Curves and Arcs

The results of the previous section can be generalized to the case of compact admissible sets $E \subset \mathbb{C}$ with multiply connected complement $\Omega := \overline{\mathbb{C}} \setminus E$. We consider the situation where Ω is regular and one of the connected components S of E is bounded by a curve or an arc, and prove the analogous assertions to Theorems 2.1 and 2.4. We restrict ourselves to the case of sufficiently smooth ∂S, and show how the reasoning of the previous section can be simplified in this case.

For $\sigma > 0$, we denote by

$$L_\sigma = \{z \in \Omega : G(z) = \log(1+\sigma)\}$$

the level lines of Green's function $G(z)$ and again set

$$E_\sigma = E \cup \{z \in \Omega : 0 < G(z) < \log(1+\sigma)\}.$$

Every level line L_σ is analytic at each point z where grad $G(z) \neq 0$. Moreover, L_σ consists of a finite number of Jordan curves that are mutually exterior except for a finite number of critical points, i.e., points where grad $G(z) = 0$ [181, pp. 65–68].

Lemma 4.1. *Let $S \subset E$ be a continuum that has positive distance to $E \setminus S$. If M_σ denotes the component of E_σ containing S, then there is a $\sigma_0 > 0$ such that*

$$E \cap M_\sigma = S \tag{4.1}$$

for all $0 < \sigma \leq \sigma_0$.

Proof. Let

$$M := \bigcap_{n \in \mathbb{N}} M_{1/n}.$$

Thus $S \subseteq M \subseteq E$. We show that M is connected. Otherwise, there are disjoint open sets U_1, U_2 with

$$U_j \cap M \neq \emptyset, \quad j = 1, 2,$$
$$M \subset (U_1 \cup U_2).$$

Since each $M_{1/n}$ is connected, we can pick $z_n \in (\mathbb{C} \setminus (U_1 \cup U_2)) \cap M_{1/n}$. Every limit point z of the sequence $\{z_n\}$ is in $(\mathbb{C} \setminus (U_1 \cup U_2)) \cap M$. This contradicts $M \subset (U_1 \cup U_2)$.

Since M is connected and S is a connected component of E, we get $S = M$. Since S has positive distance to $E \setminus S$, there is a number $n \in \mathbb{N}$ with $M_{1/n} \cap E = S$. Thus, $M_\sigma \cap E = S$ for $\sigma \leq 1/n$. $\square$

If E consists of several mutually exterior components, the most obvious difficulty is that Ω is not conformally equivalent to the exterior of a circle. Nevertheless, there is a standard analogue of $\Phi(z)$ in this case. In fact, if $\widetilde{G}$ is the harmonic conjugate of the Green function G, we have for a continuum E,

$$\Phi(z) = \exp\{G(z) + i\widetilde{G}(z)\} \tag{4.2}$$

(except possibly for a constant factor of absolute value one, since $\widetilde{G}$ is unique up to a constant). In the more general case, there is the problem that Φ defined by (4.2) is multiple-valued.

Let E, S, and M_{σ_0} be as in Lemma 4.1, and let $\Gamma \subset M_{\sigma_0}$ be a smooth curve enclosing S. By virtue of Gauss's theorem (see [156, p. 83]) and (1.1.24),

$$\mu_E(S) = \frac{1}{2\pi} \int_\Gamma \frac{\partial}{\partial \mathbf{n}_\zeta} G(\zeta)\, |d\zeta|.$$

The $\mathbf{n}_\zeta$ appearing in the integral denotes the normal at $\zeta \in \Gamma$ directed into the exterior of Γ. An easy consequence of the last relation and the Cauchy–Riemann equations (1.1.2) is the following fact:

$$\Delta_\Gamma \widetilde{G} = 2\pi \mu_E(S),$$

where $\Delta_\Gamma \widetilde{G}$ is the *net change* of $\widetilde{G}(z)$ as z runs along Γ. Indeed,

$$2\pi\mu_E(S) = \int_\Gamma \frac{\partial}{\partial \mathbf{n}_\zeta} G(\zeta)\, |d\zeta| = \int_\Gamma \frac{\partial}{\partial \mathbf{t}_\zeta} \widetilde{G}(\zeta)\, |d\zeta| = \Delta_\Gamma \widetilde{G}.$$

Here $\mathbf{t}_\zeta$ denotes the tangent vector obtained by rotating $\mathbf{n}_\zeta$ counterclockwise by an angle of $\pi/2$.

Therefore, the function

$$\Phi_b(z) := \Phi(z)^b := \exp\{b\, G(z) + i\, b\, \widetilde{G}(z)\}, \quad b := 1/\mu_E(S),$$

is a conformal mapping of the curvilinear annulus $M_{\sigma_0} \setminus E$ onto the annulus $N_{\sigma_0} := \{w : 1 < |w| \le (1+\sigma_0)^b\}$. This fact allows us to use conformal mappings to describe the properties of $G(z)$ in a neighborhood of S. We formulate the appropriate analogues of Lemmas 1.2.10 and 1.2.11 without proof, because the proofs of these lemmas were given in such a way that they can be transferred without any essential changes.

Let $p = p_n \in \mathbb{M}_n$ be a monic polynomial and let E be admissible with regular complement Ω. We denote again by ν_p the normalized zero counting measure of p and by μ_E the equilibrium measure of E. Set

$$\varepsilon_{p,E} := \sup_{z \in \mathbb{C}} U^{\mu_E - \nu_p}(z). \tag{4.3}$$

First, we consider the case where $S = L'$ is an arc. Let $\sigma_0 > 0$ be as in Lemma 4.1. For $0 < \sigma \le \sigma_0$ let

$$L'_\sigma := \partial M_\sigma = \{z : |\Phi_b(z)| = (1+\sigma)^b\},$$
$$E'_\sigma := M_\sigma = \operatorname{int} L'_\sigma,$$

and for an arbitrary subarc J of L',

$$A_\sigma(J) := \{\zeta \in E'_\sigma : \zeta_{L'} \in J\}, \tag{4.4}$$

where we use the notation

$$\zeta_{L'} := \Psi_{b,j}\left(\frac{\Phi_{b,j}(\zeta)}{|\Phi_{b,j}(\zeta)|}\right), \quad \zeta \in \overline{\Omega}_{b,j}.$$

Note that (4.4) coincides with the definition of a σ-neighborhood for J in (2.30) if we choose $L = L'$.

Theorem 4.2. *Let $L' \subset E$ be a Dini-smooth arc with positive distance to $E \setminus L'$ and $p \in \mathbb{M}_n$. Then*

$$\nu_p(\mathbb{C} \setminus E_{\sqrt{\varepsilon_{p,E}}}) \le 2\sqrt{\varepsilon_{p,E}}, \tag{4.5}$$

and there is a constant $c > 0$, depending on E and L' only, such that for all subarcs J on L',

$$|(\mu_E - \nu_p)(A_{\sqrt{\varepsilon_{p,E}}}(J))| \le c\sqrt{\varepsilon_{p,E}}. \tag{4.6}$$

In the proof we need some geometric considerations that we shall formulate first as lemmas.

Lemma 4.3. *Let ∂S be a Dini-smooth curve. Then there exists a constant $\varepsilon = \varepsilon(S, \sigma_0, b) > 0$ such that for all points ζ_1 and $\zeta_2 \in \Omega$ with $\operatorname{dist}(\zeta_j, S) \le \varepsilon$, $j = 1, 2$, and their images $\tau_j := \Phi_b(\zeta_j)$, the inequalities*

$$c_1|\tau_1 - \tau_2| \le |\zeta_1 - \zeta_2| \le c_2|\tau_1 - \tau_2| \tag{4.7}$$

hold with some constants $c_i = c_i(S, \varepsilon, b) > 0$, $i = 1, 2$.

Now let $L' := \partial S$ be an arc. Denote by z_1 and z_2 the endpoints of L'. Let $\Psi_b := \Phi_b^{-1}$,

$$t_j := \Phi_b(z_j), \quad j = 1, 2,$$
$$\Delta'_1 := \{w : 1 < |w| < (1+\sigma_0)^b, \arg t_1 < \arg w < \arg t_2\},$$
$$\Delta'_2 := \{w : 1 < |w| < (1+\sigma_0)^b\} \setminus \overline{\Delta}'_1,$$
$$\Omega_{b,i} := \Psi_b(\Delta'_i), \quad i = 1, 2,$$

and let $\Phi_{b,i}$, $i = 1, 2$, be the restriction of the function Φ_b to $\overline{\Omega}_{b,i}$. Set $\Psi_{b,i} := \Phi_{b,i}^{-1}$.

2.4 Erdős–Turán Type Theorems on Compact Sets

Lemma 4.4. *Let $L' := \partial S$ be a Dini-smooth arc. Then there exists a constant $\varepsilon = \varepsilon(S, \sigma_0, b) > 0$ such that for any $i = 1, 2$, for all points $z \in L'$, $\zeta \in \overline{\Omega}_{b,i}$ with*

$$\operatorname{dist}(\zeta, L') < \varepsilon,$$
$$|\zeta - z| \le \min_{j=1,2} |z - z_j| =: |z - z_1|,$$

and for their images $\tau_i := \Phi_{b,i}(\zeta), w_i := \Phi_{b,i}(z)$, the inequalities

$$c_1 \left| \frac{w_i - t_1}{w_i - \tau_i} \right| \le \left| \frac{z - z_1}{z - \zeta} \right| \le c_2 \left| \frac{w_i - t_1}{w_i - \tau_i} \right|, \tag{4.8}$$

$$c_3 |w_i - t_1|^2 \le |z - z_1| \le c_4 |w_i - t_1|^2 \tag{4.9}$$

hold with some constants $c_k = c_k(S, \varepsilon, b) > 0$, $k = 1, 2, 3, 4$.

Now we are in position to prove Theorem 4.2.

Proof of Theorem 4.2. The first inequality is a simple consequence of Lemma 1.4.3 and the elementary estimate

$$\log(1 + x) \ge x/2, \quad 0 < x < 1.$$

To prove (4.6) it is sufficient to consider only a subarc J of L' that contains one of the endpoints of L', say z_1. The typical situation arising in the construction process below is represented in Figure 2.7.

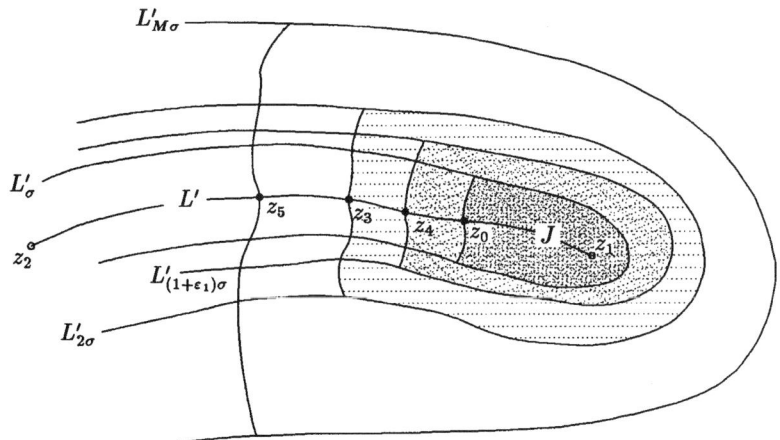

Fig. 2.7

Let z_1 and $z_0 \ne z_2$ be the endpoints of J. First, we show that

$$(\mu - \nu)(A_\sigma(J)) \ge -c_1 \sigma, \tag{4.10}$$

where $\mu := \mu_E$, $\nu := \nu_p$, $\varepsilon := \varepsilon_{p,E}$, $\sigma := \sqrt{\varepsilon}$.

We can assume that $\sigma \leq \min(1, \sigma_0)$ with σ_0 as in Lemma 4.1 and $\mu(L' \setminus J) \geq c_2 \sigma$ with an arbitrary (but fixed) constant $c_2 > 1$, which will be chosen below. For ζ_1 and $\zeta_2 \in L'$ denote by $L'(\zeta_1, \zeta_2)$ the sub-arc of L' between these points. Choose a point $z_3 \in L'(z_2, z_0)$ such that $\mu(L'(z_3, z_0)) = \sigma$.

Let $J' := L'(z_3, z_1)$ and consider the function

$$h(z) := \begin{cases} 1, & \text{if } z \in A_{2\sigma}(J'), \\ 0, & \text{otherwise in } \mathbb{C}. \end{cases}$$

We denote by $\rho(z)$, $z \in \mathbb{C}$, a regularized distance to $A_\sigma(J)$ (see Appendix D), i.e., a function with the following properties:

$$\frac{1}{8} \operatorname{dist}(z, A_\sigma(J)) \leq \rho(z) \leq 2 \operatorname{dist}(z, A_\sigma(J)), \qquad z \in \mathbb{C}, \tag{4.11}$$

$$\rho(z) \in C^\infty(\mathbb{C}), \tag{4.12}$$

$$\left| \frac{\partial^{j+k}}{\partial x^j \partial y^k} \rho(x+iy) \right| \leq c \rho(x+iy)^{1-j-k}, \qquad 1 \leq j+k \leq 2. \tag{4.13}$$

Next, we average the function h in the following way:

$$g(z) := \begin{cases} \dfrac{16}{\rho(z)^2} \displaystyle\int_{\mathbb{C}} h(\zeta) K\left(\dfrac{4|\zeta - z|}{\rho(z)}\right) dm(\zeta), & \text{if } z \in \mathbb{C} \setminus A_\sigma(J), \\ 1, & \text{if } z \in A_\sigma(J), \end{cases} \tag{4.14}$$

where $K(\zeta)$ is the averaging kernel from Section 1.5.3.

Note that according to (4.12), $g \in C^\infty(\mathbb{C})$. By Lemma 4.4 there exists a constant $\varepsilon_1 > 0$ and a point $z_4 \in L'(z_0, z_3)$ such that $\mu(L'(z_0, z_4)) = \varepsilon_1 \sigma$ and, for $z \in A_{(1+\varepsilon_1)\sigma}(L'(z_1, z_4))$,

$$\operatorname{dist}(z, A_\sigma(J)) < \frac{1}{2} \operatorname{dist}(z, \partial A_{2\sigma}(J')).$$

Therefore, using (4.11) we obtain

$$g(z) = 1, \qquad z \in A_{(1+\varepsilon_1)\sigma}(L'(z_1, z_4)). \tag{4.15}$$

Further, by Lemma 4.4 there exists a constant $c_3 > 0$ such that for $z \in \operatorname{ext} L_{c_3 \sigma}$,

$$\operatorname{dist}(z, A_{2\sigma}(J')) > \frac{1}{2} \operatorname{dist}(z, A_\sigma(J)).$$

Therefore, by (4.11) for such z,

$$\rho(z) \leq 2 \operatorname{dist}(z, A_\sigma(J)) < 4 \operatorname{dist}(z, A_{2\sigma}(J')).$$

Moreover, c_3 can be chosen so large that $g(z) = 0$ for $z \in \text{ext } L_{c_3\sigma}$.

The same reasoning shows that for a suitably large constant $c_4 > 0$ and for $z \in E'_{c_3\sigma}$ with

$$z_L \in L'(z_2, z_3), \quad \mu(L'(z_L, z_3)) \geq c_4 \sigma$$

we have $g(z) = 0$, too. Certainly, we can take $c_2 > c_3 + c_4 + 1$. As a result we can construct a constant $M(= c_3 + c_4)$ and a point $z_5 \in L'(z_2, z_3)$ with $\mu(L'(z_5, z_3)) = M\sigma$ and

$$g(z) = 0, \quad z \in \mathbb{C} \setminus A_{M\sigma}(L'(z_5, z_1)). \tag{4.16}$$

Our next aim is to estimate $\Delta g(z)$. It is obvious that

$$\Delta g(z) = 0, \quad z \in A_{(1+\varepsilon_1)\sigma}(L'(z_1, z_4)) \cup (\mathbb{C} \setminus A_{M\sigma}(L'(z_5, z_1))).$$

Let $z \in A_{M\sigma}(L'(z_1, z_4)) \setminus A_{(1+\varepsilon_1)\sigma}(L'(z_1, z_4))$. By Lemma 4.4,

$$\rho(z) \asymp \text{dist}(z, L).$$

Differentiating (4.14) and using (4.13) we get for such z,

$$|\Delta g(z)| \preceq \rho(z)^{-2} \asymp (\text{dist}(z, L))^{-2}. \tag{4.17}$$

Further, for $z \in A_{M\sigma}(L'(z_4, z_5))$ by Lemma 4.4,

$$\rho(z) \asymp |z_4 - z_5| \asymp |z_0 - z_3|,$$

and reasoning as in the previous case we obtain

$$|\Delta g(z)| \preceq |z_3 - z_0|^{-2}, \quad z \in A_{M\sigma}(L'(z_4, z_5)). \tag{4.18}$$

We can assume that $M\sigma$ is also sufficiently small, i.e.,

$$\text{dist}(z, L') < \text{dist}(z, L'_{\sigma_0}), \quad z \in L'_{M\sigma}.$$

By Lemma 1.2.3 we have for $z \in L'_t$, $0 < t \leq M\sigma$, that

$$\frac{\partial G(z)}{\partial \mathbf{n}_z} = \frac{1}{b} \frac{|\Phi'_b(z)|}{|\Phi_b(z)|} \asymp \frac{|\Phi_b(z)| - 1}{\text{dist}(z, L')}.$$

Therefore, taking into account Lemma 4.4, (4.17), and (4.18), we obtain

$$|\Delta g(z)| \preceq \frac{1}{\sigma^2} \left(\frac{\partial G(z)}{\partial \mathbf{n}_z}\right)^2, \quad z \in L'_t, \ 0 < t \leq M\sigma. \tag{4.19}$$

Since by virtue of (4.15) and (4.16)

$$\int g \, d(\mu - \nu) \leq (\mu - \nu)(A_\sigma(J)) + (M+1)\sigma, \tag{4.20}$$

84   2. Zero Distribution of Polynomials

we merely need to estimate the integral on the left-hand side of (4.20) in order to prove the desired estimate (4.10).

Using inequality (4.19) together with Fubini's theorem we have as in the proof of Theorem 2.1 (after changing coordinates with $G(\zeta)$)

$$\left| \int g\, d(\mu - \nu) \right| \leq \frac{1}{2\pi} \int (U^{\nu-\mu}(\zeta) + \varepsilon) |\Delta g(\zeta)|\, dm(\zeta)$$

$$\leq \frac{1}{2\pi} \int_0^{M\sigma} \int_{L'_t} (U^{\nu-\mu}(\zeta) + \varepsilon) |\Delta g(\zeta)| \left(\frac{\partial G(\zeta)}{\partial \mathbf{n}_\zeta} \right)^{-1} |d\zeta|\, dt$$

$$\preceq \frac{1}{\sigma^2} \int_0^{M\sigma} \int_{L_t} (U^{\nu-\mu}(\zeta) + \varepsilon) \frac{\partial G(\zeta)}{\partial \mathbf{n}_\zeta} |d\zeta|\, dt.$$

For all functions f superharmonic in Ω, we have

$$\frac{1}{2\pi} \int_{L_t} f(\zeta) \frac{\partial G(\zeta)}{\partial \mathbf{n}_\zeta} |d\zeta| \leq f(\infty)$$

(see [178, p. 22]). Thus, since $U^{\nu-\mu} + \varepsilon$ is superharmonic in Ω,

$$\left| \int g\, d(\mu - \nu) \right| \preceq \frac{\varepsilon}{\sigma} = \sigma. \tag{4.21}$$

Combining (4.20) and (4.21) we get (4.10).

The proof of

$$(\mu - \nu)(A_\sigma(J)) \leq c_5 \sigma \tag{4.22}$$

follows the same lines. This time we can assume that $\mu(J) \geq c_6 \sigma$, where c_6 is sufficiently large. Choose a point $z_3 \in L'(z_0, z_1)$ such that $\mu(L'(z_0, z_3)) = c_6 \sigma / 2$, set $J' := L'(z_3, z_1)$, and consider the function

$$h(z) := \begin{cases} 1, & \text{if } z \in A_{\sigma/c_6}(J'), \\ 0, & \text{otherwise in } \mathbb{C}. \end{cases}$$

Let $\rho(z)$ be a regularized distance to $A_{\sigma/c_6}(J'')$, where $J'' := L'(z_4, z_1)$ with the point $z_4 \in L'(z_3, z_1)$ being defined in such way that it satisfies $\mu(L'(z_3, z_4)) = \sigma$. Let the function g be given by (4.14) (with new $h(z)$, $\rho(z)$ and with $A_{\sigma/c_6}(J'')$ instead of $A_\sigma(J)$). From Lemma 4.4 we derive that c_6 can be chosen in such a way that

$$g(z) = 0, \quad z \in \mathbb{C} \setminus A_\sigma(J)),$$

$$g(z) = 1, \quad z \in A_{\sigma/c_6}(J'').$$

Then (4.20) is replaced by

$$(\mu - \nu)(A_\sigma(J)) \le \int g \, d(\mu - \nu) + \mu(J \backslash J'')$$
$$= \int g \, d(\mu - \nu) + (1 + c_6/2)\,\sigma.$$

The remaining proof proceeds as above. Comparing (4.10) and (4.22) we obtain (4.6). □

If the component S of E is a closed Jordan domain, we continue in an obvious way the previous reasoning in the proofs of Theorem 2.1 and 4.2 with the properties of the Green function described by Lemma 4.3 and obtain an analogue of Theorem 4.2.

Let $L' := \partial S$, $G = \operatorname{int} L'$. then we fix a point $z_0 \in G$ and introduce the conformal mapping φ of G onto $\mathbb{D}$ with $\varphi(z_0) = 0$, $\varphi'(z_0) > 0$. Set $\psi = \varphi^{-1}$ and σ_0 as in Lemma 4.1. Then as in Section 2.2, for $\zeta \in E_{\sigma_0} \backslash \{z_0\}$ let

$$\zeta_{L'} := \begin{cases} \psi\left(\dfrac{\Phi(\zeta)}{|\Phi(\zeta)|}\right), & \text{if } \zeta \in \Omega \cap E_{\sigma_0}, \\ \psi\left(\dfrac{\varphi(\zeta)}{|\varphi(\zeta)|}\right), & \text{if } \zeta \in G \backslash \{z_0\}, \\ \zeta, & \zeta \in L'. \end{cases}$$

Moreover, let

$$L'^{-}_r := \{\zeta : |\varphi(\zeta)| = 1 - r\}, \qquad 0 < r < 1.$$

Let J be a subarc of L' and $0 < \sigma \le \sigma_0$. Then we define

$$A_{\sigma,\sigma}(J) := (\operatorname{ext} L'^{-}_\sigma) \cap (\operatorname{int} L'_\sigma)$$

as the (σ, σ)-neighborhood of J with respect to L'.

Moreover, we define

$$\delta_{p,E} := \varepsilon_{p,E} - U^{\mu_E - \nu_p}(z_0), \tag{4.23}$$

where $\varepsilon_{p,E}$ is defined by (4.3).

Theorem 4.5. *Let $S \subset E$ be bounded by a Dini-smooth curve L' with positive distance to $E \backslash S$ and let $z_0 \in S^\circ$ be fixed and define $\gamma := \gamma_{p,E} := \sqrt{\varepsilon_{p,E} + \delta_{p,E}}$. Then*

$$\nu_p\left((\mathbb{C} \backslash E_\gamma) \cup \operatorname{int} L'^{-}_\gamma\right) \le 2\sqrt{\varepsilon_{p,E} + \delta_{p,E}}, \tag{4.24}$$

and there is a constant $c > 0$, depending on E and L' only, such that for all subarcs J of L',

$$|(\mu_E - \nu_p)(A_{\gamma,\gamma}(J))| \le c\sqrt{\varepsilon_{p,E} + \delta_{p,E}}, \tag{4.25}$$

where $A_{\gamma,\gamma}(J)$ denotes the (γ, γ)-neighborhood of J with respect to L'.

86   2. Zero Distribution of Polynomials

The proof will be omitted, since it follows the same lines as the proof of Theorem 2.1 and Theorem 4.2.

2.5 Erdős–Turán Type Theorems on a System of Intervals

The proof of Theorem 4.2 shows that the constant c in (4.6) depends on the geometry of E. Hence, it seems to be interesting that in the special case $E \subset \mathbb{R}$ the constant c can be estimated as an absolute constant independent of E.

Theorem 5.1. *Let E be a compact subset of $\mathbb{R}$ consisting of a finite number of closed intervals, and let $p \in \mathbb{M}_n$, $n \in \mathbb{N}$. Then for all intervals $I \subset \mathbb{R}$ and the corresponding strip domain $S(I) := \{z \in \mathbb{C} : \text{Re}\{z\} \in I\}$ the inequality*

$$|(\mu_E - \nu_p)(S(I))| \leq 8\sqrt{2\,\varepsilon_{p,E}/\pi} + 8\,\varepsilon_{p,E} \tag{5.1}$$

holds. Consequently,

$$|(\mu_E - \nu_p)(S(I))| \leq 8\sqrt{\varepsilon_{p,E}}, \tag{5.2}$$

where $\varepsilon_{p,E}$ is given by (4.3).

Before giving the proof we establish the following elementary fact.

Lemma 5.2. *For all $0 = \tau_0 < \tau_1 < \cdots < \tau_m = 1$ and $\eta > 0$, there is a function $\psi \in C^2(\mathbb{R})$, $\psi : \mathbb{R} \to [0,1]$ and a constant $\kappa_0 > 0$ such that for all $\kappa < \kappa_0$,*

(i) $\psi(x) = 1$ for $x \leq \kappa$,

(ii) $\psi(x) = 0$ for $x \geq 1 - \kappa$,

(iii) $\psi^{(n)}(x) = 0$ for $n = 1, 2$, $x \in [\tau_\nu - \kappa, \tau_\nu + \kappa]$, $\nu = 0, \ldots, m$,

(iv) $|\psi'(x)| \leq (2 + \eta)m$ for all $x \in \mathbb{R}$,

(v) $|\psi''(x)| \leq (4 + \eta)m$ for all $x \in \mathbb{R}$.

Proof. Let $\kappa < \frac{1}{6}$ and consider u_κ given by

$$u_\kappa(x) := \begin{cases} 0 & \text{for } x \leq \kappa, \\ -1 & \text{for } x \in \left[2\kappa, \frac{1}{2} - \kappa\right], \\ +1 & \text{for } x \in \left[\frac{1}{2} + \kappa, 1 - 2\kappa\right], \\ 0 & \text{for } x \geq 1 - \kappa, \end{cases}$$

2.5 Erdős–Turán Type Theorems on a System of Intervals   87

and linear in between. Define

$$\psi_1(x) := \frac{4}{(1-4\kappa)(1-2\kappa)} \int_1^x \int_1^y u_\kappa(t)\,dt\,dy.$$

Then, for κ sufficiently small, the function $\psi(x) := \psi_1(x)$ satisfies (i)–(v) for $m = 1$. For $m > 1$, let $\delta_i := \tau_i - \tau_{i-1}$ and

$$\eta_i := \frac{\delta_i^2}{\sum_{\nu=1}^m \delta_\nu^2}$$

for $i = 1, \ldots, m$. Now define $\psi(x) := 1$ for $x \leq 0$, $\psi(x) := 0$ for $x \geq 1$, and

$$\psi(x) := 1 - \sum_{\nu=1}^i \eta_\nu + \eta_i \psi_1\left(\frac{1}{\delta_i}(x - \tau_{i-1})\right)$$

for $x \in (\tau_{i-1}, \tau_i)$ (κ chosen as above). Then ψ satisfies (i), (ii), and (iii) as well as

$$|\psi''(x)| \leq (4+\eta)\frac{\eta_i}{\delta_i^2} = \frac{4+\eta}{\sum_{\nu=1}^m \delta_\nu^2}.$$

Since $\sum_{\nu=1}^m \delta_\nu = 1$, we get from Hölder's inequality that

$$1 = \left(\sum_{\nu=1}^m \delta_\nu\right)^2 \leq m \sum_{\nu=1}^m \delta_\nu^2,$$

and therefore (v) is true. Inequality (iv) can be proved analogously. □

Proof of Theorem 5.1. We set

$$\alpha := \inf E, \quad \beta := \sup E.$$

Since the polynomials $p(z) = \prod_{j=1}^n (z - z_j)$ and $\widetilde{p}(z) = \prod_{j=1}^n (z - \operatorname{Re}\{z_j\})$ have the same distribution of zeros with respect to the strip domains $S(I)$ and since $\varepsilon_{\widetilde{p}} \leq \varepsilon_{p,E}$, it suffices to assume that all zeros of p are real and to prove

$$(\mu - \nu)(I) \leq 4\sqrt{2\varepsilon/\pi} + 4\varepsilon \tag{5.3}$$

for any interval $I = (-\infty, b]$, where $b \leq \beta$, $\mu := \mu_E$, $\nu := \nu_p$, $\varepsilon := \varepsilon_{p,E} < \frac{1}{4}$. We may also assume that $\mu(I) > 4\sqrt{2\varepsilon/\pi}$, since otherwise, (5.3) is obvious.

Let $b' \in E \cap I$ be such that $\mu([b', b]) = a\sqrt{\varepsilon}$, where $a \leq 4\sqrt{2/\pi}$ will be specified later. We consider the Green function $G^*(z)$ for $\overline{\mathbb{C}}\setminus E^*$ with pole at infinity where

$$E^* := E \cup [b', b].$$

88   2. Zero Distribution of Polynomials

The set E^* is regular, and by the maximum principle for harmonic functions (cf. Theorem 1.1.2),

$$\mu_{E^*}([b',b]) \geq \mu_E([b',b]) = \mu([b',b]). \tag{5.4}$$

The remaining part of the proof now follows the above ideas. Let $\chi(z)$ be a harmonic conjugate of $G^*(z)$ on $\mathbb{C}\setminus[\alpha,\infty)$. Then $\chi(z)$ extends continuously to $[\alpha,\infty)$ from the upper half-plane. Moreover, (5.4) yields

$$\chi(b') - \chi(b) \geq a\pi\sqrt{\varepsilon}.$$

Fix $\eta > 0$. By Lemma 5.2 there is a twice continuously differentiable function $g_1 : \mathbb{R} \to [0,1]$ and $\kappa > 0$ such that $g_1(\chi) = 1$ for $\chi \geq \chi(b') - \kappa$, $g_1(\chi) = 0$ for $\chi \leq \chi(b) + \kappa$, and

$$\left|\frac{d^2}{d\chi^2} g_1\right| \leq \frac{4+\eta}{\varepsilon(a\pi)^2}.$$

Next, let

$$g_2(z) := \psi_1\left(\frac{e^{G^*(z)} - 1}{a\pi\sqrt{\varepsilon}}\right)$$

($\psi_1 = \psi$ from Lemma 5.2 with $m = 1$, $\eta > 0$ fixed). Then we set

$$g(z) := \begin{cases} g_1(\chi(z))\, g_2(z), & \text{if } \operatorname{Im}(z) \geq 0, \\ g_1(\chi(\bar{z}))\, g_2(z), & \text{if } \operatorname{Im}(z) < 0. \end{cases}$$

Due to the symmetry of $G^*(z)$, i.e., $G^*(z) = G^*(\bar{z})$, g is continuous on $\mathbb{C}$. Moreover, g has compact support and

$$\frac{\partial}{\partial \mathbf{n}_+} g(z) = \frac{\partial}{\partial \mathbf{n}_-} g(z) = 0, \qquad z \in [\alpha,\beta], \tag{5.5}$$

where $\mathbf{n}_+ = i$ and $\mathbf{n}_- = -i$.

Next, we prove the representation

$$g(z) = \frac{1}{2\pi}\int \Delta g(\zeta)\log|z-\zeta|\,dm(\zeta), \qquad z \in \mathbb{C}, \tag{5.6}$$

which coincides with (1.1.11) for $g \in C^2(\mathbb{C})$ with compact support. Since $|\Delta g(z)|$ is bounded on $\mathbb{C}\setminus E$, both sides of (5.6) are continuous in z and we need only to prove (5.6) for $z \notin E$. We take $\rho > 0$ so small that the circle $C(z,\rho)$ of radius ρ around z does not intersect E and apply Green's formula (1.1.9) to the region

$$R_\rho := \mathbb{C}\setminus(D(z,\rho)\cup E),$$

where $D(z,\rho)$ is the interior of $C(z,\rho)$. Since g has compact support and is continuous on E and since (5.5) holds, we get

$$\frac{1}{2\pi} \int \Delta g(\zeta) \log|\zeta - z| \, dm(\zeta)$$

$$= \lim_{\rho \to 0} \left(\frac{1}{2\pi} \int_{C(z,\rho)} \frac{\partial}{\partial \mathbf{n}_1} g(\zeta) \log|\zeta - z| \, |d\zeta| \right.$$

$$\left. - \frac{1}{2\pi} \int_{C(z,\rho)} \frac{\partial}{\partial \mathbf{n}_1} \log|\zeta - z| \, g(\zeta) \, |d\zeta| \right)$$

$$= \lim_{\rho \to 0} \left(\frac{1}{2\pi} \log \rho \int_{D(z,\rho)} \Delta g(\zeta) \, dm(\zeta) + \frac{1}{2\pi\rho} \int_{C(z,\rho)} g(\zeta) \, |d\zeta| \right),$$

where $\mathbf{n}_1$ denotes the inner normal on $C(z,\rho)$ with respect to $D(z,\rho)$. For $\rho \to 0$ the first term vanishes and the second term tends to $g(z)$.

Further, introducing Green coordinates (r,χ) such that $G^*(z) = \log r$ (see Appendix D), we have

$$|\Delta g(z)| = r^2 \left(\frac{\partial G^*}{\partial \mathbf{n}}\right)^2 \left| g_1 \frac{\partial^2 g_2}{\partial r^2} + \frac{1}{r} g_1 \frac{\partial g_2}{\partial r} + \frac{1}{r^2} g_2 \frac{\partial^2 g_1}{\partial \chi^2} \right|$$

$$\leq r^2 \left(\frac{\partial G^*}{\partial \mathbf{n}}\right)^2 \left| 1 + \frac{r-1}{r} + \frac{1}{r^2} \right| \frac{4+\eta}{(a\pi)^2 \varepsilon}$$

$$\leq r^2 \left(\frac{\partial G^*}{\partial \mathbf{n}}\right)^2 \frac{8+2\eta}{(a\pi)^2 \varepsilon}$$

for all $z \in \mathbb{C} \setminus [\alpha, \infty)$, where we have used that

$$\frac{\partial g_2}{\partial r}(r) = \int_1^r \frac{\partial^2 g_2(r)}{\partial r^2} \, dr.$$

Again, the estimation starts with

$$(\mu - \nu)(I) \leq \int g \, d(\mu - \nu) + \mu([b', b]).$$

Proceeding as in the proof of Theorem 4.2 (with the new Green function $G^*(z)$ instead of $G(z)$) we obtain, with $\eta \to 0$,

$$\int g \, d(\mu - \nu) \leq \frac{4}{(a\pi)^2} \left(2a\pi\sqrt{\varepsilon} + (a\pi)^2 \varepsilon \right).$$

Thus,
$$(\mu - \nu)(I) \le \left(\frac{8}{\pi a} + a\right)\sqrt{\varepsilon} + 4\varepsilon.$$

Choosing $a = \sqrt{8/\pi}$, inequality (5.3) follows. Furthermore, for $\sqrt{\varepsilon} < \frac{1}{8}$, the estimate (5.1) yields (5.2). But for $\sqrt{\varepsilon} \ge \frac{1}{8}$, (5.2) is obvious. □

For the case $E = [-1, 1]$ it is more common to study the distribution of zeros of $p \in \mathbb{M}_n$ with respect to some sets in $\mathbb{C}$ defined by level lines of the appropriate Green function and the function conjugate to it (in the spirit of Theorems 2.1 to 4.5). The same can be done for the compact set from Theorem 5.1. However, our estimate will then depend on the number of components of E. Let us therefore assume that

$$E = I_1 \cup I_2 \cup \cdots \cup I_k \subset \mathbb{R}$$

is a (finite) union of compact, disjoint intervals. Since E is regular, each interval I_j consists of more than a single point.

Next, we introduce *Green lines* (see Appendix D), i.e., orthogonal trajectories to the family of level lines of Green's function. Hence through every noncritical point z of Ω, i.e., where grad $G(z) \ne 0$, there exists a unique Green line. If z is a critical point, then four Green lines have z as a limit point. For $z \in \mathbb{R}$, we define $\Gamma(z)$ as the union of all Green lines with z as a limit point. Then, for an interval $I \subset \mathbb{R}$ and $\sigma > 0$, we set

$$\widetilde{A}_\sigma(I) := \left(\bigcup_{z \in J} \overline{\Gamma(z)}\right) \cap E_\sigma.$$

$\widetilde{A}_\sigma(I)$ is called σ-*neighborhood for the interval I with respect to the compact set $E \subset \mathbb{R}$*. If E itself is an interval and $I \subset E$, then $\widetilde{A}_\sigma(I) = A_\sigma(I)$ with the notion of (2.30).

As in the proof of Theorem 5.1, let

$$\alpha := \inf_{z \in E} z \quad \text{and} \quad \beta := \sup_{z \in E} z.$$

Theorem 5.3. *Let E be the union of k real compact intervals, let $p \in \mathbb{M}_n$ be a monic polynomial, and let $a > 0$. Then for all intervals $I \subset [\alpha, \beta]$,*

$$|(\mu_E - \nu_p)(\widetilde{A}_\sigma(I))| \le \left(\frac{18}{\pi a} + 2a\sqrt{k}\right)\sqrt{\varepsilon_{p,E}} + 9\,\varepsilon_{p,E} \tag{5.7}$$

holds for arbitrary $\sigma \ge \pi a\sqrt{\varepsilon_{p,E}}$, where $\varepsilon_{p,E}$ is defined by (4.3). In particular,

$$|(\mu_E - \nu_p)(\widetilde{A}_\sigma(I))| \le 8\,k^{1/4}\sqrt{\varepsilon_{p,E}} \tag{5.8}$$

for any $\sigma \ge 3\,k^{-1/4}\sqrt{\pi\,\varepsilon_{p,E}}$.

2.5 Erdős–Turán Type Theorems on a System of Intervals

Proof. As before, let $\mu := \mu_E$, $\nu := \nu_p$, $\varepsilon := \varepsilon_{p,E} < \frac{1}{4}$. First we prove that

$$(\mu - \nu)(\widetilde{A}_\sigma(I)) \leq \left(\frac{8}{\pi a} + a\sqrt{k}\right)\sqrt{\varepsilon} + 4\varepsilon \qquad (5.9)$$

for any interval $I = [\alpha, b]$, where $b \leq \beta$. We may assume $\mu([\alpha, b]) > a\sqrt{k}\,\varepsilon$. Define $b' \in E \cap I$ such that

$$\mu([b', b]) = a\sqrt{k}\,\varepsilon.$$

The arguments that follow depend only on $G(z)$, i.e., the role of $G^*(z)$ in the proof of Theorem 5.1 is taken over by the function $G(z)$ itself.

In $\mathbb{C}\setminus[\alpha, \infty]$ define a harmonic conjugate $\chi(z)$ of $G(z)$. We extend $\chi(z)$ continuously to $[\alpha, \beta]$ from the upper half-plane and remark that $\chi(z)$ is constant on each interval $S_i := [\sup I_i, \inf I_{i+1}]$. Let $m - 1$ be the number of intervals S_i in (b', b), say

$$b' < S_{j+1} < S_{j+2} < \cdots < S_{j+m-1} < b.$$

Then we set

$$\tau_0 := \chi(b'), \; \tau_1 := \chi(S_{j+1}), \; \ldots, \; \tau_{m-1} := \chi(S_{j+m-1}), \; \tau_m := \chi(b),$$

and note that $\tau_m - \tau_0 = a\pi\sqrt{k}\,\varepsilon$. For this partition $\tau_0, \ldots, \tau_m$ and $\eta > 0$, Lemma 5.2 yields a function $g_1 : \mathbb{R} \to [0, 1]$ such that

$$\left|\frac{d^2}{d\chi^2}g_1\right| \leq \frac{4+\eta}{(a\pi)^2\varepsilon}\frac{m}{k} \leq \frac{4+\eta}{(a\pi)^2\varepsilon}.$$

In addition, we set

$$g_2(z) := \psi_1\left(\frac{e^{G(z)} - 1}{a\pi\sqrt{\varepsilon}}\right)$$

and

$$g(z) := \begin{cases} g_1(\varphi(z))\,g_2(z), & \text{if } \operatorname{Im}\{z\} \geq 0, \\ g_1(\varphi(\overline{z}))\,g_2(z), & \text{if } \operatorname{Im}\{z\} < 0. \end{cases}$$

The same estimates as in the proof of Theorem 5.1 yield

$$|\Delta g(z)| \leq r^2\left(\frac{\partial G}{\partial n}\right)^2\frac{8 + 2\eta}{(a\pi)^2\,\varepsilon}, \quad z \in \mathbb{C}\setminus[\alpha, \infty).$$

Proceeding as in that proof we get (5.9).

Of course, (5.9) holds for any interval of the form $[c, \beta]$, too. Furthermore, with some simple calculations we deduce from Lemma 1.4.3 that

$$\nu(\mathbb{C}\setminus E_{\pi a\sqrt{\varepsilon}}) \leq \frac{(e-1)\sqrt{\varepsilon}}{\pi a} + \varepsilon.$$

Observing $(\nu - \mu)(\mathbb{C}) = 0$, we obtain (5.7). Now we choose $a = 3/(\sqrt{\pi}\,k^{1/4})$ in (5.7). This yields (5.8). $\square$

Remark 5.4. The proof of Theorem 5.3 shows that (5.8) can be replaced by
$$|(\mu_E - \nu_p)(\widetilde{A}_\sigma(I))| \leq 8\sqrt{\varepsilon_{p,E}}$$
for all $\sigma \geq 3\sqrt{\pi \varepsilon_{p,E}}$ if all critical points of $G(z)$ are outside of the interior of $E_{3\sqrt{\pi \varepsilon_{p,E}}}$.

Indeed, we note that all components of $E_{3\sqrt{\pi \varepsilon_{p,E}}}$ contain exactly one subinterval of E. We fix $b' \in E$ such that
$$\mu_E([b',b]) = a\sqrt{\varepsilon_{p,E}}.$$
If $[b',b] \subset E$, we proceed as in the proof of Theorem 5.1 (respectively as in the proof of Theorem 5.3 with $k = 1$). If $[b',b] \not\subset E$, we have to modify $g(z)$ in a straightforward way. Let I_j be the interval with $b' \in I_j$ and let
$$\kappa := \max_{z \in I_j} z.$$
For $\mathrm{Im}(z) \geq 0$ we define
$$g(z) := \begin{cases} g_2(z), & \text{if } \chi(z) \leq \chi(\kappa), \\ 0, & \text{elsewhere in } \mathbb{C}, \end{cases}$$
and extend $g(z)$ to the lower half-plane by setting $g(z) := g(\bar{z})$.

2.6 Historical Comments

P. Bloch and G. Pólya [40] first investigated the question of giving an upper bound of the number R of real zeros of polynomials
$$p(z) = a_0 + a_1 z + \cdots + a_n z^n$$
whenever
$$|a_0| \geq \mu', \ |a_n| \geq \mu', \ |a_\nu| \leq \mu \ (1 \leq \nu \leq n-1).$$
They proved that the number R is bounded by
$$R < A_1(\mu, \mu') \frac{n \log \log n}{\log n},$$
where $A_1(\mu, \mu')$ depends only on μ and μ'. A few years later Erhardt Schmidt [157] proved the sharper inequality
$$R^2 \leq A_2(\mu, \mu') n \log \frac{\mu}{\mu'}$$

and the still sharper one

$$R^2 \leq A_3 \, n \, \log P, \tag{6.1}$$

where

$$P = \frac{|a_0| + \cdots + |a_n|}{\sqrt{a_0 a_n}}.$$

His detailed proof was never published because I. Schur [158] found shortly thereafter an elementary proof of it; his method furnishes the estimate

$$R^2 \leq 4n \log P$$

with the best possible constant 4. Further, G. Szegő found refinements of this inequality, and discovered that Schur's extremal polynomials are essentially Jacobi polynomials [172].

Erdős and Turán could derive the inequality (0.1) of Szegő and the inequality (6.1) of Schur from a common source, namely from the inequality (0.2). The result (0.2) of Erdős and Turán is sharp up to the constant 16 in front. T. Ganelius [67] improved this constant to $2, 5619\ldots$, which was again slightly improved by M. Mignotte [7]. Moreover, in [67] various types of distributions of zeros for sequences of polynomials and exponential sums are investigated that beyond the scope of this book.

Section 2.1 is presented in accordance with [76]. Theorem 1.1 is due to R. Grothmann, whereas Theorem 1.7, which plays a central role in applications, was first proved by Blatt, Saff, and Simkani [38]. For further discussion of applications of Theorem 1.7 we refer to the papers of Grothmann [77], [78].

The results of Section 2.2 are taken from [20].

In Section 2.3 we discuss the results of Blatt and Grothmann [31] and their natural generalizations. In some cases we obtain more general results by modifying the original technique in the proofs.

Some potential theoretical aspects of the Erdős–Turán type theorems can also be found in [151]. In higher dimensions P. Sjörgen [164] proved the discrepancy estimates for measures from their potentials and energies.

3
Discrepancy Theorems via Two–Sided Bounds for Potentials

In potential theory it is well known that a mass distribution σ is uniquely determined by its potential U^σ. In this chapter we shall consider different ways of making this fact more precise for signed measures $\sigma = \sigma^+ - \sigma^-$ that are supported on curves or arcs L.

In Section 3.1 we derive global discrepancy results between σ^+ and σ^- on quasiconformal curves and arcs. Typically, one of the measures (σ^+ in our terminology) has to take over the role of an approximate of the equilibrium measure μ_L. The estimates are formulated in terms of global bounds for $|U^\sigma|$ on a level line L_δ, $\delta > 0$, of Green's function for $\operatorname{ext} L$. Of course, practically relevant estimates can be obtained only if the level line L_δ is not too far away from the curve or arc L. On the other hand, in practical applications the role of σ^- is taken over by discrete zero counting measures of polynomials. Hence, in this case global bounds make sense only if the level line L_δ is not too near to E. Altogether, the parameter δ has to be chosen appropriately for practically relevant estimates.

Section 3.2 is devoted to the special case where all measures are supported on the interval $[-1, 1]$. In contrast to the previous section the starting point is no longer a global bound of $|U^\sigma|$ on a whole level line L_δ of Green's function. Instead, starting from a fixed subinterval $\Gamma \subset [-1, 1]$ bounds for $|U^\sigma|$ on neighboring sets of Γ will play an essential role for obtaining *local* discrepancy estimates on the subarc Γ. This approach is generalized to Dini-smooth Jordan arcs and curves in Section 3.3.

3.1 Estimates for Quasiconformal Curves and Arcs

Let $L \subset \mathbb{C}$ be a bounded Jordan curve or a Jordan arc.

Definition 1.1. The *discrepancy of a signed (Borel) measure σ on L* is defined by
$$D[\sigma] := \sup_J |\sigma(J)|,$$
where the supremum is taken over all subarcs $J \subset L$.

Obtaining estimates of this discrepancy $D[\sigma]$ in terms of the logarithmic potential
$$U^\sigma(z) = \int \log \frac{1}{|z-\zeta|} \, d\sigma(\zeta)$$
is the objective of this section.

Let Φ denote the Riemann function that maps the domain
$$\Omega := \begin{cases} \overline{\mathbb{C}} \setminus L, & \text{if } L \text{ is an arc,} \\ \text{ext} L, & \text{if } L \text{ is a curve,} \end{cases}$$
conformally and univalently onto the exterior $\Delta := \overline{\mathbb{C}} \setminus \overline{\mathbb{D}}$ of the unit disk $\mathbb{D}$, where Φ is normalized by the conditions (1.2.1).

For $\delta > 0$, we use again the level lines
$$L_\delta = \{z \in \Omega : |\Phi(z)| = 1 + \delta\}$$
and define
$$\varepsilon_\sigma(\delta) := \|U^\sigma\|_{L_\delta}, \tag{1.1}$$
where $\|\cdot\|_A$ always denotes the supremum norm on a subset $A \subset \mathbb{C}$. The main result of this subsection is the following.

Theorem 1.2. *Let L be a K-quasiconformal curve, $\sigma = \sigma^+ - \sigma^-$ a signed measure on L with positive part $\sigma^+ \in \mathcal{M}(L)$ and negative part $\sigma^- \in \mathcal{M}(L)$. Moreover, let $M > 0$ and $0 < \beta \leq 1$ be constants such that for all subarcs $J \subset L$,*
$$\sigma^+(J) \leq M \, \mu_L(J)^\beta. \tag{1.2}$$

Then there exists a constant $c > 0$, depending only on K, such that for $0 < \delta < 1$,
$$D[\sigma] \leq c \left(\varepsilon_\sigma(\delta) \log \frac{1}{\delta} + \delta^{1/(2K^2)} + M\delta^{\beta/2} \right), \tag{1.3}$$
where $\varepsilon_\sigma(\delta)$ is defined by (1.1).

Before we give the proof of this theorem we shall outline its ideas and discuss some special constructions.

Let $l \subset L$ be a subarc of the curve L,
$$\gamma := \Phi(l) = \{e^{i\theta} : \theta_1 \leq \theta \leq \theta_2\}.$$

For $r > 0$ set
$$\gamma_r := \{(1+r)e^{i\theta} : \theta_1 \leq \theta \leq \theta_2\}, \quad l_r := \Psi(\gamma_r).$$

Fix
$$a := \delta + \delta^{K^2}$$

and set
$$h_{l,a}(z) := \omega(z, \mathrm{int}\, L_a, l_a), \quad z \in \mathrm{int}\, L_a.$$

Defining
$$\omega_l(z) := \omega(z, \Omega, l), \quad z \in \Omega,$$

and applying the Green formulas (1.1.10) and (1.1.13), we may write for $z \in L$,

$$h_{l,a}(z) = \omega_l(\infty) + \frac{1}{2\pi} \int_{L_\delta} \left[\left(\frac{\partial}{\partial \mathbf{n}} \omega_l(\zeta) - \frac{\partial}{\partial \mathbf{n}} h_{l,a}(\zeta) \right) \log|\zeta - z| \right.$$
$$\left. + (h_{l,a}(\zeta) - \omega_l(\zeta)) \frac{\partial}{\partial \mathbf{n}} \log|\zeta - z| \right] |d\zeta|,$$

where $\partial/\partial \mathbf{n}$ is the operator of differentiation with respect to the outward normal to the curve L_δ at the point ζ.

Integrating the last relation we get

$$\int_L h_{l,a}(z) \, d\sigma(z) = \frac{1}{2\pi} \int_{L_\delta} U^\sigma(\zeta) \left(\frac{\partial}{\partial \mathbf{n}} h_{l,a}(\zeta) - \frac{\partial}{\partial \mathbf{n}} \omega_l(\zeta) \right) |d\zeta|$$
$$+ \frac{1}{2\pi} \int_{L_\delta} (\omega_l(\zeta) - h_{l,a}(\zeta)) \frac{\partial}{\partial \mathbf{n}} U^\sigma(\zeta) |d\zeta|. \qquad (1.4)$$

This formula will be the base of our next reasoning: Since $h_{l,a}$ is an approximation of the characteristic function of l, the left-hand side of (1.4) represents an approximation of $\sigma(l)$. To obtain the inequality (1.3) we have to estimate the right-hand side in an appropriate way.

For sufficiently large δ the estimate (1.3) is trivial. Henceforth, we may assume without loss of generality that $\delta < \frac{1}{2}$.

Since L is a K-quasiconformal curve, the mappings Φ and $\Psi := \Phi^{-1}$ can be extended to K^2-quasiconformal automorphisms of $\overline{\mathbb{C}}$ (see [3, Chapter IV] or Theorem B.2.1). Set $z_0 := \Psi(0)$ and denote by φ_a the conformal mapping of int L_a onto the unit disk $\mathbb{D}$ with the normalization $\varphi_a(z_0) = 0$, $\varphi'_a(z_0) > 0$.

Consider the function $g := \varphi_a \circ \Psi$ that maps the disk $\{w : |w| < 1 + a\}$ K^2-quasiconformally onto $\mathbb{D}$ with $g(0) = 0$. We note that g is conformal in the annulus $\{w : 1 < |w| < 1 + a\}$ and, moreover, g can be extended to a K^2-quasiconformal mapping $g : \overline{\mathbb{C}} \to \overline{\mathbb{C}}$ if for $|w| > 1 + a$ we set

$$g(w) := \left[\overline{g\left(\frac{(1+a)^2}{\overline{w}}\right)} \right]^{-1}. \tag{1.5}$$

The last fact gives us the possibility to use Theorem 1.2.7 with $F := g$ or $F := g^{-1}$.

Let $u := g^{-1}$ and set

$$\Gamma = \Gamma(\delta) := \varphi_a(L_\delta) = \{z \in \mathbb{D} : |u(z)| = 1 + \delta\}.$$

Denote by $|\gamma|$ the *length* of a Jordan arc or curve γ.

Lemma 1.3. *Let $z, \zeta \in \Gamma$ be arbitrary points and let $\Gamma(z, \zeta)$ denote the shortest component of $\Gamma \setminus \{z, \zeta\}$. Then the inequality*

$$|\Gamma(z, \zeta)| \leq c|z - \zeta|$$

holds with some constant $c = c(K) > 0$.

Proof. Set $w := u(z)$, $\tau := u(\zeta)$,

$$\Gamma'(w, \tau) := u(\Gamma(z, \zeta)), \quad E := \{t \in \mathbb{D} : 1 < |u(t)| < 1 + a\}$$

(see Figure 3.1). By virtue of Theorem 1.2.7, we have $d(z, \partial E) \asymp 1 - |z|$. Suppose first that $|w - \tau| \leq \delta^{K^2}/32$. According to (1.2.8),

$$\frac{1}{16} \frac{d(z, \partial E)}{\delta^{K^2}} |\tau - w| \leq |z - \zeta| \leq \frac{d(z, \partial E)}{2}.$$

Since by the same reasoning

$$|z - \xi| \leq \frac{d(z, \partial E)}{2}, \quad \xi \in \Gamma(z, \zeta),$$

we obtain by Lemma 1.2.3 that for any $t \in \Gamma'(w, \tau)$,

$$|g'(t)| \asymp |g'(w)| \asymp \frac{1 - |z|}{\delta^{K^2}}.$$

Therefore,

$$|\Gamma(z, \zeta)| = \int_{\Gamma'(w, \tau)} |g'(t)| \, |dt| \asymp \frac{1 - |z|}{\delta^{K^2}} |w - \tau| \preceq |z - \zeta|.$$

3.1 Estimates for Quasiconformal Curves and Arcs   99

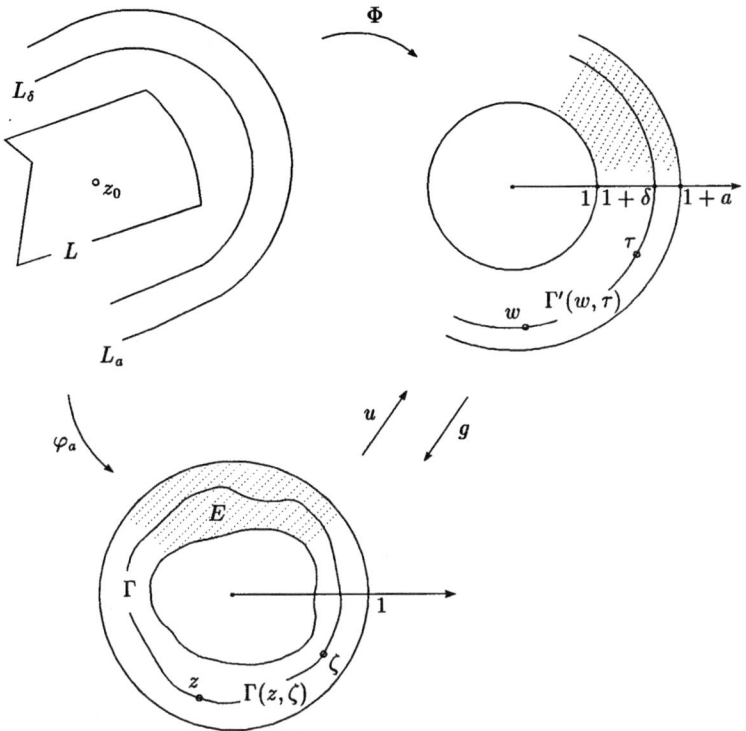

Fig. 3.1

Now let $|w - \tau| > \delta^{K^2}/32$. We subdivide the arc $\Gamma'(w, \tau)$ by points $t_1 := w, \ldots, t_{m+1} := \tau$ such that

$$\frac{\delta^{K^2}}{64} \leq |t_k - t_{k+1}| \leq \frac{\delta^{K^2}}{32}, \quad k = 1, \ldots, m.$$

Set

$$s_k := t_k \frac{1+a}{1+\delta}, \quad \xi_k := g(s_k), \quad \eta_k := g(t_k).$$

Using the previous result and Lemma 1.2.3, we find that

$$|\Gamma(\eta_k, \eta_{k+1})| \preceq |\eta_k - \eta_{k+1}| \asymp |\eta_k - \xi_k| \asymp |\xi_{k+1} - \xi_k|.$$

Hence,

$$|\Gamma(z, \zeta)| = \sum_{k=1}^{m} |\Gamma(\eta_k, \eta_{k+1})| \preceq \sum_{k=1}^{m} |\xi_k - \xi_{k+1}| \preceq |\xi_1 - \xi_{m+1}| \asymp |z - \zeta|.$$

$\square$

100   3. Discrepancy Theorems via Two–Sided Bounds for Potentials

The estimate of the right-hand side of (1.4) is now given by the following lemma.

Lemma 1.4. *For an arbitrary signed measure σ on L and $\delta < \frac{1}{2}$ the inequality*

$$\left| \int_L h_{l,a}(z)\, d\sigma(z) \right| \leq c\varepsilon_\sigma \left(\frac{\delta}{2} \right) \log \frac{1}{\delta} \tag{1.6}$$

holds with some constant $c = c(K) > 0$.

Proof. In the representation (1.4) we try to estimate the first integral on the right-hand side from above. Therefore, we want to establish the inequalities

$$\int_{L_\delta} \left| \frac{\partial}{\partial \mathbf{n}} h_{l,a}(\zeta) \right| |d\zeta| \preceq \log \frac{1}{\delta}, \tag{1.7}$$

$$\int_{L_\delta} \left| \frac{\partial}{\partial \mathbf{n}} \omega_l(\zeta) \right| |d\zeta| \preceq \log \frac{1}{\delta}. \tag{1.8}$$

The estimate (1.8) is a consequence of Lemma 1.1.5. In fact, since the harmonic measure is a conformal invariant, we obtain

$$\int_{L_\delta} \left| \frac{\partial}{\partial \mathbf{n}} \omega_l(\zeta) \right| |d\zeta| \leq \int_{|w|=1+\delta} |\text{grad } \omega(w, \Delta, \gamma)| |dw|$$

$$\preceq \sum_{j=1}^2 \int_{|w|=1+\delta} \frac{|dw|}{|w - w_j|} \preceq \log \frac{1}{\delta},$$

where w_1 and w_2 are the endpoints of the arc $\gamma := \Phi(l)$.

In order to prove (1.7) we consider the conformal mapping φ_a defined above. Let τ_1 and τ_2 be the endpoints of the arc $\varphi_a(l_a)$. Using (1.1.20), we find that for $\tau \in \mathbb{D}$,

$$|\text{grad } \omega(\tau, \mathbb{D}, \varphi_a(l_a))| \preceq \frac{1}{|\tau - \tau_1|} + \frac{1}{|\tau - \tau_2|}.$$

Denote by $\mu_j(\delta)$, where $j = 1, 2$ and $0 < \delta < 2$, the length of the portion of Γ lying in the open disk with center at τ_j and radius δ. According to Lemma 1.3, we have $\mu_j(\delta) \preceq \delta$.

Further, recalling the definition of the function g from (1.5) and applying to it Theorem 1.2.7, we obtain

$$d(\Gamma, \mathbb{T}) \preceq \delta^{K^4}.$$

Therefore, integrating by parts, we get

$$\int_{L_\delta} \left|\frac{\partial}{\partial \mathbf{n}} h_{l,a}(\zeta)\right| |d\zeta| \leq \int_\Gamma |\text{grad } \omega(\tau, \mathbb{D}, \varphi_a(l_a))| |d\tau|$$

$$\preceq \sum_{j=1}^{2} \int_{c\delta K^4}^{2} \frac{d\mu_j(x)}{x} = \sum_{j=1}^{2} \left(\frac{\mu_j(2)}{2} + \int_{c\delta K^4}^{2} \frac{\mu_j(x)}{x^2} dx \right)$$

$$\preceq \log \frac{1}{\delta},$$

which is (1.7).

In order to estimate the second integral on the right-hand side of (1.4), we set for $w \in \Delta$,

$$\widetilde{U}(w) := U^\sigma(\Psi(w)),$$
$$\widetilde{h}(w) := h_{l,a}(\Psi(w)),$$
$$\widetilde{\omega}(w) := \omega_l(\Psi(w))$$

and

$$\widetilde{\chi}(w) := \begin{cases} 1, & \text{if } w \in \gamma_\delta, \\ 0, & \text{if } w \notin \gamma_\delta. \end{cases}$$

For $|w| = 1 + \delta$ we have by Schwarz's formula (cf. (1.1.8))

$$|\text{grad } \widetilde{U}(w)| \leq \frac{1}{\pi} \int_{|\tau - w| = \delta/2} \frac{|\widetilde{U}(\tau)|}{|\tau - w|^2} |d\tau| \leq \frac{4\varepsilon_\sigma(\delta/2)}{\delta}. \qquad (1.9)$$

Since

$$\int_{L_\delta} |\omega_l(\zeta) - h_{l,a}(\zeta)| \left|\frac{\partial}{\partial \mathbf{n}} U^\sigma(\zeta)\right| |d\zeta|$$

$$\leq \int_{|w|=1+\delta} |\widetilde{\omega}(w) - \widetilde{h}(w)| |\text{grad } \widetilde{U}(w)| |dw|, \qquad (1.10)$$

we want to show that

$$\int_{|w|=1+\delta} |\widetilde{\omega}(w) - \widetilde{\chi}(w)| |dw| \preceq \delta \log \frac{1}{\delta}, \qquad (1.11)$$

$$\int_{|w|=1+\delta} |\widetilde{h}(w) - \widetilde{\chi}(w)| |dw| \preceq \delta \log \frac{1}{\delta}. \qquad (1.12)$$

Now, it follows from (1.1.19), rewritten for the exterior of the unit disk, that

$$|\widetilde{\omega}(w) - \widetilde{\chi}(w)| \preceq \delta \left(\frac{1}{|w - w_1|} + \frac{1}{|w - w_2|} \right), \qquad (1.13)$$

where $|w| = 1 + \delta$, and w_1 and w_2 are the endpoints of the arc γ.

Thus, (1.11) is a simple consequence of (1.13).

In order to establish (1.12) suppose that $|w| = 1 + \delta$ and denote by t_1 and t_2 the endpoints of the arc γ_a. Applying Lemma 1.2.9 to the restriction of the mapping Φ to $\operatorname{int} L_a$ and Lemma 1.1.5, we have

$$|\widetilde{h}(w) - \widetilde{\chi}(w)| \preceq \left(\frac{\delta^{K^2}}{\min_{j=1,2} |w - t_j|} \right)^{K^{-2}}$$

$$\leq \delta \left(|w - t_1|^{-K^{-2}} + |w - t_2|^{-K^{-2}} \right).$$

By integrating the last inequality, we get (1.12).

Finally, combining (1.4), (1.7)–(1.12), we obtain (1.6). □

Proof of Theorem 1.2. To see that $D[\sigma]$ is appropriately estimated, we have only to consider an arbitrary arc $J \subset L$ and to establish the inequality

$$\sigma(J) \geq -c_1 \left(\varepsilon_\sigma \left(\frac{\delta}{2} \right) \log \frac{1}{\delta} + \delta^{1/(2K^2)} + M \, \delta^{\beta/2} \right). \qquad (1.14)$$

Moreover, if
$$J' := \Phi(J) = \{e^{i\theta} : a \leq \theta \leq b\},$$
we need to study only the case $b - a \leq \pi$.

Set
$$\gamma := \{e^{i\theta} : a - \delta^{1/2} \leq \theta \leq b + \delta^{1/2}\}, \quad l := \Psi(\gamma),$$
$$\gamma_1 := \{e^{i\theta} : a - 2\delta^{1/2} \leq \theta \leq b + 2\delta^{1/2}\}, \quad l_1 := \Psi(\gamma_1).$$

With this choice we get from Lemma 1.4 that the function $h_{l,a}(z)$ satisfies

$$\int_L h_{l,a}(z) \, d\sigma(z) \geq c_2 \, \varepsilon_\sigma(\delta/2) \log \delta. \qquad (1.15)$$

On the other hand,

$$\int_L h_{l,a}(z) \, d\sigma(z) \leq \sigma(J) + \int_J [1 - h_{l,a}(z)] \, d\sigma^-(z)$$

$$+ \sigma^+(l_1 \setminus J) + \int_{L \setminus l_1} h_{l,a}(z) \, d\sigma^+(z). \qquad (1.16)$$

3.1 Estimates for Quasiconformal Curves and Arcs

Let us estimate each of the integrals on the right-hand side of (1.16) from above.

Using (1.1.19) and Lemma 1.2.9, we find that

$$1 - h_{l,a}(z) \preceq \delta^{1/(2K^2)}, \quad z \in J,$$
$$h_{l,a}(z) \preceq \delta^{1/(2K^2)}, \quad z \in L \setminus l_1.$$

Therefore,

$$\int_J [1 - h_{l,a}(z)]\, d\sigma^-(z) + \int_{L \setminus l_1} h_{l,a}(z)\, d\sigma^+(z) \preceq \delta^{1/(2K^2)}. \tag{1.17}$$

Furthermore, by our assumption (1.2),

$$\sigma^+(l_1 \setminus J) \le 2M(2\delta^{1/2})^\beta,$$

which, in view of (1.15)–(1.17), yields (1.3). $\square$

The results of Theorem 1.2 can be transformed in a straightforward manner to the case of a quasiconformal arc.

Theorem 1.5. *Let L be a quasiconformal arc, $\sigma = \sigma^+ - \sigma^-$ a signed measure with positive part $\sigma^+ \in \mathcal{M}(L)$ and negative part $\sigma^- \in \mathcal{M}(L)$. Moreover, let $M > 1$, $0 < \beta \le 1$ be constants such that for all subarcs $J \subset L$,*

$$\sigma^+(J) \le M \mu_L(J)^\beta. \tag{1.18}$$

Then there exist constants $c > 0$ and $K \ge 1$, depending on L and β, such that for $0 < \delta \le 1$,

$$D[\sigma] \le c \left(\varepsilon_\sigma(\delta) \log \frac{1}{\delta} + M \delta^{\beta/(2K^2)} \right). \tag{1.19}$$

Proof. Let L be a K_1-quasiconformal arc, $K_1 \ge 1$. Then by (1.2.24),

$$\mu_L(J) \le c_1 (\mathrm{diam}\ J)^{1/K_2}$$

with some constant $c_1 > 0$ and $K_2 \ge 1$, depending on L.

Moreover, there exists a bounded K_3-quasiconformal curve $\widetilde{L}$ with $L \subset \widetilde{L}$, $K_3 \ge 1$. Then (1.2.22) implies

$$c_2 (\mathrm{diam}\ J)^{K_3^2} \le \mu_{\widetilde{L}}(J)$$

with a positive constant c_2, depending on L. Hence,

$$\mu_L(J) \le c_3\, \mu_{\widetilde{L}}(J)^{1/(K_2 K_3^2)}$$

104   3. Discrepancy Theorems via Two–Sided Bounds for Potentials

with some constant $c_3 > 0$, depending on L. Therefore, (1.18) implies

$$\sigma^+(J) \leq c_3 \, M \, \mu_{\widetilde{L}}(J)^{\beta/(K_2 K_3^2)} \tag{1.20}$$

for all subarcs L of J.

If $\widetilde{G}(z)$ is Green's function of ext $\widetilde{L}$, then it is well known that $\widetilde{G}(z) < G(z)$ for all $z \in \text{ext } \widetilde{L}$. Hence, the level line

$$\widetilde{L}_\delta := \{ z \in \mathbb{C} : \widetilde{G}(z) = \log(1 + \delta) \}$$

surrounds $\widetilde{L}$ and lies in ext L_δ. The maximum principle for harmonic functions yields

$$\widetilde{\varepsilon}_\sigma(\delta) := \| U^\sigma \|_{\widetilde{L}_\sigma} \leq \varepsilon_\sigma(\delta).$$

Hence, with (1.20) and Theorem 1.2 we obtain

$$D[\sigma] \leq c \left(\varepsilon_\sigma(\delta) \log \frac{1}{\delta} + \delta^{1/(2K_3^2)} + M \delta^{\beta/(2K_2 K_3^2)} \right).$$

Choosing $K = K_2 K_3$, we obtain (1.19). □

Remark 1.6. Let L be a bounded quasiconformal arc or curve and define

$$\mathcal{L} := \{ \widetilde{L} \subset \mathbb{C} : \widetilde{L} = aL + b \text{ with } a, b \in \mathbb{C}, \, a \neq 0 \}.$$

Then the constant c in the inequality (1.3), and the constants c and K in the inequality (1.19) can be chosen fixed for all $\widetilde{L} \in \mathcal{L}$.

In applications the parameter δ will be appropriately chosen, depending on the size of $\varepsilon_\sigma(\delta)$, to obtain a discrepancy estimate merely in terms of an upper bound ε of $\varepsilon_\sigma(\delta)$.

Theorem 1.7. Let L be a quasiconformal curve or arc, $\sigma = \sigma^+ - \sigma^-$ a signed measure on L with positive part $\sigma^+ \in \mathcal{M}(L)$ and negative part $\sigma^- \in \mathcal{M}(L)$. Moreover, let

$$\sigma^+(J) \leq M \mu_L(J)^\beta$$

for any subarc $J \subset L$, where $M > 0$, $0 < \beta \leq 1$. Let $0 < \varepsilon \leq 1/e$. Then there exist positive constants $c = c(L, M)$ and $\kappa = \kappa(L, M) > 1$ such that

$$D[\sigma] \leq c \varepsilon \log \frac{1}{\varepsilon} \tag{1.21}$$

for all ε with $\varepsilon \geq \varepsilon_\sigma(\varepsilon^\kappa)$.

Proof. If L is a K-quasiconformal curve, we set $\kappa := 2K^2 + 2/\beta$ and $\delta := \varepsilon^\kappa$. We get by Theorem 1.2,

$$D[\sigma] \leq c_1 \left(\kappa \varepsilon_\sigma(\delta) \log \frac{1}{\varepsilon_\sigma(\delta)} + \varepsilon + M\varepsilon \right), \tag{1.22}$$

where c_1 depends only on K. Since $\varepsilon_\sigma(\varepsilon^\kappa) \leq \varepsilon \leq 1/e$ and the function

$$f(x) = x \log \frac{1}{x}$$

is monotonically increasing in $0 < x \leq 1/e$, we obtain (1.21) with c depending only on K and M.

If L is a quasiconformal arc, there exist constants $c > 0$ and $K > 1$ such that the discrepancy estimate (1.19) holds; i.e.,

$$D[\sigma] \leq c_1 \left(\varepsilon_\sigma(\delta) \log \frac{1}{\delta} + M \delta^{\beta/(2K^2)} \right), \qquad (1.23)$$

where c_1 and K depend on L only. Set $\kappa_1 := 2K^2$ and $\delta := \varepsilon_1^\kappa$. Then again we obtain the inequality (1.21). $\square$

The following consequence of Theorem 1.2 is especially useful.

Theorem 1.8. *Let L be a quasiconformal curve or arc, and let $p = p_n$ be a monic polynomial of degree $n \geq 2$ with simple zeros $z_j \in L$, $j = 1, \ldots, n$, such that*

$$\|p\|_L \leq A_n (\operatorname{cap} L)^n,$$
$$|p'(z_j)| \geq B_n^{-1} (\operatorname{cap} L)^n,$$
$$C_n := \max(A_n, B_n, n) \leq e^{n/e}.$$

Let ν_p denote the zero counting measure of p. Then

$$D[\mu_L - \nu_p] \leq c \frac{\log C_n}{n} \log \frac{n}{\log C_n}, \qquad (1.24)$$

where $c > 0$ is a constant depending only on L.

Proof. The positive part μ_L of the signed measure $\sigma = \mu_L - \nu_p$ satisfies (1.2) with $M = \beta = 1$. Let $\kappa > 1$ be the constant of Theorem 1.7 and let

$$\delta_n = \left(\frac{1}{n}\right)^\kappa.$$

Then by Lemmas 1.4.2 and 1.4.5 we obtain

$$\varepsilon_\sigma(\delta_n) \leq c_1 \frac{\log C_n}{n},$$

where $c_1 = c_1(L)$ is a positive constant. We may assume $c_1 \geq 1$ and $n \geq 3$. Set

$$\delta := \left(c_1 \frac{\log C_n}{n} \right)^\kappa.$$

Then $\delta > \delta_n$ and
$$\varepsilon_\sigma(\delta)^\kappa \leq \varepsilon_\sigma(\delta_n)^\kappa \leq \left(c_1 \frac{\log C_n}{n}\right)^\kappa = \delta = \varepsilon^\kappa$$
with $\varepsilon := c_1 \log C_n / n$.

If $\varepsilon \geq 1/e$, then
$$\frac{1}{e} \geq \frac{\log C_n}{n} \geq \frac{1}{c_1 e}$$
and
$$\frac{\log C_n}{n} \log \frac{n}{\log C_n} \geq \frac{1}{c_1 e}.$$

Therefore, (1.24) is trivial in this case. If $\varepsilon \leq 1/e$, then this discrepancy estimate follows from Theorem 1.7. □

Finally, we would like to discuss the requirement of quasiconformality in our results. More precisely, we claim that in the case of curves and arcs with cusps, in general, a universal estimate for the discrepancy like (1.24) with some constant c does not exist. We will try to explain this fact with an illustrating example to Theorem 1.8 and domains of the following form:
$$E = E(f) := \{z = x + iy : 0 < x < 1, 0 < y < f(x)\},$$
where f is a positive monotonically increasing function on $[0,1]$ such that $f(0) = 0$ and there exists an inverse function $g := f^{-1}$ (with the same properties).

For $n \geq 2$, we consider the Jordan arc
$$S = S(n) := L \setminus (0, g(n^{-2})),$$
where $L := \partial E$ (see Figure 3.2).

Let $S_{2/n}$ denote the corresponding level curve of the conformal mapping from $\overline{\mathbb{C}} \setminus S$ onto Δ with standard normalization at ∞. Since for any $z \in S_{2/n}$ and $n \in \mathbb{N}$ large enough
$$d(z, S) \geq \frac{\operatorname{diam} S}{4} \frac{(2/n)^2}{1 + 2/n} > n^{-2}$$
(see Lemma 1.2.5), we conclude that
$$E \subset \operatorname{int} S_{2/n},$$
and consequently,
$$\operatorname{cap} S \leq \operatorname{cap} L \leq \left(1 + \frac{2}{n}\right) \operatorname{cap} S.$$

Further, let $z_1, \ldots, z_n$ be the points of an nth Fekete point set of S.

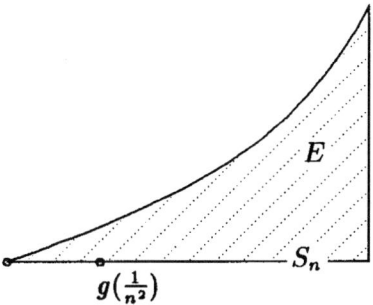

Fig. 3.2

According to Lemma 1.1.13 and Theorem 1.3.3, the monic Fekete polynomial

$$q_n(z) := \prod_{j=1}^{n}(z - z_j)$$

satisfies

$$|q_n'(z_j)| \geq (\text{cap } S)^n \succeq (\text{cap } L)^n, \quad j = 1, \ldots, n,$$

and

$$\|q_n\|_L \preceq \|q_n\|_S \preceq n^2(\text{cap } S)^n \leq n^2(\text{cap } L)^n.$$

Hence, the right-hand side of estimate (1.24) has order

$$\frac{\log C_n}{n} \log \frac{n}{\log C_n} \asymp \frac{(\log n)^2}{n}.$$

On the other hand, it is obvious that

$$D[\mu_L - \nu_n] \succeq g(n^{-2}).$$

Since g can tend to zero arbitrarily slowly as $x \to 0$, it is impossible to obtain either an estimate of the same type as in Theorem 1.8 or any other estimate with a universal right-hand side.

Note that Theorem 1.8 is best possible even for the interval $L = [-1, 1]$.

Theorem 1.9. *Let $\{C_n\}$ be an arbitrary sequence with the property that $n \leq C_n \leq e^{n/2}$. Then there are monic polynomials $p_n \in \mathbb{M}_n, n \in \mathbb{N}$, with all zeros on $[-1, 1]$ such that*

$$\|p_n\|_{[-1,1]} \leq C_n \, 2^{-n},$$

$$|p_n'(x_{i,n})| \geq \frac{1}{C_n} 2^{-n}, \quad 1 \leq i \leq n,$$

for every zero $x_{i,n}$ of p_n, and such that the estimate

$$D[\nu_{p_n} - \mu_{[-1,1]}] \geq c \frac{\log C_n}{n} \log \frac{n}{\log C_n}$$

holds with some $c > 0$ independent of n.

For a proof, see [177] or [156, p. 427].

3.2 Local Estimates for Intervals

In this section we explain how to derive a local version of the above results. We begin our investigation with the special case $L := I = [-1, 1]$ to prepare some insight into the methods for the more general case of Dini-smooth arcs and curves. Moreover, the results of this case are sharper than the estimates for Dini-smooth arcs. Especially, the results are well suited for applications to orthogonal and extremal polynomials (cf. Chapter 5).

Let $L \subset \mathbb{C}$ be a bounded Jordan arc or Jordan curve, and let $\Gamma \subset L$ be a subarc of L.

Definition 2.1. The *local discrepancy of a signed measure σ on the subarc Γ* is defined by

$$D_\Gamma[\sigma] := \sup_J |\sigma(J)|, \qquad (2.1)$$

where the supremum is taken over all subarcs J of Γ.

Let $\Gamma = [a, b]$ and let us define for $\delta > 0$,

$$B(\delta, \Gamma) := \{z = x + iy : a - \delta \leq x \leq b + \delta, |y| \geq \delta^3\} \qquad (2.2)$$

(see Figure 3.3) and

$$\varepsilon_\sigma(\delta, \Gamma) := \max_{z \in B(\delta, \Gamma)} |U^\sigma(z)|. \qquad (2.3)$$

In the following theorem we give an estimate of $D_\Gamma[\sigma]$ in terms of $\varepsilon_\sigma(\delta, \Gamma)$ and δ. Furthermore, if Γ is an interval with $\sigma^-(\Gamma) = 0$, then we want to get some insight to bound $\sigma^+(\Gamma)$. For this reason we need the interval

$$\Gamma_{1/2} := \left[\frac{a+b}{2} - \frac{b-a}{4}, \frac{a+b}{2} + \frac{b-a}{4}\right],$$

which has the same center as Γ but half the length of Γ. We remind that $|J|$ denotes the length of the interval J.

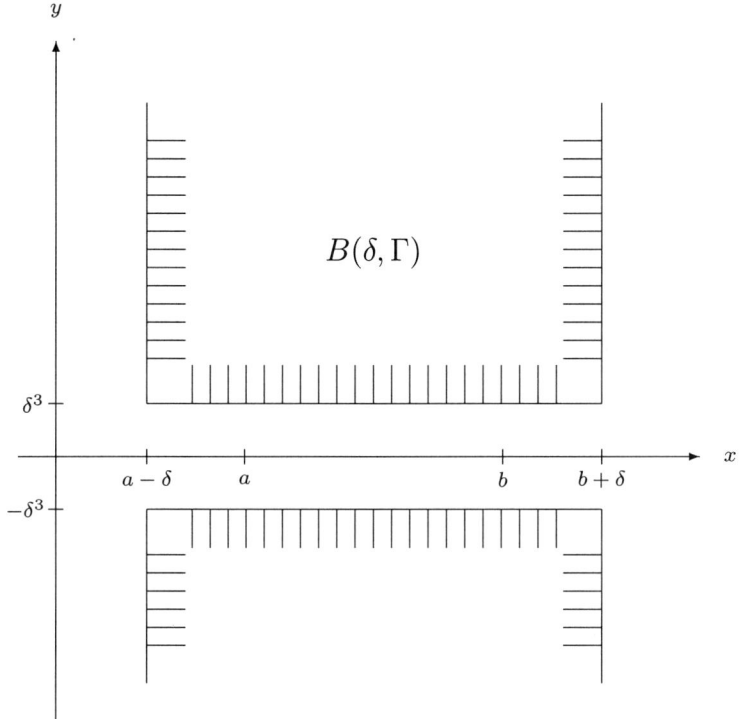

Fig. 3.3

Theorem 2.2. *Let $\sigma = \sigma^+ - \sigma^-$ be a signed measure with σ^+, $\sigma^- \in \mathcal{M}(I)$ and let $\Gamma = [a,b]$ be a closed subinterval of I. Let $\delta \leq |\Gamma|/4$. Then*

$$\sigma^+(\Gamma_{1/2}) \leq 3\left(\varepsilon_\sigma(\delta,\Gamma) + \delta \log \frac{1}{\delta}\right) + \sigma^-(\Gamma). \tag{2.4}$$

Moreover, if $M > 0$, $0 < \gamma \leq 1$ are constants such that

$$\sigma^+(J) \leq M|J|^\gamma \tag{2.5}$$

for all subintervals $J \subset [a-\delta, b+\delta]$, then for $0 < \delta \leq \tfrac{1}{2}$,

$$D_\Gamma[\sigma] \leq 4\varepsilon_\sigma\left(2 + \log \frac{|\Gamma|}{\delta}\right) + 21\,\delta^2 \log \frac{1}{\delta} + 2M\,\delta^\gamma. \tag{2.6}$$

Proof. For abbreviation let

$$\varepsilon := \varepsilon_\sigma(\delta,\Gamma).$$

We prove (2.6) first. We may assume that $\delta < 1/21$, since otherwise, (2.6) is trivial.

Let $f : \mathbb{R} \to [0,1]$ be a monotone function such that $f(x) = 1$ for $-\infty < x \leq 0$ and $f(x) = 0$ for $1 \leq x < \infty$. Moreover, f can be chosen in such a way that f is twice continuously differentiable and $|f''(x)| \leq 4$ for all $x \in \mathbb{R}$.

Let J be a subinterval of Γ. First, let us show that

$$\sigma(J) \geq -\left(4\varepsilon\left(2 + \log\left(1 + \frac{|\Gamma|}{\delta}\right)\right) + 21\,\delta^2 \log\frac{1}{\delta} + 2\,M\,\delta^\gamma\right). \tag{2.7}$$

We introduce the function

$$g(z) = g(x+iy) := f\left(\frac{2}{\delta}(x-b)\right) f\left(\frac{2}{\delta}(a-x)\right), \quad z \in \mathbb{C},$$

which is twice continuously differentiable in $\mathbb{R}^2 \cong \mathbb{C}$ and all partial derivatives up to second order vanish in $\mathbb{C} \setminus S_1$ and S_2, where S_1 and S_2 are the strips

$$S_1 := \{x+iy : a - \delta/2 \leq x \leq b + \delta/2\}$$

and

$$S_2 := \{x+iy : a \leq x \leq b\}.$$

Let us define the polygonal line l consisting of the line segments

$$l_1 := \left[a - \frac{\delta}{2} + i\delta,\ a + i\delta\right],$$

$$l_2 := \left[a + i\delta,\ \frac{b+a}{2} + i\left(\delta - \frac{b-a}{2}\right)\right],$$

$$l_3 := \left[\frac{b+a}{2} + i\left(\delta - \frac{b-a}{2}\right),\ b + i\delta\right],$$

$$l_4 := \left[b + i\delta,\ b + \frac{\delta}{2} + i\delta\right].$$

Here and in what follows, $[z_1, z_2]$ stands for the line segment between the points z_1 and z_2. Moreover, let $\bar{l} = \{\bar{z} : z \in l\}$ denote the reflection of l at the real axis. We define V as the closed subset of S_1 bounded by l and $\bar{l}$ (see Figure 3.4).

By Green's formula, for $z \in I$,

$$g(z) = -\frac{1}{2\pi} \int_{\partial V} g(\zeta) \frac{\partial}{\partial \mathbf{n}_\zeta} \log \frac{1}{|\zeta - z|} |d\zeta| - \frac{1}{2\pi} \int_V \Delta g(\zeta) \log \frac{1}{|\zeta - z|}\, dx\, dy,$$

where $\partial/\partial \mathbf{n}_\zeta$ denotes the operator of differentiation with respect to the outward normal to the curve ∂V at the point $\zeta \in \partial V$. This normal derivative exists at all points different from the corners of the polygon ∂V.

Keeping in mind that $\Delta g(\zeta) = 0$ for $\zeta \in S_2$ and that $g(\zeta)$ and $U^\sigma(\zeta)$ are

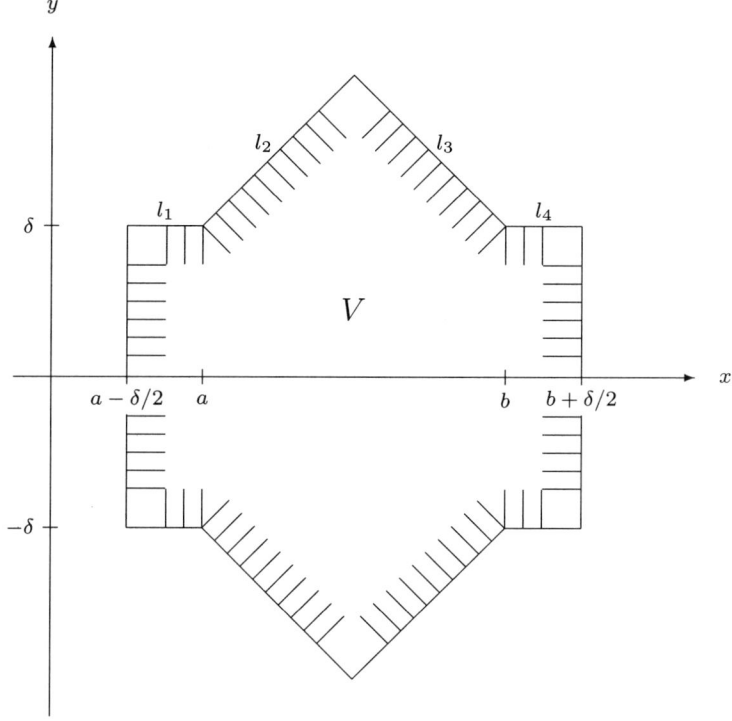

Fig. 3.4

symmetric with respect to the real axis, integration with respect to σ and Fubini's theorem imply that

$$\int g\, d\sigma = -\frac{1}{\pi}\int_l g(\zeta)\frac{\partial}{\partial \mathbf{n}_\zeta}U^\sigma(\zeta)\,|d\zeta| - \frac{1}{\pi}\int_{V_1 \cup V_2}\Delta g(\zeta)\, U^\sigma(\zeta)\,dx\,dy, \quad (2.8)$$

where

$$V_1 := \{x+iy : y \in [0, \delta^3],\ x \in [a-\delta/2, a] \cup [b, b+\delta/2]\}$$

and

$$V_2 := \{x+iy : y \in [\delta^3, \delta],\ x \in [a-\delta/2, a] \cup [b, b+\delta/2]\}.$$

To estimate the first integral on the right-hand side of (2.8) we need bounds for $|\operatorname{grad} U^\sigma(\zeta)|$. Hence, let us use in the disk $|z - \zeta| \leq r$ with $r < \operatorname{dist}(\zeta, I)$ Schwarz's formula

$$h(z) = \frac{1}{2\pi i}\int_{|t-\zeta|=r}\frac{t+z-2\zeta}{t-z}U^\sigma(t)\frac{dt}{t-\zeta},$$

where h is analytic in $|z - \zeta| < r$ with $U^\sigma(z)$ as real part. Differentiation yields

$$h'(z) = \frac{1}{\pi i} \int_{|t-\zeta|=r} U^\sigma(t) \frac{dt}{(t-z)^2}.$$

For $z \in B = B(\delta, J)$ with $|\operatorname{Im} z| \geq 2\delta^3$ we may choose $r = d/2$ with $d = \operatorname{dist}(z, \partial B)$ and obtain

$$|\operatorname{grad} U^\sigma(z)| \leq \frac{4\varepsilon}{d}. \tag{2.9}$$

Since $\operatorname{dist}(l, \partial B) > \delta/2$, we obtain

$$\left| \int_{l_1 \cup l_4} g(\zeta) \frac{\partial}{\partial \mathbf{n}_\zeta} U^\sigma(\zeta) \, |d\zeta| \right| \leq \int_{l_1 \cup l_4} |\operatorname{grad} U^\sigma(\zeta)| \, |d\zeta| \leq 8\varepsilon \tag{2.10}$$

and

$$\left| \int_{l_2 \cup l_3} g(\zeta) \frac{\partial}{\partial \mathbf{n}_\zeta} U^\sigma(\zeta) \, |d\zeta| \right| = \left| \int_{l_2 \cup l_3} \frac{\partial}{\partial \mathbf{n}_\zeta} U^\sigma(\zeta) \, |d\zeta| \right|$$
$$\leq \int_{l_2 \cup l_3} |\operatorname{grad} U^\sigma(\zeta)| \, |d\zeta| \leq 8\sqrt{2}\,\varepsilon \log\left(1 + \frac{|\Gamma|}{\delta}\right). \tag{2.11}$$

Since $\sigma(I) = 0$, for $z \in \mathbb{C} \setminus I$ we have the obvious inequality

$$|U^\sigma(z)| = \left| \int \log \frac{|\zeta - z|}{d(z, I)} \, d\sigma(\zeta) \right| \leq \log\left(1 + \frac{2}{d(z, I)}\right). \tag{2.12}$$

Hence,

$$\left| \int_{V_1} \Delta g(\zeta) U^\sigma(\zeta) \, dx \, dy \right| \leq \frac{16}{\delta^2} \delta \int_0^{\delta^3} \log\left(1 + \frac{2}{y}\right) dy \leq 64\,\delta^2 \log\frac{1}{\delta}, \tag{2.13}$$

where we note for further use that

$$\int_0^{\delta^3} \log\left(1 + \frac{2}{y}\right) dy \leq 4\,\delta^3 \log\frac{1}{\delta}, \qquad 0 < \delta \leq \frac{1}{6}. \tag{2.14}$$

Next,

$$\left| \int_{V_2} \Delta g(\zeta) U^\sigma(\zeta) \, dx \, dy \right| \leq 16\,\varepsilon. \tag{2.15}$$

Taking into account (2.8), (2.10)–(2.13) and (2.15) we obtain

$$\left| \int g \, d\sigma \right| \leq 4\varepsilon \left(2 + \log \left(1 + \frac{|\Gamma|}{\delta} \right) \right) + 21 \, \delta^2 \log \frac{1}{\delta},$$

which together with the relation

$$\int g \, d\sigma \leq \sigma(J) + \sigma^+ \left([a - \frac{\delta}{2}, a] \right) + \sigma^+ \left([b, b + \frac{\delta}{2}] \right) \leq \sigma(J) + 2 \, M \, \delta^\gamma$$

implies (2.7).

In order to prove that

$$\sigma(J) \leq 4\varepsilon \left(2 + \log \left(1 + \frac{|\Gamma|}{\delta} \right) \right) + 21 \, \delta^2 \log \frac{1}{\delta} + 2 \, M \, \delta^\gamma \qquad (2.16)$$

we may assume that $|\Gamma| \geq 2\,\delta$. Then we repeat the same reasoning with the function

$$g(x + iy) := f \left(\frac{2}{\delta} (x - b + \delta) \right) f \left(\frac{2}{\delta} (a + \delta - x) \right). \qquad (2.17)$$

Combining (2.7) with (2.16) yields (2.6).

To prove (2.4) we note first that for $\frac{1}{2} \geq \delta \geq \frac{1}{6}$ the right-hand side is greater than 1. Hence, we may assume that $\delta < \frac{1}{6}$. We define for $\beta := |\Gamma|/4$ the functions

$$g_1(x) = f \left(\frac{1}{\beta} (x - b + \beta) \right) f \left(\frac{1}{\beta} (a + \beta - x) \right), \qquad (2.18)$$

$$g_2(y) = f \left(\frac{1}{\beta} y \right) f \left(-\frac{1}{\beta} y \right), \qquad (2.19)$$

and

$$g(x + iy) = g_1(x) \, g_2(y).$$

Then g is twice continuously differentiable in $\mathbb{R}^2$, and all partial derivatives up to second order vanish in $\mathbb{C} \setminus S$, where S is the rectangle

$$S = \{ x + iy : x \in \Gamma, \, |y| \leq \beta \}.$$

Again we obtain by Green's formula

$$\int g \, d\sigma = -\frac{1}{2\pi} \int_S \Delta g(\zeta) \, U^\sigma(\zeta) \, dx \, dy.$$

We subdivide S into $S = S_1 \cup S_2$ with

$$S_1 := \{ x + iy : x \in \Gamma, \, |y| \leq \delta^3 \}$$

and
$$S_2 := \{x + iy : x \in \Gamma,\ \delta^3 \leq |y| \leq \beta\}.$$

Since $|\Delta g(\zeta)| \leq 2/\beta^2$ for all $\zeta \in \mathbb{C}$, we use again (2.14) and obtain, analogously to (2.13),

$$\left|\int_{S_1} \Delta g(\zeta)\, U^\sigma(\zeta)\, dx\, dy\right| \leq \frac{4}{\beta^2}|\Gamma|\int_0^{\delta^3} \log\left(1+\frac{2}{y}\right) dy \leq \frac{16}{\beta^2}|\Gamma|\, \delta^3 \log\frac{1}{\delta}$$
$$\leq 11\, \delta \log\frac{1}{\delta}.$$

Moreover,
$$\left|\int_{S_2} \Delta g(\zeta)\, U^\sigma(\zeta)\, dx\, dy\right| \leq 16\,\varepsilon.$$

Summarizing,
$$\sigma^+(\Gamma_{1/2}) - \sigma^-(\Gamma) \leq \int g\, d\sigma < 3\left(\varepsilon + \delta \log\frac{1}{\delta}\right).$$

$\square$

Concerning the sharpness of the theorem we refer to the investigation of Pollaczek polynomials in Chapter 7, especially to Theorem 7.7.5 and Theorem 7.7.8.

3.3 Local Estimates for Dini–Smooth Arcs and Curves

Let us turn to the case of a general bounded Jordan arc L with subarc Γ. Denote by z_1 and z_2 the endpoints of L. Since the function Φ can be extended continuously to these points, we set $\Psi := \Phi^{-1}$ and for $j = 1, 2$ and $\delta > 0$,

$$e^{i\theta_j} = t_j := \Phi(z_j)$$
$$\Delta_1 := \{t = re^{i\theta} : r > 1,\ \theta_1 < \theta < \theta_2\},$$
$$\Delta_2 := \Delta \setminus \overline{\Delta}_1,\quad \Omega_j := \Psi(\Delta_j),$$
$$L_\delta := \{\zeta \in \Omega : |\Phi(\zeta)| = 1+\delta\}.$$

The restrictions Φ_j of the function Φ to the regions Ω_j can be extended continuously to $\partial\Omega_j$, and we keep the previous notation for these extensions. Denote by $E_\delta := \text{int } L_\delta$ the bounded component of $\mathbb{C}\setminus L_\delta$.

Further, define according to (2.2.30) the δ-neighborhood for F with respect to the arc L by
$$A_\delta(F) := \{\zeta \in E_\delta : \zeta_L \in F\}, \qquad F \subset L,$$
where we use the notation $\Psi_j := \Phi_j^{-1}$,
$$\zeta_L := \Psi_j \left(\frac{\Phi_j(\zeta)}{|\Phi_j(\zeta)|}\right), \qquad \zeta \in \overline{\Omega}_j.$$
Let $\varepsilon > 0$ be sufficiently small, ξ_1 and ξ_2 the endpoints of Γ, and let
$$\Phi_j(\xi_k) = e^{i\theta_j^k}, \qquad j, k = 1, 2.$$
Without loss of generality we assume that
$$\theta_1^1 < \theta_1^2 \le \theta_2^2 < \theta_2^1 < \theta_1^1 + 2\pi.$$
Set
$$B_\varepsilon^*(\Gamma) := \{t = re^{i\theta} : r > 1, \theta_1^1 - \varepsilon < \theta < \theta_1^2 + \varepsilon \text{ or } \theta_2^2 - \varepsilon < \theta < \theta_2^1 + \varepsilon\},$$
and
$$B_\varepsilon = B_\varepsilon(\Gamma) := \Psi(B_\varepsilon^*(\Gamma))$$
(see Figure 3.5). The quantity
$$a(\delta) = a(\delta, \sigma, \varepsilon, \Gamma) := \sup_{z \in B_\varepsilon \cap L_\delta} |U^\sigma(z)| \qquad (3.1)$$
plays a key role in our main results.

The next essential point is the geometrical structure of L in a neighborhood of Γ. In what follows, we use the following notation:
$$J_{\delta,j} := \{\zeta \in L_\delta \cap \overline{\Omega}_j : \zeta_L \in J\} \quad \text{for } J \subset L, \delta > 0 \text{ and } j = 1, 2,$$
$$J_\delta := J_{\delta,1} \cup J_{\delta,2}.$$

Theorem 3.1. *Let L be a Jordan arc, and let $\Gamma \subset L$ be a subarc such that $\Gamma \cap \overline{B}_\varepsilon$ is Dini-smooth for some $\varepsilon > 0$. Suppose that $\sigma = \sigma^+ - \sigma^-$ is a signed measure on L with positive part σ^+, negative part σ^-, and $\sigma^+(L) = \sigma^-(L) = 1$. Moreover, for all subarcs $J \subset L \cap \overline{B}_\varepsilon$ let*
$$\sigma^+(J) \le c\, \mu_L(J)^\beta \qquad (3.2)$$
with some constants $c > 0$ and $0 < \beta \le 1$.

Then there exist constants $c_1, c_2 > 0$ depending only on $L, \Gamma, \varepsilon, c$, and β such that
$$D_\Gamma[\sigma] \le c_1 \inf_{0 < \delta < \varepsilon} \left(\frac{1}{\delta} \sup_{\delta \le t < 3\delta} a(t) + \left(\delta \log \frac{1}{\delta}\right)^\beta\right), \qquad (3.3)$$
$$D_\Gamma[\sigma] \le c_2 \inf_{0 < \delta < \varepsilon} \left(\sup_{\delta \le t \le \varepsilon} a(t) \log \frac{1}{\delta} + \left(\delta \log \frac{1}{\delta}\right)^\beta\right), \qquad (3.4)$$
where $a(t)$ is defined by (3.1).

3. Discrepancy Theorems via Two–Sided Bounds for Potentials

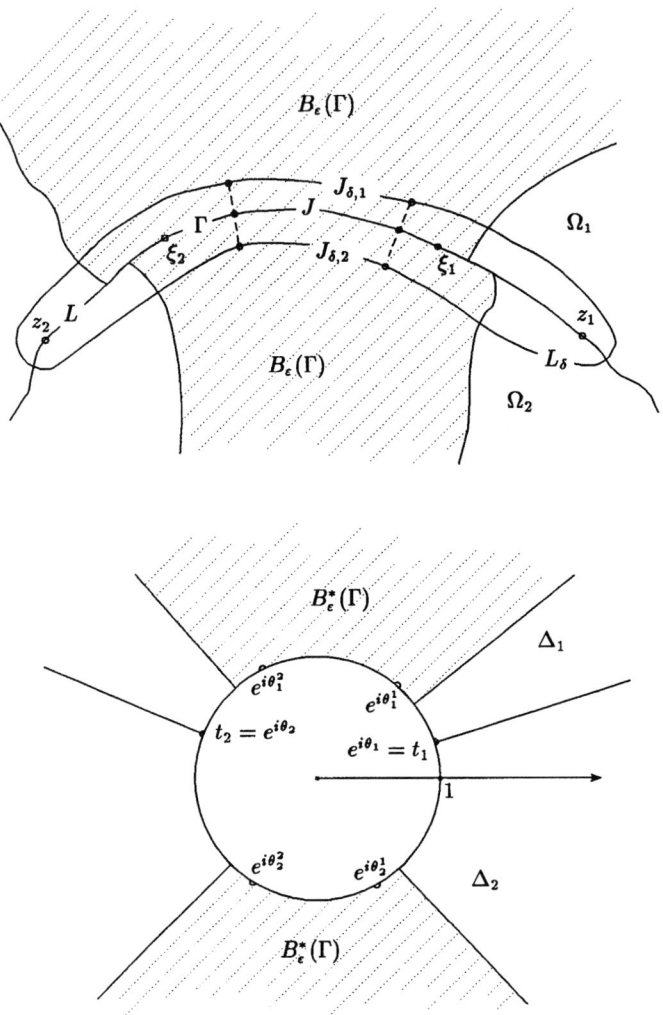

Fig. 3.5

Before giving the proof of Theorem 3.1 we make some remarks. One of the immediate consequences of (1.2.7) is the following estimate for $0 < \delta \leq 1$:

$$|L_\delta| = \int_{|w|=1+\delta} |\Psi'(w)|\,|dw| \leq \frac{16\pi \operatorname{diam} L_1}{\delta} = \frac{c}{\delta}. \tag{3.5}$$

The next fact is also a consequence of (1.2.7), however, not so obvious. Let G be a Jordan domain and suppose $S \subset \partial G$ consists of at most two subarcs of ∂G. Denote by $\omega(z) := \omega(z, G, S)$ the harmonic measure of S at the point $z \in G$ with respect to G.

Lemma 3.2. *For any $z \in G$,*

$$|\operatorname{grad} \omega(z)| \leq 8 \, \frac{\min\{\omega(z), 1 - \omega(z)\}}{d(z, \partial G)}. \tag{3.6}$$

Proof. Denote by $\varphi = \varphi_z$ the conformal mapping of G onto the unit disk $\mathbb{D}$ with the normalization $\varphi(z) = 0$, $\varphi'(z) > 0$. Note that by (1.2.7),

$$|\operatorname{grad} \omega(z)| = |\operatorname{grad} \omega^*(0)| \, |\varphi'(0)| \leq \frac{4}{d} \, |\operatorname{grad} \omega^*(0)|, \tag{3.7}$$

where $d := d(z, \partial G)$ and

$$\omega^*(w) := \omega(\varphi^{-1}(w)), \quad w \in \mathbb{D}.$$

Further, by the Schwarz formula (cf. (1.1.8)),

$$|\operatorname{grad} \omega^*(0)| = \frac{1}{\pi} \left| \int_0^{2\pi} \omega^*(e^{i\theta}) \, \frac{d\theta}{e^{i\theta}} \right| \leq 2 \min\{\omega^*(0), 1 - \omega^*(0)\}. \tag{3.8}$$

Comparing (3.7) and (3.8) we get (3.6). □

Let L and $\Gamma \subset L$ be as in Theorem 3.1. It is obvious that locally in a neighborhood of the smooth part of L the distortion properties of the mappings Φ and Ψ are the same as for the case of a Dini-smooth arc. We restrict ourselves to the formulation of only some of the consequences (for details, see Lemmas 1.2.10–1.2.12). More precisely, let $z \in \Omega_j \cap B_{\varepsilon/2}(\Gamma)$. Denote by z_1 the endpoint of L, that is nearest to z_L. Then

$$\left| \frac{z_L - z}{z_1 - z} \right| \asymp \left| \frac{\Phi_j(z_L) - \Phi_j(z)}{\Phi_j(z_1) - \Phi_j(z)} \right| = \frac{|\Phi(z)| - 1}{|\Phi(z) - \Phi(z_1)|}, \tag{3.9}$$

$$|z_L - z| \asymp d(z, L), \tag{3.10}$$

$$d(z, L_{3\delta}) \asymp d(z, L) \asymp \delta \, \frac{|z - z_1|}{|\Phi(z) - \Phi(z_1)|}, \quad z \in B_{\varepsilon/2}(\Gamma) \cap L_{2\delta}. \tag{3.11}$$

We complete this remark with two direct consequences of the definition of a potential of a measure σ on L with $\sigma(L) = 0$: for $z \in \Omega \setminus \{\infty\}$,

$$|U^\sigma(z)| = \left| \int \log \frac{|\zeta - z|}{d(z, L)} \, d\sigma \right| \leq \log \left(1 + \frac{\operatorname{diam} L}{d(z, L)}\right), \tag{3.12}$$

$$|\operatorname{grad} U^\sigma(z)| \leq \frac{2}{d(z, L)}. \tag{3.13}$$

The most complicated situation for our studies is the case $\overline{B}_\varepsilon \not\supset L$ (otherwise, we have global estimates for the appropriate potentials and are in a position to use known results and constructions from Section 3.1). Hence, we will assume that

$$\theta_2^2 - \theta_2 > \varepsilon, \quad \theta_2 - \theta_1^2 > \varepsilon.$$

Therefore, the endpoint z_2 of L is always outside of $\overline{B}_\varepsilon$, and for z_1 there are two possibilities: $z_1 \in \overline{B}_\varepsilon$ (more complicated case) and $z_1 \notin \overline{B}_\varepsilon$ (simpler case). Below, our constructions are mainly directed to the first of these cases. However, the results are true for both.

We assume that $\delta > 0$ is sufficiently small and consider an arbitrary subarc $J \subset \Gamma$. If $\overline{B}_\varepsilon$ includes one of the endpoints of L (we already denoted it by z_1), we can extend Γ to the arc with the same properties but with one of the endpoints moving to z_1. In this case we assume, in addition, that one of the endpoints of J also coincides with z_1.

Consider the arc

$$\gamma := \Phi_1(J) = \{e^{i\theta} : \eta_1 \leq \theta \leq \eta_2\}$$

and set for $k = 1, 2$,

$$\gamma^k := \left\{ e^{i\theta} : \eta_1 - kM\delta \log \frac{1}{\delta} \leq \theta \leq \eta_2 + kM\delta \log \frac{1}{\delta} \right\} \cap \partial \Delta_1,$$

where M is a constant whose choice will be made below, and define

$$J^k := \Psi_1(\gamma^k).$$

The function

$$g(z) := \omega(z, E_{3\delta}, J_{3\delta}^1)$$

will play a key role in our reasoning.

Lemma 3.3. *There exist sufficiently large constants M, c_j, $j = 1, \ldots, 4$, independent of J and δ such that the function g possesses the following properties:*

$$1 - g(z) \leq c_1 \delta^4, \quad z \in A_{3\delta}(J), \tag{3.14}$$

$$g(z) \leq c_2 \delta^4, \quad z \in A_{3\delta}(L\backslash J^2), \tag{3.15}$$

$$|\operatorname{grad} g(z)| \leq c_3 \delta^2, \quad z \in L_{2\delta}, \ z_L \notin J^2\backslash J, \tag{3.16}$$

$$|\operatorname{grad} g(z)| \leq \frac{c_4}{\delta} \frac{|\Phi(z) - \Phi(z_1)|}{|z - z_1|}, \quad z \in L_{2\delta}, \ z_L \in J^2\backslash J, \tag{3.17}$$

where z_1 is the appropriate endpoint of L.

Proof. We begin with a property of the harmonic measure with respect to the region $E_{3\delta}$. We assert that the inequality

$$g(z) \preceq \exp\left(-c_5 \frac{d(\Phi_1(z_L), \gamma^1)}{\delta}\right), \quad z \in A_{3\delta}(L \setminus J^1), \tag{3.18}$$

3.3 Local Estimates for Dini–Smooth Arcs and Curves

holds.

Indeed, without loss of generality we assume that $d := d(\Phi_1(z_L), \gamma^1) \geq 2\delta$ and $z \in (B_{\varepsilon/2} \cap E_{3\delta}) \setminus A_{3\delta}(J^1)$. Let ξ_3 be the endpoint of J^1 nearest to z_L along L. Denote by $Q \subset E_{3\delta}$ the quadrilateral whose boundary consists of

$$\gamma_1 := \{\zeta \in E_{3\delta} : \zeta_L = \xi_3\},$$
$$\gamma_2 := \{\zeta \in E_{3\delta} : \zeta_L = z_L\},$$

and of the corresponding two subarcs of $L_{3\delta}$ (see Figure 3.6).

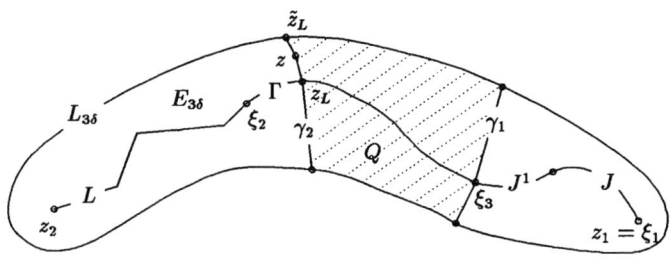

Fig. 3.6

Let $m(Q)$ be the module of Q, i.e., the module of the family of all crosscuts of Q separating the sides γ_1 and γ_2 in Q. According to (1.2.17) and the comparison principle (1.2.11),

$$g(z) \preceq \exp\{-\pi m(Q)\}. \tag{3.19}$$

Further, we apply Rengel's inequality (B.1.5) to find that

$$m(Q) \geq \frac{s(Q)^2}{A(Q)}, \tag{3.20}$$

where $A(Q)$ is the Euclidean area of Q and $s(Q)$ is the distance between γ_1 and γ_2 in Q, i.e.,

$$s(Q) := \inf_{\gamma \in \Gamma_Q(\gamma_1, \gamma_2)} |\gamma|,$$

where $\Gamma_Q(\gamma_1, \gamma_2)$ denotes the set of all crosscuts of Q that join γ_1 and γ_2. Since by Theorem 1.2.7,

$$s(Q) \succeq |z_L - \xi_3|,$$
$$A(Q) \preceq |z_L - \xi_3| |z_L - \tilde{z}_L|,$$

where
$$\tilde{z}_L := \Psi\left((1+3\delta)\frac{\Phi(z)}{|\Phi(z)|}\right),$$
we see that by virtue of analogues of Lemmas 1.2.11 and 1.2.12 as well as (3.20),
$$m(Q) \succeq \left|\frac{z_L - \xi_3}{z_L - \tilde{z}_L}\right| = \left|\frac{z_L - \xi_3}{z_L - z_1}\right|\left|\frac{z_L - z_1}{z_L - \tilde{z}_L}\right|$$
$$\asymp \left|\frac{\Phi_1(z_L) - \Phi_1(\xi_3)}{\Phi_1(z_L) - \Phi_1(z_1)}\right|\left|\frac{\Phi_1(z_L) - \Phi_1(z_1)}{\Phi_1(z_L) - \Phi_1(\tilde{z}_L)}\right| \asymp \frac{d}{\delta}. \qquad (3.21)$$

Combining (3.19) and (3.21) we get (3.18).

Repeating step by step the reasoning above we get for $z \in A_{3\delta}(J^1)$ the inequality
$$1 - g(z) \preceq \exp\left(-c_6 \frac{d(\Phi_1(z_L), \mathbb{T}\setminus\gamma^1)}{\delta}\right). \qquad (3.22)$$

The inequalities (3.14) and (3.15) (for sufficiently large M) are immediate consequences of (3.22) and (3.18), respectively.

Estimate (3.16) follows simply from (3.14) and (3.15) by using Lemma 3.2 and the relation
$$d(L_{2\delta}, L_{3\delta}) \succeq \delta^2$$
(see Lemma 1.2.5).

Finally, inequality (3.17) can also be directly derived from Lemma 3.2 if we take (3.11) into account. $\square$

Proof of Theorem 3.1. Our first objective is to show (using the function g investigated above) that
$$\sigma(J) \geq -c\left[(a(2\delta) + c(2\delta)\,\delta)\log\frac{1}{\delta} + \left(\delta\log\frac{1}{\delta}\right)^\beta + b(2\delta)\right], \qquad (3.23)$$
where for $t > 0$ we have used the following notation:
$$b_j(t) := \sup_{\zeta_1,\zeta_2 \in L_t \cap \Omega_j \cap B_{\varepsilon/2}} |\tilde{U}^\sigma(\zeta_1) - \tilde{U}^\sigma(\zeta_2)|, \quad j=1,2,$$
$$b(t) := \max_{j=1,2} b_j(t),$$
$$c(t) := \sup_{\zeta \in L_t \cap B_{\varepsilon/2}} \frac{|\text{grad } U^\sigma(\zeta)|}{|\Phi'(\zeta)|},$$
and $\tilde{U}^\sigma$ is a harmonic conjugate of U^σ in Ω.

By Green's formula (1.1.10) for $z \in L$,
$$g(z) = \frac{1}{2\pi}\int_{L_{2\delta}} \left(\log\frac{1}{|\zeta - z|}\frac{\partial g(\zeta)}{\partial \mathbf{n}} - g(\zeta)\frac{\partial}{\partial \mathbf{n}_\zeta}\log\frac{1}{|\zeta - z|}\right)|d\zeta|,$$

where $\partial/\partial \mathbf{n} = \partial/\partial \mathbf{n}_\zeta$ denotes the operator of differentiation with respect to the outward normal to the curve $L_{2\delta}$ at the point ζ.
Therefore,

$$\int g\, d\sigma = \frac{1}{2\pi} \int\limits_{L_{2\delta}} \left(\frac{\partial g(\zeta)}{\partial \mathbf{n}} U^\sigma(\zeta) - g(\zeta) \frac{\partial}{\partial \mathbf{n}} U^\sigma(\zeta) \right) |d\zeta|. \qquad (3.24)$$

Our next purpose is to estimate the integrals on the right-hand side of (3.24). Let

$$S := J_{2\delta}^2 \setminus J_{2\delta},$$

and let τ_j, $j = 1,2$, be the endpoints of J. The set S consists of 2 or 4 disjoint subarcs of $L_{2\delta}$. To be specific, let us consider only the more complicated case of 4 subarcs. Denote by S_k^j the one that has $\Psi_k((1 + 2\delta)\Phi_k(\tau_j)) =: \tau_k^j$ as endpoint. For $z \in S_k^j$ set $w := \Phi(z)$, $\tilde{g}(w) := g(\Psi(w))$, $w_1 := \Phi_1(z_1) = \Phi_2(z_1)$. For such z, by (3.17), (1.2.7), (3.9), and (3.10)

$$\frac{d(z, L)}{|z - z_1|} \asymp \frac{\delta}{|w - w_1|},$$

$$|\operatorname{grad} \tilde{g}(w)| = |\operatorname{grad} g(z)||\Psi'(w)| \preceq \frac{1}{\delta} \frac{|w - w_1|}{|z - z_1|} \frac{d(z, L)}{\delta} \asymp \frac{1}{\delta}.$$

As a result we obtain according to (3.5), (3.12), (3.13), and (3.16) the following estimate:

$$\left| \int\limits_{L_{2\delta}} \frac{\partial g(\zeta)}{\partial \mathbf{n}} U^\sigma(\zeta) |d\zeta| \right| \preceq a(2\delta) \int\limits_S |\operatorname{grad} g(\zeta)| |d\zeta|$$

$$+ \log \frac{1}{d(L_{2\delta}, L)} \int\limits_{L_{2\delta} \setminus S} |\operatorname{grad} g(\zeta)| |d\zeta|$$

$$\preceq a(2\delta) \int\limits_{\Phi(S)} |\operatorname{grad} \tilde{g}(w)| |dw| + \left(\log \frac{1}{\delta} \right) \delta^2 |L_{2\delta}|$$

$$\preceq (a(2\delta) + \delta) \log \frac{1}{\delta}. \qquad (3.25)$$

Further,

$$\left| \int_{L_{2\delta}} g(\zeta) \frac{\partial}{\partial \mathbf{n}} U^\sigma(\zeta) |d\zeta| \right| \leq \sum_{j=1}^{2} \left| \int_{J_{2\delta,j}} \frac{\partial}{\partial \mathbf{n}} U^\sigma(\zeta) |d\zeta| \right| + c(2\delta) |\Phi(S)|$$

$$+ \sum_{j=1}^{2} \int_{J_{2\delta,j}} (1 - g(\zeta)) \left| \frac{\partial}{\partial \mathbf{n}} U^\sigma(\zeta) \right| |d\zeta|$$

$$+ \int_{L_{2\delta} \setminus J_{2\delta}^2} g(\zeta) \left| \frac{\partial}{\partial \mathbf{n}} U^\sigma(\zeta) \right| |d\zeta|. \quad (3.26)$$

To estimate the integrals on the right-hand side of (3.26), we write

$$\left| \int_{J_{2\delta,j}} \frac{\partial}{\partial \mathbf{n}} U^\sigma(\zeta) |d\zeta| \right| = \left| \int_{J_{2\delta,j}} \frac{\partial}{\partial \mathbf{t}_\zeta} \widetilde{U}^\sigma(\zeta) |d\zeta| \right| \leq b_j(2\delta),$$

where $\mathbf{t}_\zeta$ is the tangent vector obtained by a rotation of $\mathbf{n} = \mathbf{n}_\zeta$ counterclockwise by an angle of $\pi/2$. Thus, continuing (3.26) and using Lemma 3.3, and (3.5) as well as (3.13) we obtain

$$\left| \int_{L_{2\delta}} g(\zeta) \frac{\partial}{\partial \mathbf{n}} U^\sigma(\zeta) |d\zeta| \right| \preceq b(2\delta) + c(2\delta) \delta \log \frac{1}{\delta} + \delta. \quad (3.27)$$

Comparing (3.24), (3.25), and (3.27) we get

$$\left| \int g \, d\sigma \right| \preceq (a(2\delta) + c(2\delta)\delta + \delta) \log \frac{1}{\delta} + b(2\delta). \quad (3.28)$$

Further, taking into account (3.2), (3.14), (3.15), (3.28), and the relations

$$\int g \, d\sigma = \int_J d\sigma - \int_J (1-g) \, d\sigma + \int_{J^2 \setminus J} g \, d\sigma + \int_{L \setminus J^2} g \, d\sigma$$

$$\leq \sigma(J) + \int_J (1-g) \, d\sigma^- + \sigma^+(J^2 \setminus J) + \int_{L \setminus J^2} g \, d\sigma^+,$$

we obtain (3.23).

Moreover, changing our reasoning in an obvious way, we have quite the same estimate for the quantity $\sigma(L \setminus J)$. Taking into account the relation $\sigma(J) = -\sigma(L \setminus J)$ we get the estimate

$$|\sigma(J)| \preceq (a(2\delta) + c(2\delta)\delta) \log \frac{1}{\delta} + \left(\delta \log \frac{1}{\delta} \right)^\beta + b(2\delta). \quad (3.29)$$

3.3 Local Estimates for Dini-Smooth Arcs and Curves

To obtain the required inequalities (3.3) and (3.4) we have to study the relations of the quantities $a(2\delta)$, $b(2\delta)$, and $c(2\delta)$. For $w \in \Delta$, $z \in \Omega$ let

$$u^\sigma(w) := U^\sigma(\Psi(w)),$$
$$\widetilde{u}^\sigma(w) := \widetilde{U}^\sigma(\Psi(w)),$$
$$v^\sigma(w) := u^\sigma(w) + i\widetilde{u}^\sigma(w), \quad V^\sigma(z) := v^\sigma(\Phi(z)).$$

For w with $|w| \geq 1 + 2\delta$, $z := \Psi(w)$, and $r := |w| - 1 - \delta$ we have, according to the Schwarz formula (cf. (1.1.8)),

$$|\text{grad } u^\sigma(w)| = |(v^\sigma)'(w)| \leq \frac{1}{\pi r} \int_0^{2\pi} |u^\sigma(w + re^{i\theta})| \, d\theta$$

$$\leq \frac{2}{r} \sup_{0 \leq \theta < 2\pi} |u^\sigma(w + re^{i\theta})|. \tag{3.30}$$

Since

$$|\text{grad } U^\sigma(z)| = |\text{grad } u^\sigma(\Phi(z))| \, |\Phi'(z)|,$$

we have from (3.30),

$$c(2\delta) \preceq \frac{1}{\delta} \sup_{\delta \leq t \leq 3\delta} a(t). \tag{3.31}$$

Now let $\zeta_1, \zeta_2 \in L_{2\delta} \cap B_{\varepsilon/2} \cap \Omega_j$, $w_k := \Phi(\zeta_k) = (1 + 2\delta) e^{i\alpha_k}$, and suppose that $\alpha_1 < \alpha_2$. To prove Theorem 3.1 we can assume that δ is so small that

$$2M\delta \log \frac{1}{\delta} < \frac{\varepsilon}{4},$$

where M is the constant appearing in the construction of the arcs J^1 and J^2. Consider the arc

$$l := l_1 \cup l_2 \cup l_3, \tag{3.32}$$

where

$$l_1 := \left\{ w = \left(1 + \frac{\varepsilon}{4}\right) e^{i\theta} : \alpha_1 \leq \theta \leq \alpha_2 \right\}$$

and

$$l_k := \left\{ w = re^{i\alpha_k}, 2\delta \leq r - 1 \leq \frac{\varepsilon}{4} \right\}, \quad k = 1, 2.$$

This arc joins the points w_1 and w_2. For $w \in l$ we have, by virtue of (3.30),

$$|\text{grad } \widetilde{u}^\sigma(w)| = |\text{grad } u^\sigma(w)| \leq \frac{2}{|w| - 1 - \delta} \sup_{\delta \leq t \leq \varepsilon} a(t).$$

Therefore, a simple computation yields

$$|\widetilde{U}^\sigma(\zeta_2) - \widetilde{U}^\sigma(\zeta_1)| \leq \int_l |\text{grad } \widetilde{u}^\sigma(w)| \, |dw| \preceq \left(\log \frac{1}{\delta}\right) \sup_{\delta \leq t \leq \varepsilon} a(t).$$

Hence,
$$b(2\delta) \preceq \left(\log \frac{1}{\delta}\right) \sup_{\delta \leq t \leq \varepsilon} a(t). \tag{3.33}$$

If we take in our construction instead of the arc l given by (3.32) the simpler arc
$$\{w = (1 + 2\delta) e^{i\theta} : \alpha_1 \leq \theta \leq \alpha_2\},$$
then by repeating step by step the reasoning above we get the inequality
$$b(2\delta) \preceq \frac{1}{\delta} \sup_{\delta \leq t \leq 3\delta} a(t). \tag{3.34}$$

Comparing (3.29), (3.31), (3.33), and (3.34) we obtain (3.3) and (3.4). □

The sharpness of the estimates (3.3) and (3.4) (up to the logarithmic term) and a principal difference between local and global discrepancies can be seen from the next result.

Theorem 3.4. Let $L := [-1, 1]$, $\Gamma := [-\frac{1}{4}, \frac{1}{4}]$. For each $0 < \alpha < 0.01$ there exists a signed measure $\sigma = \sigma_\alpha$, satisfying inequality (3.2) with a constant c independent of α and with $\beta = 1$, such that
$$D[\sigma] = D_L[\sigma] \geq c_1 \sqrt{\alpha}, \tag{3.35}$$
$$D_\Gamma[\sigma] \geq c_2 \alpha, \tag{3.36}$$

and moreover,
$$\inf_{0 < \delta < \varepsilon} \left(\frac{1}{\delta} \sup_{\delta \leq t \leq 3\delta} a(t) + \delta \log \frac{1}{\delta}\right) \leq c_3 \alpha \sqrt{\log \frac{1}{\alpha}}, \tag{3.37}$$
$$\inf_{0 < \delta < \varepsilon} \left(\delta + \sup_{\delta \leq t \leq \varepsilon} a(t)\right) \log \frac{1}{\delta} \leq c_4 \alpha \log \frac{1}{\alpha}, \tag{3.38}$$

where $\varepsilon > 0, c_j > 0$, $j = 1, \ldots, 4$, are independent of α.

Proof. Denote by σ^+ the equilibrium measure for $[-1 + 2\alpha, 1] =: S_+$ and by σ^- the corresponding one for the interval $[-1, 1 - 2\alpha] =: S_-$.

We are going to prove that the signed measure $\sigma := \sigma^+ - \sigma^-$ satisfies (3.35)–(3.38).

At first note that for any subarc $J \subset [-\frac{1}{2}, \frac{1}{2}]$, each of $\sigma^\pm$ satisfies the inequality
$$\sigma^\pm(J) \preceq |J| \preceq \mu_L(J).$$
Hence, (3.2) is fulfilled (with $\beta = 1$).

Inequality (3.35) is also trivial:
$$D_L[\sigma] \geq \sigma^+([1 - 2\alpha, 1]) \succeq \sqrt{\alpha}.$$

3.3 Local Estimates for Dini–Smooth Arcs and Curves

Further, denote by $\Psi_\pm$ the conformal mapping of Δ onto $\overline{\mathbb{C}}\setminus S_\pm$ given by formula

$$\Psi_\pm(w) := \frac{1-\alpha}{2}\left(w + \frac{1}{w}\right) \pm \alpha, \qquad w \in \Delta. \qquad (3.39)$$

The inverse mapping $\Phi_\pm := \Psi_\pm^{-1}$ is defined as follows:

$$\Phi_\pm(z) := \frac{1}{1-\alpha}\left(z \mp \alpha + \sqrt{(z \mp \alpha)^2 - (1-\alpha)^2}\right), \qquad z \in \overline{\mathbb{C}}\setminus S_\pm,$$

where $\sqrt{(z \mp \alpha)^2 - (1-\alpha)^2}$ will denote the branch of the square root that is asymptotically equal to $(z \mp \alpha)$ near infinity.

For the respective potentials, we have

$$U^{\sigma^\pm}(z) = -\log|\Phi_\pm(z)| - \log\frac{1-\alpha}{2}, \qquad z \in \Omega := \mathbb{C}\setminus[-1,1].$$

Our next purpose is to estimate (from below) the discrepancy $D_\Gamma[\sigma]$. Let $J := [0, \frac{1}{4}]$ and let $0 < \eta_- < \eta_+ < \theta_- < \theta_+ < \pi$ be such that

$$\Psi_\pm(e^{i\theta_\pm}) = 0, \qquad \Psi_\pm(e^{i\eta_\pm}) = \frac{1}{4}.$$

Then

$$|\sigma(J)| = \frac{1}{\pi}\left|(\eta_+ - \eta_-) - (\theta_+ - \theta_-)\right|. \qquad (3.40)$$

Since

$$(e^{i\theta_+} - e^{i\theta_-})(1-\alpha) = -2\alpha,$$

$$(e^{i\eta_+} - e^{i\eta_-})(1-\alpha) = -2\alpha + i\left(\sqrt{(1-\alpha)^2 - \left(\frac{1}{4}-\alpha\right)^2}\right.$$

$$\left. - \sqrt{(1-\alpha)^2 - \left(\frac{1}{4}+\alpha\right)^2}\right)$$

$$= -2\alpha + iB(\alpha),$$

where

$$B(\alpha) := \frac{\alpha}{\sqrt{(1-\alpha)^2 - (\frac{1}{4}-\alpha)^2} + \sqrt{(1-\alpha)^2 - (\frac{1}{4}+\alpha)^2}} \geq \frac{\alpha}{2(1-\alpha)}$$

and

$$\theta_+ - \theta_- = 2\arcsin\alpha,$$

$$\eta_+ - \eta_- = 2\arcsin\left(\alpha\sqrt{1 + \left(\frac{B(\alpha)}{2\alpha}\right)^2}\right),$$

126   3. Discrepancy Theorems via Two–Sided Bounds for Potentials

we obtain according to (3.40) after an elementary computation that
$$|\sigma([0,1/4])| \succeq \alpha.$$
Thus, (3.36) is established.

The next step is connected with an estimation of the quantity
$$U^\sigma(z) = \log\left|\frac{\Phi_-(z)}{\Phi_+(z)}\right| = \log\left|\frac{z+\alpha+\sqrt{(z+\alpha)^2-(1-\alpha)^2}}{z-\alpha+\sqrt{(z-\alpha)^2-(1-\alpha)^2}}\right|$$
for $z = x+iy$ with $-\frac{1}{2} \le x \le \frac{1}{2}$, $0 < |y| < 1$. Since for the values z under consideration
$$\sqrt{(z+\alpha)^2-(1-\alpha)^2} - \sqrt{(z-\alpha)^2-(1-\alpha)^2} = O(\alpha),$$
we have
$$|U^\sigma(z)|$$
$$\le \left|\log\left(1 + \frac{2\alpha + \sqrt{(z+\alpha)^2-(1-\alpha)^2} - \sqrt{(z-\alpha)^2-(1-\alpha)^2}}{z-\alpha+\sqrt{(z-\alpha)^2-(1-\alpha)^2}}\right)\right|$$
$$= O(\alpha).$$
Therefore, taking $\delta = \alpha$ we obtain the estimate (3.38).

Next, since
$$U^\sigma(z) = \log\left(1 + \frac{|\Phi_-(z)|^2 - |\Phi_+(z)|^2}{(|\Phi_-(z)|+|\Phi_+(z)|)\,|\Phi_+(z)|}\right),$$
for the proof of (3.37) it is enough to show that for $|y| \asymp \alpha/\sqrt{\log(1/\alpha)}$ and $-\frac{1}{2} \le x \le \frac{1}{2}$,
$$|\Phi_+(z)|^2 - |\Phi_-(z)|^2 = O(\alpha^2), \quad z = x+iy. \tag{3.41}$$
(Then we need only to set $\delta = \alpha/\sqrt{\log(1/\alpha)}$.) To do this, we assume for definiteness that $y > 0$ and write
$$\sqrt{(x+iy\pm\alpha)^2-(1-\alpha)^2} = \sqrt{-(1-x^2)+2(\alpha+ixy\pm x\alpha)+O(\alpha^2)}$$
$$= i\sqrt{1-x^2}\left(1 - \frac{\alpha+ixy\pm x\alpha}{1-x^2}\right) + O(\alpha^2).$$
Therefore,
$$(1-\alpha)^2\,|\Phi_\pm(x+iy)|^2$$
$$= \left(x\mp\alpha+\frac{xy}{\sqrt{1-x^2}}\right)^2 + \left(y+\sqrt{1-x^2}-\frac{\alpha(1\mp x)}{\sqrt{1-x^2}}\right)^2 + O(\alpha^2)$$
$$= x^2 \mp 2x\alpha + 2\frac{x^2 y}{\sqrt{1-x^2}} + 1 - x^2 + 2y\sqrt{1-x^2} - 2\alpha \pm 2\alpha x + O(\alpha^2)$$
$$= 1 + 2y\left(\frac{x^2}{\sqrt{1-x^2}}+\sqrt{1-x^2}\right) - 2\alpha + O(\alpha^2).$$

Estimate (3.41) follows immediately from the last inequality. □

We note that the simple example of the measure $\sigma := \mu_{[-1,1]} - \mu_{[0,1]}$ shows that a local version of Theorem 2.2.4, which concerns global estimates for the discrepancy related to the Erdős–Turán type theorems, is impossible.

Finally, we want to point out how the results for the arc can be transformed to the case of a Jordan curve.

Let Φ denote again the Riemann mapping of the domain $\Omega = \operatorname{ext} L$ onto Δ normalized by $\Phi(\infty) = \infty$, $\Phi'(\infty) \geq 0$, and $\Psi = \Phi^{-1}$. Denote by $w = \varphi(z)$ the conformal mapping of $G = \operatorname{int} L$ onto $\mathbb{D}$ and normalized by $\varphi(z_0) = 0$, $\varphi'(z_0) > 0$, where $z_0 \in G$. Set $\psi = \varphi^{-1}$.

Furthermore, we use some notation that we have introduced already in Section 2.2.1. For $\zeta \in \mathbb{C} \setminus \{z_0\}$ set

$$\zeta_L := \begin{cases} \Psi\left(\dfrac{\Phi(\zeta)}{|\Phi(\zeta)|}\right), & \text{if } \zeta \in \Omega \setminus \{\infty\}, \\ \psi\left(\dfrac{\varphi(\zeta)}{|\varphi(\zeta)|}\right), & \text{if } \zeta \in G \setminus \{z_0\}, \\ \zeta, & \text{if } \zeta \in L, \end{cases}$$

$$L_r^+ := \{\zeta : |\Phi(\zeta)| = 1 + r\}, \qquad r > 0,$$

and

$$L_r^- := \{\zeta : |\varphi(\zeta)| = 1 - r\}, \qquad 0 < r < 1.$$

Let Γ be a fixed subarc of L with endpoints ζ_1 and ζ_2 and let

$$\Phi(\zeta_k) = e^{i\theta_k}, \qquad k = 1, 2.$$

Without loss of generality we may assume that

$$\theta_1 < \theta_2 < \theta_1 + 2\pi.$$

Let $\varepsilon > 0$ and define

$$B_\varepsilon^+ = B_\varepsilon^+(\Gamma) := \{z = \Psi(t) : t = re^{i\varphi} \ (r > 1,\ \theta_1 - \varepsilon < \theta < \theta_2 + \varepsilon)\},$$
$$B_\varepsilon^- = B_\varepsilon^-(\Gamma) := \{z = \psi(t) : t = re^{i\varphi} \ (r < 1,\ \theta_1 - \varepsilon < \theta < \theta_2 + \varepsilon)\},$$

and

$$\Gamma_\varepsilon = L \cap \overline{(B_\varepsilon^+ \cup B_\varepsilon^-)}.$$

Then the quantity

$$b(\delta) = b(\delta, \sigma, \varepsilon, \Gamma) := \max\left(\sup_{z \in B_\varepsilon^+ \cap L_\delta^+} |U^\sigma(z)|,\ \sup_{z \in B_\varepsilon^- \cap L_\delta^-} |U^\sigma(z)|\right) \qquad (3.42)$$

will play the same role as $a(\delta)$ of (3.1) in the case of a Jordan curve.

Theorem 3.5. *Let L be a Jordan curve, and let $\Gamma \subset L$ be a subarc such that Γ_ε is Dini-smooth for some $\varepsilon > 0$. Suppose that $\sigma = \sigma^+ - \sigma^-$ is a signed measure with $\sigma^+, \sigma^- \in \mathcal{M}(L)$ and for all subarcs $J \subset \Gamma_\varepsilon$ the inequality*

$$\sigma^+(J) \leq c\,\mu_L(J)^\beta$$

holds with constants $c > 0$ and $0 < \beta \leq 1$.

Then there exist constants $c_1, c_2 > 0$ depending on L, Γ, ε, c, β, and z_0 such that

$$D_\Gamma[\sigma] \leq c_1 \inf_{0 < \delta < \varepsilon} \left(\frac{1}{\delta} \sup_{\delta \leq t < 3\delta} b(t) + \left(\delta \log \frac{1}{\delta} \right)^\beta \right),$$

$$D_\Gamma[\sigma] \leq c_2 \inf_{0 < \delta < \varepsilon} \left(\sup_{\delta \leq t \leq \varepsilon} b(t) \log \frac{1}{\delta} + \left(\delta \log \frac{1}{\delta} \right)^\beta \right),$$

where $b(t)$ is defined by (3.42).

The proof follows the lines of the proof of Theorem 3.1 and is therefore omitted.

3.4 Historical Comments

First results concerning the distribution of simple zeros of polynomials on intervals or the unit circle (analogous to Theorem 1.8) were obtained by H.–P. Blatt in [28] using two-sided bounds for the difference of logarithmic potentials $U^{\mu_L - \nu_p}$ and function-theoretical arguments. These results were refined and sharpened by V. Totik [177].

Blatt and Mhaskar [34] generalized these discrepancy estimates to signed measures on curves and arcs of Hölder class C^{1+}.

The discrepancy estimates of Section 3.1 for quasiconformal curves and arcs were established in [19]. The main results of Section 3.2 for local estimates in the case of an interval are proved by Andrievskii, Blatt, and Mhaskar in [21].

In the case of the unit circle F. Amoroso and M. Mignotte [7] showed that (1.24) can be made more precise, namely

$$D[\mu_L - \nu_p] \leq 13 \max\left(1, \log \frac{2n}{\log C_n}\right) \frac{\log C_n}{n},$$

by using methods of harmonic analysis.

In $\mathbb{R}^d$, $d \geq 3$, M. Götz [71, 72] investigated the discrepancy of signed measures in terms of two-sided bounds of Newtonian and Green potentials. His results generalize the discrepancy theorems in the plane of this chapter to higher dimensions.

4
Discrepancy Theorems via One-Sided Bounds for Potentials

In Chapter 2 we obtained discrepancy estimates for the zero distribution of a polynomial p in connection with the equilibrium measure μ_L of a Jordan curve or arc L. The basic quantities involved have been the two terms

$$\varepsilon_{p,L} = \max_{z \in \mathbb{C}} U^{\mu_L - \nu_p}(z)$$

and

$$\delta_{p,L} = \varepsilon_{p,L} - U^{\mu_L - \nu_p}(z_0),$$

where $z_0 \in \text{int } L$ is fixed if L is a curve. In Section 2.3 we have outlined that it is possible to restrict the essential quantities to the *outer bounds*

$$\varepsilon_{p,L}(r) = \max_{z \in L_r} U^{\mu_L - \nu_p}(z), \qquad r > 0,$$

in the case of a Jordan arc. If L is a curve, we replace $\delta_{p,L}$ by the smaller *inner bound*

$$\delta_{p,L}(r) = \max_{z \in L_r^-} U^{\mu_L - \nu_p}(z) - U^{\mu_L - \nu_p}(z_0), \qquad 0 < r < 1,$$

where L_r^- is the level line of the conformal mapping $\varphi(z)$ of int L onto $\mathbb{D}$ normalized by $\varphi(z_0) = 0$, $\varphi'(z_0) > 0$, as in (1.4.5). Then the discrepancy estimates can be formulated in terms of $\varepsilon_{p,L}(r) + \delta_{p,L}(r)$. In this chapter we shall discuss this approach carefully for general signed measures.

In Section 4.1 we consider the situation that L is a piecewise Dini-smooth curve or arc and $\sigma = \sigma^+ - \sigma^-$ is a signed measure with probability measures

130   4. Discrepancy Theorems via One-Sided Bounds for Potentials

σ^+ and σ^- supported on L. The discrepancy estimates are based on *outer one-sided bounds* for $U^\sigma(z)$.

The problem for more general Jordan curves is investigated in Section 4.2. But we have to pay for the more general situation by taking into account not only outer but also *inner one-sided bounds* for $U^\sigma(z)$.

Section 4.3 is concerned with analytic Jordan curves L and $\sigma = \mu_L - \nu$, where ν is a probability measure on L. It turns out that weak inner bounds for

$$\delta_\nu(\rho) := \max_{z \in L_\rho^-} U^{\mu_L - \nu}(z) - U^{\mu_L - \nu}(z_0),$$

where ρ ($0 < \rho < 1$) is fixed, are already sufficient to obtain discrepancy estimates. For example, in the case where $\nu = \nu_p$ is the zero counting measure of a polynomial p of degree n with all zeros on L and if $\delta_{\nu_p}(\rho)$ is not too large, e.g., when $\delta_{\nu_p}(\rho) \le e^{\sqrt{n}}$, the discrepancy estimate is of order $1/\log n$.

In the final section we consider again a Dini-smooth arc L and obtain discrepancy estimates of type analogous to the case of analytic Jordan curves.

4.1 Outer Bounds for Potentials

Let L be a Jordan curve or arc, $\sigma = \sigma^+ - \sigma^-$ a signed measure on L with positive part $\sigma^+ \in \mathcal{M}(L)$ and negative part $\sigma^- \in \mathcal{M}(L)$. Moreover, for all subarcs J of L the inequality

$$\sigma^+(J) \le c\mu_L(J)^\beta \qquad (1.1)$$

holds with constants $c > 0$ and $0 < \beta \le 1$. As before, we consider the normalized conformal mapping Φ of the unbounded component Ω of $\overline{\mathbb{C}} \setminus L$ onto Δ and its level curves

$$L_\delta = \{z \in \Omega : |\Phi(z)| = 1 + \delta\}, \qquad \delta > 0.$$

Our first result will be formulated in terms of the *outer bounds*

$$a_{\sup}(\delta) := \sup_{z \in L_\delta} U^\sigma(z), \qquad (1.2)$$

$$a_{\inf}(\delta) := -\inf_{z \in L_\delta} U^\sigma(z). \qquad (1.3)$$

Note that $a_{\sup}(\delta) \ge 0$, $a_{\inf}(\delta) \ge 0$ and let

$$a(\delta) := \min\{a_{\sup}(\delta), a_{\inf}(\delta)\}. \qquad (1.4)$$

Theorem 1.1. *Let L be a piecewise Dini-smooth curve with all inner angles less than or equal to π, or let L be an arbitrary piecewise Dini-smooth* arc. Let $\sigma = \sigma^+ - \sigma^-$ be a signed measure with $\sigma^+, \sigma^- \in \mathcal{M}(L)$ and let σ^+ satisfy (1.1). Then there exists a constant $c_1 = c_1(L, c) > 0$ such that for $\delta > 0$,

$$D[\sigma] \leq c_1 \left(\frac{a(\delta)}{\sqrt{\delta}} + \delta^{\beta/2} \right), \tag{1.5}$$

where $a(\delta)$ is the outer bound in (1.4).

Remark 1.2. If $\sigma > 0$, then $a(\delta) > 0$ and $a(\delta)$ is monotonically decreasing for $\delta > 0$. Hence, there exists a unique $\delta^* > 0$ such that

$$\delta^* = a(\delta^*)^{2/(\beta+1)}.$$

Therefore, if $\delta \leq a(\delta)^{2/(\beta+1)}$, we obtain from (1.5) that

$$D[\sigma] \leq 2\,c_1\,a(\delta)^{\beta/(\beta+1)}. \tag{1.6}$$

In the case where ε is a global upper bound of U^σ, i.e.,

$$\varepsilon \geq \sup_{z \in \mathbb{C}} U^\sigma(z),$$

then (1.6) yields the estimate

$$D[\sigma] \leq 2c_1\, \varepsilon^{\beta/(\beta+1)}.$$

Proof of Theorem 1.1. First, let L be an arc. Since L is quasiconformal, we can use the constructions and results from Section 1.2.5, i.e., we introduce the domains Ω_j and Δ_j as well as the functions Φ_j and Ψ_j, $j = 1, 2$, that are the restrictions of the conformal mappings Φ and Ψ to $\overline{\Omega}_j$ and $\overline{\Delta}_j$, respectively. Important is the fact that by Lemma 1.2.8 for all these functions we can use the assertion of Theorem 1.2.7.

We restrict ourselves to the case $a(\delta) = a_{\sup}(\delta)$, i.e.,

$$U^\sigma(z) \leq a(\delta), \qquad z \in \text{ext } L_\delta. \tag{1.7}$$

The other case $a(\delta) = a_{\inf}(\delta)$ can be handled in the same way.

Consider an arbitrary subarc $J \subset L$. The discrepancy estimate (1.5) is equivalent to

$$D[\sigma] \leq c_2 \left(\frac{a(\delta/2)}{\sqrt{\delta}} + \delta^{\beta/2} \right)$$

with some constant $c_2 = c_2(L, c)$. Moreover, it is sufficient to establish the inequality

$$\sigma(J) \geq -c_2 \left(\frac{a(\delta/2)}{\sqrt{\delta}} + \delta^{\beta/2} \right) \tag{1.8}$$

132   4. Discrepancy Theorems via One-Sided Bounds for Potentials

for sufficiently small δ in the case where one of the endpoints of J coincides with one of the endpoints of L.

By our assumption, $L = \cup_{j=1}^{m} l_j$, where each l_j is a Dini-smooth arc. As before, denote by z_j and z_{j+1} the endpoints of l_j and by ζ_1 and ζ_2 the endpoints of J. We can assume that $\zeta_1 = z_1$.

Let $\varepsilon_1 > 0$ be a fixed constant such that

$$|z_j - z_{j+1}| \geq 4\varepsilon_1, \qquad j = 1, \ldots, m. \tag{1.9}$$

We distinguish two cases.

Case A: ζ_2 is "far away" from the corners, that is,

$$|\zeta_2 - z_j| \geq \varepsilon_1, \qquad j = 2, \ldots, m.$$

Case B: There exists $2 \leq j_0 \leq m$ such that $|\zeta_2 - z_{j_0}| \leq \varepsilon_1$.

Discussion of Case A. The most delicate situation occurs if $\zeta_2 \in l_m$. Therefore, let us investigate this case in detail (from the reasoning below it will be quite clear how to handle the other cases). The typical situation arising in the constructions below is shown in Figure 4.1. Since by Lemma 1.2.12,

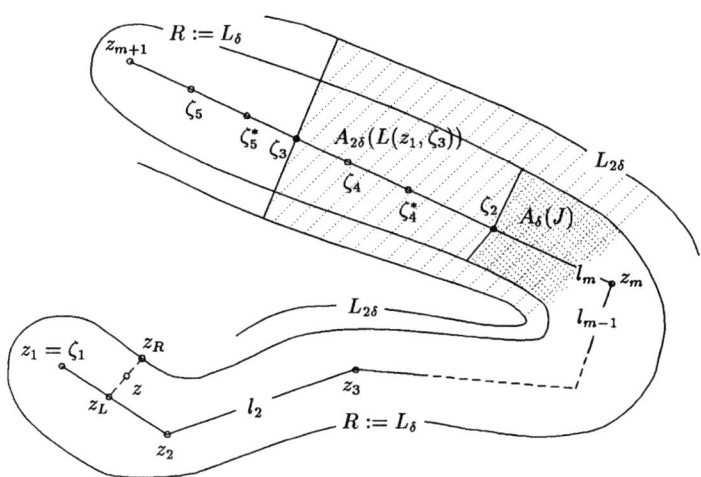

Fig. 4.1

$$-\sigma(J) = \sigma(L\backslash J) \leq \sigma^+(L\backslash J) \leq c_1 |\zeta_2 - z_{m+1}|^{\beta/2},$$

we may assume without loss of generality that

$$|\zeta_2 - z_{m+1}| \geq c_2 t^2, \tag{1.10}$$

where
$$t := \sqrt{\delta} \tag{1.11}$$

and c_2 is a constant that is large enough for our reasoning.

Set $R := L_\delta$, $E := \text{int } R$, $R' := \{w : |w| = 1 + \delta\}$. Moreover, for any point $z \in \Omega$ define the following points: $w := \Phi(z)$, $w_L := w/|w|$, $w_R := (1+\delta)w_L$, $z_L := \Psi(w_L)$, $z_R := \Psi(w_R)$. Let $e^{i\theta_{m+1}} := \Phi(z_{m+1})$, and let $e^{i\eta_1}$ as well as $e^{i\eta_2}$ be points such that $\eta_1 < \theta_{m+1} < \eta_2 < \eta_1 + 2\pi$, $\Psi(e^{i\eta_1}) = \Psi(e^{i\eta_2}) = \zeta_2$.

By Theorem 1.2.7 and (1.10),
$$\eta_1 - \theta_{m+1} \asymp \theta_{m+1} - \eta_2,$$
$$|\eta_j - \theta_{m+1}| \geq c_3 t, \quad j = 1, 2,$$

where the constant c_3 can be made sufficiently large by an appropriate choice of c_2.

Next, we use the construction from the proof of Theorem 2.4.2. Namely, we introduce the point $\zeta_3 := \Phi_1(e^{i(\eta_1+t)})$ and consider the function
$$h(z) := \begin{cases} 1, & \text{if } z \in A_{2\delta}(L(z_1,\zeta_3)), \\ 0, & \text{otherwise in } \mathbb{C}, \end{cases} \tag{1.12}$$

where $L(z,\zeta)$ denotes the subarc of L between the points z and $\zeta \in L$ and
$$A_r(\Gamma) := \{z : |\Phi(z)| \leq 1 + r, z_L \in \Gamma\}, \quad r > 0, \Gamma \subset L$$

is the r-neighborhood for Γ with respect to L.

We average this function $h(z)$ as in (2.4.14). Let $\rho(z), z \in \mathbb{C}$, be a regularized distance to $A_\delta(J)$, that is, a function satisfying (2.4.11)–(2.4.13) (with $\sigma := \delta$). Consider the function $g(z)$, given by (2.3.12) (with $\sigma := \delta$). In the proof of Theorem 2.4.2 we have established some properties of this function, whence we can derive the following consequences. If c_3 is large enough, there exist constants c_4 and c_5 with $c_4 < 1, 2 < c_5 < c_3$ such that for the points $\zeta_j := \Psi_1(e^{i(\eta_1+c_j t)})$, $\zeta_j^* := \Psi_1(e^{i(\eta_1+c_j t/2)})$, $j = 4, 5$, and for the restriction g_R of g to R we have
$$0 \leq g_R(z) \leq 1, \quad z \in R,$$
$$g_R(z) = 1, \quad z \in R, z_L \in L(z_1, \zeta_4),$$
$$g_R(z) = 0, \quad z \in R, z_L \in L(\zeta_5^*, z_{m+1}).$$

Further, if $z = x + iy$, $\xi = x_0 + iy_0 \in R$, $z_L, \xi_L \in L(\zeta_2, \zeta_5)$, then
$$g_R(z) = g_R(\xi) + A(\xi)(x - x_0) + B(\xi)(y - y_0) + r(z, \xi), \tag{1.13}$$

where
$$|A(\xi)| + |B(\xi)| \preceq |\zeta_2 - \zeta_3|^{-1}, \qquad (1.14)$$
$$|r(z,\xi)| \preceq \frac{|z-\xi|^2}{|\zeta_2 - \zeta_3|^2}. \qquad (1.15)$$

Let $f_+(z)$ be the harmonic function in ext R and let $f_-(z)$ be the harmonic function in int R that satisfy the boundary conditions
$$f_+(z) = f_-(z) = g_R(z), \qquad z \in R.$$
Set
$$\widetilde{f}_+(w) := f_+[\Psi(w)], \qquad |w| \geq 1+\delta,$$
$$\widetilde{f}_-(w) := f_-[\Psi(w)], \qquad 1 \leq |w| \leq 1+\delta,$$
$$\widetilde{f}(w) := \begin{cases} \widetilde{f}_+(w), & \text{if } |w| \geq 1+\delta, \\ \widetilde{f}_-(w), & \text{if } 1 \leq |w| < 1+\delta. \end{cases}$$

Lemma 1.3. *There exists a constant $c_6 > 0$ such that for $1 \leq |w| \leq 1+2\delta$,*
$$|\widetilde{f}(w) - \widetilde{f}(w_R)| \leq c_6 \, t. \qquad (1.16)$$

The proof of Lemma 1.3 will be given later.

Returning to the proof of (1.8), we average the functions f_+ and f_- in some special way. Let $K(z), z \in \mathbb{C}$, be an arbitrary averaging kernel, i.e., $K \in C^\infty(\mathbb{C})$ and K satisfies (1.5.3)–(1.5.5). Consider in Δ the function
$$\widetilde{\kappa}(w) := \begin{cases} \dfrac{16}{\delta^2} \int \widetilde{f}(\tau) K\left(\dfrac{4(\tau - w)}{\delta}\right) dm(\tau), & \text{if } |w| > 1 + \tfrac{1}{2}\delta, \\ \widetilde{f}_-(w), & \text{if } 1 \leq |w| \leq 1 + \tfrac{1}{2}\delta. \end{cases}$$

If $1 \leq |w| < 1 + \tfrac{3}{4}\delta$ or $|w| > 1 + \tfrac{5}{4}\delta$, the value of $\widetilde{\kappa}(w)$ is an average of the harmonic function $\widetilde{f}_-(w)$ or $\widetilde{f}_+(w)$, respectively. Hence, by (1.5.6),
$$\widetilde{\kappa}(w) := \begin{cases} \widetilde{f}_+(w), & \text{if } |w| > 1 + \tfrac{5}{4}\delta, \\ \widetilde{f}_-(w), & \text{if } 1 \leq |w| < 1 + \tfrac{3}{4}\delta, \end{cases}$$
and therefore $\Delta\widetilde{\kappa}(w) = 0$ for such w. Note that $\widetilde{\kappa} \in C^\infty(\Delta)$ and
$$0 \leq \widetilde{\kappa}(w) \leq 1. \qquad (1.17)$$
Moreover, according to (1.16) and (1.5.7),
$$|\Delta\widetilde{\kappa}(w)| \preceq \frac{t}{\delta^2}, \qquad 1 + \frac{3}{4}\delta \leq |w| \leq 1 + \frac{5}{4}\delta. \qquad (1.18)$$

4.1 Outer Bounds for Potentials

Finally, the function

$$\kappa(z) := \begin{cases} \widetilde{\kappa}[\Phi(z)], & \text{if } z \in \Omega, \\ f_-(z), & \text{if } z \in L, \end{cases}$$

has in $\mathbb{C}$ partial derivatives of all orders, and by Green's formula (1.1.9) and (1.1.12),

$$\int \Delta \kappa(z)\, dm(z) = 0. \tag{1.19}$$

Next, we will establish the inequality

$$\left| \int \kappa\, d\sigma \right| \preceq \frac{a\, t}{\delta}, \tag{1.20}$$

where $a := a(\delta/2)$. In fact, by Green's formula (1.1.10) and (1.1.13), κ can be represented as

$$\kappa(z) = \kappa(\infty) + \frac{1}{2\pi} \int \Delta \kappa(\zeta) \log|z - \zeta|\, dm(\zeta), \qquad z \in \mathbb{C}.$$

Since

$$\int \Delta \kappa(\zeta)\, dm(\zeta) = 0$$

(cf. (2.2.13)), we obtain according to (1.7), (1.18), (1.19), and (1.1.3) that

$$\left| \int \kappa\, d\sigma \right| = \frac{1}{2\pi} \left| \int (a - U^\sigma(\zeta))\, \Delta \kappa(\zeta)\, dm(\zeta) \right|$$

$$\leq \frac{1}{2\pi} \int_\Omega \left(a - \widetilde{U}^\sigma(w) \right) |\Delta \widetilde{\kappa}(w)|\, dm(w)$$

$$\preceq \frac{t}{\delta^2} \int_{1+\frac{3}{4}\delta}^{1+\frac{5}{4}\delta} r \int_0^{2\pi} \left(a - \widetilde{U}^\sigma(re^{i\theta}) \right) d\theta\, dr \preceq \frac{a\, t}{\delta},$$

where $\widetilde{U}^\sigma(w) := U^\sigma(\Psi(w))$.

The conditions (1.16), (1.17), and (1.20) imply

$$(-\sigma)(J) \leq - \int \kappa\, d\sigma + \sigma^+(\Gamma) + \int_{L \setminus (J \cup \Gamma)} \kappa\, d\sigma^+$$

$$+ \int_J (1 - \kappa)\, d\sigma^- \preceq \left(\frac{a\, t}{\delta} + t^\beta \right),$$

where
$$\Gamma := L(\zeta_2, \zeta_5).$$

Hence (1.8) is proved.

Discussion of Case B. It is enough to consider the situation that
$$|\zeta_2 - z_2| \leq \varepsilon_1.$$

In this case our reasoning will be the same as before. However, the construction of the auxiliary function $\widetilde{f}$ needs some modification.

Let L have an exterior angle $\alpha\pi$ at the point z_2 with respect to the region Ω_1, $0 < \alpha \leq 1$, and let $e^{i\eta_j} := \Phi_j(\zeta_2)$, $e^{i\theta_2} = w_2 := \Phi_1(z_2)$. We fix two points $\xi_1 \in l_1$ and $\xi_2 \in l_2$ such that
$$|\xi_1 - z_1| = \varepsilon_1, \quad |\xi_2 - z_3| = \varepsilon_1, \tag{1.21}$$

where ε_1 is the constant from (1.9), and set
$$\Phi_k(\xi_j) =: t_j^k = e^{i\beta_j^k}, \quad k, j = 1, 2$$

(see Figure 4.2). The points $(1+\delta)\, t_j^2$ divide R' into two subarcs. Denote by S' the subarc that includes the point $(1+\delta)\, w_2$. Without loss of generality we assume that
$$\eta_2 < \beta_1^2 < \beta_1^1 < \eta_1.$$

We choose a constant c, $1 \leq c \leq 10$, such that
$$\theta_2 \notin [\eta_1 + ct, \eta_1 + 4ct], \tag{1.22}$$

where as before, $t = \sqrt{\delta}$. The construction of the function g follows almost the same line as in the previous case. We introduce the point $\zeta_3 := \Psi_1(e^{i(\eta_1 + 2ct)})$ as well as the sets $A_\delta(J)$ and $A_{2\delta}(L(z_1, \zeta_3))$. Let the function $h(z)$ be given by (1.12). We average this function by using the regularized distance $\rho(z)$ to $A_\delta(J)$ (see Appendix D) in the following way:

$$g(z) := \begin{cases} \dfrac{16}{\rho_1(z)^2} \displaystyle\int_{\mathbb{C}} h(\zeta)\, K\left(\dfrac{4|\zeta - z|}{\rho_1(z)}\right) dm(\zeta), & \text{if } z \in \mathbb{C} \setminus A_\sigma(J), \\ 1, & \text{if } z \in A_\sigma(J), \end{cases} \tag{1.23}$$

where $\rho_1(z) := \varepsilon_2\, \rho(z)$, and the sufficiently small constant $\varepsilon_2 > 0$ is such that $g(z)$ has the following properties: There is a constant c_2, $c_2 < c$, such that for the points
$$\zeta_4 := \Psi_1(e^{i(\eta_1 + (c+c_2)t)}), \quad \zeta_5 := \Psi_1(e^{i(\eta_1 + (4c - c_2)t)}),$$
$$\zeta_4^* := \Psi_1(e^{i(\eta_1 + (c+c_2/2)t)}), \quad \zeta_5^* := \Psi_1(e^{i(\eta_1 + (4c - 2c_2)t)}),$$

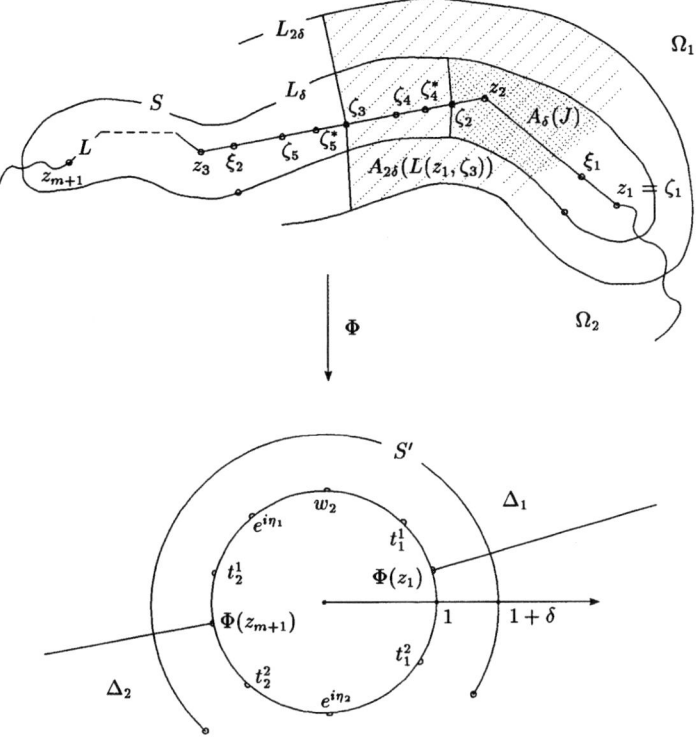

Fig. 4.2

and for the restriction g_S of g to $S := \Psi(S')$, we have

$$0 \leq g_S(z) \leq 1, \quad z \in S,$$
$$g_S(z) = 1, \quad z \in S,\ z_L \in L(z_1, \zeta_4),$$
$$g_S(z) = 0, \quad z \in S,\ z_L \in L(\zeta_5^*, z_{m+1}).$$

Further, if $z = x + iy$, $\xi = x_0 + iy_0 \in S$, $z_L, \xi_L \in L(\zeta_2, \zeta_5)$, then

$$g_S(z) = g_S(\xi) + \widehat{A}(\xi)(x - x_0) + \widehat{B}(\xi)(y - y_0) + \widehat{r}(z, \xi), \qquad (1.24)$$

where

$$|\widehat{A}(\xi)| + |\widehat{B}(\xi)| \preceq |\zeta_2 - \zeta_3|^{-1}, \qquad (1.25)$$

$$|\widehat{r}(z, \xi)| \preceq \frac{|z - \xi|^2}{|\zeta_2 - \zeta_3|^2}. \qquad (1.26)$$

Let f be the function harmonic in $\overline{\mathbb{C}} \setminus S$ with continuous boundary values
$$f(z) := g_S(z), \qquad z \in S.$$
For $w \in \overline{\Delta}$, set $\widetilde{f}(w) := f(\Psi(w))$. Then the following estimate holds.

Lemma 1.4. *Let $w \in S'_\delta := \{w \in \overline{\Delta} : \mathrm{dist}\,(w, S') \leq \delta\}$. Then*
$$|\widetilde{f}(w) - \widetilde{f}(w^*)| \leq c_1 t, \tag{1.27}$$
where $w^* \in S'$ is the point of S' nearest to w.

The proof of Lemma 1.4 will be given later. Using this lemma, the proof of (1.8) is completed by repeating step by step our reasoning from the previous case with the new function f.

If L is a curve, the proof of (1.5) is simpler than in the previous case of an arc, because we don't need to take care of endpoints. Here is a short sketch of the proof: As before, we assume that (1.7) holds. Obviously, it is enough to prove (1.8) for an arbitrary subarc $J \subset L$ with endpoints ζ_1, ζ_2 such that
$$J \subset l_j \cup l_{j+1}, \ |\zeta_k - z_i| \geq \varepsilon_3, \qquad k = 1, 2, \quad i = j, j+2,$$
$$\zeta_1 \in l_j, \ |\zeta_1 - z_{j+1}| \geq \varepsilon_3,$$
where $\varepsilon_3 > 0$ is fixed and l_j, $j = 1, \ldots, m$, are the Dini-smooth arcs with endpoints z_j, z_{j+1} that constitute the curve L. We may restrict ourselves to the case $j = 1$.
Let
$$e^{i\theta_2} := \Phi(z_2), \quad e^{i\eta_j} := \Phi(\zeta_j), \quad j = 1, 2.$$
We assume that $\eta_1 < \eta_2$ and choose a constant c, $1 \leq c \leq 10$, such that
$$\theta_2 \notin [\eta_2 + ct, \eta_2 + 4ct].$$
Then we construct two auxiliary points
$$\zeta_1^* := \Psi(e^{i(\eta_1 - t)}), \quad \zeta_2^* := \Psi(e^{i(\eta_2 + 2ct)}),$$
where $t := \sqrt{\delta}$, and denote by J_1 the subarc of L between the points ζ_1^* and ζ_2^*, that includes J. Next, we introduce the two sets
$$\widetilde{A}_\delta(J) := \{\zeta \in (\mathrm{int}\, L_\delta) \setminus (\mathrm{int}\, L) : \zeta_L \in J\},$$
$$\widetilde{A}_{2\delta}(J_1) := \{\zeta \in (\mathrm{int}\, L_{2\delta}) \setminus (\mathrm{int}\, L) : \zeta_L \in J_1\},$$
the function
$$h(z) := \begin{cases} 1, & \text{if } z \in \widetilde{A}_{2\delta}(J_1), \\ 0, & \text{otherwise in } \mathbb{C}, \end{cases}$$

and the function $\rho(z)$, which is a regularized distance to $\widetilde{A}_\delta(J)$ (see Appendix D). We average the function $h(z)$ as in (1.23) and obtain the function $g(z)$ and its restriction $g_R(z)$ to $R := L_\delta$. Then the same arguments as before lead to the estimate (1.8). □

We have to append the proof of the two lemmas that we have used above.

Proof of Lemma 1.3. We begin with the proof of (1.16) for $1 + \delta < |w| < 1 + 2\delta$. Let $\zeta := \Psi(w)$. If $\zeta_L \notin L(\zeta_4^*, \zeta_5)$, then by construction of the function g_R, there is a sufficiently small constant $c_1 > 0$ such that for τ with

$$|\tau - w_R| \leq c_1 t, \quad |\tau| = 1 + \delta,$$

the function $g_R(\Psi(\tau))$ is equal to 0 or 1. Therefore, inequality (1.16) follows simply from Lemma 1.1.5 (see (1.1.19)).

The reasoning above also remains true for ζ with $\zeta_L \in L(\zeta_6, \zeta_5)$, where

$$\zeta_6 := \Psi_1 \left(\exp \left\{ \frac{1}{2} i \left(\arg \Phi_1(\zeta_5^*) + \arg \Phi_1(\zeta_5) \right) \right\} \right). \quad (1.28)$$

Hence, the nontrivial case appears when $\zeta_L \in L(\zeta_4^*, \zeta_6)$. Set $\xi := \zeta_R$ and consider the representation of the function g_R in the form (1.13). Theorem 1.2.7 and Lemma 1.2.12 imply that

$$\left| \frac{\zeta - \zeta_R}{\zeta_2 - \zeta_3} \right| \preceq \left| \frac{\zeta_L - \zeta_R}{\zeta_L - \zeta_2} \right| \asymp \frac{\delta}{t} = t. \quad (1.29)$$

In the following, denote by the same symbol $r(z, \xi)$ the harmonic extension of the function appearing in (1.13) to $\text{ext } R \setminus \{\infty\}$ as follows:

$$r(z, \xi) := f(z) - g_R(\xi) - A(\xi)(x - x_0) - B(\xi)(y - y_0),$$

and set

$$\widetilde{r}(\tau) := r(\Psi(\tau), \xi), \quad |\tau| \geq 1 + \delta. \quad (1.30)$$

Since by Lemmas 1.2.12 and 1.2.3 for $z \in R$ with $z_L \in L(\zeta_2, \zeta_5)$,

$$\left| \frac{z - \xi}{\zeta_2 - \zeta_3} \right| \preceq \begin{cases} \left| \dfrac{z - \zeta_L}{\zeta_2 - \zeta_L} \right|, & \text{if } |z - \xi| \geq |\zeta_L - \zeta_R|, \\ \left| \dfrac{z - \zeta_R}{\zeta_L - \zeta_R} \right| \left| \dfrac{\zeta_L - \zeta_R}{\zeta_L - \zeta_2} \right|, & \text{if } |z - \xi| < |\zeta_L - \zeta_R|, \end{cases}$$

$$\preceq \frac{|\Phi(z) - w_R|}{t}, \quad (1.31)$$

we have according to (1.15),

$$|\widetilde{r}(\tau)| \preceq \frac{|\tau - w_R|^2}{t^2}, \quad |\tau| = 1 + \delta, \, |\tau - w_R| \leq c_2 t. \quad (1.32)$$

140   4. Discrepancy Theorems via One-Sided Bounds for Potentials

At the same time, by the definition of the function $\tilde{r}(\tau)$ and (1.13)–(1.15), relation (1.32) remains true for τ such that $|\tau| > 1 + \delta$, $|\tau - w_R| = c_2 t$.

Further, direct computation yields

$$|\tilde{r}(w)| \preceq t. \tag{1.33}$$

Indeed, let us introduce the auxiliary function $\widetilde{R}(\tau)$ that is the harmonic extension to $|\tau| \geq 1 + \delta$ of the function

$$\widetilde{R}(\tau) := \begin{cases} |\tilde{r}(\tau)|, & \text{if } |\tau| = 1 + \delta, \ |\tau - w_R| \leq c_2 t, \\ c_3, & \text{otherwise for } |\tau| = 1 + \delta. \end{cases}$$

It is clear that for sufficiently large c_3 we have

$$|\tilde{r}(\tau)| \leq \widetilde{R}(\tau)$$

on the boundary of the domain

$$\{\tau : |\tau| > 1 + \delta, \ |\tau - w_R| < c_2 t\}.$$

Therefore, by the maximum principle for harmonic functions (see Theorem 1.1.2), the Poisson formula (1.1.5), and (1.32) we obtain

$$|\tilde{r}(w)| \leq \widetilde{R}(w) = \widetilde{R}(|w|e^{i\theta})$$

$$= \frac{1}{2\pi} \int_0^{2\pi} \widetilde{R}((1+\delta)e^{i\eta}) \frac{|w|^2 - (1+\delta)^2}{|w|^2 - 2|w|(1+\delta)\cos(\theta - \eta) + (1+\delta)^2} \, d\eta$$

$$\preceq \delta \left(\frac{1}{t^2} \int_{\theta - c_2 t}^{\theta + c_2 t} d\eta + \int_{\theta + c_2 t}^{\theta + \pi} \frac{d\eta}{(\eta - \theta)^2} + \int_{\theta - \pi}^{\theta - c_2 t} \frac{d\eta}{(\eta - \theta)^2} \right) \preceq \frac{\delta}{t} = t.$$

Comparing (1.13), (1.14), (1.29), and (1.33) we get (1.16) for $1 + \delta < |w| < 1 + 2\delta$.

Now we assume that $\zeta := \Psi(w)$, $1 < |w| < 1 + \delta$. Reasoning as in the proof of Lemma 3.3 above (see inequalities (3.18) and (3.22)) we see that (1.16) is true if $\zeta_L \notin L(\zeta_4^*, \zeta_6)$. Therefore, without loss of generality we assume that $\zeta_L \in L(\zeta_4^*, \zeta_6)$. As before, we represent the function g_R in the form (1.13) and set $\xi := \zeta_R$. Our next purpose is to study the behavior of $r(z, \xi)$, where now $r(z, \xi)$ also denotes the extension of the respective function to $E := \text{int } R$.

Let $A := A_\delta(L(\zeta_2, \zeta_5))$. By (1.13)–(1.15) and (1.31), the function $r(z, \xi)$ satisfies for $z \in \partial A$ the inequalities

$$r(z, \xi) \preceq \left| \frac{z - \zeta_R}{\zeta_2 - \zeta_3} \right|^2 \preceq \left| \frac{z_L - \zeta_R}{\zeta_2 - \zeta_3} \right|^2 \preceq \left| \frac{z_L - \zeta_L}{\zeta_2 - \zeta_L} \right|^2 + \left| \frac{\zeta_L - \zeta_R}{\zeta_L - \zeta_2} \right|^2$$

$$\preceq \frac{|\Phi_1(z_L) - \Phi_1(\zeta_L)|^2}{t^2} + \delta. \tag{1.34}$$

4.1 Outer Bounds for Potentials 141

Next, in order to introduce one special measure, we proceed as follows. Let

$$\Phi_1(\zeta_2) =: e^{i\eta_1}, \ \Phi_1(\zeta_5) =: e^{i\eta_5}, \ \Phi_1(\zeta_L) =: e^{i\theta},$$
$$\eta_1 < \theta < \eta_5 < \eta_1 + 2\pi.$$

As we know,
$$\theta - \eta_1 \asymp \eta_5 - \theta \asymp t.$$

For η with $\eta_1 \leq \eta \leq \eta_5$, the points ξ_1 and ξ_2, defined by the conditions

$$\xi_k \in R, \ \Phi_1((\xi_k)_L) = e^{i\eta}, \qquad k = 1, 2,$$

divide ∂A into two parts. We denote by B_η the one of them that includes the point ζ_2. Let

$$u(\eta) := \omega(\zeta, A, B_\eta), \quad \eta_1 \leq \eta \leq \eta_5,$$

be the appropriate harmonic measure. Reasoning as in the proof of (3.18) and (3.22), we have

$$u(\eta) \leq c_1 \exp\left\{c_2 \frac{\eta - \theta}{\delta}\right\}, \qquad \eta_1 \leq \eta \leq \theta, \qquad (1.35)$$

$$1 - u(\eta) \leq c_1 \exp\left\{c_2 \frac{\theta - \eta}{\delta}\right\}, \qquad \theta \leq \eta \leq \eta_5. \qquad (1.36)$$

Now by virtue of (1.34), (1.35), and (1.36) we obtain

$$|r(\zeta, \zeta_R)| \preceq \int_{\eta_1}^{\eta_5} \left(\frac{(\eta - \theta)^2}{t^2} + \delta\right) du(\eta) + e^{-c_3 t/\delta}$$

$$\preceq \delta + \frac{1}{t^2} \left(\int_{\eta_1}^{\theta} + \int_{\theta}^{\eta_5}\right) (\eta - \theta)^2 \, du(\eta)$$

$$\leq \delta + \frac{2}{t^2} \left(\int_{\eta_1}^{\theta} (\theta - \eta) u(\eta) \, d\eta + \int_{\theta}^{\eta_5} (\eta - \theta)(1 - u(\eta)) \, d\eta\right)$$

$$\preceq \delta + \frac{1}{t}\left(\int_{\eta_1}^{\theta} e^{c_2 \frac{\eta-\theta}{\delta}} \, d\eta + \int_{\theta}^{\eta_5} e^{c_2 \frac{\theta-\eta}{\delta}} \, d\eta\right) \preceq \frac{\delta}{t} = t.$$

At last, taking into account (1.13), (1.14), (1.29) (which is true in this case, too) and the last inequality, we have

$$|\widetilde{f}(w) - \widetilde{f}(w_R)| = |A(\zeta_R)\operatorname{Re}\{\zeta - \zeta_R\} + B(\zeta_R)\operatorname{Im}\{\zeta - \zeta_R\} + r(\zeta, \zeta_R)|$$

$$\preceq \left|\frac{\zeta - \zeta_R}{\zeta_2 - \zeta_3}\right| + |r(\zeta, \zeta_R)| \preceq t.$$

$\square$

Proof of Lemma 1.4. According to Beurling's theorem (see [4, p. 43]), for all $\zeta \in V := \overline{\mathbb{C}} \setminus S$ with the properties

$$|\zeta - z_2| \geq 2\varepsilon_1, \quad \operatorname{dist}(\zeta, S) \leq \varepsilon_1,$$

where ε_1 is the constant of (1.9), we have

$$|f(\zeta) - f(\zeta^*)| \leq c_1 \sqrt{\operatorname{dist}(\zeta, S)}, \tag{1.37}$$

where $\zeta^* \in S$ is the point of S nearest to ζ.

Therefore, in the case that w and $\zeta := \Psi(w)$ satisfy

$$1 < |w| \leq 1 + \delta, \quad |\zeta - z_2| \geq 2\varepsilon_1,$$

inequality (1.27) is a simple consequence of (1.37) and the maximum principle for harmonic functions (see Theorem 1.1.2). Furthermore, applying (1.37) or Lemma 1.1.5 we obtain (1.27) for w with $1 + \delta < |w| \leq 1 + 2\delta$, $\operatorname{dist}(w, S') < \delta$, $\zeta_L \notin L(\zeta_4^*, \zeta_6)$, where ζ_6 is defined by (1.28).

If $1 + \delta < |w| \leq 1 + 2\delta$, $\operatorname{dist}(w, S') < \delta$, and $\zeta_L \in L(\zeta_4^*, \zeta_6)$, we set $\xi := \zeta_R$ and write the function g_S in the form (1.24). Reasoning as in the proof of (1.16) (see the previous Lemma 1.3) we also get (1.27).

Thus, we have only one nontrivial case, namely

$$|\zeta - z_2| \leq 2\varepsilon_1, \quad \zeta \in \Omega_1 \cap (\operatorname{int} R). \tag{1.38}$$

Next, we discuss the following auxiliary result. Let $\kappa(z)$, $z \in \overline{\mathbb{C}}$, be a continuous function that is harmonic in V. Set

$$\widetilde{\kappa}(\tau) := \kappa(\Psi(\tau)), \quad \tau \in \Delta.$$

For ζ as in (1.38) we want to estimate the quantity

$$|\widetilde{\kappa}(w) - \widetilde{\kappa}(w_R)| = |\kappa(\zeta) - \kappa(\zeta_R)|$$

by some expressions using the notion of harmonic measure and some special crosscuts of the domain V.

It follows from Theorem 1.2.7 that

$$\operatorname{diam}\left(\{z \in (\operatorname{int} R) \cap \Omega_1 : z_L = \zeta_L\}\right) =: d \asymp |\zeta_L - \zeta_R|. \tag{1.39}$$

Without loss of generality we assume that δ and, consequently, d are sufficiently small. For $d < r < \varepsilon := \varepsilon_1/2$ we denote by $\gamma(r) = \gamma_\zeta(r)$ the arc of the intersection $\{z : |z - \zeta_L| = r\} \cap V$ that separates the point ζ from ∞. Define $l(r) = l_\zeta(r)$ to be the subarc of S that has the same endpoints as $\gamma(r)$ (see Figure 4.3). It is obvious that

$$\operatorname{diam} l(r) \preceq r.$$

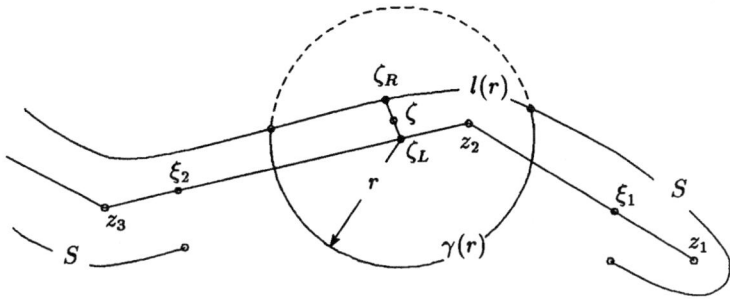

Fig. 4.3

Consider the Jordan domain $W = W_s \subset V$, $d < s \leq \varepsilon$, that includes the point ζ and is bounded by $\partial W = l(s) \cup \gamma(s)$. For $r > 0$ define

$$u(r) := \sup_{\substack{z \in \partial W, \\ |z-\zeta_R| \leq r}} |\kappa(z) - \kappa(\zeta_R)|,$$

$$m := u(\operatorname{diam} W).$$

Let $v(r) := \omega(\zeta, W, \partial W \setminus l(r))$, $d < r \leq s$, be the harmonic measure of the arc $\partial W \setminus l(r)$ at the point ζ with respect to W. For convenience, let $v(d) := 1$.

Choosing a natural number k such that

$$\frac{s}{2} \leq 2^k d < s,$$

we get by the maximum principle for harmonic functions (assuming that $s > 2d$)

$$|\kappa(\zeta) - \kappa(\zeta_R)| \leq u(\operatorname{diam} l(d)) + \sum_{j=0}^{k-1} v(2^j d)\, u(\operatorname{diam} l(2^{j+1} d)) + m\, v(s/2)$$

$$\leq u(cd) + 2 \int_d^s \frac{u(cr)}{r} v\left(\frac{r}{2}\right) dr + m\, v(s/2).$$

Denote by $\Gamma(r)$, $d < r < s$, the family of all crosscuts of W that separate the point ζ from $\partial W \setminus l(r)$. By (1.2.17) we obtain

$$v(r) \preceq \exp\{-\pi\, m(\Gamma(r))\}, \tag{1.40}$$

144   4. Discrepancy Theorems via One-Sided Bounds for Potentials

where $m(\Gamma(r))$ is the module of the family $\Gamma(r)$. By the integrated version of the composition laws (1.2.15),

$$m(\Gamma(r)) \geq \int_d^r \frac{dx}{|\gamma(x)|},$$

where $|\gamma(x)|$ denotes the length of $\gamma(x)$. Together with (1.40) this implies

$$|\kappa(\zeta) - \kappa(\zeta_R)| \preceq u(2cd) + \int_{2d}^s \frac{u(cr)}{r} \exp\left(-\pi \int_d^{r/2} \frac{dx}{|\gamma(x)|}\right) dr$$

$$+ m \exp\left(-\pi \int_d^{s/2} \frac{dx}{|\gamma(x)|}\right).$$

By (1.2.27) and Theorem 1.2.7 we have for $d < x < \varepsilon$ and some $0 < \beta < 1$,

$$\varepsilon_2 \, x \leq |\gamma(x)| \leq \pi \left(x + c_1 \, x \, h(c_2 x) + c_3 \, d^\beta x^{1-\beta}\right),$$

where h is the function from the definition of a Dini-smooth curve. Hence, by (1.2.26),

$$\pi \int_d^{r/2} \frac{dx}{|\gamma(x)|} = \int_d^{r/2} \frac{dx}{x} + \int_d^{r/2} \frac{\pi \, x - |\gamma(x)|}{x \, |\gamma(x)|} \, dx$$

$$\geq \log \frac{r}{2d} - \frac{c_1}{\varepsilon_2} \int_0^{r/2} \frac{h(c_2 x)}{x} \, dx - \frac{c_3}{\varepsilon_2} d^\beta \int_d^{r/2} \frac{x^{1-\beta}}{x^2} \, dx$$

$$\geq \log \frac{r}{d} - c_4,$$

and, consequently,

$$|\kappa(\zeta) - \kappa(\zeta_R)| \preceq u(2cd) + d \int_{2d}^s \frac{u(cr)}{r^2} \, dr + m \frac{d}{s}. \qquad (1.41)$$

It is simple to prove (1.27) if in addition to (1.38), $\zeta_L \notin L(\zeta_4^*, \zeta_6)$. Indeed, now let $\kappa := f$, $s := \varepsilon_3 \, |\zeta_L - \zeta_3|$, where the sufficiently small constant ε_3 is chosen such that $u(cs) = 0$. Therefore, by virtue of (1.41) and the obvious inequality

$$\left|\frac{\zeta_L - \zeta_R}{\zeta_L - \zeta_3}\right| \preceq \frac{\delta}{t} = t \qquad (1.42)$$

(which is a direct consequence of Lemma 1.2.10) we get (1.27).

The situation is more complicated if in addition to (1.38), $\zeta_L \in L(\zeta_4^*, \zeta_6)$. In this case we represent the function g_S in the form (1.24) with $\xi := \zeta_R$ and set $\kappa(z) := \widehat{r}(z, \xi)$ (i.e., $\kappa(z)$ is the harmonic extension of $\widehat{r}(z, \xi)$ from S to V), $s := \varepsilon_4 |\zeta_2 - \zeta_3|$, where ε_4 is chosen so small that for $z \in l(s)$ the function $\kappa(z)$ satisfies (1.26). Since by (1.24) and (1.25),

$$|\kappa(z)| \preceq 1, \quad z \in \gamma(s),$$

we have according to (1.26) and (1.41)

$$|\widehat{r}(\zeta, \zeta_R)| \preceq \frac{d^2}{s^2} + \frac{d}{s^2} \int_{2d}^{s} dr + \frac{d}{s} \asymp \frac{d}{s} \asymp \left|\frac{\zeta_L - \zeta_R}{\zeta_2 - \zeta_3}\right|. \tag{1.43}$$

Comparing (1.43), (1.24), and (1.25) we get

$$|f(\zeta) - f(\zeta_R)| \preceq \left|\frac{\zeta_L - \zeta_R}{\zeta_2 - \zeta_3}\right| \asymp \left|\frac{\zeta_L - \zeta_R}{\zeta_L - \zeta_2}\right| \preceq \frac{\delta}{t} = t,$$

where in the last inequality we have used the analogue of (1.42), which also follows simply from Lemma 1.2.10. $\square$

Next, consider an application of Theorem 1.1 to the distribution of zeros of monic polynomials.

Theorem 1.5. *Let L be as in Theorem 1.1. Suppose that the monic polynomial $p = p_n \in \mathbb{M}_n$, $n \geq 2$, with all zeros on L satisfies at least one of the following two conditions:*

(i) $\|p\|_L \leq A_n (\operatorname{cap} L)^n$ *with* $1 \leq A_n < e^n$;

(ii) *all zeros $z_1, \ldots, z_n$ of p are simple and*

$$|p'(z_j)| \geq \frac{1}{A_n} (\operatorname{cap} L)^n, \quad j = 1, \ldots, n,$$

with $n \leq A_n < e^n$.

Then there exists a constant $c > 0$ depending only on L such that

$$D[\mu_L - \nu_p] \leq c\sqrt{\delta_n}, \tag{1.44}$$

where

$$\delta_n := \frac{\log A_n}{n},$$

and ν_p is the zero counting measure of p.

146   4. Discrepancy Theorems via One-Sided Bounds for Potentials

Proof. By (1.4.2) we get

$$a_{\sup}(\delta/2) \leq \frac{1}{n} \log A_n, \qquad 0 < \delta < 1,$$

for the polynomial p satisfying condition (i), and by Lemma 1.4.5,

$$a_{\inf}(\delta/2) \leq \frac{c_1}{n} \left(\log A_n + \log \frac{1}{\delta} + n\delta \right), \qquad 0 < \delta < 1,$$

for p satisfying (ii), where $c_1 = c_1(L) > 0$. Hence, (1.5) with $\delta := \delta_n$ yields (1.44). □

There is a simple way to extend the results of this theorem in the case (ii) to arbitrary quasiconformal curves and arcs. However, the estimate (1.44) becomes slightly weaker (up to logarithmic terms).

Theorem 1.6. *Let L be a quasiconformal curve or arc and let the monic polynomial $p = p_n \in \mathbb{M}_n$, $n > 1$, with all zeros on L satisfy condition (ii) from Theorem 1.5 with $n \leq A_n \leq e^{\sqrt{n}/e}$. Then there exists a constant $c > 0$ depending only on L such that*

$$D[\mu_L - \nu_p] \leq c \, \frac{\log A_n}{n^{1/2}} \log \frac{n^{1/2}}{\log A_n}. \tag{1.45}$$

Proof. Let $Z^1 := \{z_1, \ldots, z_n\}$ be the zeros of p and $w_1(z) := p(z)$. Next, for $j \geq 2$ we define inductively

$$M_{j-1} := \|w_{j-1}\|_L = |w_{j-1}(z_{n+j-1})|,$$
$$Z^j := Z^{j-1} \cup \{z_{n+j-1}\},$$
$$w_j(z) := w_{j-1}(z)(z - z_{n+j-1}).$$

By construction, we get for the *Vandermonde expression*

$$V(S) := \left\{ \prod_{\substack{z,t \in S, \\ z \neq t}} |z - t| \right\}^{1/2}, \qquad S \subset \mathbb{C} \text{ finite},$$

the relation

$$V(Z^{j+1}) = V(Z^j) \, M_j.$$

For simplicity, let us now assume that $\operatorname{cap} L = 1$. We claim that for sufficiently large n there exists an index i, $1 \leq i \leq \sqrt{n} + 1$, such that

$$\log M_i \leq 2 n^{1/2} \log A_n. \tag{1.46}$$

Let us assume that (1.46) is false. Then condition (ii) yields

$$\log V(Z^1) \geq -\frac{n}{2} \log A_n.$$

We fix $k := [\sqrt{n} + 1]$. Then

$$\log V(Z^{k+1}) = \log V(Z^1) + \sum_{j=1}^{k} \log M_j$$

$$\geq -\frac{n}{2} \log A_n + 2n \log A_n > n \log n. \qquad (1.47)$$

On the other hand, by Theorem 1.3.3 we have the upper bound

$$\log V_m \leq (m/2) \log(4 e^{-1} \log m + 4) + (m/2) \log m, \qquad (1.48)$$

where V_m is the Vandermonde expression for an mth Fekete point set in L. For $m = n+1+k$, the upper bound in (1.48) is equal to $n \log n(1/2+o(1))$. Comparing this upper bound with the lower bound (1.47) we obtain a contradiction for sufficiently large n. Hence, (1.46) is true.

Now,

$$w_1(z) = \frac{w_i(z)}{(z - z_{n+1}) \cdots (z - z_{n+i-1})}.$$

For $z \in L_{1/n}$ we obtain, using Lemma 1.2.5,

$$|(z - z_{n+1}) \cdots (z - z_{n+i-1})| \geq (c_1 n^{-2})^{i-1},$$

and therefore

$$\log |w_1(z)| \leq c_2 n^{1/2} \log A_n, \quad z \in L_{1/n},$$

where the constant $c_2 > 0$ is independent of n. The same estimate is fulfilled for $z \in L$. If $c_2 \log A_n \leq \sqrt{n}/e$, then

$$\|w_1\|_L \leq e^{n/e},$$

and Corollary 3.1.8 yields (1.45). If $c_2 \log A_n > \sqrt{n}/e$ then

$$\frac{\log A_n}{\sqrt{n}} \log \frac{\sqrt{n}}{\log A_n} \geq \frac{1}{c_2},$$

since $\log A_n \leq \sqrt{n}/e$. Hence in this case the inequality (1.45) is trivial. □

We have shown in Section 2.3 that Theorem 1.5 (and henceforth Theorem 1.1) is sharp for the case (i). We complete this section by showing the sharpness of Theorem 1.5 in the case (ii).

Theorem 1.7. *Let L be an arbitrary Jordan curve with* cap $L = 1$. *For any $n \in \mathbb{N}$ and $n < A_n \leq e^n$ there exists a monic polynomial $p \in \mathbb{M}_n$ with zeros on L, satisfying condition (ii) of Theorem 1.5, such that*

$$D[\mu_L - \nu_p] \geq c \sqrt{\frac{\log A_n}{n}}, \quad c = \frac{1}{\sqrt{32 \pi^3}}. \qquad (1.49)$$

148   4. Discrepancy Theorems via One-Sided Bounds for Potentials

Proof. First we consider the subarc of L given by the formula

$$S = S_a := \{z \in L : a \leq \arg \Phi(z) \leq 2\pi\},$$

where $a \in (0, \pi/8)$ is a parameter, the choice of which will be made more precise later.

Next, we introduce the conformal mapping Ψ_S of Δ onto $\overline{\mathbb{C}} \setminus S$ normalized by $\Psi_S(\infty) = \infty$, $\Psi'_S(\infty) > 0$.

Let $\varepsilon > 0$ be arbitrary, $R = R(\varepsilon) > 1$ be sufficiently large and such that for ζ with $|\zeta| = R$,

$$|(\Phi \circ \Psi_S)(\zeta)| \leq (1+\varepsilon) R \, \Psi'_S(\infty) =: A.$$

We define two families of curves. Denote by Γ_1 the family of all locally rectifiable curves in the annulus

$$\{z : 1 < |z| < R\}$$

that separate its boundary components.

Let Γ_2 be the family of all locally rectifiable curves and arcs (with endpoints on $\{e^{it} : 0 < t < a\}$) in the annulus

$$\{w : 1 \leq |w| \leq A\}$$

that separate the circle $\{w : |w| = A\}$ from the arc $\{e^{i\theta} : a \leq \theta \leq 2\pi\}$ (see Figure 4.4).

By virtue of the comparison principle (1.2.11) and conformal invariance of the module (1.2.11) we have

$$\frac{1}{2\pi} \log R = m(\Gamma_1) \leq m(\Gamma_2). \tag{1.50}$$

Our next aim is to estimate from above the module $m(\Gamma_2)$. To do this, we consider the functions

$$\rho_1(w) := \begin{cases} \dfrac{1}{2\pi|w|}, & \text{if } 1 \leq |w| \leq A, \\ 0, & \text{otherwise}, \end{cases}$$

$$\rho_2(w) := \begin{cases} \dfrac{1}{\pi}, & \text{if } |w| \geq 1, \, |w - 1| \leq 2a, \\ 0, & \text{otherwise}, \end{cases}$$

$$\rho(w) := \max\{\rho_1(w), \rho_2(w)\}.$$

It is clear that for each $\gamma \in \Gamma_2$,

$$\int_\gamma \rho(w) \, |dw| \geq 1.$$

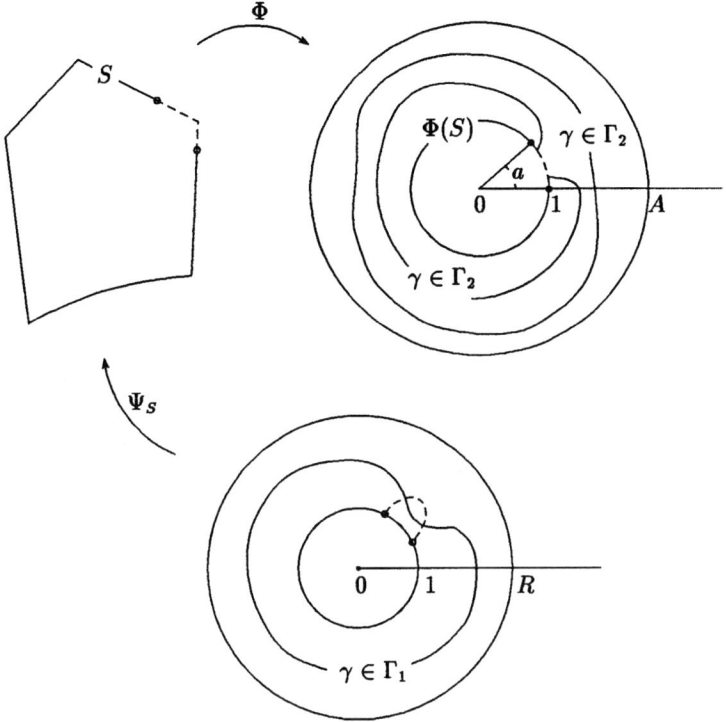

Fig. 4.4

Therefore, by the L-definition of the module,

$$m(\Gamma_2) \leq \int_{\mathbb{C}} \rho_1^2(w)\, dm(w) + \int_{\mathbb{C}} \rho_2^2(w)\, dm(w) \leq \frac{1}{2\pi} \log A + 4\,a^2.$$

Comparing the last estimate with (1.50) we get

$$0 \leq \frac{1}{2\pi} \log[(1+\varepsilon)\,\Psi_S'(\infty)] + 4\,a^2.$$

Since ε was arbitrary, we have

$$\exp\{c_1 a^2\}\, \mathrm{cap} S \geq 1,$$

where $c_1 := 8\pi$. Now consider the nth Fekete polynomial q_n for the arc S. According to Lemma 1.1.13 and Theorem 1.3.3, at its zeros $z_j \in S$, $j = 1, \ldots, n$,

$$|q_n'(z_j)| \geq (\mathrm{cap} S)^{n-1} \geq \exp\{-c_1 a^2 n\}.$$

If
$$a := \sqrt{\frac{\log A_n}{c_1 n}},$$
then the polynomial q_n (considered on L) satisfies condition (ii) of Theorem 1.5. At the same time,
$$D[\mu - \nu] \geq \frac{a}{2\pi} = c\sqrt{\frac{\log A_n}{n}},$$
where c is given by (1.49). □

4.2 Inner and Outer Bounds for Potentials

We establish an analogue of Theorem 1.1 for general domains with piecewise smooth boundary. As before, consider the conformal mapping Φ and its level curves L_δ, $\delta > 0$, and introduce the additional conformal mapping φ of int L onto the unit disk $\mathbb{D}$ with $\varphi(z_0) = 0$, $\varphi'(z_0) > 0$, where $z_0 \in$ int L is a fixed point. Let
$$L_\delta^+ := L_\delta$$
and
$$L_\delta^- := \{z : |\varphi(z)| = 1 - \delta\}, \quad 0 < \delta < 1.$$

Our next results will be formulated in terms of the *outer bounds*
$$a_{\sup}^+(\delta) := a_{\sup}(\delta) = \sup_{z \in L_\delta^+} U^\sigma(z),$$
$$a_{\inf}^+(\delta) := a_{\inf}(\delta) = -\inf_{z \in L_\delta^+} U^\sigma(z),$$

and the *inner bounds*
$$a_{\sup}^-(\delta) := \sup_{z \in L_\delta^-} U^\sigma(z) - U^\sigma(z_0), \tag{2.1}$$
$$a_{\inf}^-(\delta) := -\inf_{z \in L_\delta^-} U^\sigma(z) + U^\sigma(z_0). \tag{2.2}$$

Note that $a_{\sup}^-(\delta) \geq 0$ and $a_{\inf}^-(\delta) \geq 0$.

Theorem 2.1. *Let L be a piecewise Dini-smooth curve, and let $z_0 \in$ int L be fixed. Moreover, let $\sigma = \sigma^+ - \sigma^-$ be a signed measure with $\sigma^+, \sigma^- \in \mathcal{M}(L)$ and let σ^+ satisfy inequality (1.1). Then there exists a constant $c_1 = c_1(L, c, z_0) > 0$ such that for $\delta > 0$,*
$$D[\sigma] \leq c_1 \left(\frac{a_{\sup}^+(\delta) + a_{\sup}^-(\delta)}{\sqrt{\delta}} + \delta^{\beta/2} \right) \tag{2.3}$$

and
$$D[\sigma] \le c_1 \left(\frac{a_{\inf}^+(\delta) + a_{\inf}^-(\delta)}{\sqrt{\delta}} + \delta^{\beta/2} \right). \tag{2.4}$$

Proof. We restrict ourselves to prove (2.3), and it is sufficient to obtain the estimate
$$\sigma(J) \ge -c \left(\frac{a_{\sup}^+(\delta/2) + a_{\sup}^-(\delta/2)}{\sqrt{\delta}} + \delta^{\beta/2} \right) \tag{2.5}$$
for a subarc $J \subset L$ such that
$$J \subset l_1 \cup l_2, \ |\zeta_2 - z_j| \ge 2\varepsilon_1, \quad j = 1, 3,$$
$$\zeta_1 \in l_1, \ |\zeta_1 - z_2| = 2\varepsilon_1,$$
where ζ_1 and ζ_2 are the endpoints of J, and $\varepsilon_1 > 0$ is fixed and satisfies (1.9). Let L have an interior angle $\alpha\pi$ at the point z_2 with respect to the domain $G := \operatorname{int} L$, $0 < \alpha < 2$.

The proof, i.e., the construction of the auxiliary functions $f_\pm$, depends essentially on the value of α. More precisely, we have to distinguish two cases.

(a) $\alpha \le 1$: Consider fixed points $\xi_1 \in l_1$ and $\xi_2 \in l_2$ satisfying
$$|\xi_1 - z_1| = |\xi_2 - z_3| = \varepsilon_1.$$
Let $\varphi(\xi_j) =: \tau_j$. The points $(1-\delta)\tau_1$ and $(1-\delta)\tau_2$ divide the circle $\{\tau : |\tau| = 1-\delta\}$ into two parts. Denote by s' the one that does not include the point $(1-\delta)\,\varphi(z_2)$. Let $s := \psi(s')$, where $\psi := \varphi^{-1}$. As in the proof of Theorem 1.1 for the case of a curve we construct the function g_R, $R := L_\delta$.

Let f_+ be the harmonic extension of g_R to $\operatorname{ext} R$. By f_- we denote the harmonic extension of the function that is equal to g_R on R and equal to 0 on s to $(\operatorname{int} R)\backslash s$.

Thus, the construction of the function f_- is essentially different from that in the proof of Theorem 1.1 for curves. Nevertheless, the corresponding analogue of Lemma 1.4 can be proved, too.

Now consider the functions
$$\widetilde{f}(w) := \begin{cases} f_+[\Psi(w)], & \text{if} \quad |w| \ge 1+\delta, \\ f_-[\Psi(w)], & \text{if} \quad 1 < |w| < 1+\delta, \\ f_-[\psi(w)], & \text{if} \quad 0 \le |w| < 1, \end{cases}$$

$$\widetilde{\kappa}(w) := \begin{cases} \dfrac{16}{\delta^2} \int \widetilde{f}(\tau) K\left(\dfrac{4(\tau-w)}{\delta}\right) dm(\tau), & \text{if } 1+\tfrac{3}{4}\delta \le |w| \le 1+\tfrac{5}{4}\delta, \\ \dfrac{16}{\delta^2} \int \widetilde{f}(\tau) K\left(\dfrac{4(\tau-w)}{\delta}\right) dm(\tau), & \text{if } 1-\tfrac{5}{4}\delta \le |w| \le 1-\tfrac{3}{4}\delta, \\ \widetilde{f}(w), & \text{otherwise}, \end{cases}$$

152   4. Discrepancy Theorems via One-Sided Bounds for Potentials

and
$$\kappa(z) := \begin{cases} \widetilde{\kappa}(\Phi(z)), & \text{if } z \in \Omega, \\ \widetilde{\kappa}(\varphi(z)), & \text{if } z \in G, \\ f_-(z), & \text{if } z \in L. \end{cases}$$

To obtain (2.5) we have to repeat the reasoning of the proof of Theorem 1.1 with the new function κ. Following arguments analogous to those used in Section 2.2, the quantity $a^+_{\sup}(\delta/2) + a^-_{\sup}(\delta/2)$ takes over the role of $\varepsilon_{p,L} + \delta_{p,L}$ in the proof of Theorem 2.2.1, and we obtain the estimate (2.5).

(b) $\alpha > 1$: We repeat our reasoning from the case (a), but now we interchange the roles of the domains G and Ω, the functions φ and Φ, and so forth. □

Like Theorem 1.1, Theorem 2.1 has a direct application to the distribution of zeros of monic polynomials.

Theorem 2.2. *Let L be as in Theorem 2.1 and let $p = p_n$ be a monic polynomial in $\mathbb{M}_n$, $n \geq 2$, with all zeros on L, that satisfies one of the conditions (i),(ii) from Theorem 1.5. Then inequality (1.44) holds with*

$$\delta_n := \frac{1}{n} \log \frac{A_n^2 (\operatorname{cap} L)^n}{|p_n(z_0)|}$$

in the case (i), and

$$\delta_n := \frac{1}{n} \log \frac{A_n^4 |p_n(z_0)|}{(\operatorname{cap} L)^n}$$

in the case (ii).

The proof of this assertion follows the same lines if we take into account Lemma 1.4.5.

4.3 Inner Bounds for Potentials of Signed Measures on Analytic Jordan Curves

We begin our discussion with the case of the unit disk $\mathbb{D}$ and a monic polynomial

$$p_n(z) = a_0 + a_1 z + \cdots + z^n = \prod_{j=1}^{n} (z - z_j), \qquad a_0 \neq 0. \qquad (3.1)$$

The results of Theorem 2.2.1, concerning the zero counting measure ν_{p_n} can be used, and these lead to estimates in terms of

$$N_0 = N_0(p_n) = \frac{\|p_n\|_{C_1}}{\sqrt{|a_0|}},$$

where $C_r := \{z : |z| = r\}$ for $r \geq 0$. Hence the quantity N_0 describes the distribution of the zeros of p_n in the different sectors, which was already proved by Erdős and Turán [57].

It is natural to ask whether by restricting only

$$N_\rho = N_\rho(p_n) = \frac{\|p_n\|_{C_{1-\rho}}}{\sqrt{a_0}}$$

with fixed ρ ($0 < \rho < 1$) a similar equidistribution theorem can be deduced. It is easy to see that this is not the case. Indeed, let

$$\chi_n(z) := \sum_{j=0}^{n} \frac{z^j}{j!}, \quad p_n(z) := z^n \chi_n\left(\frac{\sqrt[n]{n!}}{z}\right).$$

For $|z| = 1/(2e)$ we have by Stirling's formula

$$|w| := \frac{\sqrt[n]{n!}}{|z|} > 2n,$$

i.e., the terms $|w|^n/n!$, $|w|^{n-1}/(n-1)!$, ..., $|w|$ decrease more rapidly than a geometric progression with quotient $\frac{1}{2}$. Therefore,

$$\left|\chi_n\left(\frac{\sqrt[n]{n!}}{z}\right)\right| < 2\frac{|w|^n}{n!} = \frac{2}{|z|^n},$$

and

$$N_\rho \leq 2, \quad \rho \leq \frac{1}{2e}.$$

On the other hand, as Szegő [171] showed, we have

$$\lim_{n\to\infty} \nu_{\chi_n}(\{z : \text{Re}\{z\} \leq 0\}) = \frac{1}{2} + \frac{1}{e\pi},$$

i.e., the zeros of p (which have the same angular distribution as the conjugate numbers of the zeros of χ_n) are not uniformly distributed in the different sectors.

So without imposing any further conditions on p we can not hope to get an equidistribution theorem similar to that of the previous sections. However, we shall show that the simple additional condition that all zeros of p are outside of $\mathbb{D}$ saves the situation.

There is a simple method to reduce the general case to the case where all the zeros of p lie on the unit circle. In fact, let $z = \rho e^{i\theta}$ be a fixed point, and let $\zeta = re^{i\tau}$ move along the line $\{\zeta : \arg \zeta = \tau\}$ (τ is fixed). Then

$$\frac{|z-\zeta|^2}{|\zeta|} = \frac{\rho^2}{r} + r - 2\rho\cos(\theta - \tau).$$

If r moves from $+\infty$ to 1, the expression on the right decreases monotonically. Hence,

$$\frac{|z-\zeta|^2}{|\zeta|} \geq |z - e^{i\tau}|^2. \tag{3.2}$$

Let now p be as in (3.1),

$$z_j = r_j e^{i\tau_j}, \quad r_j \geq 1 \quad \text{for } j = 1, \ldots, n.$$

Then (3.2) gives for $z \in \mathbb{D}$

$$\frac{|z - z_j|^2}{|z_j|} \geq |z - e^{i\tau_j}|^2, \quad j = 1, \ldots, n.$$

Multiplying these inequalities we obtain

$$\frac{|p_n(z)|^2}{|a_0|} \geq \left| \prod_{j=1}^n (z - e^{i\tau_j}) \right|^2,$$

i.e.,

$$N_\rho(\widetilde{p}_n) \leq N_\rho(p_n), \tag{3.3}$$

where $\widetilde{p}_n(z) := \prod_{j=1}^n (z - e^{i\tau_j})$.

Since the number of zeros of $\widetilde{p}_n$ in the different angles $\{z = re^{i\theta} : r \geq 0, \alpha \leq \theta \leq \beta\}, 0 \leq \alpha < \beta \leq 2\pi$, is identical with that of p_n the estimate (3.3) allows us to rewrite the inequalities describing the behavior of $\nu_{\widetilde{p}}$ in terms of $N_\rho(p)$.

We generalize the situation above in the following way. Let $L \subset \mathbb{C}$ be a bounded Jordan curve and let $\sigma = \mu_L - \nu$ be a signed measure on L with $\nu \in \mathcal{M}(L)$. As before, let $\Phi : \Omega := \text{ext } L \to \Delta$ and $\varphi : G := \text{int } L \to \mathbb{D}$ denote the associated Riemann mapping functions (with standard normalization at ∞ and at some fixed point $z_0 \in G$, respectively). Further, set

$$\Psi := \Phi^{-1}, \quad \psi := \varphi^{-1},$$

and

$$l_r := L_{1-r}^- = \{\zeta : |\varphi(\zeta)| = r\}, \quad 0 < r \leq 1.$$

Our next result will be formulated in terms of the *inner bound*

$$\delta_\nu(1-\rho) := \sup_{z \in l_\rho}(U^{\mu_L - \nu}(z)) - U^{\mu_L - \nu}(z_0) = -\inf_{z \in l_\rho}(U^\nu(z)) + U^\nu(z_0) \tag{3.4}$$

of the logarithmic potential $U^{\mu_L - \nu}$.

4.3 Signed Measures on Analytic Jordan Curves   155

If $\nu = p_n$ denotes the zero counting measure of $p_n \in \mathbb{M}_n$, then

$$\delta_\nu(1-\rho) = \frac{1}{n} \frac{\log \|p_n\|_{l_\rho}}{|p(z_0)|}. \qquad (3.5)$$

Theorem 3.1. *Let L be an analytic curve, that is there exists $r, 0 < r < 1$, such that Φ can be continued analytically and univalently to $\operatorname{ext} l_r$ and*

$$\Phi(z) \neq 0, \qquad z \in \operatorname{ext} l_r,$$

$$\frac{1}{M_r} \leq |\Phi'(z)| \leq M_r, \qquad z \in \operatorname{ext} l_r,$$

$$\frac{1}{m_r} \leq |\varphi'(z)| \leq m_r, \qquad z \in l_r,$$

with some $m_r, M_r \geq 1$. Then for $m \in \mathbb{N}$, $r < \rho \leq 1$, and $\nu \in \mathcal{M}(L)$,

$$D[\mu_L - \nu] \leq c \left(\frac{1}{m} + \delta_\nu(1-\rho) t(r)^{-2m} \left(\frac{m}{t(r)} m_r M_r r + \frac{\rho}{\rho - r} \right) \right),$$

where $c > 0$ is an absolute constant and

$$t(r) := \min_{|w|=r} |\Phi(\psi(w))|.$$

Proof. We begin with a special construction based on the notion of the *Jackson kernel* (see Appendix C).

Let $m \in \mathbb{N}$ be sufficiently large ($m \geq 100$) and a be a parameter subjected only to the restriction

$$\frac{2}{m} \leq a < 1.$$

We consider the even trigonometric polynomial

$$\pi_m(\theta, a) := \frac{1}{b} \int_{-a\pi}^{a\pi} \left(\frac{\sin \frac{m+1}{2}(t-\theta)}{\sin \frac{t-\theta}{2}} \right)^4 dt = a_0(a) + \sum_{j=1}^{2m} a_j(a) \cos j\theta,$$

where

$$b := \int_{-\pi}^{\pi} \left(\frac{\sin \frac{m+1}{2}(t-\theta)}{\sin \frac{t-\theta}{2}} \right)^4 dt = \int_{-\pi}^{\pi} \left(\frac{\sin \frac{m+1}{2} t}{\sin \frac{t}{2}} \right)^4 dt \asymp m^3.$$

A simple calculation gives the following relations:

$$a_0(a) = a, \qquad (3.6)$$

$$0 \leq \pi_m(\theta, a) \leq 1, \qquad \theta \in \mathbb{R}, \qquad (3.7)$$

$$\pi_m(\theta, a) \geq c_1, \qquad -a\pi \leq \theta \leq a\pi, \qquad (3.8)$$

$$\pi_m(\theta, a) \preceq \frac{1}{m^3(\theta - a\pi)^3}, \qquad a\pi < \theta \leq \pi, \qquad (3.9)$$

$$\pi_m(\theta, a) \preceq \frac{1}{m^3 |\theta + a\pi|^3}, \qquad -\pi \leq \theta < -a\pi. \qquad (3.10)$$

156   4. Discrepancy Theorems via One-Sided Bounds for Potentials

Next, we introduce the rational function

$$\Pi_m(w) = \Pi_m(w, a, \theta_0) := a_0(a) + \sum_{j=1}^{2m} \frac{a_j(a)}{w^j} e^{i\theta_0 j}, \qquad w \neq 0, w \in \mathbb{C},$$

where $\theta_0 \in \mathbb{R}$, so that

$$\pi_m(\theta - \theta_0, a) = \text{Re}\left\{\Pi_m(e^{i\theta})\right\}, \qquad \theta \in \mathbb{R}.$$

We need some information about the derivative of Π_m. For w with $|w| = 1 + 1/m$, the Schwarz formula (1.1.8) gives

$$|\Pi'_m(w)| \leq \frac{1}{\pi} \int_0^{2\pi} \frac{\pi_m(\theta - \theta_0, a)}{|e^{i\theta} - w|^2} d\theta \preceq m.$$

Therefore, by the maximum principle applied to $w^{2m+1}\Pi'_m(w)$ in $|w| < 1 + 1/m$, we have for w with $|w| = t$, $0 < t \leq 1$,

$$|\Pi'_m(w)| \preceq m \left(\frac{1+1/m}{t}\right)^{2m+1} \preceq m\, t^{-(2m+1)}. \tag{3.11}$$

Let

$$Q_m(w) := \text{Re}\{\Pi_m(w)\}, \qquad w \neq 0,$$

and

$$\widetilde{\Pi}_m(z) := \Pi_m(\Phi(z)), \quad \widetilde{Q}_m(z) := Q_m(\Phi(z)), \qquad z \in \text{ext } l_r.$$

Since for any $w = re^{i\tau} \in \mathbb{D}$ by (3.11)

$$|Q_m(w)| \leq Q_m(e^{i\tau}) + \int_r^1 |\Pi'_m(te^{i\tau})|\, dt$$

$$\leq 1 + c_2\, m \int_r^1 \frac{dt}{t^{2m+1}} \leq 1 + \frac{c_2}{2} r^{-2m},$$

we have

$$|Q_m(w)| \preceq |w|^{-2m}, \qquad 0 < |w| < 1. \tag{3.12}$$

Applying the Green formula (1.1.10) and (1.1.13) to the function $\widetilde{Q}_m(z)$ we may write for $z \in L$,

$$\widetilde{Q}_m(z) = \widetilde{Q}_m(\infty) + \frac{1}{2\pi} \int_{l_r} \left(\frac{\partial \widetilde{Q}_m(\zeta)}{\partial \mathbf{n}_\zeta} \log|\zeta - z| - \widetilde{Q}_m(\zeta) \frac{\partial}{\partial \mathbf{n}_\zeta} \log|\zeta - z|\right) |d\zeta|,$$

4.3 Signed Measures on Analytic Jordan Curves

where $\partial/\partial \mathbf{n}_\zeta$ is the operator of differentiation with respect to the outward normal to the curve l_r at the point ζ.

Integrating the last relation, we get

$$B = B(\nu, a, \theta_0, m, L) := \int_L \widetilde{Q}_m(z)\, d\nu(z) - a$$

$$= -\frac{1}{2\pi} \int_{l_r} \left(U^\nu(\zeta) \frac{\partial \widetilde{Q}_m(\zeta)}{\partial \mathbf{n}_\zeta} - \widetilde{Q}_m(\zeta) \frac{\partial}{\partial \mathbf{n}_\zeta} U^\nu(\zeta) \right) |d\zeta|. \tag{3.13}$$

Our next task is to estimate the quantity B in two different ways. Let

$$\alpha_\nu(r) := -\inf_{z \in l_r} U^\nu(z). \tag{3.14}$$

Since

$$\int_{l_r} \frac{\partial \widetilde{Q}_m(\zeta)}{\partial \mathbf{n}_\zeta} |d\zeta| = 0,$$

we obtain that

$$\left| \int_{l_r} U^\nu(\zeta) \frac{\partial \widetilde{Q}_m(\zeta)}{\partial \mathbf{n}_\zeta} |d\zeta| \right| \leq \int_{l_r} (\alpha_\nu(r) + U^\nu(\zeta)) \left| \frac{\partial \widetilde{Q}_m(\zeta)}{\partial \mathbf{n}_\zeta} \right| |d\zeta|$$

$$\preceq m\, t(r)^{-(2m+1)}\, m_r\, M_r \int_{|w|=r} (\alpha_\nu(r) + U^\nu(\psi(w)))\, |dw|$$

$$\preceq 2\pi r\, m\, t(r)^{-(2m+1)} m_r\, M_r\, \delta_\nu(1-r). \tag{3.15}$$

Further, by (3.12),

$$\left| \int_{l_r} \widetilde{Q}_m(\zeta) \frac{\partial}{\partial \mathbf{n}_\zeta} U^\nu(\zeta) |d\zeta| \right| \preceq t(r)^{-2m} \int_{l_r} |\operatorname{grad} U^\nu(\zeta)|\, |d\zeta|$$

$$= t(r)^{-2m} \int_{|w|=r} |\operatorname{grad} U^\nu(\psi(w))|\, |dw|. \tag{3.16}$$

As a simple consequence of the Schwarz formula (cf. (1.1.8)) we have for $r < \rho \leq 1$,

$$\int_{|w|=r} |\operatorname{grad} U^\nu(\psi(w))|\, |dw| = \int_{|w|=r} |\operatorname{grad} (\alpha_\nu(\rho) + U^\nu(\psi(w)))|\, |dw|$$

$$\leq \frac{1}{\pi} \int\limits_{|w|=r} \int\limits_{|t|=\rho} (\alpha_\nu(\rho) + U^\nu(\psi(t))) \frac{|dt|\,|dw|}{|t-w|^2}$$

$$= \frac{1}{\pi} \int\limits_{|t|=\rho} (\alpha_\nu(\rho) + U^\nu(\psi(t))) \,|dt| \int\limits_{|w|=r} \frac{|dw|}{|t-w|^2}$$

$$\preceq \frac{\rho}{\rho-r} \delta_\nu(1-\rho). \tag{3.17}$$

Hence, by virtue of (3.13)–(3.17),

$$|B| \preceq V, \tag{3.18}$$

where

$$V = V(\nu, m, \rho, r, L) := \delta_\nu(1-\rho) t(r)^{-2m} \left(\frac{m}{t(r)} m_r M_r r + \frac{\rho}{\rho-r} \right).$$

Next, we want to show that

$$(\nu - \mu)(J) \geq -c \left(\frac{1}{m} + V \right) \tag{3.19}$$

for all subarcs $J \subset L$. Let $J \subset L$ be an arbitrary arc and let $e^{i\theta_0}$ be the midpoint of $J' := \Phi(J)$. Without loss of generality we assume that

$$J' = \Phi(J) = \left\{ e^{i\theta} : -a\pi \leq \theta \leq a\pi \right\},$$

i.e., $\theta_0 = 0$.

We begin with the case $\mu(J) \leq 2/m$. Then (3.19) is trivial. Taking $a = 2/m$ in (3.13) we have by (3.7) and (3.8)

$$c_1 \nu(J) \leq \int_L \widetilde{Q}_m(z)\, d\nu(z) = B + \frac{2}{m}$$

and consequently by (3.18)

$$\nu(J) \preceq \frac{1}{m} + V. \tag{3.20}$$

Now let $2/m < \mu_L(J) =: a < 1$. From (3.18) we obtain

$$\int_L \widetilde{Q}_m(z)\, d\nu(z) \geq a - c_3 V. \tag{3.21}$$

To estimate the left-hand term in (3.21) we construct the following division of the arc $\{|w| = 1\} \setminus J'$.

We distinguish two cases.
If $1 - a \leq 1/m$, then
$$\Gamma_1 := \{e^{i\theta} : a\pi \leq \theta \leq \pi\}.$$

If $1 - a > 1/m$, then
$$\Gamma_{k+1} := \left\{e^{i\theta} : \left(a + \frac{k}{m}\right)\pi \leq \theta \leq \left(a + \frac{k+1}{m}\right)\pi\right\}$$
$$(k = 0, \ldots, k_0 := [m(1-a)] - 1),$$
$$\Gamma_{k_0+2} := \left\{e^{i\theta} : \left(a + \frac{k_0+1}{m}\right)\pi \leq \theta \leq \pi\right\}.$$

In both cases let
$$\Gamma_{-k} := \{e^{i\theta} : e^{-i\theta} \in \Gamma_k\},$$
$$\Gamma'_{\pm k} := \Psi(\Gamma_{\pm k}).$$

Since by (3.20)
$$\nu(\Gamma'_{\pm k}) \preceq \frac{1}{m} + V,$$
we have owing to (3.9) and (3.10)
$$\int_L \widetilde{Q}_m(z)\, d\nu(z) \leq \nu(J) + \nu(\Gamma'_1) + \nu(\Gamma'_{-1})$$
$$+ c_4 \sum_{k=1}^{k_0+1} \frac{1}{k^3} \left(\nu(\Gamma'_{k+1}) + \nu(\Gamma'_{-(k+1)})\right)$$
$$\leq \nu(J) + c_5 \left(\frac{1}{m} + V\right). \tag{3.22}$$

Comparing (3.20)–(3.22), we have for any subarc $J \subset L$,
$$(\nu - \mu_L)(J) \geq -c_6 \left(\frac{1}{m} + V\right). \tag{3.23}$$

To get the upper estimate of $(\nu - \mu_L)(J)$ we need, due to the relation
$$(\nu - \mu_L)(J) = -(\nu - \mu_L)(L \setminus J),$$
merely to apply (3.23) to the arc $L \setminus J$. □

Next we formulate some corollaries.

Corollary 3.2. *Let $0 < r < \rho < 1$ in Theorem 3.1 be fixed and let moreover $\delta_\nu(1-\rho) \leq \frac{1}{2}$. Then the inequality*
$$D[\mu_L - \nu] \leq c_1 \left(\log \frac{1}{\delta_\nu(1-\rho)}\right)^{-1} \tag{3.24}$$
holds with some constant $c_1 = c_1(L, r, \rho, z_0) > 0$.

Indeed, we can assume that

$$m := \left[\frac{1}{4} \frac{\log \frac{1}{\delta_\nu(1-\rho)}}{\log \frac{1}{t(r)}}\right] > 1, \qquad (3.25)$$

where $[a]$ denotes the integral part of $a > 0$. Then choosing in Theorem 3.1 m as in (3.25) yields (3.24).

It is natural that for the unit circle $L = \mathbb{T}$ and $z_0 = 0$ we can say more about the quantity c_1 in (3.24). Indeed, inserting $r = \rho/2$ in Theorem 3.1 and taking into account that $t(r) = r$, $m_r = M_r = 1$, we have in this case with the choice of m as in (3.25)

$$D[\mu_L - \nu] \le c_2 \log \frac{2}{\rho} \left(\log \frac{1}{\delta_\nu(1-\rho)}\right)^{-1}, \qquad (3.26)$$

where $c_2 > 0$ is an absolute constant.

Estimate (3.24) reveals the surprising fact that the condition concerning the estimates for the potential $U^\nu(z)$ on a level line l_ρ, where $0 < \rho < 1$ is fixed, which is a much weaker condition than that on L, can ensure the equidistribution of the zeros.

Remark 3.3. Though the error term in (3.24) is worse than that of Theorems 1.1 and 2.1, its order is best possible for every $0 < \rho < 1$.

Indeed, let $2 \le h \le n$ and $0 < \rho < 1$ be given (n is sufficiently large). Write

$$\rho =: e^{-c}, \quad c > 0,$$

and put

$$l := \left[\frac{1}{c} \log h\right], \quad k := \left[\frac{cn}{\log h}\right].$$

The polynomial

$$\widehat{p}_n(z) := (1 - z^l)^k$$

has a zero of multiplicity bigger than

$$\frac{cn}{\log h} - 1 > \frac{c}{2} \frac{n}{\log h}.$$

Therefore, for $L = \mathbb{T}$ and the zero counting measure $\nu := \nu_{\widehat{p}_n}$ it holds that

$$D[\mu_L - \nu] \succeq \frac{1}{\log h}. \qquad (3.27)$$

At the same time, on the circle $\{z : |z| = \rho\}$ we have

$$|\widehat{p}_n(z)| \le \left(1 + \exp\left\{-c\left(\frac{1}{c}\log h - 1\right)\right\}\right)^{cn/\log h} < \exp\left\{\frac{cn}{\log h} \frac{e^c}{h}\right\}.$$

Thus,
$$\delta_\nu(1-\rho) = \alpha_\nu(\rho) \preceq \frac{1}{h \log h}. \tag{3.28}$$

Relations (3.27) and (3.28) show the sharpness of (3.24).

Corollary 3.4. *Let L be analytic. Then*
$$D[\mu_L - \nu] \le c_3 \sqrt{\delta_\nu}, \tag{3.29}$$
where $c_3 = c_3(L, z_0) > 0$ and $\delta_\nu := \delta_\nu(0)$.

Indeed, from the distortion properties of the conformal mappings ψ and Φ (see Lemma 1.2.10) we conclude that
$$t\left(1 - \frac{1}{m}\right) \ge 1 - \frac{c_4}{m}, \quad c_4 = c_4(L, z_0) > 0.$$
Consequently, choosing in Theorem 3.1
$$m := \left[\frac{1}{\sqrt{\delta_\nu}}\right], \quad \rho = 1, \quad r = 1 - \frac{1}{m},$$
we obtain (3.29), which is of the same kind as (2.3).

4.4 Another Approach for Dini-Smooth Arcs

In this section we prove an analogue of Theorem 3.1 for the case of a Jordan arc L. Since L does not have interior points, it is natural to state the results in terms of bounds of the potential $U^{\mu_L - \nu}$ along level lines of the conformal mapping Φ (instead of φ as in Theorem 3.1), where $\nu \in \mathcal{M}(L)$. Thus, our basic result will be formulated in terms of the outer bounds
$$a_{\sup}(\delta) := \sup_{z \in L_\delta} U^{\mu_L - \nu}(z), \tag{4.1}$$
$$a_{\inf}(\delta) := - \inf_{z \in L_\delta} U^{\mu_L - \nu}(z), \tag{4.2}$$
and
$$a(\delta) := \min(a_{\sup}(\delta), a_{\inf}(\delta)). \tag{4.3}$$

Theorem 4.1. *Let L be a Dini-smooth arc, $\nu \in \mathcal{M}(L)$, and μ_L the equilibrium measure for L. Then for any $m \in \mathbb{N}$ and $\delta \ge 0$ the inequality*
$$D[\mu_L - \nu] \le c \left(\frac{1}{m} + m(1+\delta)^m a(\delta)\right) \tag{4.4}$$
holds with a constant $c > 0$ depending only on L, where $a(\delta)$ is defined by (4.3).

Before proving Theorem 4.1 we recall some properties of the conformal mappings Φ and $\Psi := \Phi^{-1}$ in the case of a Dini-smooth arc L.

We denote by z_1 and z_2 the endpoints of L. Since the function Φ can be extended continuously to these points, for $j = 1, 2$, we set

$$e^{i\theta_j} := \Phi(z_j), \quad \theta_1 < \theta_2 < \theta_1 + 2\pi,$$
$$\Delta_1 := \{t = re^{i\theta} : r > 1, \theta_1 < \theta < \theta_2\},$$
$$\Delta_2 := \Delta \setminus \overline{\Delta_1}, \quad \Omega_j := \Psi(\Delta_j).$$

According to Lemma 1.2.8 and (B.2.1) the restrictions Φ_j, $j = 1, 2$, of the function Φ to the region Ω_j can be extended to K_1-quasiconformal mappings of $\overline{\mathbb{C}}$ onto itself for some $K_1 = K_1(L) \geq 1$. We preserve the same notation Φ_j for these extensions.

At first we formulate some corollaries that follow directly from Lemma 1.2.12.

Let L be a Dini-smooth arc and let $L(z, \zeta)$ denote the subarc of L between the points z and $\zeta \in L$.

Corollary 4.2. *For any points $z, \zeta \in L$ and their images*

$$w_j := \Phi_j(z), \quad t_j := \Phi_j(\zeta), \qquad j = 1, 2,$$

we have

$$|w_1 - t_1| \asymp |w_2 - t_2| \asymp \mu_L(L(z, \zeta)).$$

Further, set

$$\rho_u(z) := \operatorname{dist}(z, L_u), \quad u > 0.$$

Corollary 4.3. *Let $0 < u \leq a < \theta_2 - \theta_1$ and $\theta_1 \leq \theta \leq \theta_2$. Consider the arcs*

$$\widetilde{\gamma} := [\theta - a, \theta + a] \cap [\theta_1, \theta_2],$$
$$\gamma := \{\zeta \in L : \zeta = \Psi(e^{i\tau}) : \tau \in \widetilde{\gamma}\}.$$

Then

$$\frac{\rho_u(z)}{\operatorname{diam} \gamma} \asymp \frac{u}{a}.$$

In the proof of Theorem 4.1 we are going to approximate certain functions given on an arc by complex polynomials. In connection with this problem we are going to use the Dzjadyk polynomial kernel $K^{(j)}_{r,m,k,n}(\zeta, z)$ introduced in Appendix C.

For $\zeta \in L$, $j = 1, 2$, $m \in \mathbb{N}$, let

$$\widetilde{\zeta}^{(j)} = \widetilde{\zeta}^{(j)}(m) := \Psi\left(\Phi_j(\zeta)\left(1 + \frac{1}{m}\right)\right).$$

4.4 Another Approach for Dini–Smooth Arcs

Using Theorem 1.2.7 we obtain for $z, \zeta \in L$, $j = 1, 2$,

$$\left| \frac{\widetilde{\zeta}^{(j)} - \zeta}{\widetilde{\zeta}^{(j)} - z} \right| \preceq 1,$$

$$\frac{1}{|z - \widetilde{\zeta}^{(j)}|} \preceq \frac{1}{|z - \widetilde{z}^{(j)}| + |z - \zeta|} \asymp \frac{1}{\rho_{1/m}(z) + |z - \zeta|},$$

and

$$\left| \frac{\widetilde{\zeta}^{(j)} - \zeta}{\widetilde{\zeta}^{(j)} - z} \right|^{K_1^2} \preceq \left| \frac{\Phi_j(\widetilde{\zeta}^{(j)}) - \Phi_j(\zeta)}{\Phi_j(\widetilde{\zeta}^{(j)}) - \Phi_j(z)} \right|^{K_1} = \left| \frac{\Phi_j(\widetilde{z}^{(j)}) - \Phi_j(z)}{\Phi_j(\widetilde{z}^{(j)}) - \Phi_j(\zeta)} \right|^{K_1}$$

$$\preceq \left| \frac{\widetilde{z}^{(j)} - z}{\widetilde{z}^{(j)} - \zeta} \right| \preceq \frac{\rho_{1/m}(z)}{|z - \zeta|}.$$

Hence, we have by Theorem C.2.4 for some value of k (which will be considered fixed from now on) and for $z, \zeta \in L$, $j = 1, 2$, $m \in \mathbb{N}$,

$$\left| \frac{1}{\zeta - z} - K^{(j)}_{0,2,k,m}(\zeta, z) \right| \leq c_1 \frac{(\rho_{1/m}(z))^2}{|\zeta - z|^3}, \quad (4.5)$$

$$\left| K^{(j)}_{0,2,k,m}(\zeta, z) \right| \leq \frac{c_2}{\rho_{1/m}(z) + |\zeta - z|}, \quad (4.6)$$

where $c_l = c_l(L, k) > 0$, $l = 1, 2$. Moreover, with a suitable orientation of L,

$$\sum_{j=1}^{2} \frac{(-1)^{j+1}}{2\pi i} \int_L K^{(j)}_{0,2,k,m}(\zeta, z) \, d\zeta = 1 - \frac{a}{m^{2k}} =: \gamma_{m,k}, \quad 0 < a \leq c_3(k) \quad (4.7)$$

(cf. Lemma C.2.7).

Note that the function $K^{(j)}_{0,2,k,m}(\zeta, z)$ is a polynomial with respect to z of degree at most

$$2(k+2)(m-1)k - 1.$$

Consider an arbitrary subarc J of L and denote by ζ_1 and ζ_2 its endpoints. Our aim is to construct a polynomial approximation of the characteristic function

$$f(z) = f(z, J, L) := \begin{cases} 1, & \text{if } z \in J, \\ 0, & \text{if } z \in L \setminus J. \end{cases}$$

Without loss of generality we suppose that

$$\theta_1 < \tau_1^1 < \tau_2^1 \leq \theta_2 \leq \tau_1^2 < \tau_1^2 < \theta_1 + 2\pi,$$

where

$$\Psi(e^{i\tau_j^k}) = \zeta_j, \quad j, k = 1, 2.$$

164   4. Discrepancy Theorems via One-Sided Bounds for Potentials

Lemma 4.4. *For any $m \in \mathbb{N}$ there exists a complex polynomial $P_m(z) = P_m(z, J, L)$ of degree at most m such that*

$$0 \leq \operatorname{Re}\{P_m(z)\} \leq 1, \qquad z \in L, \tag{4.8}$$

$$|P_m(z)| \leq 2, \qquad z \in L, \tag{4.9}$$

$$\operatorname{Re}\{P_m(z)\} \geq \frac{1}{4}, \qquad z \in J, \tag{4.10}$$

$$\left| \int P_m \, d\mu - \mu(J) \right| \leq \frac{c_1}{m}, \tag{4.11}$$

$$|P_m(z) - f(z)| \leq \frac{c_2}{(m\,\mu(L(z,\widehat{z})))^2}, \qquad z \in L, \tag{4.12}$$

where $\mu := \mu_L$, $\widehat{z}$ is whichever of the points ζ_1 and ζ_2 nearest to z and where the positive constants c_j, $j = 1, 2$, are independent of m and J.

Proof. We are going to use a standard procedure for the construction of polynomials approximating continuous functions on an arc (see [50, 17]). Since the function $f(z)$ is not continuous and since in addition to the purely approximating properties (4.11) and (4.12) the desired polynomial $P_m(z)$ must satisfy conditions (4.8)–(4.10), we will perform the appropriate correction of this scheme.

First we introduce an auxiliary functions as follows. For $\theta_1 \leq \theta \leq \theta_2$ set

$$\widetilde{g}(\theta) := \begin{cases} \dfrac{1}{2}\left(\dfrac{b}{\tau_1^1 - \theta + b}\right)^2, & \text{if } \theta_1 \leq \theta \leq \tau_1^1, \\[2mm] 1 - \dfrac{1}{2}\left(\dfrac{b}{\theta - \tau_1^1 + b}\right)^2, & \text{if } \tau_1^1 \leq \theta \leq \dfrac{\tau_1^1 + \tau_2^1}{2}, \\[2mm] 1 - \dfrac{1}{2}\left(\dfrac{b}{\tau_2^1 - \theta + b}\right)^2, & \text{if } \dfrac{\tau_1^1 + \tau_2^1}{2} \leq \theta \leq \tau_2^1, \\[2mm] \dfrac{1}{2}\left(\dfrac{b}{\theta - \tau_2^1 + b}\right)^2, & \text{if } \tau_2^1 \leq \theta \leq \theta_2, \end{cases}$$

where $b := c/m$ and the sufficiently large constant $c > 1$, which plays the role of a parameter, will be chosen later.

Further, set $g(z) := \widetilde{g}(\theta)$, where $\theta_1 \leq \theta \leq \theta_2$ is such that $\Phi_1(z) = e^{i\theta}$. The function $g(z)$ is close to $f(z)$ in the integral sense, i.e., by Corollary 4.2,

$$\int |f - g|\,d\mu \leq 8 \int_0^\infty \left(\dfrac{b}{\varepsilon_1\theta + b}\right)^2 d\theta \preceq b. \tag{4.13}$$

We denote by $p_m(z)$ the polynomial of degree at most

$$m' := 2\,(k+2)\,(m-1)\,k - 1 \asymp m$$

4.4 Another Approach for Dini–Smooth Arcs

(k is fixed) defined by

$$p_m(z) := \frac{1}{2\pi i \gamma_{m'}} \int_L g(z) \left(K_{m'}^{(1)}(\zeta, z) - K_{m'}^{(2)}(\zeta, z) \right) d\zeta, \qquad z \in L, \quad (4.14)$$

where

$$K_{m'}^{(j)}(\zeta, z) := K_{0,2,k,m}^{(j)}(\zeta, z) \quad j = 1, 2,$$

is the polynomial kernel of degree m' satisfying (4.5)–(4.7) and $\gamma_{m'} := \gamma_{m,k}$ is the number given by formula (4.7). In order to investigate the behavior of the quantity $|g(z) - p_m(z)|$ we note that

$$\frac{1}{2\pi i \gamma_{m'}} \int_L \left(K_{m'}^{(1)}(\zeta, z) - K_{m'}^{(2)}(\zeta, z) \right) d\zeta = 1 \qquad (4.15)$$

and introduce for $z \in L$ an arbitrary point $z^* \in \Omega$ satisfying

$$|z - z^*| \leq m^{-4}.$$

The following relations follow easily from (4.14) and (4.15):

$$g(z) - p_m(z) = \frac{1}{2\pi i \gamma_{m'}} \int_L (g(z) - g(\zeta)) \left(K_{m'}^{(1)}(\zeta, z) - K_{m'}^{(2)}(\zeta, z) \right) d\zeta$$

$$= \frac{1}{2\pi i \gamma_{m'}} \sum_{j=1}^{2} (-1)^j \int_L (g(z) - g(\zeta)) \left(\frac{1}{\zeta - z^*} - K_{m'}^{(j)}(\zeta, z) \right) d\zeta$$

$$=: \frac{1}{2\pi i \gamma_{m'}} \sum_{j=1}^{2} \int_L A_j(\zeta, z) \, d\zeta.$$

If we show that the constant c can be chosen such that

$$|g(z) - p_m(z)| \leq \frac{1}{2} |f(z) - g(z)|, \qquad z \in L, \qquad (4.16)$$

then by virtue of (4.13) the polynomial $P_m(z) := p_{[\varepsilon m]}(z)$, $\varepsilon := (2(k+2)k)^{-1}$, will satisfy conditions (4.8)–(4.12).

In order to prove (4.16) we consider an arbitrary point $z \in L$, say $\Phi_1(z) = e^{i\theta}$ with $\theta_1 \leq \theta \leq \theta_2$. We will study in detail only the case

$$\theta_1 \leq \theta \leq \tau_1^1$$

(according to the symmetric character of the function $\widetilde{g}(\theta)$, the other cases can be handled in quite the same way).

Set $a := \sqrt{c}/m$ and consider the arcs

$$\widetilde{\gamma} := [\theta - a, \theta + a] \cap [\theta_1, \theta_2],$$

$$\gamma := \{\zeta \in L : \zeta = \Psi(e^{i\tau}), \tau \in \widetilde{\gamma}\}.$$

4. Discrepancy Theorems via One-Sided Bounds for Potentials

With this notation we have

$$g(z) - p_m(z) = \frac{1}{2\pi i \gamma_{m'}} \sum_{j=1}^{2} \left(\int_{\gamma} A_j(\zeta, z) \, d\zeta + \int_{L \setminus \gamma} A_j(\zeta, z) \, d\zeta \right). \quad (4.17)$$

Since by (4.6) and Corollary 4.3 for $j = 1, 2$,

$$\int_{\gamma} \left| K_{m'}^{(j)}(\zeta, z) \right| |d\zeta| \preceq \int_{\gamma} \frac{|d\zeta|}{\rho_{1/m}(z) + |z - \zeta|} \preceq \log \frac{\operatorname{diam} \gamma}{\rho_{1/m}(z)}$$

$$\asymp \log \frac{\sqrt{c}/m}{1/m} = \frac{1}{2} \log c, \quad (4.18)$$

we have

$$\left| \sum_{j=1}^{2} \int_{\gamma} A_j(\zeta, z) \, d\zeta \right| \leq \sum_{j=1}^{2} \sup_{\zeta \in \gamma} |g(z) - g(\zeta)| \int_{\gamma} \left| K_{m'}^{(j)}(\zeta, z) \right| |d\zeta|$$

$$\preceq (\log c) \sup_{\zeta \in \gamma} |g(z) - g(\zeta)|. \quad (4.19)$$

In order to estimate the second integral on the right-hand side of (4.17) we note that by Lemma 1.2.12 for $\zeta \in L \setminus \gamma$,

$$|\zeta - z| \asymp |\zeta - z^*|$$

and

$$\rho_{1/m}(z) \succeq m^{-2}$$

(cf. Lemma 1.2.5). Thus, for $\zeta \in L \setminus \gamma$ we may use the obvious inequality

$$\left| \frac{1}{\zeta - z^*} - \frac{1}{\zeta - z} \right| = \frac{|z^* - z|}{|\zeta - z||\zeta - z^*|} \preceq \frac{\rho_{1/m}^2(z)}{|\zeta - z|^3}$$

and the property (4.5) of Dzjadyk's kernel $K_{m'}^{(j)}(\zeta, z)$ to write

$$\left| \int_{L \setminus \gamma} A_j(\zeta, z) \, d\zeta \right| \preceq (\rho_{1/m}(z))^2 \int_{L \setminus \gamma} |g(z) - g(\zeta)| \frac{|d\zeta|}{|z - \zeta|^3}. \quad (4.20)$$

Next we consider two particular possibilities for the location of the point z.

(i) Let $0 \le \tau_1^1 - \theta \le b$. Then we have by Corollary 4.3,

$$|g(\zeta) - g(z)| \preceq \frac{a}{b} = \frac{1}{\sqrt{c}}, \qquad \zeta \in \gamma, \tag{4.21}$$

$$\left| \int_{L \setminus \gamma} A_j(\zeta, z) \, d\zeta \right| \preceq (\rho_{1/m}(z))^2 \int_{L \setminus \gamma} \frac{|d\zeta|}{|z - \zeta|^3}$$

$$\preceq \left(\frac{\rho_{1/m}(z)}{\operatorname{diam} \gamma} \right)^2 \preceq \left(\frac{1/m}{\sqrt{c}/m} \right)^2 = \frac{1}{c}, \tag{4.22}$$

$$|f(z) - g(z)| = g(z) \ge \frac{1}{8}. \tag{4.23}$$

Therefore, comparing (4.17), (4.19), (4.21)–(4.23) and choosing c sufficiently large we obtain inequality (4.16).

(ii) Let $\tau_1^1 - \theta \ge b$. In this case

$$|f(z) - g(z)| = g(z) \asymp \left(\frac{b}{\tau_1^1 - \theta} \right)^2. \tag{4.24}$$

Further, if $\zeta \in \gamma$, then

$$|g(\zeta) - g(z)| \preceq \frac{a \, b^2}{(\tau_1^1 - \theta)^3}.$$

Therefore, by (4.19) and (4.24),

$$\left| \sum_{j=1}^{2} \int_{\gamma} A_j(\zeta, z) \, d\zeta \right| \preceq \frac{a \, b^2}{(\tau_1^1 - \theta)^3} \log c \le \frac{\log c}{\sqrt{c}} |f(z) - g(z)|. \tag{4.25}$$

The estimation of the second integral on the right-hand side of (4.17) is based on the following reasoning. Let $\zeta \in L \setminus \gamma$, $\zeta = \Psi(e^{i\tau})$, $\theta_1 \le \tau \le \theta_2$. Consider the sets

$$\tilde{\gamma}_1 := \left(\left[\tau - \frac{\tau_1^1 - \tau}{2}, \tau + \frac{\tau_1^1 - \tau}{2} \right] \cap [\theta_1, \tau_1^1] \right) \setminus \tilde{\gamma},$$
$$\gamma_1 := \{ \zeta \in L : \zeta = \Psi(e^{i\tau}), \tau \in \tilde{\gamma}_1 \}.$$

Since for $\zeta \in \gamma_1$,

$$|g(\zeta) - g(z)| \le g(\zeta) + g(z) \asymp g(z),$$

we deduce from (4.20), (4.24), and Corollary 4.3 that

$$\left| \int_{L \setminus \gamma} A_j(\zeta, z) \, d\zeta \right|$$

$$\preceq \left(\rho_{1/m}(z) \right)^2 g(z) \int_{\gamma_1} \frac{|d\zeta|}{|\zeta - z|^3} + \left(\rho_{1/m}(z) \right)^2 \int_{L \setminus \gamma_1} \frac{|d\zeta|}{|\zeta - z|^3}$$

$$\preceq g(z) \left(\frac{\rho_{1/m}(z)}{\operatorname{diam} \gamma} \right)^2 + \left(\frac{\rho_{1/m}(z)}{|z - \zeta_1|} \right)^2$$

$$\preceq g(z) \left(\frac{1/m}{\sqrt{c/m}} \right)^2 + \left(\frac{1/m}{\tau_1^1 - \theta} \right)^2$$

$$\preceq \frac{1}{c} |f(z) - g(z)|. \tag{4.26}$$

Comparing (4.17), (4.25) and (4.26) we again obtain (4.16) by choosing c sufficiently large. □

Proof of Theorem 4.1. In the following we again write $\mu := \mu_L$. Let $J \subset L$ be an arbitrary subarc, $r := 1 + \delta \geq 1$, $m \in \mathbb{N}$. We consider the polynomial $P_m(z) = P_m(z, J, L)$ from Lemma 4.4 and set

$$Q_m(z) := \operatorname{Re}\{P_m(z)\}.$$

By (4.9) and the Bernstein–Walsh lemma (see Lemma 1.1.11)

$$|Q_m(z)| \leq |P_m(z)| \leq 2 |\Phi(z)|^m, \quad z \in \Omega, \tag{4.27}$$

which implies that for $R > r$ and $z \in L_{R-1}$,

$$|\operatorname{grad} Q_m(z)| = |P_m'(z)| \leq \frac{1}{2\pi} \int_{|\zeta - z| = d(z, L_{R(1+1/m)-1})} \frac{|P_m(\zeta)|}{|\zeta - z|^2} |d\zeta|$$

$$\leq \frac{e \, R^m}{d(z, L_{R(1+1/m)-1})}. \tag{4.28}$$

Applying the Green formula (1.1.10) to the harmonic polynomial $Q_m(z)$ we have for $z \in L$,

$$Q_m(z) = \frac{1}{2\pi} \int_{L_{R-1}} \left(\frac{\partial Q_m(\zeta)}{\partial \mathbf{n}_\zeta} \log \frac{1}{|\zeta - z|} - Q_m(\zeta) \frac{\partial}{\partial \mathbf{n}_\zeta} \log \frac{1}{|\zeta - z|} \right) |d\zeta|,$$

where $\partial/\partial \mathbf{n}_\zeta$ denotes the operator of differentiation with respect to the outward normal to the curve L_{R-1} at the point $\zeta \in L_{R-1}$.

4.4 Another Approach for Dini–Smooth Arcs

Integrating the last relation, we get

$$\int Q_m \, d\nu - \mu(J) + \int (f - Q_m) \, d\mu = \int Q_m (d\nu - d\mu)$$
$$= \frac{1}{2\pi} \int_{L_{R-1}} \left(U^{\nu-\mu}(\zeta) \frac{\partial Q_m(\zeta)}{\partial \mathbf{n}_\zeta} - Q_m(\zeta) \frac{\partial}{\partial \mathbf{n}_\zeta} U^{\nu-\mu}(\zeta) \right) |d\zeta|. \quad (4.29)$$

Note that by virtue of (4.11),

$$\left| \int (f - Q_m) \, d\mu \right| \preceq \frac{1}{m}. \quad (4.30)$$

Our next task is to estimate the integrals on the right-hand side of (4.29). We may assume without loss of generality that

$$a := a(\delta) = \sup_{\zeta \in L_\delta} U^{\mu-\nu}(\zeta).$$

Moreover, we set $R := r(1 + 1/m)$. Since

$$\frac{R-1}{R\left(1 + \frac{1}{m}\right) - R} \geq \frac{m}{m+1} \geq \frac{1}{2},$$

we have by Lemma 1.2.3

$$\frac{d(z, L)}{d(z, L_{R(1+1/m)-1})} \leq 16 \, m \, \frac{R-1}{R}.$$

Therefore, a routine argument involving Lemma 1.2.3, (1.1.3), (1.1.12), and (4.28) shows that

$$\left| \int_{L_{R-1}} U^{\nu-\mu}(\zeta) \frac{\partial Q_m(\zeta)}{\partial \mathbf{n}_\zeta} |d\zeta| \right| \leq \int_{L_{R-1}} (a + U^{\nu-\mu}(\zeta)) \left| \frac{\partial Q_m(\zeta)}{\partial \mathbf{n}_\zeta} \right| |d\zeta|$$
$$\preceq R^m \int_{L_{R-1}} (a + U^{\nu-\mu}(\zeta)) |\Phi'(\zeta)| \frac{d(z, L)}{d(z, L_{R(1+1/m)-1})} \frac{|d\zeta|}{R-1}$$
$$\preceq m \, R^{m-1} \int_{|w|=R} (a + U^{\nu-\mu}(\Psi(w))) \, |dw|$$
$$= 2\pi \, m \, R^m a \preceq m \, r^m a. \quad (4.31)$$

Further, by (4.27),

$$\left| \int_{L_R} Q_m(\zeta) \frac{\partial}{\partial \mathbf{n}_\zeta} U^{\nu-\mu}(\zeta) |d\zeta| \right|$$

$$\leq 2 R^m \int_{L_{R-1}} \left| \operatorname{grad} U^{\nu-\mu}(\zeta) \right| |d\zeta|$$

$$= R^m \int_{|w|=R} \left| \operatorname{grad} U^{\nu-\mu}(\Psi(w)) \right| |dw|. \tag{4.32}$$

As a simple consequence of the Schwarz formula (cf. (1.1.8)) we have

$$\int_{|w|=R} \left| \operatorname{grad} U^{\nu-\mu}(\Psi(w)) \right| |dw| = \int_{|w|=R} \left| \operatorname{grad} \left(a + U^{\nu-\mu}(\Psi(w)) \right) \right| |dw|$$

$$\leq \frac{1}{\pi} \int_{|w|=R} \int_{|t|=r} \left(a + U^{\nu-\mu}(\Psi(t)) \right) \frac{|dt| \, |dw|}{|t-w|^2}$$

$$= \frac{1}{\pi} \int_{|t|=r} \left(a + U^{\nu-\mu}(\Psi(t)) \right) \left(\int_{|w|=R} \frac{|dw|}{|t-w|^2} \right) |dt|$$

$$\preceq \frac{a r}{R - r} = a m. \tag{4.33}$$

Hence, by virtue of (4.29)–(4.33) the quantity

$$B = B(\nu, m, J, L) := \int Q_m d\nu - \mu(J)$$

satisfies the relation

$$|B| \preceq M = M(m, r, a) := \frac{1}{m} + m r^m a. \tag{4.34}$$

In order to complete the proof of Theorem 4.1 we have to establish the inequality

$$|\nu(J) - \mu(J)| \preceq M. \tag{4.35}$$

We begin with the case $\mu(J) \leq 1/m$. According to (4.10),

$$\frac{1}{4} \nu(J) \leq \int_J Q_m \, d\nu + \int_{L \setminus J} Q_m \, d\nu = \int Q_m \, d\nu \leq B + \frac{1}{m}.$$

Hence, (4.35) follows directly from (4.34).

4.4 Another Approach for Dini–Smooth Arcs

Now let $1/m < \mu(J) =: l < 1$. Without loss of generality we may assume that one of the endpoints of J coincides with the endpoint of L, say $\zeta_1 = z_1$. We divide $L(\zeta_2, z_2) = L \setminus J$ by certain points

$$\xi_1 := \zeta_2, \ldots, \xi_k := z_2$$

moving along L successively from ζ_2 to z_2 and such that

$$\mu(L(\xi_1, \xi_2)) = \cdots = \mu(L(\xi_{k-2}, \xi_{k-1})) = \frac{1}{m}, \quad \mu(L(\xi_{k-1}, \xi_k)) \leq \frac{1}{m}.$$

Since

$$\nu(L(\xi_{j-1}, \xi_j)) \preceq M, \quad j = 2, \ldots, k,$$

we have owing to (4.8) and (4.12),

$$\int Q_m d\nu = \int_J Q_m \, d\nu + \sum_{j=1}^{k-1} \int_{L(\xi_j, \xi_{j+1})} Q_m \, d\nu$$

$$\leq \nu(J) + c_1 \nu(L(\xi_1, \xi_2)) + c_2 \sum_{j=2}^{k-1} \frac{\nu(L(\xi_j, \xi_{j+1}))}{(m\mu(L(\xi_1, \xi_j)))^2}$$

$$\leq \nu(J) + c_3 M \sum_{j=2}^{k-1} \frac{1}{j^2} \leq \nu(J) + c_4 M. \tag{4.36}$$

Comparing (4.34) and (4.36) we obtain

$$\nu(J) - \mu(J) \geq -c_5 M. \tag{4.37}$$

To get the upper estimate we need, due to the relation

$$\nu(J) - \mu(J) = -\left(\nu(L \setminus J) - \mu(L \setminus J)\right),$$

merely to apply (4.37) to the arc $L \setminus J$. □

From Theorem 4.1 we derive two corollaries in order to show the connection of this result with the previous ones.

Corollary 4.5. *For any $\delta > 0$ with $a(\delta) < 1$ we have*

$$D[\mu_L - \nu] \leq c_1 \log(2(1+\delta)) \frac{1}{|\log a(\delta)|}, \tag{4.38}$$

where $c_1 = c_1(L) > 0$.

Indeed, nontrivial is only the case where $a(\delta) \leq \frac{1}{2}$ and

$$m := \left[\frac{1}{2} \frac{|\log a(\delta)|}{\log(2(1+\delta))}\right] > 1,$$

where $[a]$ denotes the integral part of $a \geq 0$. Inserting this value of m in (4.4), we get (4.38) after an elementary calculation.

The estimate (4.38) is an obvious analogue of inequality (3.24).

Corollary 4.6. *For $0 \leq \delta \leq 1$ we have*

$$D[\mu_L - \nu] \leq c_2 \sqrt{\delta + a(\delta)}, \tag{4.39}$$

where $c_2 = c_2(L) > 0$ (cf. Theorem 1.1).

Indeed, we can assume that

$$m := \left[\frac{1}{\sqrt{\delta + a(\delta)}} \right] \geq 1. \tag{4.40}$$

Then taking m as in (4.40) in (4.4) we obtain (4.39).

4.5 Historical Comments

The results of Section 4.1 are presented in accordance with [18], but the estimates here are sharper then those in [18]. For smooth curves and arcs of class C^{2+} the results were proved by J. Hüsing [85].

In the proof of Theorem 1.6 we apply a scheme from [31], where it was applied for alternation points of polynomials of best uniform approximation. This scheme was pointed out to the second author by E.B. Saff and V. Totik. In [34] the same scheme was applied to extreme points of best approximants on smooth arcs and curves.

The special case of Section 4.3, when $\nu = \nu_p$ is a zero counting measure on the unit disk, was investigated by Erdős and Turán [56]. Corollary 3.2 represents their result. Ganelius [68] showed that this result as well as the discrepancy estimate in [55] and [57] can be derived from a lemma on Fourier coefficients of bounded integrable functions.

The reduction of the zeros to the boundary of the unit disk at the beginning of Section 4.3 is an artifice due to I. Schur [158] that was successfully used by Erdős and Turán in their papers [55, 56, 57].

5
Discrepancy Theorems via Energy Integrals

If E is a compact set in $\mathbb{C}$ with cap $E > 0$, then the equilibrium measure μ_E minimizes the energy expression

$$I[\sigma] = \int U^\sigma(z)\, d\sigma(z)$$

with respect to all positive Borel measures σ supported on E with $\sigma(E) = 1$. Since $I[\mu_E] = -\log \operatorname{cap} E$, the energy $I[\sigma]$ is positive for all positive Borel measures σ on E if $0 < \operatorname{cap} E < 1$. Consequently, for such E,

$$\sigma(E)^2 \leq \frac{I[\sigma]}{|\log \operatorname{cap} E|}.$$

Thus, the total mass $\sigma(E)$ can be estimated by its energy.

In this chapter we consider the question of whether discrepancy estimates for signed measures σ on quasiconformal curves or arcs can be formulated in terms of the energy integral associated with σ. The proofs require some estimates for the modulus of a doubly split extended complex plane (Section 5.1). The main theorem is formulated in Section 5.2 for closed quasidisks E with boundary L and signed measures $\sigma = \sigma^+ - \sigma^-$ supported on L. Both measures σ^+ and σ^- must have finite energy. For the case that $\sigma^+ = \mu_L$, the discrepancy estimate can be stated impressively as

$$D[\sigma] \leq c\sqrt{I[\sigma] \log \frac{1}{I[\sigma]}}$$

for sufficiently small $I[\sigma]$, where c is a positive constant depending only on L.

5.1 Modulus of a Doubly Split Plane

We consider the following special geometric configuration: Let L be a bounded Jordan curve consisting of four closed arcs L^1, J^1, L^2, J^2, disjoint up to their endpoints, as is shown in Figure 5.1. Further, we consider

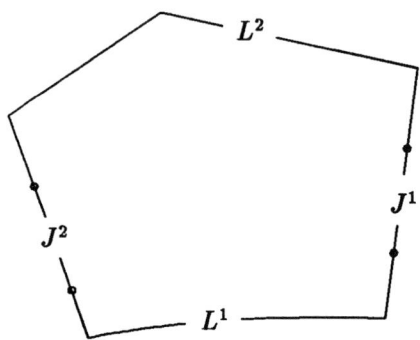

Fig. 5.1

the doubly connected domain $U := \overline{\mathbb{C}} \setminus (L^1 \cup L^2)$ and a conformal mapping f of U onto the annulus

$$B_R := \{w : 1 < |w| < R\}$$

satisfying the boundary correspondence

$$\lim_{z \to L^1} |f(z)| = 1, \quad \lim_{z \to L^2} |f(z)| = R$$

(cf. [75, p. 208]). The quantity $R = R(U)$ is called the *modulus* of the domain U.

First, let $L := \mathbb{T}$ and

$$J^1 := \{e^{i\theta} : |\theta| \leq \delta_1\},$$
$$J^2 := \{e^{i\theta} : |\theta - \theta_1| \leq \delta_2\},$$

where $0 < \delta_1 \leq \delta_2$, $\delta_1 < \pi/4$, $\delta_1 + \delta_2 < \theta_1 < 2\pi - \delta_1 - \delta_2$.

Lemma 1.1. *Under the above assumptions,*

$$\frac{1}{R(U) - 1} \leq 4 + \frac{8}{\pi} \log \frac{2\pi}{\delta_1}. \tag{1.1}$$

5.1 Modulus of a Doubly Split Plain

Proof. We consider the curve

$$\gamma := [0, 2] \cup [0, 2e^{i\theta_1}] \cup \{2e^{i\theta} : 0 \leq \theta \leq \theta_1\}$$

(see Figure 5.2) and the integrals

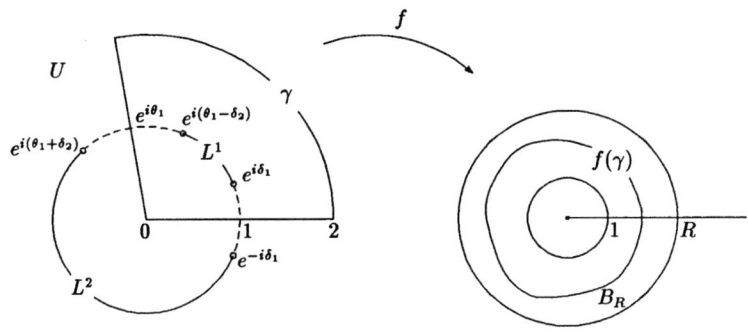

Fig. 5.2

$$I_1 := \int_\gamma \frac{|d\zeta|}{\text{dist }(\zeta, \partial U)}, \quad I_2 := \int_{f(\gamma)} \frac{|dw|}{\text{dist }(w, \partial B_R)}.$$

The first of them can be estimated from above as follows:

$$I_1 \leq \left(\int_{\{2e^{i\theta}: 0 \leq \theta \leq \theta_1\}} \cdots + 4 \int_{[0,1]} \cdots \right)$$

$$\leq 4\pi + 4 \int_0^1 \frac{dx}{|x - e^{i\delta_1}|}$$

$$\leq 4\pi + 8 \left(\int_0^{\cos \delta_1} \frac{dx}{\sin \delta_1 + \cos \delta_1 - x} + \int_{\cos \delta_1}^1 \frac{dx}{x - \cos \delta_1 + \sin \delta_1} \right)$$

$$= 4\pi + 8 \left(\log \frac{\sin \delta_1 + \cos \delta_1}{\sin \delta_1} + \log \frac{1 - \cos \delta_1 + \sin \delta_1}{\sin \delta_1} \right)$$

$$\leq 4\pi + 8 \log \frac{2\pi}{\delta_1}. \tag{1.2}$$

Since dist $(w, \partial B_R) \leq (R-1)/2$ for each $w \in f(\gamma)$, we have that

$$I_2 \geq 2 \int_0^{2\pi} \frac{d\theta}{R-1} = \frac{4\pi}{R-1}. \tag{1.3}$$

To compare the integrals introduced above we apply Lemma 1.2.3 with $G_1 = U \setminus \{\infty\}$ to obtain

$$I_2 \leq \int_\gamma \frac{|f'(\zeta)||d\zeta|}{\text{dist }(f(\zeta), \partial B_R \cup \{f(\infty)\})} \leq 4 \int_\gamma \frac{|d\zeta|}{\text{dist }(\zeta, \partial U)} = 4 I_1.$$

Hence, (1.2) and (1.3) imply (1.1). □

Next, we assume that the Jordan curve L and the respective subarcs satisfy

$$\text{diam } J^1 = \text{diam } J^2 =: \delta \leq \frac{1}{2} \text{ diam } L, \tag{1.4}$$

and

$$\min\{\text{diam } L^1, \text{diam } L^2\} \geq \frac{1}{12} \text{ diam } L. \tag{1.5}$$

Lemma 1.2. *Under the assumptions (1.4) and (1.5) the modulus of U satisfies*

$$R(U) \leq c = e^{288\pi^2}. \tag{1.6}$$

Proof. Let Γ denote the family of all Jordan curves $\gamma \subset U$ that separate L^1 and L^2. It is well known that

$$m(\Gamma) = m(f(\Gamma)) = \frac{1}{2\pi} \log R \tag{1.7}$$

(cf. (1.2.10) and (A.1.8)). Hence, for the proof of (1.6) it is enough to find an upper estimate for $m(\Gamma)$. To this end, we fix a point $z^* \in L$ and consider the function

$$\rho(z) := \begin{cases} \dfrac{6}{\text{diam } L}, & \text{if } |z - z^*| \leq 2 \text{ diam } L, \\ 0, & \text{if } |z - z^*| > 2 \text{ diam } L. \end{cases}$$

By our assumption (1.5),

$$\int_\gamma \rho(z) |dz| \geq 1, \qquad \gamma \in \Gamma.$$

Therefore, by the L-definition of the module of a family of curves (cf. Appendix A.1),
$$m(\Gamma) \le \int_{\mathbb{C}} \rho^2 \, dm = 144\,\pi.$$

Comparing (1.7) with the last inequality we obtain (1.6). □

Lemma 1.3. *Let L be K-quasiconformal and satisfy (1.4) and (1.5). Then*
$$\frac{1}{\log R(U)} \le c_1 + c_2 \log \frac{\operatorname{diam} L}{\delta} \tag{1.8}$$
holds with some constants c_j, $j = 1, 2$, that depend only on K.

Proof. Let Φ be the conformal mapping of $\Omega := \operatorname{ext} L$ onto Δ normalized as in (2.1). Since L is K-quasiconformal, Φ can be extended to a K^2-quasiconformal mapping of $\overline{\mathbb{C}}$ onto itself (see [3, Chapter IV] or Theorem B.2.1). Therefore, by Theorem 1.2.7,
$$\operatorname{diam} \Phi(J^j) \ge c_3 \left(\frac{\delta}{\operatorname{diam} L} \right)^{K^2}$$
with some $c_3 = c_3(K) > 0$. At the same time, according to the quasi-invariance of the module under quasiconformal mappings (see (1.2.20) and (1.7)), we have
$$\frac{1}{\log R(U)} \le \frac{K^2}{\log R(\Phi(U))}.$$

Comparing these estimates with Lemma 1.1, Lemma 1.2, and the inequality
$$\frac{1}{\log(1+x)} \le \frac{c}{\log(1+c)} \frac{1}{x}, \qquad 0 < x \le c < \infty,$$
we obtain (1.8). □

Finally, we recall a result due to Totik [177, Lemma 3.1].

Lemma 1.4. *For any $x \in \mathbb{R}$ and $0 < \eta < \theta$,*
$$\left| \int_\eta^\theta \log \left| \frac{x+t}{x-t} \right| \frac{dt}{t} \right| \le 10.$$

For a proof, see [156, p. 434].

5.2 Estimates of Mass Distributions from Their Energies

Let $L \subset \mathbb{C}$ be a Jordan curve, and let $\sigma = \sigma^+ - \sigma^-$ be a signed measure such that the Borel measures σ^+, σ^- are in $\mathcal{M}(L)$. We suppose that each of the measures σ^+ and σ^- has finite logarithmic energy. In particular, this implies that $\sigma^+(B) = \sigma^-(B) = 0$ for any Borel set $B \subset L$ of capacity 0 (see [178, Theorem III.7]). Our aim is to estimate the discrepancy $D[\sigma]$ in terms of the *energy integral*

$$I[\sigma] := \int U^\sigma d\sigma = \iint \log \frac{1}{|\zeta - z|} d\sigma(z) \, d\sigma(\zeta), \qquad (2.1)$$

which according to the formula

$$\int |\text{grad } U^\sigma(\zeta)|^2 \, dm(\zeta) = 4\pi^2 I[\sigma] \qquad (2.2)$$

(see [108, Theorem 1.20]) is nonnegative (cf. also [156, p. 29]).

Let

$$\alpha_+(\delta) := \inf_{\substack{J \subset L \\ \sigma^+(J) \geq \delta}} \text{diam } J, \qquad \delta > 0.$$

Theorem 2.1. *Let $L \subset \mathbb{C}$ be a K-quasiconformal curve and $\sigma = \sigma^+ - \sigma^-$ a signed measure, where $\sigma^+, \sigma^- \in \mathcal{M}(L)$ have finite energy. Then*

$$D[\sigma]^2 \leq c_1 I[\sigma] \log \frac{c_2 \, \text{diam } L}{\alpha_+(D[\sigma]/16)} \qquad (2.3)$$

with constants $c_1 > 0$ and $c_2 \geq 1$ depending only on K.

Proof. Let $d := D[\sigma]$. There exists a closed subarc $J \subset L$ such that

$$\text{diam } J \geq \frac{1}{4} \text{diam } L, \quad \text{diam } (\overline{L \setminus J}) \geq \frac{1}{4} \text{diam } L, \qquad (2.4)$$

and

$$\sigma(J) \geq \frac{d}{4}. \qquad (2.5)$$

This partition can be obtained as follows: First we choose a subarc $J_1 \subset L$ such that $|\sigma(J_1)| \geq d/2$ and diam $J_1 \geq (\text{diam } L)/2$. Each point $z^* \in J_1$ divides J_1 in two subarcs $J_{1,1}$ and $J_{1,2}$. We fix z^* such that

$$\text{diam } J_{1,j} \geq \frac{1}{2} \text{diam } J_1 \geq \frac{1}{4} \text{diam } L, \qquad j = 1, 2.$$

Without loss of generality we can assume that $|\sigma(J_{1,1})| \geq d/4$. Hence, setting $J := J_{1,1}$ or $J := \overline{L \setminus J_{1,1}}$ we obtain an arc with the desired properties (2.4) and (2.5).

5.2 Estimates of Mass Distributions from Their Energies

In the following we set $L^2 := \overline{(L \setminus J)}$.

Next, we divide J into three closed subarcs J^1, L^1, and J^2, mutually disjoint except for their endpoints, such that

$$\text{diam } J^1 = \text{diam } J^2 = \frac{1}{2}\alpha_+\left(\frac{d}{16}\right), \qquad (2.6)$$

where J^1 and J^2 include the endpoints of J (see Figure 5.1). By virtue of (2.5) we have

$$\sigma(L^2) \leq -\frac{d}{4}, \quad \sigma^+(J^1) \leq \frac{d}{16}, \quad \sigma^+(J^2) \leq \frac{d}{16}. \qquad (2.7)$$

Finally, (2.4) and (2.6) yield

$$\text{diam } L^1 \geq \text{diam } J^1 \geq \frac{1}{12}\text{diam } L.$$

Hence, we have constructed a subdivision of L in four subarcs satisfying (1.4) and (1.5) with

$$\delta := \frac{1}{2}\alpha_+\left(\frac{d}{16}\right).$$

In a next step we introduce the doubly connected domain $U := \overline{\mathbb{C}} \setminus \{L^1 \cup L^2\}$, the annulus B_R and the conformal mapping $f : U \to B_R$, as above. For $0 < \varepsilon < 1$ sufficiently small, we set $w_0 := f(\infty)$ and define

$$g_1(w) := \begin{cases} 1, & \text{if } |w| > R - 2\varepsilon, \\ 0, & \text{if } |w| \leq 1 + 2\varepsilon \text{ or } |w - w_0| \leq 2\varepsilon, \\ \dfrac{\log|w|}{\log R}, & \text{otherwise in } \mathbb{C}. \end{cases}$$

Next, we average this function as follows:

$$g_2(w) := \frac{1}{\varepsilon^2} \int_{\mathbb{C}} g_1(\tau) K\left(\frac{\tau - w}{\varepsilon}\right) dm(\tau), \quad w \in \mathbb{C},$$

where $K(t)$, $t \in \mathbb{C}$, is an arbitrary averaging kernel, i.e., $K(t) \in C^\infty(\mathbb{C})$ and $K(t)$ satisfies (1.5.3)–(1.5.5).

Taking into account (1.5.6) and the analogue of (1.5.7) for the first derivatives of g_2 we obtain the following properties of g_2:

$$g_2 \in C^\infty(\mathbb{C}),$$

$$g_2(w) = g_1(w) \quad \text{for} \quad \begin{cases} w \in D(1+\varepsilon), \\ w \in \mathbb{C} \setminus D(R-\varepsilon), \\ w \in D(w_0, \varepsilon), \\ w \in D(R-3\varepsilon) \setminus D(1+3\varepsilon) \setminus D(w_0, 3\varepsilon), \end{cases}$$

and

$$|\text{grad } g_2(w)| \leq \frac{1}{\varepsilon}, \quad w \in \mathbb{C}.$$

Therefore,

$$\int_{\mathbb{C}} |\text{grad } g_2(w)|^2 \, dm(w)$$

$$\leq \frac{1}{\varepsilon} \int_{1+\varepsilon \leq |w| \leq 1+3\varepsilon} dm(w) + \frac{1}{\varepsilon} \int_{R-3\varepsilon \leq |w| \leq R-\varepsilon} dm(w)$$

$$+ \frac{1}{\varepsilon} \int_{\varepsilon \leq |w-w_0| \leq 3\varepsilon} dm(w) + \frac{2\pi}{(\log R)^2} \int_1^R \frac{dr}{r}$$

$$= 2\pi \left(2 + 2R + 4\varepsilon + \frac{1}{\log R}\right).$$

Next, we introduce the function

$$g(z) := g_2(f(z)), \quad z \in U,$$

extended by 0 to L^1 and by 1 to L^2. Note that

$$\int_{\mathbb{C}} |\text{grad } g(z)|^2 \, dm(z) = \int_{\mathbb{C}} |\text{grad } g_2(w)|^2 \, dm(w) \qquad (2.8)$$

$$\leq 2\pi \left(6 + 2R + \frac{1}{\log R}\right).$$

By Green's formula (1.1.11) we have for $z \in L$ and $\zeta = \xi + i\eta \in \mathbb{C}$,

$$g(z) = -\frac{1}{2\pi} \int_{\mathbb{C}} \left(\frac{\partial}{\partial \xi} g(\zeta) \frac{\partial}{\partial \xi} \log |\zeta - z| + \frac{\partial}{\partial \eta} g(\zeta) \frac{\partial}{\partial \eta} \log |\zeta - z|\right) dm(\zeta).$$

Therefore, making of use (2.2), (2.7) as well as (2.8), Fubini's theorem and Hölder's inequality imply

$$\frac{d}{8} \leq \left| \int g(z) \, d\sigma(z) \right|$$

$$= \left| \frac{1}{2\pi} \int \int \left(\frac{\partial}{\partial \xi} g(\zeta) \frac{\partial}{\partial \xi} \log |\zeta - z| \right. \right.$$

$$\left. \left. + \frac{\partial}{\partial \eta} g(\zeta) \frac{\partial}{\partial \eta} \log |\zeta - z| \right) d\sigma(z) \, dm(\zeta) \right|$$

$$\leq \frac{1}{2\pi} \int |\mathrm{grad}\, g(\zeta)| \, |\mathrm{grad}\, U^\sigma(\zeta)| \, dm(\zeta)$$

$$\leq \left(\int |\mathrm{grad}\, g(\zeta)|^2 \, dm(\zeta) \right)^{1/2} \left(\frac{1}{4\pi^2} \int |\mathrm{grad}\, U^\sigma(\zeta)|^2 \, dm(\zeta) \right)^{1/2}$$

$$\leq \sqrt{2\pi} \left(6 + 2R + \frac{1}{\log R} \right)^{1/2} I[\sigma]^{1/2}.$$

Comparing this with Lemma 1.2 and Lemma 1.3 we obtain (2.3). $\square$

Remark 2.2. The same estimate (2.3) holds also for quasiconformal arcs. The proof can be based on a reduction to the case of a curve and can be found in the dissertation of J. Hüsing [86].

The next example shows that Theorem 2.1 is sharp up to the constants c_1 and c_2.

Example 2.3. Let $0 < \alpha < 1$ be some constant. For $0 < \varepsilon < 1$ we define

$$v_\varepsilon(t) := \begin{cases} \dfrac{\varepsilon^{2(1-\alpha)}}{t}, & \text{if } \varepsilon^2 \leq |t| \leq \varepsilon, \\ 0, & \text{otherwise}, \end{cases}$$

and consider the measure

$$d\nu_\varepsilon := v_\varepsilon(t) \, dt.$$

Then we define the signed measure $\sigma := \sigma^+ - \sigma^-$ on $[-1, 1]$ by

$$d\sigma^+ := c_\alpha \, |x|^{-\alpha} \, dx \quad \text{and} \quad d\sigma^- := c_\alpha \left(|x|^{-\alpha} \, dx + d\nu_\varepsilon \right)$$

with

$$c_\alpha := \frac{1-\alpha}{2}.$$

It is obvious that $\sigma^+, \sigma^- \in \mathcal{M}([-1, 1])$. We remark that

$$D[\sigma] = c_\alpha \int_{\varepsilon^2}^{\varepsilon} v_\varepsilon(t) \, dt = c_\alpha \int_{\varepsilon^2}^{\varepsilon} \frac{\varepsilon^{2(1-\alpha)}}{t} \, dt = c_\alpha \, \varepsilon^{2(1-\alpha)} \log \frac{1}{\varepsilon}. \qquad (2.9)$$

To find an upper estimate for $I[\sigma]$ we notice that for each $x \in [-1,1]$

$$|U^\sigma(x)| = |c_\alpha U^{\nu_\varepsilon}(x)|$$

$$= c_\alpha \left| \int_{\varepsilon^2}^{\varepsilon} \log\frac{1}{|x-t|} \frac{\varepsilon^{2(1-\alpha)}}{t} dt + \int_{-\varepsilon}^{-\varepsilon^2} \log\frac{1}{|x-t|} \frac{\varepsilon^{2(1-\alpha)}}{t} dt \right|$$

$$= c_\alpha \, \varepsilon^{2(1-\alpha)} \left| \int_{\varepsilon^2}^{\varepsilon} \log\frac{|x+t|}{|x-t|} \frac{1}{t} dt \right| \leq 10 \, c_\alpha \, \varepsilon^{2(1-\alpha)}, \qquad (2.10)$$

where we have used Lemma 1.4 for the last inequality. From (2.9) and (2.10) we obtain

$$I[\sigma] = \int U^\sigma(x) \, d\sigma(x)$$

$$\leq 20 \, c_\alpha \, \varepsilon^{2(1-\alpha)} |\sigma([\varepsilon^2, \varepsilon])| = 20 \, c_\alpha^2 \, \varepsilon^{4(1-\alpha)} \log \frac{1}{\varepsilon}. \qquad (2.11)$$

Since

$$\alpha_+(\delta) = 2 \, \delta^{1/(1-\alpha)}, \quad 0 < \delta \leq 1,$$

using (2.9) and (2.11) we obtain for ε small enough that

$$D[\sigma]^2 \geq c \, I[\sigma] \log \frac{2}{\alpha_+(D[\sigma]/16)},$$

where $c > 0$ depends only on α.

Finally, we would like to discuss why the curve L has to be quasiconformal in Theorem 2.1. More precisely, we will show with the following example that in the case of curves with cusps the estimate (2.3) does not hold in general.

Example 2.4. Let $\beta > 1$. We define a Jordan curve L as follows:

$$L^1 := \{z = x + ix^\beta \,:\, 0 \leq x \leq 1\},$$
$$L^2 := \{z = x - ix^\beta \,:\, 0 \leq x \leq 1\},$$
$$L^3 := \left\{z = \sqrt{2} e^{i\theta} \,:\, |\theta| \leq \frac{\pi}{4}\right\},$$
$$L := L^1 \cup L^2 \cup L^3$$

(see Figure 5.3). It is obvious that L possesses a cusp at the origin, and according to the Ahlfors criterion (cf. 1.2.18) L is not quasiconformal. In addition, for $0 < \varepsilon < \frac{1}{2}$, we consider the subarcs

$$L^+ := L \setminus \{z = x + ix^\beta \,:\, x \in (0, \varepsilon)\},$$
$$L^- := L \setminus \{z = x - ix^\beta \,:\, x \in (0, \varepsilon)\}$$

5.2 Estimates of Mass Distributions from Their Energies   183

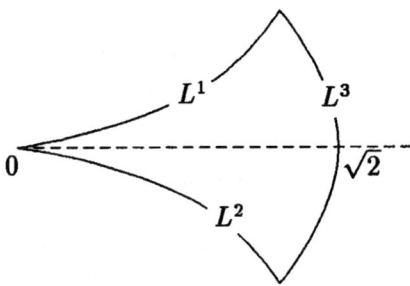

Fig. 5.3

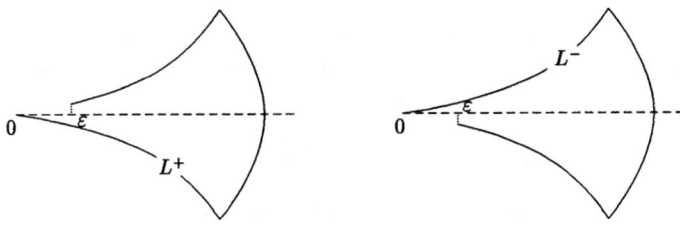

Fig. 5.4

(see Figure 5.4) and define $\sigma := \mu_{L^+} - \mu_{L^-}$, where μ_{L^+} and μ_{L^-} denote the equilibrium measures of L^+ and L^-, respectively. In the following we will show that inequality (2.3) is not valid for σ, though σ satisfies the assumptions of Theorem 2.1.

To get the desired result we prove some auxiliary estimates. First we show that

$$|U^\sigma(z)| \leq c_1 \varepsilon^{\beta/2}, \qquad z \in L_\pm^\varepsilon := L \setminus L^\pm \tag{2.12}$$

with $c_1 = c_1(\beta) > 0$.

Indeed, let $\Phi_\pm : \overline{\mathbb{C}} \setminus L^\pm \to \Delta$ be the conformal mappings such that $\Phi_\pm(\infty) = \infty$ and $\Phi'_\pm(\infty) > 0$. Note that

$$U^\sigma(z) = -\log|\Phi_\pm(z)|, \qquad z \in \mathbb{C} \setminus L^\pm.$$

Taking into account that the capacity of $L^\pm$ is at least (diam $L^\pm)/4 > \frac{1}{4}$ and the inequality

$$\text{dist }(z, L^\pm) \geq \text{cap } L^\pm \frac{(|\Phi_\pm(z)| - 1)^2}{|\Phi_\pm(z)|}, \qquad z \in \mathbb{C} \setminus L^\pm$$

(see Lemma 1.2.5), we obtain (2.12).

The equilibrium measure of a subarc J of a Jordan arc or curve E is the harmonic measure of this subarc at infinity (cf. (1.2.3) and (1.2.4)). Hence, by virtue of Theorem 1.1.4, it is a monotonic function of both J and L.

Using, in addition, distortion properties of conformal mappings of domains with piecewise smooth boundary onto the disk (see [147, p. 52]) we find that

$$D[\sigma] \geq \mu_{L^-}(L_+^\varepsilon) \geq \mu_L(L_+^\varepsilon) \geq c_2 \sqrt{\varepsilon}. \tag{2.13}$$

Further, we claim that for any subarc $J \subset L$,

$$\mu_{L^+}(J) \leq c_3 \sqrt{\text{diam } J}. \tag{2.14}$$

In order to prove (2.14) we can assume that J is a subarc of one of the smooth arcs L^2, L^3, or $L^1 \setminus L_+^\varepsilon$, which we denote by L^*. Then using Lemma 1.2.11 and Lemma 1.2.12 we have

$$\mu_{L^+}(J) \leq \mu_{L^*}(J) \leq c_4 \sqrt{\text{diam } J}.$$

Moreover, the analysis of the proof of these lemmas (more precisely, of Lemma 1.2.10) shows that the constant c_4 depends only on β.

Inequality (2.14) implies that

$$\alpha_+(\delta) \geq \frac{\delta^2}{c_3^2}, \qquad 0 < \delta < \frac{1}{2}. \tag{2.15}$$

Now (2.12) and (2.14) yield

$$I[\sigma] = 2 \int_{L_-^\varepsilon} |U^\sigma(z)| \, d\mu_{L^+}(z) \leq c_5 \varepsilon^{(1+\beta)/2}.$$

Thus, by (2.15),

$$I[\sigma] \log \frac{\text{diam } L}{\alpha_+(D[\sigma]/16)} \leq c_6 \varepsilon^{(1+\beta)/2} \log \frac{1}{\varepsilon},$$

and according to (2.13) the estimate of the form (2.3) is violated for any $\beta > 1$.

Finally, we want to present a reformulation of the inequality (2.3) such that the discrepancy $D[\sigma]$ is estimated by $I[\sigma]$ alone.

5.2 Estimates of Mass Distributions from Their Energies

Corollary 2.5. *Let L be a quasiconformal curve and $\sigma = \mu_L - \sigma^-$, where $\sigma^- \in \mathcal{M}(L)$ has finite energy. Then there exists a constant $c > 0$ such that*

$$D[\sigma] \le c\sqrt{I[\sigma]\log\frac{1}{I[\sigma]}} \qquad (2.16)$$

if $0 < I[\sigma] < 1/e$.

Proof. Since L is K-quasiconformal, we know from (1.2.22) that there exists a constant $c_1 > 0$ such that

$$\mu_L(J) \le c_1 (\operatorname{diam} J)^{1/K^2}$$

for all subarcs J of L. Hence,

$$\alpha_+(D[\sigma]/16) \ge c_2 D[\sigma]^{K^2},$$

and by (2.3),

$$D[\sigma]^2 \le c_3 \left(I[\sigma]\log\frac{1}{D[\sigma]} + I[\sigma] \right) \qquad (2.17)$$

with some constant $c_3 \ge 1$. Next, we assert that

$$D[\sigma] \le \sqrt{2c_3 I[\sigma]\log\frac{1}{I[\sigma]}}. \qquad (2.18)$$

If $D[\sigma] \ge 1/e$, then from (2.17) we obtain

$$D[\sigma]^2 \le 2c_3 I[\sigma],$$

and (2.18) is true.

If $D[\sigma] < 1/e$, we prove (2.18) by contradiction. Let us assume that (2.18) is false, i.e.,

$$D[\sigma] > \sqrt{2c_3 I[\sigma]\log\frac{1}{I[\sigma]}}.$$

Then

$$\frac{D[\sigma]^2}{\log D[\sigma]^{-1}} \ge \frac{2c_3 I[\sigma]\log I[\sigma]^{-1}}{\log I[\sigma]^{-1}/2} = 4c_3 I[\sigma].$$

Hence,

$$D[\sigma]^2 \ge 4c_3 I[\sigma]\log D[\sigma]^{-1} \ge c_3 \left(I[\sigma]\log D[\sigma]^{-1} + 3 I[\sigma] \right),$$

which contradicts (2.17). $\square$

5.3 Historical Comments

Theorem 2.1 was first proved by W. Kleiner [94] for the case of smooth L and under some additional restrictions on the measure σ. Note that the original proof of Kleiner is very technical and difficult to follow. The presented version of Theorem 2.1 and both examples in Section 6.3 are due to J. Hüsing [86]. The construction of Example 2.4 uses essentially an idea of V. Totik [177].

W. Kleiner [95] has developed a method to apply discrepancy theorems in terms of energy integrals to the zero distribution of polynomials with simple zeros. Since the zero counting measures have infinite energies, Kleiner proposed to replace the discrete mass at a point z_0 by its balayage measure on small circles with center z_0. Using this approach, J. Hüsing [86, 87] could prove discrepancy estimates for the zero counting measure of polynomials not only for polynomials with simple zeros, but also with multiple zeros.

The analogue of Theorem 2.1 for quasiconformal arcs is investigated in detail by J. Hüsing [86].

For higher dimensions, $n \geq 3$, discrepancy estimates via energy integrals have been derived by W. Kleiner [96] and P. Sjögren [164].

6
Applications of Jentzsch–Szegő and Erdős–Turán Type Theorems

In this chapter we consider applications of Jentzsch–Szegő type and Erdős–Turán type theorems that were derived in Chapter 2.

In Section 6.1 it is shown that the zeros of best-approximating polynomials to a function f on a compact set E have the characteristic property that every boundary point of E is a limit point of their zeros, provided that f is not analytic on E and f does not vanish identically on any component of $E°$. Moreover, the zero distributions of the best approximants are weakly convergent to the equilibrium measure of E for a subsequence of the best-approximating polynomials. In general, the weak convergence does not hold for the whole sequence of best approximants.

Section 6.2 considers the analogous problems for near-best approximants. The situation is completely different from the case of best approximants. Only under special conditions (Theorem 2.4) does there exist at least one limit point of zeros of near-best approximants.

Maximally convergent polynomials again show a characteristic phenomenon: The zero distributions of maximally convergent polynomials converge at least for a subsequence to the equilibrium measure of the maximal Green's level curve, in whose interior the function f is analytic. Moreover, in Section 6.3 we investigate maximally convergent polynomials defined by interpolation and derive necessary conditions on the distribution of the nodes if the interpolating polynomials converge maximally at least for one single function.

For Jordan domains G we investigate orthogonal polynomials on G and obtain discrepancy estimates concerning the distribution of their a-values compared with the equilibrium distribution of the boundary of G (Section

6.4). The zeros of the orthogonal polynomials show exceptional behavior in contrast to all other a-values, $a \neq 0$.

The final Section 6.5 is concerned with the zero distribution of Bieberbach polynomials and their derivatives.

6.1 Polynomials of Best Uniform Approximation

Let $E \subset \mathbb{C}$ be an admissible compact set i.e., the complement $\Omega := \overline{\mathbb{C}} \setminus E$ of E with respect to the extended complex plane is connected and cap $E > 0$.

We denote by $p_n^* = p_n^*(f)$ the polynomial in $\mathbb{P}_n$ of best uniform approximation to $f \in A(E)$ on E, that is,

$$\|f - p_n^*\|_E = E_n(f, E) := \inf_{p \in \mathbb{P}_n} \|f - p\|_E.$$

Writing

$$p_n^*(z) = a_n^* z^n + \cdots,$$

it will be shown that the coefficients a_n^* carry the information about whether the function f is analytic on E.

Proposition 1.1. *Let $f \in A(E)$, where E is admissible and has a regular boundary. Then the following assertions are equivalent:*

(i) $\limsup_{n \to \infty} |a_n^*|^{1/n} = 1/\operatorname{cap} E$;

(ii) *f is not analytic on E.*

Proof. Let $E_n(f) := E_n(f, E)$. Since $f \in A(E)$, we remark that Mergelyan's theorem implies

$$\lim_{n \to \infty} E_n(f) = 0 \tag{1.1}$$

(see (1.5.1)). By Lemma 1.1.12 and the fact that $\{\|p_n^*\|_E\}$ is a bounded sequence we have

$$\rho \limsup_{n \to \infty} |a_n^*|^{1/n} \leq \limsup_{n \to \infty} \|p_n^*\|_E^{1/n} \leq 1, \tag{1.2}$$

where $\rho := \operatorname{cap} E$.

Now suppose that (i) holds. We have to show that f cannot be analytically extended to ∂E. Assume, to the contrary, that f is analytic on E. For each $\delta > 0$, we consider the set

$$E_\delta := E \cup \{z \in \Omega : G(z) < \log(1 + \delta)\}.$$

Then f is analytic in $\overline{E}_\delta$ for some (sufficiently small) $\delta > 0$, and by Theorem 1.5.3,

$$\limsup_{n \to \infty} E_n(f)^{1/n} < \frac{1}{1 + \delta}.$$

Standard reasoning, involving the last inequality and Lemma 1.1.11 (cf. proof of Theorem C.3.7), shows that the series

$$p_1^* + \sum_{j=1}^{\infty} (p_{j+1}^* - p_j^*)$$

converges uniformly on $\overline{E}_\delta$. Hence, from (1.2) we get (with E replaced by $\overline{E}_\delta$)

$$\limsup_{n \to \infty} |a_n^*|^{1/n} \leq \frac{1}{(1+\delta)\rho} < \frac{1}{\rho},$$

which contradicts (i).

Next, assume that f is not analytic on E. If (i) does not hold, then from (1.2) we have

$$\limsup_{n \to \infty} |a_n^*|^{1/n} < 1/\rho. \tag{1.3}$$

Let $T_n \in \mathbb{M}_n, n \in \mathbb{N}$, be the Chebyshev polynomials for E and define

$$\widetilde{p}_{n-1} := p_n^* - a_n^* T_n.$$

It follows from (1.3) and Lemma 1.1.14 that

$$\limsup_{n \to \infty} \|p_n^* - \widetilde{p}_{n-1}\|_E^{1/n} < 1. \tag{1.4}$$

Moreover, from the extremal property of p_{n-1}^*, we have

$$E_{n-1}(f) = \|f - p_{n-1}^*\|_E \leq \|f - \widetilde{p}_{n-1}\|_E$$
$$\leq \|f - p_n^*\|_E + \|p_n^* - \widetilde{p}_{n-1}\|_E$$
$$= E_n(f) + \|p_n^* - \widetilde{p}_{n-1}\|_E,$$

and hence, from (1.4) we find that

$$\limsup_{n \to \infty} (E_{n-1}(f) - E_n(f))^{1/n} \leq \limsup_{n \to \infty} \|p_n^* - \widetilde{p}_{n-1}\|_E^{1/n} < 1.$$

Since $\lim_{n \to \infty} E_n(f) = 0$, we obtain

$$E_n(f) = \sum_{k=n}^{\infty} (E_k(f) - E_{k+1}(f))$$

and consequently

$$\limsup_{n \to \infty} E_n^{1/n} < 1. \tag{1.5}$$

By virtue of Theorem 1.5.3, inequality (1.5) implies that f is analytic on E, which is the desired contradiction. □

Proposition 1.1 and Theorem 2.1.7 lead to the following statement.

190   6. Jentzsch–Szegő and Erdős–Turán Type Theorems

Theorem 1.2. *Let f and E be as in Theorem 1.1. Assume further that f is not analytic on E and does not vanish identically on any component of $E°$. Let $\{p_n^*\}_{n\in\mathbb{N}}$ be the sequence of polynomials of best uniform approximation to f on E. Then the zero counting measures $\nu_{p_n^*}$ converge weakly to μ_E as $n \to \infty$ through some subsequence $\Lambda = \Lambda(f) \subset \mathbb{N}$.*

Proof. Set $p_n = p_n^*/a_n^*$ and apply Theorem 2.1.7 with Λ chosen such that

$$|a_n^*|^{1/n} \longrightarrow 1/\operatorname{cap} E \quad \text{as } n \to \infty, \quad n \in \Lambda.$$

With Proposition 1.1 we get the desired result. □

Our next goal is to show that Theorem 1.2 does not hold for the whole sequence $\{p_n^*\}_{n\in\mathbb{N}}$ in general.

Theorem 1.3. *Let $E \subset \mathbb{C}$ be an admissible compact set. Then there is a function $f \in A(E)$ that is not analytic on E and an increasing sequence $\{k(n)\}_{n\in\mathbb{N}}$ of integers such that*

$$\lim_{n\to\infty} \nu_{p_{k(n)}^*}(S) = 0 \tag{1.6}$$

for all bounded $S \subset \mathbb{C}$.

Proof. For $n \in \mathbb{N}$ let

$$p_n(z) := \frac{1}{n^2 \|T_n\|_E} T_n(z),$$

where by $T_n(z)$ we denote the nth Chebyshev polynomial on E.

We construct an increasing sequence $\{k(n)\}_{n\in\mathbb{N}_0}$ of natural numbers, a sequence $\{m(n)\}_{n\in\mathbb{N}_0}$ of natural numbers, and a sequence $\{\alpha_n\}_{n\in\mathbb{N}}$ of real numbers by induction. The desired function

$$f(z) := \sum_{n=0}^{\infty} \left(p_{k(n)+m(n)}(z) + \alpha_{n+1} p_{k(n+1)}(z) \right) \tag{1.7}$$

will be the limit of

$$S_N(z) := \sum_{n=0}^{N} \left(p_{k(n)+m(n)}(z) + \alpha_{n+1} p_{k(n+1)}(z) \right), \quad N \in \mathbb{N}.$$

We will require that $k(n) + m(n) < k(n+1)$ and $0 < \alpha_{n+1} < 1$. This ensures that (1.7) converges uniformly on E, i.e., $f \in A(E)$.

We start with $k(0) = 0$ and choose $m(0)$ such that

$$\sum_{j=m(0)}^{\infty} \frac{1}{j^2} < 1.$$

Assume now that $k(n), m(n)$ have been constructed for $0 \leq n \leq N$ and α_n for $1 \leq n \leq N$. Since S_{N-1} is the best approximant to $S_{N-1} + p_{k(N)+m(N)}$ in $\mathbb{P}_{k(N)}$, by the continuity of the best approximation operator (cf. Theorem 1.5.1) and Hurwitz's theorem, there is an $\varepsilon_N > 0$ such that for all $g \in A(E)$ with

$$\|g - (S_{N-1} + p_{k(N)+m(N)})\|_E < \varepsilon_N, \tag{1.8}$$

$p_{k(N)}^*(g)$ is a polynomial of exact degree $k(N)$ and does not have more zeros in $D(N) := \{z : |z| < N\}$ than S_{N-1} has in $D(2N)$.

Choose $k(N+1)$ such that

$$k(N+1) > (k(N) + m(N))^2,$$

$$\sum_{j=k(N+1)}^{\infty} \frac{1}{j^2} < \varepsilon_N/2.$$

By Hurwitz's theorem, we can also choose $0 < \alpha_{N+1} < \varepsilon_N/2$ such that S_N has at most $k(N) + m(N)$ zeros in $D(2N+2)$ and the function f in (1.7) satisfies (1.8) with $g = f$.

Next, we define $m(N+1)$. Again $S_{N-1} + p_{k(N)+m(N)}$ is the best approximation to S_N in $\mathbb{P}_{k(N)+m(N)}$. Its leading coefficient is

$$\beta_N := (k(N) + m(N))^{-2} \|T_{k(N)+m(N)}\|^{-1}. \tag{1.9}$$

By the continuity of the best approximation operator we can choose $\delta_N > 0$ so small that for all $g \in A(E)$ with

$$\|g - S_N\|_E \leq \delta_N, \tag{1.10}$$

the leading coefficient γ_N of $p_{k(N)+m(N)}^*(g)$ satisfies

$$|\gamma_N| > \beta_N/2. \tag{1.11}$$

We now choose $m(N+1)$ such that

$$\sum_{j=k(N+1)+m(N+1)}^{\infty} \frac{1}{j^2} < \delta_N/2. \tag{1.12}$$

By (1.12), for f defined by (1.7) we have (1.10) with $g = f$. Hence, by (1.9) and (1.11),

$$\limsup_{n \to \infty} |a_n^*|^{1/n} \geq \limsup_{n \to \infty} \|T_n\|_E^{-1/n} = (\operatorname{cap} E)^{-1}.$$

According to Proposition 1.1 and (1.2) the function f is not analytic on E.

Finally, for the zero-counting measure we have

$$\nu_{p_{k(N)}^*}(D(N)) \leq \frac{k(N-1) + m(N-1)}{k(N)} < \frac{1}{k(N)^{1/2}}.$$

For bounded $S \subset \mathbb{C}$, this estimate implies (1.6). □

Furthermore, a quantified version of Theorem 1.2 can be proved.

Theorem 1.4. *Let E be a closed quasidisk or a quasiconformal arc satisfying conditions (2.2.4) or (2.2.29). Assume that the function $f \in A(E)$ is not infinitely often differentiable on the boundary of E. Then, for all $0 < \sigma_0 < 1$ there exist infinitely many n such that for any subarc J of ∂E and $\sigma > \sigma_0$,*

$$|(\mu_E - \nu_{p_n^*})(A_{\sigma,\sigma}(J))| \leq c\left(\frac{\log n}{n}\right)^{\alpha/(1+\alpha)}$$

if E is bounded by a Jordan curve and $z_0 \in \operatorname{int} L$ is an arbitrary fixed point with $f(z_0) \neq 0$ (or

$$|(\mu_E - \nu_{p_n^*})(A_\sigma(J))| \leq c\left(\frac{\log n}{n}\right)^{\alpha/(1+\alpha)}$$

in the case of a Jordan arc E), where the constant $c > 0$ depends on f but not on J, n, and σ.

Proof. We may assume $\|f\|_E = 1$, $\operatorname{cap} E = 1$. Thus, $\|p_n^*\|_E \leq 2$ for all $n \in \mathbb{N}$. Let

$$p_{n+1}^* - p_n^* = a_{n+1}^* \Phi_{n+1} + p_n,$$

where $p_n \in \mathbb{P}_n$ and $\Phi_{n+1} \in \mathbb{M}_{n+1}$ is the Faber polynomial of degree $n+1$ on E. Then

$$|a_{n+1}^*| \|\Phi_{n+1}\|_E = \|f - (p_n^* + p_n) - (f - p_{n+1}^*)\|_E.$$

Since

$$\|\Phi_n\|_E \leq cn, \quad n \in \mathbb{N},$$

with some constant $c > 0$ (cf. Theorem 1.3.2),

$$|a_{n+1}^*| \geq \frac{E_n(f) - E_{n+1}(f)}{cn}, \quad E_n(f) := E_n(f, E).$$

Next, consider the monic polynomial $P_n := p_n^*/a_n^*$, whenever $a_n^* \neq 0$. For such n,

$$\|P_n\|_E \leq \frac{2cn}{E_{n-1}(f) - E_n(f)}. \tag{1.13}$$

Now we show that there exists a subsequence $\Lambda \subset \mathbb{N}$ and a constant $s > 0$ such that

$$E_{n-1}(f) - E_n(f) \geq n^{-s}, \quad n \in \Lambda. \tag{1.14}$$

Let us assume that (1.14) is false and fix a real number $s > 5$. Then there exists an $n_0 \in \mathbb{N}$ such that

$$E_{n-1}(f) - E_n(f) < n^{-s}, \qquad n > n_0,$$

and for all $n \geq n_0$,

$$E_n(f) = \sum_{j=1}^{\infty}(E_{n+j-1}(f) - E_{n+j}(f)) \leq \sum_{j=1}^{\infty} \frac{1}{(n+j)^s} \leq \int_n^{\infty} \frac{ds}{x^s}$$
$$= \frac{1}{(s-1)\,n^{s-1}}.$$

By the inverse theorems of approximation theory (cf. Theorem C.3.7), this means that f has $[s/2] - 2$ continuous derivatives on E. Since s was arbitrary, this contradicts our assumption and proves (1.14).

Let us first assume that E is a Jordan arc. By (1.13) and (1.14),

$$\|P_n\|_E \leq 2c\,n^{s+1}, \qquad n \in \Lambda,$$

and our statement is an immediate consequence of Theorem 2.2.4.

In the case of a Jordan curve we fix z_0 in the interior of E such that $f(z_0) \neq 0$ and apply Theorem 2.2.1 as above. □

6.2 Polynomials of Near-Best Approximation

Let us now consider the behavior of polynomials of near-best approximation. We say that a sequence of polynomials $\{p_n\}_{n \in \mathbb{N}}$ is of *near-best approximation* to f on $E \subset \mathbb{C}$ if $p_n \in \mathbb{P}_n$, $n \in \mathbb{N}$, and there is a constant $c \geq 1$ such that

$$\|f - p_n\|_E \leq c\,E_n(f, E), \qquad n \in \mathbb{N}. \qquad (2.1)$$

As is shown in the next examples, we can no longer expect that every point on ∂E is a limit point of zeros of the p_n, not to mention that the zero distributions $\nu_n := \nu_{p_n}$ associated with the p_n have a subsequence converging weakly to μ_E.

Example 2.1. Let $f(z) := \sqrt{z}$ for $\operatorname{Re} z \geq 0$ and set

$$D_a := \{z : |z - a| \leq a\}, \qquad a > 0.$$

Then, for $P_n \in \mathbb{P}_n$,

$$\sup_{z \in D_1} |f(z) - P_n(z)| = \frac{1}{\sqrt{a}} \sup_{w \in D_a} \left|f(w) - \sqrt{a}\,P_n\left(\frac{w}{a}\right)\right|.$$

Thus, if $\{p_n\}_{n\in\mathbb{N}}$ is the sequence of best uniform approximants to f on D_2, it satisfies (2.1) with $E = D_1$. But no point of $\partial D_1 \setminus \{0\}$ is a limit point of zeros of the p_n, since the p_n converge to $\sqrt{z}$ uniformly on D_2.

This example can be simply generalized. Namely, let G_1 and G_2 be Jordan domains bounded by Dini-smooth curves and such that

$$G_1 \subset G_2, \quad \partial G_1 \cap \partial G_2 = \{z_0\}.$$

For $0 < \alpha < 1$ we introduce the function

$$f_\alpha(z) := (z - z_0)^\alpha,$$

which is defined such that $f_\alpha \in A(\overline{G_2})$. For the modulus of continuity of f_α along ∂G_j, $j = 1, 2$, we have

$$\omega_j(f_\alpha, \delta) \asymp \delta^\alpha, \quad 0 < \delta \leq 1.$$

Then by Theorem C.4.1 we have

$$E_n(f_\alpha, \overline{G}_j) \asymp n^{-\alpha}.$$

Therefore, near-best approximants p_n, $n \in \mathbb{N}$, to f on $\overline{G}_2$ are also near-best approximants to f on $\overline{G}_1$. At the same time, no point of $\partial G_1 \setminus \{z_0\}$ is a limit point of zeros of the p_n.

Notice that in Example 2.1, the set $E = D_1$ has nonempty interior. An analogous example can be constructed for $E = [-1, 1]$.

Example 2.2. Set $E_1 := [-1, 1]$,

$$E_2 := \{z = x + iy : |x| \leq 2, |y| \leq x^2\},$$

and define

$$f(z) := \begin{cases} z, & \text{if Re } z \geq 0, \\ -z, & \text{if Re } z < 0, \end{cases}$$

so that $f(x) = |x|$ for $x \in \mathbb{R}$. If we denote by $E_{j,n}$ the error of the best uniform approximation to f on E_j with respect to $\mathbb{P}_n$, then according to the results of Section 8.1,

$$E_{2,n} \leq c E_{1,n}, \quad n \in \mathbb{N},$$

with some constant $c > 1$.

This result yields another example of a sequence $\{p_n\}_{n\in\mathbb{N}}$ satisfying (2.1), where $E = [-1, 1]$ and p_n is the best uniform approximation to f on E_2. Notice that no point $x \in [-1, 1] \setminus \{0\}$ can be a limit point of zeros of such p_n.

6.2 Polynomials of Near-Best Approximation   195

The above discussion naturally raises the question as to whether near-best approximants (in the sense of (2.1)) necessarily have at least one limit point of zeros on ∂E when f is not analytic on E. The next theorem shows that the answer is no; that is, it may happen that no point of ∂E attracts zeros of the whole sequence of near-best approximants.

Let $E \subset \mathbb{C}$ be an arbitrary continuum with connected complement $\Omega := \overline{\mathbb{C}} \setminus E$. Denote as before by $\Phi : \Omega \to \Delta$ the Riemann mapping function with the standard normalization $\Phi(\infty) = \infty$, $\Phi'(\infty) > 0$. Let $L := \partial E$,

$$L_r := \{\zeta : |\Phi(\zeta)| = 1 + r\}, \quad E_r := \text{int}\, L_r, \quad r > 0.$$

Theorem 2.3. *Let E and $r > 0$ be given. There exists a function $f \in A(\overline{E_r})$ that is not analytic on $\overline{E}_r$ and a sequence of polynomials $\{p_n\}_{n \in \mathbb{N}}$ of near-best approximation to f on $\overline{E}_r$ such that for any $R > r$ there is an $N = N(R)$ such that p_n has no zeros in $\overline{E}_R$ for any $n \geq N$.*

Before giving the proof of this statement we recall some auxiliary facts from the theory of Faber polynomials (for details, see [165, Chapter 2]). We denote by $\Phi_k(z)$, $k \in \mathbb{N}$, the kth Faber polynomial for E defined in Section 1.3. It is known that

$$\Phi_k(z) = \Phi(z)^k + h_k(z), \quad z \in \Omega, \qquad (2.2)$$

where $h_k(z)$ is a function analytic in Ω and satisfying $h_k(\infty) = 0$ (see [165, p. 131]). Let $0 < \rho < r < \infty$ be given. Then

$$|h_k(\zeta)| \leq c(\rho, E)(1 + \rho)^k, \qquad \zeta \in L_\rho \qquad (2.3)$$

(see [165, pp. 132, 134]). We introduce the polynomials $\widetilde{\Phi}_k(z) := \Phi_k(z)/(1+r)^k$. Let k be large enough. By (2.2) and (2.3),

$$\|\widetilde{\Phi}_k\|_{\overline{E}_r} \asymp 1. \qquad (2.4)$$

Further, since for z with $\text{dist}\,(z, L_r) \leq 1/k$,

$$|h'_k(z)| = \left| \frac{1}{2\pi} \int_{|\zeta - z| = \text{dist}(L_r, L_\rho) - 1/k} \frac{h(\zeta)}{(\zeta - z)^2} d\zeta \right| \preceq c(\rho, E)(1 + \rho)^k,$$

we have for such points

$$|\widetilde{\Phi}'_k(z)| \asymp k. \qquad (2.5)$$

For any function $f \in A(\overline{E}_r)$ we introduce its modulus of continuity along L_r as follows:

$$\omega(f, \delta) = \omega(f, L_r, \delta) := \sup_{\substack{z, \zeta \in L_r, \\ |z - \zeta| \leq \delta}} |f(z) - f(\zeta)|, \qquad \delta > 0.$$

Then, as a direct consequence of (2.5), we obtain

$$\omega(\widetilde{\Phi}_k, \delta) \preceq k\delta, \qquad 0 < \delta \leq \frac{1}{k}. \tag{2.6}$$

At the same time a simple computation involving (2.2) and (2.3) shows that

$$\omega\left(\widetilde{\Phi}_k, \frac{1}{k}\right) \asymp 1. \tag{2.7}$$

Proof of Theorem 2.3. Let $M = M(r, E) > 1$ be a sufficiently large, but fixed, number whose choice will be clear from the reasoning below. Let $m_1 := 1$ and $m_{j+1} := (m_j + 1)^{m_j}$, $j \in \mathbb{N}$. Set

$$g(z) := \sum_{j=1}^{\infty} M^{-j} \widetilde{\Phi}_{m_j}(z), \qquad z \in E_r.$$

For sufficiently large j we have $m_j \geq M^j$, which implies that

$$\lim_{j \to \infty} (M^{-j})^{1/m_j} = 1.$$

Therefore, according to the well-known results on the series of Faber polynomials (cf. [165, p. 138]), the function $g(z)$ is analytic in E_r° and has a singular point on L_r.

Next, we study the modulus of continuity $\omega(g, \delta)$ of g along L_r. We begin with $\delta = 1/m_k$ and show that

$$\omega\left(g, \frac{1}{m_k}\right) \asymp M^{-k}, \qquad k \in \mathbb{N}. \tag{2.8}$$

Indeed, let

$$u_k(z) := M^{-k} \widetilde{\Phi}_{m_k}(z),$$

$$g_k(z) := \sum_{j=1}^{k} u_j(z).$$

Since

$$g(z) = g_{k-1}(z) + u_k(z) + (g(z) - g_k(z)),$$

we obtain by (2.4), (2.6), and (2.7),

$$\omega\left(g_{k-1}, \frac{1}{m_k}\right) \leq \sum_{j=1}^{k-1} M^{-j} \omega\left(\widetilde{\Phi}_{m_j}, \frac{1}{m_k}\right) \preceq \frac{1}{m_k} \sum_{j=1}^{k-1} M^{-j} m_j \preceq \frac{m_{k-1}}{m_k},$$

$$\omega\left(u_k, \frac{1}{m_k}\right) = M^{-k} \omega\left(\widetilde{\Phi}_{m_k}, \frac{1}{m_k}\right) \asymp M^{-k},$$

$$\omega\left(g - g_k, \frac{1}{m_k}\right) \leq 2 \sum_{j=k+1}^{\infty} M^{-j} \|\widetilde{\Phi}_{m_j}\|_{\overline{E}_r} \preceq M^{-k}(M-1)^{-1}.$$

Taking M sufficiently large we get (2.8).

To extend inequality (2.8) to all $0 < \delta \leq 1$ we define the function

$$\kappa(\delta) := \begin{cases} M^{-k}, & \text{if } \dfrac{1}{m_{k+1}} \leq \delta < \dfrac{1}{m_k}, \ k \in \mathbb{N}, \\ M^{-1}, & \text{if } \delta \geq 1. \end{cases}$$

Then (2.8) implies that

$$\omega(g, \delta) \asymp \kappa(\delta), \qquad 0 < \delta \leq 1. \tag{2.9}$$

The function $\kappa(\delta)$ is nondecreasing with $\kappa(+0) = 0$, and it satisfies the inequality

$$\kappa(t\,\delta) \leq c_1 \, t^\alpha \kappa(\delta), \qquad \delta > 0, \quad t > 1 \tag{2.10}$$

with some $c_1 > 0$, $0 < \alpha < 1$.

Indeed, the nontrivial case is only when $t\delta \leq \delta_0 < 1$ is sufficiently small, in particular, such that

$$\kappa(2\,M\,t\delta) \leq M\,\kappa(t\delta).$$

For $1 < t < 2M$ inequality (2.10) is fulfilled with $c_1 = M$ and an arbitrary $\alpha > 0$. Let $t \geq 2M$. We can find $s \in \mathbb{N}_0$ such that

$$(2M)^s \leq t < (2M)^{s+1}$$

and write

$$\begin{aligned}
\kappa(t\,\delta) &\leq \kappa((2M)^{s+1}\,\delta) \\
&= \frac{\kappa((2M)^{s+1}\,\delta)}{\kappa((2M)^s\,\delta)} \frac{\kappa((2M)^s\,\delta)}{\kappa((2M)^{s-1}\,\delta)} \cdots \frac{\kappa(2M\,\delta)}{\kappa(\delta)} \kappa(\delta) \\
&\preceq M^{s+1}\kappa(\delta) = M(2M)^{\alpha s}\kappa(\delta) \leq M\,t^\alpha \kappa(\delta),
\end{aligned}$$

where

$$\alpha := \frac{\log M}{\log 2M} < 1.$$

Next, we define the function $f(z) := e^{g(z)}$. Since

$$\omega(f, \delta) \asymp \omega(g, \delta), \qquad 0 < \delta \leq 1,$$

we have by virtue of (2.9), (2.10), and Theorem C.4.1,

$$E_n(f, \overline{E}_r) \asymp \kappa\left(\frac{1}{n}\right). \tag{2.11}$$

Hence, the function $f(z)$ belongs to $A(\overline{E}_r)$ and has a singularity on L_r.
Let
$$Q_k(z) := \sum_{j=0}^{m_k} \frac{1}{j!}(g_k(z))^j.$$

Note that $Q_k \in \mathbb{P}_{m_{k+1}}$. Finally, for $m_{k+1} \leq n < m_{k+2}$, we set
$$p_n(z) := Q_k(z) \in \mathbb{P}_n.$$

We claim that f and $\{p_n\}$ satisfy all the requirements of the theorem.

To show that p_n is a polynomial of near-best approximation we need to establish, in accordance with (2.11), the relation

$$\|f - Q_k\|_{\overline{E}_r} \preceq M^{-k}, \qquad k \in \mathbb{N}. \tag{2.12}$$

To prove (2.12) we observe that for sufficiently large M,
$$\|g_k\|_{\overline{E}_r} \leq 1.$$

Thus, from the obvious inequality

$$\left| e^w - \sum_{j=0}^{N} \frac{w^j}{j!} \right| = \left| \frac{1}{N!} \int_0^w (w-t)^N e^t \, dt \right| \leq |w|^{N+1} e^{|w|} \frac{1}{N!} \tag{2.13}$$

with $N = m_k$, $w = g_k(z)$, we get

$$\left| Q_k(z) - e^{g_k(z)} \right| \leq \frac{e}{m_k!} \preceq M^{-k}. \tag{2.14}$$

At the same time, for $z \in \overline{E}_r$,

$$\left| e^{g_k(z)} - e^{g(z)} \right| \preceq |g_k(z) - g(z)| \leq \sum_{j=k+1}^{\infty} M^{-j} \|\widetilde{\Phi}_{m_j}\|_{\overline{E}_r} \preceq M^{-k}. \tag{2.15}$$

Combining (2.14) and (2.15) we obtain (2.12).

Finally, we prove that for k large enough $Q_k(z)$ has no zeros in E_k. Let $|\Phi(z)| = k+1 > r+1$. Then by (2.2) and (2.3),

$$|g_k(z)| \leq c_2 \left(\frac{k+1}{r+1}\right)^{m_k} \sum_{j=1}^{k} M^{-j} \leq \frac{m_{k+1}}{5},$$

for k sufficiently large. Taking into account (2.13) it is easy to see that for $N \geq 5\,|w|$ we have

$$\left| e^w - \sum_{j=0}^{N} \frac{w^j}{j!} \right| < e^{-|w|}. \tag{2.16}$$

Indeed, using the inequality $N! > N^N e^{-N}$ we obtain

$$\left| e^w - \sum_{j=0}^{N} \frac{w^j}{j!} \right| \leq \left(\frac{|w|}{N} \right)^N |w| e^{|w|+N} \leq |w| e^{(1-\log 5)N + 2|w|} e^{-|w|}$$

$$\leq |w| e^{-(5\log 5 - 7)|w|} e^{-|w|} \leq \frac{1}{e(5\log 5 - 7)} e^{-|w|} < e^{-|w|}.$$

Setting $N = m_k$, $w = g_k(z)$ in (2.16), we obtain

$$\left| e^{g_k(z)} - Q_k(z) \right| < e^{-|g_k(z)|} \leq \left| e^{g_k(z)} \right|.$$

Thus, Rouché's theorem guarantees that Q_k has no zeros in E_k. □

The last theorem should be compared with the next statement, which says that if we require enough regularity for the error in best approximation of the function f, then at least one point of ∂E is a limit point of the zeros of near-best approximants.

Theorem 2.4. *Let E be admissible with regular boundary, and suppose that $f \in A(E)$ is not analytic on E and does not vanish identically on any component of E° or any component of E. Suppose that there exist constants $0 < \alpha < 1$ and $c > 1$ such that*

$$\liminf_{n\to\infty} \frac{E_{[\alpha n]}(f, E)}{E_n(f, E)} > c. \qquad (2.17)$$

If $p_n \in \mathbb{P}_n$, $n \in \mathbb{N}$, satisfies (2.1) (with the same constant c), then at least one point of ∂E must be a limit point of zeros of the sequence $\{p_n\}_{n\in\mathbb{N}}$.

Note that Theorem 2.4 in particular applies to the situation in Example 2.2, i.e., $E = [-1, 1]$, $f(x) = |x|$. Indeed, since

$$\frac{c_1}{n} \leq E_n(|x|, [-1, 1]) \leq \frac{c_2}{n}$$

(cf. section 8.1), we have

$$\liminf_{n\to\infty} \frac{E_{[\alpha n]}(f, E)}{E_n(f, E)} > \frac{c_1}{2\alpha c_2}.$$

Thus, choosing $\alpha = c_1/(2 c c_2)$ we obtain inequality (2.17) for any sequence $\{p_n\}_{n\in\mathbb{N}}$ with (2.1). Hence, $\{p_n\}_{n\in\mathbb{N}}$ must have at least one limit point of zeros on $[-1, 1]$.

Proof of Theorem 2.4. Assume to the contrary that no point of ∂E is a limit point of zeros of $\{p_n\}_{n\in\mathbb{N}}$. Since (2.1) implies that the p_n converge to f uniformly on E and f does not vanish identically on any component of E°, by Hurwitz's theorem the set of zeros of f in E° coincides with the set of limit points in E° of the zeros of $\{p_n\}_{n\in\mathbb{N}}$. Thus, f can have at most finitely many zeros in E°, since otherwise, either f vanishes identically on some component of E° or a point of ∂E is a limit point of zeros of $\{p_n\}_{n\in\mathbb{N}}$.

Let $z_1,\ldots,z_m$ denote the zeros of f in E°. The assumption on the zeros of p_n implies that there exists $R_0 > 1$ such that for all n large, say $n > n_0$, the set

$$E_{R_0-1} = E \cup \{z \in \mathbb{C}\setminus E : G(z) < \log R_0\}$$

contains precisely m zeros $z_{1,n},\ldots,z_{m,n}$ of p_n, where $z_{j,n} \longrightarrow z_j$ as $n \to \infty$. We claim that for each R_1, with $1 < R_1 < R_0$,

$$\limsup_{n\to\infty} \|p_n\|_{E_{R_1-1}}^{1/n} = 1. \qquad (2.18)$$

To establish (2.18) we first define

$$\widetilde{p}_n(z) := p_n(z) \left(\prod_{j=1}^m (z - z_{j,n})\right)^{-1}, \qquad n > n_0.$$

Then $\widetilde{p}_n \in \mathbb{P}_{n-m}$, and $\widetilde{p}_n$ does not have zeros in E_{R_0-1}. Furthermore, since the p_n are uniformly bounded on E, the Bernstein–Walsh lemma (see Lemma 1.1.11) implies that

$$\limsup_{n\to\infty} \|\widetilde{p}_n\|_{E_{R_0-1}}^{1/n} \leq R_0. \qquad (2.19)$$

Next, we note that E_{R_0-1} consists of finitely many simply connected components that are bounded by Jordan curves. On any such component B of E_{R_0-1} we can define a single-valued analytic function

$$\widehat{p}_n(z) := \frac{1}{n} \log \widetilde{p}_n(z).$$

Since B must contain a component C of E and f does not vanish identically in C,

$$\lim_{n\to\infty} \widetilde{p}_n(z) = f(z) \left(\prod_{j=1}^m (z - z_j)\right)^{-1} \neq 0$$

for infinitely many points in C (recall that $\overline{\mathbb{C}}\setminus E$ is regular, so E cannot contain isolated points). Of course, for such points of C we have

$$\lim_{n\to\infty} \widehat{p}_n(z) = 0, \qquad (2.20)$$

and by Vitali's theorem $\widehat{p}_n$ converges to zero uniformly on each compact subset of B. This fact implies (2.18).

Next, let $P_{[\alpha n]} \in \mathbb{P}_{[\alpha n]}$ be the Lagrange interpolant to p_n in the $N :=$

$[\alpha n]+1$ Fekete points $\zeta_1,\ldots,\zeta_N$ for the set E. By the well-known Hermite formula (see, for example, [165, p. 11]) we have for $z \in E$,

$$p_n(z) - P_{N-1}(z) = \frac{1}{2\pi i} \int_{\partial E_{R_1-1}} \frac{p_n(\zeta)}{\zeta - z} \frac{q_N(z)}{q_N(\zeta)} d\zeta,$$

where

$$q_N(z) := \prod_{k=1}^{N}(z - \zeta_k)$$

is the Fekete polynomial. Then it follows easily from (2.18) and Lemma 1.1.14 that

$$\limsup_{n\to\infty} \|p_n - P_{[\alpha n]}\|_E^{1/n} \leq R_1^{-\alpha} < 1. \tag{2.21}$$

Furthermore, by (2.1) we have for $E_n := E_n(f, E)$,

$$E_{[\alpha n]} \leq \|f - P_{[\alpha n]}\|_E \leq \|f - p_n\|_E + \|p_n - P_{[\alpha n]}\|_E$$
$$\leq c\, E_n + \|p_n - P_{[\alpha n]}\|_E,$$

and hence

$$\frac{E_{[\alpha n]}}{E_n} \leq c + \frac{\|p_n - P_{[\alpha n]}\|_E}{E_n}. \tag{2.22}$$

But since f is not analytic on E, Theorem 1.5.3 yields

$$\limsup_{n\to\infty} E_n^{1/n} = 1.$$

Let Λ denote a subsequence of $\mathbb{N}$ for which

$$E_n^{1/n} \to 1 \quad \text{as } n \to \infty,\ n \in \Lambda.$$

Then we obtain from (2.21) that

$$\lim_{\substack{n\to\infty \\ n\in\Lambda}} \frac{\|p_n - P_{[\alpha n]}\|_E}{E_n} = 0.$$

Hence, in view of (2.22), we have

$$\liminf_{n\to\infty} \frac{E_{[\alpha n]}}{E_n} \leq c,$$

which contradicts our assumption (2.17). $\square$

6.3 Polynomials of Maximal Convergence

Let E be admissible with regular boundary, $\Omega = \overline{\mathbb{C}} \setminus E$. Suppose f is analytic on E, but not entire. Then there exists a *maximal* $R > 1$ such that f has an analytic extension to

$$E_{R-1} = \{z \in \Omega \,:\, G(z) < \log R\} \cup E.$$

Then a sequence of polynomials $p_n \in \mathbb{P}_n$, $n \in \mathbb{N}$, is said to *converge maximally* to f on E if

$$\limsup_{n \to \infty} \|f - p_n\|_E^{1/n} = \frac{1}{R} \tag{3.1}$$

(cf. Theorem 1.5.3).

Such maximally convergent polynomials can be obtained by interpolation (Walsh [181]). Let

$$Z_n \,:\, z_{0,n}, z_{1,n}, \ldots, z_{n,n} \in E, \qquad n \in \mathbb{N}, \tag{3.2}$$

be a sequence of interpolating points. Then we can interpolate the function f, analytic on E, with respect to $\mathbb{P}_n$, i.e., there exists $p_n \in \mathbb{R}_n$ such that

$$p_n(z_{k,n}) = f(z_{k,n}) \qquad \text{for } k = 0, \ldots, n,$$

if all points of Z_n are pairwise distinct. In the general case of Hermite interpolation the defining equalities have to be modified according to the multiplicity of the nodes involving derivatives of f.

By Hermite's formula, the interpolating polynomial p_n has a representation

$$p_n(z) = \frac{1}{2\pi i} \int_{L_r} \frac{f(\zeta)\,(\omega_n(\zeta) - \omega_n(z))}{\omega_n(\zeta)\,(\zeta - z)} \, d\zeta, \qquad z \in E,$$

where L_r is a level line of Green's function, $0 < r < R - 1$, and

$$\omega_n(z) := \prod_{k=0}^{n} (z - z_{k,n}). \tag{3.3}$$

By Cauchy's formula, the error $f - p_n$ can be written as

$$f(z) - p_n(z) = \frac{1}{2\pi i} \int_{L_r} \frac{\omega_n(z)\, f(\zeta)}{\omega_n(\zeta)\,(\zeta - z)} \, d\zeta. \tag{3.4}$$

From these well-known facts one can deduce that $\{p_n\}$ converges maximally for all f analytic on E if

$$\lim_{n \to \infty} \|\omega_n\|_E^{1/n} = \operatorname{cap} E. \tag{3.5}$$

If all interpolation points are located on the boundary of E, the condition (3.5) is equivalent to $\nu_n \xrightarrow{*} \mu_E$ as $n \to \infty$, where ν_n denotes the normalized counting measure $\nu_n = \nu(Z_n)$ of the interpolating point set Z_n.

Let $a \in \Omega$ be fixed and let

$$f_a(z) := \frac{1}{z-a}.$$

Then the interpolating polynomial for f_a is

$$p_n(z) = \frac{1}{z-a} - \frac{\omega_n(z)}{\omega_n(a)(z-a)}.$$

Suppose $Z_n \subset \partial E$. Then it is well known (Walsh [181]) that the condition $\nu_n \xrightarrow{*} \mu_E$ as $n \to \infty$ is necessary for the maximal convergence of the interpolating polynomials on the subclass

$$\{f_a \,:\, a \in \Omega\}.$$

On the other hand, it seems to be interesting to investigate necessary conditions about the distribution of the interpolating set Z_n if the sequence $\{p_n\}$ converges maximally to a *fixed single* function f.

Theorem 3.1. *Let E be admissible with regular boundary and let f be analytic on E, but not entire. Suppose $\{p_n\}$ is a sequence of maximally convergent polynomials to f that are interpolating f on the sets*

$$Z_n \,:\, z_{0,n}, \ldots, z_{n,n} \in \partial E.$$

Then the normalized counting measures $\nu_n = \nu(Z_n)$ have the equilibrium measure μ_E as a weak limit point.

The proof will be based on the following observation: If the measures $\nu_n \in \mathcal{M}(\partial E)$ converge to μ_E, then the logarithmic potentials U converge to U^{μ_E} locally uniformly on Ω, and vice versa by Lemma 1.1.8 and Helly's selection theorem.

Let R be as in (3.1). Since $\{p_n\}$ converges maximally to f, it is well known that

$$\limsup_{n \to \infty} \|f - p_n\|_{L_r}^{1/n} = \frac{1+r}{R}, \qquad 0 < r < R - 1. \tag{3.6}$$

Proposition 3.2. *Let f and p_n be as in Theorem 3.1, $0 < s < r < R-1$, and let $\varepsilon > 0$ be such that*

$$\|U^{\mu_E} - U^{\nu_n}\|_{L_r} \geq \varepsilon.$$

Then there exists a constant $\kappa = \kappa(\varepsilon) < 1$, independent of n, such that

$$\|f - p_n\|_{L_s}^{1/n} < \kappa \frac{1+s}{R}.$$

204 6. Jentzsch–Szegő and Erdős–Turán Type Theorems

Proof. For abbreviation let $\mu := \mu_E$. We define
$$M_\varepsilon := \{\sigma \in \mathcal{M}(\partial E) \,:\, \|U^\mu - U^\sigma\|_{L_r} \geq \varepsilon\}$$
and fix the parameter r_0 such that $0 < r_0 < s < r < R - 1$.
For $\sigma \in M_\varepsilon$ we define the harmonic function g_σ on
$$U := E_r \setminus \overline{E}_{r_0}$$
that extends continuously to L_{r_0} and L_r with
$$g_\sigma(z) := \begin{cases} 0, & z \in L_r, \\ \min(U^{\mu-\sigma}(z) - c_\sigma, 0), & z \in L_{r_0}, \end{cases} \quad (3.7)$$
where
$$c_\sigma := \min_{z \in L_r} U^{\mu-\sigma}(z).$$
By the maximum principle, $c_\sigma < 0$ and
$$g_\sigma(z) \leq 0 \,(\not\equiv 0) \qquad \text{for } z \in L_s.$$
Moreover, we claim that there exists a $\delta > 0$ such that for all $\sigma \in M_\varepsilon$,
$$g_\sigma(z) < -\delta \qquad \text{for } z \in L_s. \quad (3.8)$$
Indeed, assume to the contrary that there exists $\{\sigma_n\} \subset M_\varepsilon$ such that
$$\max_{z \in L_s} g_{\sigma_n}(z) > -\frac{1}{n}.$$
By Helly's selection theorem we assume that $\sigma_n \xrightarrow{*} \sigma$ with $\sigma \in M_\varepsilon$. The uniform convergence of U^{σ_n} on L_s implies that
$$\max_{z \in L_s} g_\sigma(z) \geq 0.$$
Together with the boundary conditions this yields $g_\sigma \equiv 0$. But this contradicts the fact that $\sigma \in M_\varepsilon$. Hence, (3.8) is true.

Since (3.6) holds, there exists $N_{1,\delta} \in \mathbb{N}$ such that for $n \geq N_{1,\delta}$,
$$\frac{1}{n} \log |(f - p_n)(z)| \leq \log(1 + r) - \log R + \frac{\delta}{4}, \qquad z \in L_r. \quad (3.9)$$

Next we define the function
$$h_n(z) := \frac{1}{n} \log |(f - p_n)(z)| - G(z) + \log R.$$
This function $h_n(z)$ is subharmonic in $E_{R-1} \cap \Omega$ and thus obeys the maximum principle in this domain.

For $z \in L_r$ we have $g_{\nu_n}(z) = 0$, and with (3.9),
$$h_n(z) \leq \frac{\delta}{4} \leq g_{\nu_n}(z) + \frac{\delta}{4}, \qquad z \in L_r.$$

For $z \in L_{r_0}$ and $g_{\nu_n}(z) = 0$ we have again

$$h_n(z) \le \frac{\delta}{4} \le g_{\nu_n}(z) + \frac{\delta}{4}, \qquad z \in L_{r_0}.$$

For $z \in L_{r_0}$ (with $g_{\nu_n}(z) < 0$) we first use Cauchy's remainder formula (3.4) and obtain

$$\frac{1}{n}\log|(f-p_n)(z)| \le -U^{\nu_n}(z) + \max_{\zeta \in L_r} U^{\nu_n}(\zeta) + \frac{1}{n}\log\|f-p_n\|_{L_r} + \frac{\delta}{4}$$

for n sufficiently large, $n \ge N_{2,\delta}$.

Then for $z \in L_{r_0}$ and $g_{\nu_n}(z) < 0$ we get for $n \ge \max(N_{1,\delta}, N_{2,\delta})$ with (3.6),n

$$h_n(z) \le -U^{\nu_n}(z) + \max_{\zeta \in L_r} U^{\nu_n}(\zeta) + \log\frac{1+r}{1+r_0} + \frac{\delta}{2}$$

$$= U^{\mu-\nu_n}(z) - \min_{\zeta \in L_r} U^{\mu-\nu_n}(\zeta) + \frac{\delta}{2}$$

$$= U^{\mu-\nu_n}(z) - c_{\nu_n} + \frac{\delta}{2} \le g_{\nu_n}(z) + \frac{\delta}{2}.$$

Altogether we have obtained on the boundary of $U = E_r \setminus \overline{E}_{r_0}$ that

$$h_n(z) \le g_{\nu_n}(z) + \frac{\delta}{2}.$$

Therefore, for $z \in L_s$ the maximum principle and (3.8) yield

$$h_n(z) \le -\frac{\delta}{2},$$

or

$$\|f-p_n\|_{L_s}^{1/n} \le \frac{1+s}{R}e^{-\delta/2} = \kappa\frac{1+s}{R},$$

where $\kappa = e^{-\delta/2} < 1$. $\qquad \square$

Proof of Theorem 3.1. Let $\mu = \mu_E$, $r > 0$ fixed. Let

$$\varepsilon := \liminf_{n\to\infty} \|U^\mu - U^{\nu_n}\|_{L_r}.$$

By Helly's selection theorem there exists a subsequence $\Lambda \subset \mathbb{N}$ such that

$$\nu_n \xrightarrow{*} \sigma \text{ as } n \to \infty, \ n \in \Lambda,$$

and

$$\varepsilon = \lim_{\substack{n\to\infty\\ n\in\Lambda}} \|U^\mu - U^{\nu_n}\|_{L_r}.$$

206   6. Jentzsch–Szegő and Erdős–Turán Type Theorems

If $\varepsilon > 0$, then by the previous proposition,

$$\|f - p_n\|_{L_s}^{1/n} < \kappa \frac{1+s}{R}, \quad n \in \mathbb{N},$$

where $\kappa > 0$. Then f is analytic in a neighborhood of E_{R-1}, contradicting the definition of R.

If $\varepsilon = 0$, then $\|U^\mu - U^\sigma\|_{L_r} = 0$, and by the maximum principle,

$$U^\mu(z) = U^\sigma(z), \quad z \in \Omega.$$

Hence Lemma 1.1.8 yields $\mu = \sigma$. □

Note that ν_n may not converge for the whole sequence to μ_E. As an example, let

$$f(z) = \sum_{n=0}^{\infty} z^{n!}$$

where $\mathbb{D}$ is the domain of convergence. Then f cannot be analytically extended out of $\mathbb{D}$. Set $E = [-\frac{1}{2}, \frac{1}{2}]$ and denote by s_n the nth partial sum of the power series of f with center 0. Then

$$\lim_{n \to \infty} \|f - s_{n!}\|_E^{1/n} = 0$$

and $\nu_{n!} \overset{*}{\not\to} \mu_E$.

Finally, we consider the zero distribution of maximally convergent polynomials p_n. Since near-best approximants to a function f of the above type are maximally convergent, the following result is in some sense surprising compared with Theorem 2.3.

Theorem 3.3. *Let E admissible with regular boundary. Suppose $f \in A(E)$ does not vanish identically on any closed connected component of E. If the sequence of polynomials $\{p_n\}_{n \in \mathbb{N}}$, $p_n \in \mathbb{P}_n$, converges maximally to f with geometrical rate $1/R$, $1 < R < \infty$, then there is a subsequence $\Lambda \subset \mathbb{N}$ such that*

$$\nu_{p_n} \overset{*}{\to} \mu_{E_{R-1}} \quad \text{as } n \to \infty, n \in \Lambda. \tag{3.10}$$

Proof. We wish to apply Theorem 2.1.1 with E replaced by E_{R-1} and $k_n = n$. By (3.1) and the Bernstein–Walsh lemma (Lemma 1.1.11), we have

$$\limsup_{n \to \infty} \|p_n - p_{n+1}\|_{E_{\widetilde{R}-1}}^{1/n} \leq \frac{\widetilde{R}}{R} \quad \text{for } \widetilde{R} \geq 1. \tag{3.11}$$

Thus, for all $1 \leq \widetilde{R} < R$ the sequence $\{p_n\}_{n \in \mathbb{N}}$ converges on $E_{\widetilde{R}-1}$, and we get (2.1.3). Moreover,

$$\limsup_{n \to \infty} \|p_n\|_{E_{R-1}}^{1/n} \leq 1,$$

which implies (2.1.2).

To establish (2.1.4) we take $S := L_{R_1-1}$ for some fixed $R_1 > R$ and prove first that there exists a subsequence $\Lambda \subset \mathbb{N}$ with the following property:

$$\lim_{\substack{n \to \infty \\ n \in \Lambda}} \frac{1}{n} \log \|p_n\|_S \geq \log \frac{R_1}{R}. \tag{3.12}$$

Indeed, assume to the contrary that

$$\limsup_{n \to \infty} \frac{1}{n} \log \|p_n\|_S < \log \frac{R_1}{R}.$$

Then we get

$$b := \limsup_{n \to \infty} \frac{1}{n} \log \|p_n - p_{n+1}\|_S < \log \frac{R_1}{R}.$$

Define a harmonic function g on $E_{R_1-1} \setminus E$ that tends to 0 on ∂E and to $b - \log(R_1/R)$ on L_{R_1-1}. By the maximum principle for subharmonic functions (see Theorem 1.1.9) and (3.9),

$$\limsup_{n \to \infty} \frac{1}{n} \log |(p_n - p_{n+1})(z)| \leq g(z) + G(z) - \log R$$

for all $z \in E_{R_1-1} \setminus E$. Thus

$$\limsup_{n \to \infty} \|p_n - p_{n+1}\|_{E_{R-1}}^{1/n} < 1.$$

It follows that $\{p_n\}$ converges in $E_{\widetilde{R}-1}$ for some $\widetilde{R} > R$, which is impossible.

Thus we have (3.4); i.e., (2.1.4) holds for p_n with $n \in \Lambda$. Now Theorem 3.1 follows directly from Theorem 2.1.1. □

That the sequence Λ in Theorem 3.3 does not, in general, coincide with $\mathbb{N}$ can be seen from the following example.

Consider the function

$$f(z) := \sum_{j=0}^{\infty} (z(1+z))^{3^j} = \sum_{j=0}^{\infty} (z^{3^j} + \cdots + z^{2 \cdot 3^j}) = \sum_{k=1}^{\infty} a_k z^k,$$

and define the sequence of polynomials

$$p_n(z) := \sum_{k=1}^{n} a_k z^k, \qquad n \in \mathbb{N}. \tag{3.13}$$

Let r_0 be the radius of convergence of the series

$$\sum_{k=1}^{\infty} a_k z^k.$$

Since the set of convergence of the subsequence $\{p_{2 \cdot 3^j}\}_{j \in \mathbb{N}}$ is the interior of the lemniscate
$$\{z : |z(1+z)| = 1\},$$
we have $0 < r_0 < \infty$. Certainly, the polynomials p_n in (3.13) converge maximally on any disk $\{z : |z| \leq r\}$ with $0 < r < r_0$. On the other hand, (3.10) is violated for $\Lambda := \{2 \cdot 3^j : j \in \mathbb{N}\}$.

6.4 a-Values of Orthogonal Polynomials on Quasidisks

Let L be a bounded Jordan curve, $G := \text{int } L$, and $E = \overline{G}$.

As usual, $\Phi : \overline{C} \setminus E \longrightarrow \Delta$ denotes the Riemann mapping function normalized by $\Phi(\infty) = \infty$ and $\Phi'(\infty) = 0$ with level curves
$$L_r = \{z \in \mathbb{C} : |\Phi(z)| = 1 + r\}. \tag{4.1}$$

In the following we consider *weight functions $h(z)$ on the domain G*; i.e., $h(z)$ is nonnegative and Lebesgu measurable on E with
$$\int_G h(z)\,dm(z) > 0, \tag{4.2}$$
where $dm(z)$ denotes the 2-dimensional Lebesgue measure.

Applying the Gram–Schmidt orthogonalization method to the sequence $1, z, z^2, \ldots$ of polynomials, we obtain uniquely defined polynomials
$$Q_n(z) = Q_n(h, z) = \lambda^n z^n + \cdots, \qquad n \in \mathbb{N}_0, \tag{4.3}$$
with $\lambda_n > 0$ and
$$\int_G Q_k(z)\overline{Q_l(z)}\,dm(z) = \begin{cases} 1, & \text{if } k = l, \\ 0, & \text{if } k \neq l. \end{cases} \tag{4.4}$$

Let $d(A, B) = \text{dist}(A, B)$ for $A, B \subset \mathbb{C}$. We assume throughout this section that
$$h(z) \geq c_1 (d(z, L))^m, \qquad z \in G, \tag{4.5}$$
with some constants $m > 0$, $c_1 > 0$.

Definition 4.1. Let $a \in \mathbb{C}$. The measure $\nu_{Q_n}^a$ that associates the mass $1/n$ with each of the a-values of the polynomial Q_n, that is, with the roots of the equation
$$Q_n(z) = a,$$
is called the *a-value counting measure of Q_n*.

6.4 a-Values of Orthogonal Polynomials on Quasidisks

The aim of this section is to derive the following consequence of Theorem 2.2.1.

Theorem 4.2. *Let L be a quasiconformal bounded Jordan curve, $z_0 \in$ int L, $0 < \alpha = \alpha(L) \leq 1$, and c_0 as in Theorem 2.2.1, and let $h(z)$ satisfy (4.5). Then for each complex number $a\ (\neq 0)$,*

$$|(\mu_L - \nu_{Q_n}^a)(A_{\sigma,\tau}(J))| \leq c \left(\frac{\log n}{n}\right)^{\alpha/(1+\alpha)}, \qquad n \geq 2, \qquad (4.6)$$

for all subarcs J of L and all σ and τ such that

$$\sigma \geq \sigma_0 := c_0 \left(\frac{\log n}{n}\right)^{\alpha/(1+\alpha)} \qquad \text{and} \qquad 1 > \tau \geq \tau_0 := \sigma_0^{1/\alpha},$$

where the positive constant c depends only on L and a.

Before we give the proof of this theorem we shall make some remarks and prove some special properties of the orthogonal polynomials.

Remark 4.3. The exceptional role in Theorem 4.2 of the value $a = 0$ becomes obvious if we consider the system

$$\left\{ \left(\frac{n+1}{\pi}\right)^{1/2} z^n \right\}_{n \in \mathbb{N}_0}$$

of orthogonal polynomials on the unit disk $\mathbb{D}$ with respect to the weight function $h(z) \equiv 1$.

Remark 4.4. The example constructed below shows that the restriction (4.5) in Theorem 4.2 cannot be omitted in general.

Indeed, let $G := \mathbb{D}$,

$$h(z) = h(|z|) := \exp\left\{-\exp\left\{\frac{1}{1-|z|}\right\}\right\}, \qquad z \in \mathbb{D}.$$

It is easy to see that the function $h(z)$ does not satisfy (4.5) and $Q_n(z) = \lambda_n z^n$, where

$$\lambda_n^{-2} = \int_0^{2\pi}\int_0^1 h(r)\, r^{2n+1}\, dr\, d\theta = 2\pi \int_0^1 h(r)\, r^{2n+1}\, dr.$$

Set (for sufficiently large n) $r_n := 1 - 1/(\log n)$. Then we have

$$(2\pi \lambda_n^2)^{-1} \leq r_n^{2n+1} + \exp\left\{-\exp\left\{\frac{1}{1-r_n}\right\}\right\} \leq \exp\left\{-c_1 \frac{n}{\log n}\right\}.$$

That means that the roots $z_1, \ldots, z_n$ of the equation $Q_n(z) = 1$ satisfy

$$|z_j| = \lambda_n^{-1/n} \leq 1 - \frac{c_2}{\log n}, \qquad j = 1, \ldots, n.$$

Thus, in this case the conclusion of Theorem 4.2 is violated.

Remark 4.5. It is interesting to compare the statement of Theorem 4.2 with Picard's theorem [75, p. 72], which asserts that an analytic function assumes in an arbitrary neighborhood of its essential singularity all finite complex values with at most one possible exception.

In order to establish (4.6) we are going to apply Theorem 2.2.1 to the monic polynomial

$$p_n(z) := \frac{Q_n(z) - a}{\lambda_n}. \tag{4.7}$$

Therefore, the appropriate estimates of the quantity $\|Q_n\|$ and the leading coefficient λ_n will be the objective of our next investigation. We begin with the following assertion.

Lemma 4.6. Let $h(z)$ satisfy (4.5). Then for each point $z_0 \in G := \text{int } L$,

$$\lim_{n \to \infty} Q_n(z_0) = 0. \tag{4.8}$$

Proof. Let $\psi(w) = \psi_{z_0}(w)$ be the conformal mapping of the unit disk $\mathbb{D}$ onto G with $\psi(0) = z_0$, $\psi'(0) > 0$. The mapping ψ transforms the polynomial

$$K_n(z) := \sum_{j=0}^{n} \overline{Q_j(z_0)}\, Q_j(z)$$

into the function $\widetilde{K}_n(w) := K_n[\psi(w)]$, and for $w \in \mathbb{D}$,

$$\widetilde{K}_n(w)\, \psi'(w) = \sum_{k=0}^{\infty} a_k\, w^k,$$

where $a_0 = K_n(z_0)\, \psi'(0)$. If we take into account that

$$d(\psi(w), L) \succeq (1 - |w|)^2, \qquad w \in \mathbb{D}$$

(cf. Lemma 1.2.5), we get

$$K_n(z_0) = \sum_{j=0}^{n} |Q_j(z_0)|^2 = \int_G h(z)\, |K_n(z)|^2\, dm(z)$$

$$\succeq \int_{\mathbb{D}} (1 - |w|)^{2m} \left| \sum_{k=0}^{\infty} a_k w^k \right|^2 dm(w)$$

$$\geq 2\pi \sum_{k=0}^{\infty} |a_k|^2 \int_0^{1/2} (1 - r)^{2m}\, r^{2k+1}\, dr$$

$$\succeq |a_0|^2 = (K_n(z_0)\, \psi'(0))^2.$$

6.4 a-Values of Orthogonal Polynomials on Quasidisks

Since
$$K_n(z_0) \preceq [\psi'(0)]^{-2} \preceq 1,$$
it follows that the series
$$\sum_{j=0}^{\infty} |Q_j(z_0)|^2$$
converges, from which (4.8) follows immediately. □

Lemma 4.7. *Let E be a closed bounded Jordan domain with boundary L and level lines L_r as in (4.1). Then for any $p \in \mathbb{M}_n$, $n \in \mathbb{N}$, and $t > 0$,*
$$\int_{(\mathrm{int}\, L_t) \setminus E} |p(z)|^2 \, dm(z) \geq 2\pi (\mathrm{cap}\, E)^{2(n+1)} t. \tag{4.9}$$

Proof. For an arbitrary fixed point $z_0 \in \mathrm{int}\, L$ set
$$q(z) = q_{n,z_0}(z) := \int_{z_0}^{z} p(\zeta) \, d\zeta,$$
$$\widetilde{q}(w) := q(\Phi^{-1}(w)), \quad w \in \Delta,$$
$$I_t := \int_{\mathrm{int}\, L_t} |p(z)|^2 \, dm(z).$$

Then by the analytic Green formula (see [146, p. 15]), we may write
$$I_t = \frac{1}{2i} \int_{L_t} p(z) \overline{q(z)} \, dz = \frac{1}{2i} \int_{|w|=1+t} \widetilde{q}'(w) \overline{\widetilde{q}(w)} \, dw.$$

Using the Laurent series expansion of the function $\widetilde{q}(w)$ in a neighborhood of ∞, i.e.,
$$\widetilde{q}(w) = \frac{\rho^{n+1}}{n+1} w^{n+1} + \sum_{k=0}^{n} b_k w^k + \sum_{k=1}^{\infty} \frac{c_k}{w^k},$$
where $\rho := \mathrm{cap}\, E$, we find that
$$I_t = \pi \left[\frac{\rho^{2(n+1)}}{n+1} (1+t)^{2(n+1)} + \sum_{k=1}^{n} k |b_k|^2 (1+t)^{2k} - \sum_{k=1}^{\infty} k |c_k|^2 (1+t)^{-2k} \right].$$

Hence,
$$\int_{\mathrm{int}\, L_t \setminus E} |p(z)|^2 \, dm(z) \geq \lim_{r \to 0} (I_t - I_r)$$
$$\geq \pi \frac{\rho^{2(n+1)}}{n+1} \left[(1+t)^{2(n+1)} - 1 \right] > 2\pi \rho^{2(n+1)} t.$$

□

Lemma 4.8. *Let L be quasiconformal and let $h(z)$ satisfy (4.5). Then for the leading coefficient λ_n of the orthogonal polynomial $Q_n(z)$ the inequalities*

$$c_1 n^{-2} \leq \lambda_n (\operatorname{cap} L)^n \leq c_2 n^c, \qquad n \in \mathbb{N}, \qquad (4.10)$$

hold with some positive constants c, c_1, c_2 independent of n.

Proof. Let $q_n(z)$ be an nth Fekete polynomial for $L := \partial G$. Lemma 1.1.13 and Theorem 1.3.3 show that

$$\|q_n\|_L \preceq \rho^n n^2, \quad \rho := \operatorname{cap} L.$$

On the other hand, using the expansion

$$q_n = \lambda_n^{-1} Q_n + d_{n-1} Q_{n-1} + \cdots + d_0 Q_0,$$

we obtain

$$\int_G h(z) |q_n(z)|^2 \, dm(z) = \lambda_n^{-2} + \sum_{k=0}^{n-1} |d_k|^2 \geq \lambda_n^{-2},$$

from which the left-hand side of (4.10) follows easily. It should be pointed out that so far we did not use the assumption that L is quasiconformal.

In order to establish the right-hand side of (4.10) we extend Φ quasiconformally to $G := \int L$ and set for $0 < u < 1$,

$$L_u^* := \{\zeta \in G : |\Phi(\zeta)| = 1 - u\}.$$

Denote by Φ_u the conformal mapping of $\operatorname{ext} L_u^*$ onto Δ normalized by $\Phi_u(\infty) = \infty$, $\Phi_u'(\infty) > 0$.

For $\zeta \in L$ Lemma B.2.2 implies that

$$c_3 u \leq |\Phi_u(\zeta)| - 1 \leq c_4 u, \qquad (4.11)$$

which yields

$$\operatorname{cap} L_u^* \leq \rho \leq (1 + c_4 u) \operatorname{cap} L_u^*. \qquad (4.12)$$

We observe that according to [3, Chapter IV] (see also Theorem B.2.1) L_u^* is K^2-quasiconformal if L is K-quasiconformal, $K \geq 1$. Consequently, by the same reason the mapping Φ_u can be extended to a K^4-quasiconformal mapping of the plane onto itself, a fact that makes it possible to use Theorem 1.2.7. Hence, for

$$\zeta \in M_u := \left\{\zeta \in G : 1 \leq |\Phi_u(\zeta)| \leq 1 + \frac{c_3}{2} u\right\}$$

we have

$$d(\zeta, L) \succeq u^{K^4}.$$

Next we apply Lemma 4.7 to the closed Jordan domain $\widetilde{E} := \overline{\operatorname{int} L_u^*}$, where $u = 1/n$, and to the level parameter $t := c_3 u/2$. We obtain with (4.9)

$$\lambda_n^{-2} = \int_G h(z) \left|\frac{Q_n(z)}{\lambda_n}\right|^2 dm(z) \succeq n^{-mK^4} \int_{M_{1/n}} \left|\frac{Q_n(z)}{\lambda_n}\right|^2 dm(z)$$

$$\succeq n^{-mK^4-1} (\operatorname{cap} L_{1/n}^*)^{2(n+1)}.$$

In view of (4.12), this yields the right-hand side of (4.10). □

Proof of Theorem 4.2. Since $|Q_n(z)|^2$ is subharmonic, we obtain with (1.1.4)

$$|Q_n(z)|^2 \leq \frac{1}{\pi r^2} \int_{D(z,r)} |Q_n(\zeta)|^2 dm(\zeta).$$

For $z \in L_{1/n}^*$ and $r = d(z, L)/2$ we deduce with (4.5) that

$$|Q_n(z)| \preceq \frac{1}{r^{m+2}} \int_{D(z,r)} h(\zeta) |Q_n(\zeta)|^2 dm(\zeta) \preceq n^c$$

with some positive constant $c > 0$, independent of n. By the Bernstein–Walsh lemma (Lemma 1.1.11 and (4.11)) we have

$$\|Q_n\|_L \preceq n^c, \qquad (4.13)$$

and therefore

$$\|p_n\|_L \preceq n^c (\operatorname{cap} L)^n,$$

where p_n is defined by (4.7). In the notation of Section 2.2, we have

$$\varepsilon_{p_n} = \sup_{z \in \mathbb{C}} U^{\mu_L - \nu_{p_n}}(z) \preceq \frac{\log n}{n}. \qquad (4.14)$$

Moreover, by (4.8),

$$\lim_{n \to \infty} \lambda_n |p_n(z_0)| = |a|,$$

and with Lemma 4.8,

$$\delta_{p_n} = \frac{1}{n} \log \frac{\|p_n\|_L}{|p(z_0)|} \preceq \frac{\log n}{n}. \qquad (4.15)$$

Then the inequality (4.2) follows from Theorem 2.2.1. □

It turns out that by imposing supplementary restrictions on the geometry of L we can also derive some information about the zeros of Q_n.

Theorem 4.9. Let L, z_0, α, and c_0 be as in Theorem 4.2, and suppose that for some $k \in \mathbb{N}$ the conformal mapping φ of $G = \operatorname{int} L$ onto $\mathbb{D}$ with $\varphi(z_0) = 0$, $\varphi'(z_0) > 0$ satisfies the condition

$$\left\|\varphi^{(k)}\right\|_G = \infty.$$

Then there exist a subsequence $\Lambda \subset \mathbb{N}$ and a positive constant c depending only on z_0, c_0, and L such that the zero counting measure $\nu_{\widetilde{Q}_n}$ of the orthogonal polynomial $\widetilde{Q}_n(z) = Q_n(h,z)$, where $h(z) \equiv 1$, satisfies the inequality

$$|(\mu_L - \nu_{\widetilde{Q}_n})(A_{\sigma,\tau}(J))| \leq c \left(\frac{\log n}{n}\right)^{\alpha/(1+\alpha)}, \qquad n \in \Lambda, \qquad (4.16)$$

for any subarc J of L and all σ and τ with the properties

$$\sigma \geq \sigma_0 := c_0 \left(\frac{\log n}{n}\right)^{\alpha/(1+\alpha)}, \qquad 1 > \tau \geq \tau_0 := \sigma_n^{1/\alpha}.$$

Proof. The assumptions of this theorem coincide with the assumptions of Theorem 5.7 below. Therefore, there exists a sequence $\Lambda \subset \mathbb{N}$ such that for each $n \in \Lambda$, $\widetilde{Q}_n(z_0)$ satisfies (5.23) with $\zeta_0 = z_0$. Applying Theorem 2.2.1 to the polynomial

$$p_n(z) := \frac{\widetilde{Q}_n(z)}{\lambda_n}, \qquad n \in \Lambda,$$

and taking into account (4.10), (4.13), and (5.23), we get (4.16). $\square$

Finally, we want to show how the difficulties with the zero distribution of $Q_n(z) = Q_n(h,z)$ can be overcome if one uses for the zeros in $G = \operatorname{int} L$ the balayage of these discrete measures to the boundary curve L. Moreover, we can handle the general case of a–values for $a \in \mathbb{C}$.

Let $a \in \mathbb{C}$. Then we subdivide the a-value counting measure $\nu_{Q_n}^a$ of Q_n by

$$\nu_{Q_n}^a = \nu_{1,Q_n}^a + \nu_{2,Q_n}^a,$$

where

$$\nu_{1,Q_n}^a(B) = \nu_{Q_n}^a(B \cap \overline{\operatorname{ext} L}), \qquad B \subset \mathbb{C},$$

and

$$\nu_{2,Q_n}^a(B) = \nu_{Q_n}^a(B \cap \operatorname{int} L), \qquad B \subset \mathbb{C}.$$

Let us denote by $\widehat{\nu_{2,Q_n}^a}$ the *balayage measure* of ν_{2,Q_n}^a in G to the boundary L. Then $\widehat{\nu_{2,Q_n}^a}(L) = \nu_{2,Q_n}^a(G)$ and

$$U^{\widehat{\nu_{2,Q_n}^a}}(z) = U^{\nu_{2,Q_n}^a}(z) \qquad \text{for } z \in \overline{\operatorname{ext} L}. \qquad (4.17)$$

Next, we define
$$\widehat{\nu_{Q_n}^a} := \nu_{1,Q_n}^a + \widehat{\nu_{2,Q_n}^a} \tag{4.18}$$

and
$$A_\sigma(J) := A_{\sigma,\sigma}(J) \cap \overline{\operatorname{ext} L} \tag{4.19}$$

for any subarc J of L.

Now we want to apply the result of Theorem 4.1.1 for the case of a piecewise Dini-smooth L, $\sigma^+ = \mu_L$, and $\sigma^- = \widehat{\nu_{Q_n}^a}$. Since $\widehat{\nu_{Q_n}^a}$ is not supported on L, we cannot immediately apply Theorem 4.1.1. But some inspection of the proof of Theorem 4.1.1 shows that its proof can be slightly modified so that

$$a(n^{-\kappa}) = \sup_{z \in L_{n-\kappa}} U^{\mu_L - \widehat{\nu_{Q_n}^a}}(z) \leq c_1 \frac{\log n}{n}, \tag{4.20}$$

where $c_1 = c_1(\kappa)$ depends only on κ. Then the following result can be derived.

Theorem 4.10. *Let $a \in \mathbb{C}$ and let L be a piecewise Dini-smooth Jordan curve with all inner angles less than or equal to π and let $h(z)$ satisfy (4.5). Then there exists a constant $c > 0$, depending only on L and a, such that*

$$\left| \mu_L - \widehat{\nu_{Q_n}^a}(A_\sigma(J)) \right| \leq c \left(\frac{\log n}{n} \right)^{1/2} \tag{4.21}$$

for all subarcs J of L and all $\sigma \geq (\log n/n)^{1/2}$.

6.5 Zeros of Bieberbach Polynomials and Their Derivatives

We begin with some standard definitions and results concerning the notion of Bieberbach polynomials (for details, see [62, Chapter I, §5]).

Let G be a bounded Jordan domain, $\zeta \in G$. By the Riemann mapping theorem there exists a unique conformal mapping $w = f_\zeta(z)$ of G onto a disk $D(r_\zeta) := \{z : |z| < r_\zeta\}$ such that

$$f_\zeta(\zeta) = 0, \quad f'_\zeta(\zeta) = 1.$$

The radius r_ζ of this disk is called the *conformal radius* of G with respect to ζ. We consider the Hilbert space

$$L^2(G) := \{ g : g \text{ analytic in } G, \|g\|_{L^2(G)}^2 := (g,g) < \infty \}$$

for the inner product

$$(g,h) := \int_G g(z)\,\overline{h(z)}\,dm(z), \qquad (5.1)$$

where $dm(z)$ is the 2-dimensional Lebesgue measure. Let $K(z,\zeta)$ denote the *Bergman kernel function* of G which has the reproducing property

$$g(\zeta) = (g, K(\cdot,\zeta)), \qquad g \in L^2(G). \qquad (5.2)$$

Then it is known that $r_\zeta = (\pi K(\zeta,\zeta))^{-1/2}$ and that for $z \in G$,

$$f'_\zeta(z) = \frac{K(z,\zeta)}{K(\zeta,\zeta)}, \qquad f_\zeta(z) = \frac{1}{K(\zeta,\zeta)} \int_\zeta^z K(t,\zeta)\,dt.$$

Next, let $\widetilde{Q}_n(z) := Q_n(1,z) = \lambda_n z^n + \cdots$, $\lambda_n > 0$, be the sequence of orthogonal polynomials for the inner product (5.1), i.e.,

$$\int_G \widetilde{Q}_k(z)\,\overline{\widetilde{Q}_l(z)}\,dm(z) = \begin{cases} 1, & k=l, \\ 0, & k \neq l. \end{cases}$$

Since G is a Jordan region, it is known that $\{Q_n\}_{n \in \mathbb{N}_0}$ forms a complete orthogonal system for $L^2(G)$ and, consequently, from the reproducing property (5.2) that $K(z,\zeta)$ has the $L^2(G)$-convergent Fourier series expansion

$$K(z,\zeta) \sim \sum_{j=0}^\infty \overline{\widetilde{Q}_j(\zeta)}\,\widetilde{Q}_j(z).$$

The nth *Bieberbach polynomial*, $n \in \mathbb{N}$, can be defined as follows:

$$\pi_n(z) = \pi_n(z,\zeta) := \frac{1}{K_{n-1}(\zeta,\zeta)} \int_\zeta^z K_{n-1}(t,\zeta)\,dt,$$

where $K_{n-1}(z,\zeta)$ denotes the partial Fourier sum

$$K_{n-1}(z,\zeta) := \sum_{j=0}^{n-1} \overline{\widetilde{Q}_j(\zeta)}\,\widetilde{Q}_j(z).$$

Before stating the main results it is convenient to formulate and prove some auxiliary assertions.

Lemma 5.1. *If $R > 1$ is given, then there exists a closed Jordan domain $E \subset G$ such that the closed region $\overline{G}$ lies interior to the level curve*

$$l_R := \{z : |\Phi_{\overline{\mathbb{C}}\backslash E}(z)| = R\}, \qquad (5.3)$$

where $\Phi_{\overline{\mathbb{C}}\backslash E}$ is the conformal mapping of $\overline{\mathbb{C}}\backslash E$ onto Δ with $\Phi_{\overline{\mathbb{C}}\backslash E}(\infty) = \infty$, $\Phi'_{\overline{\mathbb{C}}\backslash E}(\infty) > 0$.

6.5 Zeros of Bieberbach Polynomials and Their Derivatives

For a proof, see [181, p. 28].

Lemma 5.2. *The orthogonal polynomials $\widetilde{Q}_n$ satisfy*

$$\lim_{n\to\infty} \left\|\widetilde{Q}_n\right\|_{\overline{G}}^{1/n} = 1. \tag{5.4}$$

Proof. Let $R > 1$ be given. By Lemma 5.1, there exists a closed Jordan region $E \subset G$ such that $\overline{G}$ lies interior to the level curve l_R given by (5.3). Let $d := \text{dist}(E, \partial G)$. Since by (1.1.4) for $z \in E$,

$$|\widetilde{Q}_n(z)|^2 \leq \frac{1}{\pi d^2} \int_{|\zeta-z|\leq d} |\widetilde{Q}_n(\zeta)|^2 \, dm(\zeta) \leq \frac{1}{\pi d^2} \int_G |\widetilde{Q}_n(\zeta)|^2 \, dm(\zeta) = \frac{1}{\pi d^2},$$

we obtain by the Bernstein–Walsh lemma (Lemma 1.1.11)

$$\left\|\widetilde{Q}_n\right\|_{\overline{G}} \leq \left\|\widetilde{Q}_n\right\|_{l_R} \leq \frac{R^n}{d\sqrt{\pi}}.$$

Thus,

$$\limsup_{n\to\infty} \left\|\widetilde{Q}_n\right\|_{\overline{G}}^{1/n} \leq R.$$

But since R is arbitrary, we obtain

$$\limsup_{n\to\infty} \left\|\widetilde{Q}_n\right\|_{\overline{G}}^{1/n} \leq 1.$$

The inequality

$$\liminf_{n\to\infty} \left\|\widetilde{Q}_n\right\|_{\overline{G}}^{1/n} \geq 1$$

is an easy consequence of the fact that the Polynomials $\widetilde{Q}_n$ are orthogonal, i.e.,

$$\left\|\widetilde{Q}_n\right\|_{\overline{G}}^2 \geq \frac{1}{m(G)} \int_G |\widetilde{Q}_n(\zeta)|^2 \, dm(\zeta) = \frac{1}{m(G)}$$

with $m(G) = \int_G dm(\zeta) > 0$. Thus, (5.4) holds. $\square$

Since the left-hand inequality of (4.6) is true for any bounded Jordan curve, as we mentioned at the beginning of the proof of Lemma 4.8, the following lemma holds.

Lemma 5.3. *The leading coefficients λ_n of the polynomials $\widetilde{Q}_n$ satisfy*

$$\liminf_{n\to\infty} \lambda_n^{1/n} \geq \frac{1}{\text{cap } \overline{G}}. \tag{5.5}$$

Lemma 5.4. *With the above notation the asymptotic relation*

$$\limsup_{n\to\infty} |\widetilde{Q}_n(\zeta)|^{1/n} = \frac{1}{R} \leq 1, \qquad \zeta \in G, \tag{5.6}$$

holds, where $R = R(f_\zeta) \geq 1$ *is the largest number such that* f_ζ *has an analytic (single-valued) extension throughout* $G_R := \operatorname{int} L_{R-1}$.

Notice that if f_ζ has a singularity on $L = \partial G$, then $R = 1$. Moreover, if $R < \infty$, then the analytic extension of f_ζ has at least one singularity on L_{R-1}.

Proof of Lemma 5.4. Since

$$K(\zeta,\zeta) f'_\zeta(z) = K(z,\zeta) = \sum_{n=0}^{\infty} \overline{\widetilde{Q}_n(\zeta)} \, \widetilde{Q}_n(z), \qquad z \in G, \tag{5.7}$$

we see that the constants $\overline{\widetilde{Q}_n(\zeta)}$ are the Fourier coefficients of the function $K(\zeta,\zeta) f'_\zeta(z)$, i.e.,

$$\overline{\widetilde{Q}_n(\zeta)} = K(\zeta,\zeta) \int_G f'_\zeta(z) \overline{\widetilde{Q}_n(z)} \, dm(z).$$

First, suppose that $R > 1$, so that f_ζ (and hence f'_ζ) is analytic in $G_R \supset \overline{G}$, and let $P_n \in \mathbb{P}_n, n \in \mathbb{N}$, denote the polynomial of best uniform approximation to f'_ζ on $\overline{G}$. From Theorem 1.5.3 we have

$$\limsup_{n\to\infty} \|f'_\zeta - P_{n-1}\|_{\overline{G}}^{1/n} \leq \frac{1}{R}.$$

Furthermore, by the orthogonality property of the polynomials $\widetilde{Q}_n$,

$$\overline{\widetilde{Q}_n(\zeta)} = K(\zeta,\zeta) \int_G (f'_\zeta(z) - P_{n-1}(z)) \overline{\widetilde{Q}_n(z)} \, dm(z). \tag{5.8}$$

Thus, from (5.8) and the Cauchy–Schwarz inequality we have

$$\limsup_{n\to\infty} |\widetilde{Q}_n(\zeta)|^{1/n} \leq \frac{1}{R}. \tag{5.9}$$

Note that in the case where f_ζ is not analytic on $\overline{G}$, i.e., when $R = 1$, the inequality (5.9) remains valid because of (5.4).

Next, let us suppose that

$$\limsup_{n\to\infty} |\widetilde{Q}_n(\zeta)|^{1/n} = \frac{1}{R_1} < \frac{1}{R}. \tag{5.10}$$

6.5 Zeros of Bieberbach Polynomials and Their Derivatives

Then from (5.4) and Lemma 1.1.11 we have for $R < r < R_1$,

$$\limsup_{n\to\infty} \left\|\widetilde{Q}_n\right\|_{\overline{G}_r}^{1/n} \leq r,$$

and by (5.10),

$$\limsup_{n\to\infty} \left\|\overline{\widetilde{Q}_n(\zeta)}\,\widetilde{Q}_n\right\|_{\overline{G}_r}^{1/n} \leq \frac{r}{R_1} < 1.$$

Thus, the series in (5.7) converges uniformly on G_r to an analytic extension of $K(\zeta,\zeta)f'_\zeta$. But this contradicts the definition of R. Hence, (5.10) is false, and (5.6) is proved. □

We can now state the first of the main results concerning zeros of Bieberbach polynomials.

Theorem 5.5. *Suppose that G is a Jordan domain, $\zeta \in G$, and let $1 \leq R = R(f_\zeta) < \infty$ be the largest number such that f_ζ has an analytic (single-valued) extension throughout $G_R := \operatorname{int} L_{R-1}$. Let $\Lambda \subset \mathbb{N}$ be a sequence for which*

$$\lim_{\substack{n\to\infty,\\ n\in\Lambda}} |\widetilde{Q}_n(\zeta)|^{1/n} = \frac{1}{R}. \tag{5.11}$$

Then in the weak-star topology of measures, the normalized counting measures for the zeros of π_n and π'_{n+1} satisfy

$$\nu_{\pi_{n+1}} \xrightarrow{*} \mu_{\overline{G}_R}, \quad \text{as } n \to \infty,\ n \in \Lambda, \tag{5.12}$$

$$\nu_{\pi'_{n+1}} \xrightarrow{*} \mu_{\overline{G}_R}, \quad \text{as } n \to \infty,\ n \in \Lambda, \tag{5.13}$$

where $\mu_{\overline{G}_R}$ is the equilibrium measure for $\overline{G}_R$.

Proof. We shall first establish (5.13). Since

$$\pi'_{n+1}(z) = \frac{1}{K_n(\zeta,\zeta)} \sum_{j=0}^{n} \overline{\widetilde{Q}_j(\zeta)}\,\widetilde{Q}_j(z),$$

we see that the leading coefficient of $\pi'_{n+1}(z) = \gamma_n z^n + \cdots$ is given by

$$\gamma_n := \frac{\overline{\widetilde{Q}_n(\zeta)}\,\lambda_n}{K_n(\zeta,\zeta)}. \tag{5.14}$$

Notice that (5.11) yields $\gamma_n \neq 0$ for sufficiently large $n \in \Lambda$. Set $p_n := \pi'_{n+1}/\gamma_n$, $n \in \Lambda$. Then we shall show that the hypotheses of Theorem 2.1.7 are valid with $E = \overline{G}_R$.

Since $K_n(\zeta,\zeta)$ converges to $K(\zeta,\zeta) > 0$ as $n \to \infty$, we deduce from (5.14), (5.5), and (5.11) that

$$\liminf_{\substack{n \to \infty, \\ n \in \Lambda}} |\gamma_n|^{1/n} \geq \frac{1}{R \operatorname{cap} \overline{G}} = \frac{1}{\operatorname{cap} \overline{G}_R}. \tag{5.15}$$

Furthermore, from (5.4) and (5.6) it follows that π'_{n+1} converges to f'_ζ uniformly on compact subsets of G_R, and therefore

$$\limsup_{n \to \infty} \|\pi'_{n+1}\|_{\overline{G}_R}^{1/n} \leq 1. \tag{5.16}$$

Thus, from (5.15) and (5.16) we have

$$\limsup_{\substack{n \to \infty, \\ n \in \Lambda}} \|p_n\|_{\overline{G}_R}^{1/n} \leq \operatorname{cap} \overline{G}_R,$$

which establishes the first property of Theorem 2.1.7. Moreover, since f'_ζ can have at most a finite number of zeros in any compact subset $A \subset G_R$, the second property of Theorem 2.1.7 follows from the theorem of Hurwitz. Hence,

$$\nu_{p_n} = \nu_{\pi'_{n+1}} \xrightarrow{*} \mu_{\overline{G}_R} \quad \text{as } n \to \infty, n \in \Lambda.$$

Finally, we note that the leading coefficient of π_{n+1} differs from that of π'_{n+1} only by the factor $1/(n+1)$, which does not affect the nth root estimates needed for applying Theorem 2.1.7. Thus, by the same reasoning as above, we get (5.12). □

In geometric function theory Bieberbach polynomials are defined in another way. Let G be a Jordan domain, and let $\zeta_0 \in G$ be fixed, $f := f_{\zeta_0}$. It is known that f minimizes the integral

$$\int_G |g'(z)|^2 \, dm(z)$$

in the class of all functions g analytic in G and normalized by the conditions $g(\zeta_0) = 0$, $g'(\zeta_0) = 1$ (see [75, p. 30]).

In order to approximate f it is natural to consider the expression

$$\int_G |p'_n(z)|^2 \, dm(z), \quad n \in \mathbb{N},$$

which is minimized in the class

$$\{p_n \in \mathbb{P}_n : p_n(\zeta_0) = 0, \, p'_n(\zeta_0) = 1\} \tag{5.17}$$

by exactly the nth *Bieberbach polynomial* π_n for the region G (cf. [165, p. 228]).

It turns out that in the class given by (5.17), π_n also minimizes the integral
$$\int_G |f'(z) - p'_n(z)|^2 \, dm(z).$$

Thus, Bieberbach polynomials provide approximations to the mapping function f in the sense that π_n converges to f locally uniformly on compact subsets of G. The latter is a direct consequence of the fact that convergence in the norm of $L^2(G)$ implies uniform convergence on each compact subset of G (cf. [165, p. 209]). Sometimes, more can be said about the behavior of $|f(z) - \pi_n(z)|$. Since we are interested in quasidisks, it is useful to keep in mind the following result.

Theorem 5.6. *Let G be a bounded quasidisk. Then there are constants $c > 0$ and $\gamma > 0$ depending only on G and ζ_0 such that*
$$\|f - \pi_n\|_{\overline{G}} \leq c \, n^{-\gamma}. \tag{5.18}$$

For a proof, see Section 8.2.

According to Rouché's theorem the last theorem means that for n large enough π_n does not have zeros in G except for ζ_0.

It was shown in Theorem 5.5 that the zero distributions of π_n and π'_n are governed by the location of the singularities of the mapping function f. We shall be interested in the case where all boundary points of G are limit points of zeros of $\{\pi_n\}$ or $\{\pi'_n\}$. By Theorem 5.5 the boundary $L = \partial G$ attracts zeros of π_n and π'_n if and only if the function f cannot be analytically extended to a neighborhood of $\overline{G}$. This fact explains the restriction (5.19) in the following theorem.

Theorem 5.7. *Let G be a quasidisk, $L = \partial G$, $z_0 = \zeta_0 \in G$, and α and c_0 as in Theorem 2.2.1. Moreover, suppose that for some $k \in \mathbb{N}$ the function $\varphi := f/r_{\zeta_0}$ satisfies the condition*
$$\|\varphi^{(k)}\|_G = \infty. \tag{5.19}$$

Then there exist a sequence $\Lambda \subset \mathbb{N}$ and constants $c_1, c_2 > 0$, depending only on z_0, c_0, and G, such that
$$|(\mu_L - \nu_{\pi'_{n+1}})(A_{\sigma,\tau}(J))| \leq c_1 \left(\frac{\log n}{n}\right)^{\alpha/(1+\alpha)}, \qquad n \in \Lambda, \tag{5.20}$$

and
$$|(\mu_L - \nu_{\pi_{n+1}})(\overline{\Omega} \cap A_{\sigma,\tau}(J))| \leq c_2 \left(\frac{\log n}{n}\right)^{\alpha/(1+\alpha)}, \qquad n \in \Lambda, \tag{5.21}$$

for any subarc J of L and all σ and τ with the properties
$$\sigma \geq \sigma_0 := c_0 \left(\frac{\log n}{n}\right)^{\alpha/(1+\alpha)} \qquad \text{and} \qquad 1 > \tau > \tau_0 := \sigma_0^{1/\alpha}.$$

Proof. We begin with the sequence of orthogonal polynomials $\widetilde{Q}_n(z)$ and claim that there exists a sufficiently large constant $k > 0$ such that

$$\limsup_{j \to \infty} |\widetilde{Q}_j(\zeta_0)| \, j^k > 0. \tag{5.22}$$

Indeed, since for any $P \in \mathbb{P}_n$, $m, n \in \mathbb{N}$,

$$\left\| P^{(m)} \right\|_{\overline{G}} \leq c(m, G) \, n^{2m} \, \|P\|_{\overline{G}}$$

(cf. Theorem C.3.6 and Lemma 1.2.5), we have according to (4.13)

$$\left\| \widetilde{Q}_n^{(m)} \right\|_{\overline{G}} \leq c_1 n^{c+2m},$$

where $c = c(G)$, $c_1 = c_1(G, m)$ are positive constants. Therefore, if we assume to the contrary that (5.22) is not valid, i.e.,

$$\limsup_{j \to \infty} |\widetilde{Q}_j(z_0)| \, j^k = 0$$

for any $k > 0$, then the function

$$f'(z) = \frac{K(z, \zeta_0)}{K(\zeta_0, \zeta_0)}$$

and all its derivatives possess bounded uniform norms along G. This contradicts our assumption (5.19).

According to (5.22) there exist constants $c > 0$, $k \in \mathbb{N}$ and a sequence $\Lambda \subset \mathbb{N}$ such that

$$|\widetilde{Q}_n(\zeta_0)| \geq c n^{-k}, \qquad n \in \Lambda. \tag{5.23}$$

Hence, for $n \in \Lambda$

$$\pi'_{n+1}(z) = \gamma_n z^n + \cdots, \qquad \gamma_n := \frac{\overline{\widetilde{Q}_n(\zeta_0)}}{K_n(\zeta_0, \zeta_0)} \lambda_n \neq 0.$$

By virtue of (4.10) and (5.23),

$$|\gamma_n| \preceq \lambda_n \preceq n^c \, (\operatorname{cap} L)^{-n}$$

and

$$|\gamma_n| \succeq \lambda_n \, n^{-k} \succeq n^{-(2+k)} \, (\operatorname{cap} L)^{-n}.$$

In the second estimate above we have used the inequality $K_n(\zeta_0, \zeta_0) \preceq 1$ from the proof of Lemma 4.6.

Now we consider the monic polynomial

$$p_n(z) := \frac{\pi'_{n+1}(z)}{\gamma_n}, \qquad n \in \Lambda. \tag{5.24}$$

For its uniform norm on $\overline{G}$ we have

$$\|p_n\|_{\overline{G}} \preceq n^2 \|\pi_{n+1}\|_{\overline{G}} \; n^{2+k} \, (\text{cap } L)^n \preceq n^{4+k} \, (\text{cap } L)^n.$$

In addition,

$$p_n(\zeta_0) = \frac{1}{\gamma_n} \succeq n^{-c} \, (\text{cap } L)^n.$$

Applying Theorem 2.2.1 to the polynomial p_n we obtain (5.20).

For the proof of (5.21) we note that $\|\pi_n\|_{\overline{G}} \preceq 1$, $|\pi_n(z'_0)| \succeq 1$ for some fixed point $z'_0 \in G$, $z'_0 \neq z_0$, and the leading coefficient of π_n differs from the corresponding one of π'_n only by a factor $1/n$, which does not affect the estimates needed for applying Theorem 2.2.1. Thus, by the same reasoning as above, applying Theorem 2.2.1 with some fixed point $z'_0 \in G$, $z'_0 \neq z_0$, instead of z_0, we get (5.21). □

6.6 Historical Comments

The theorems of Bernstein establish an exact relationship between the rate of convergence of the sequence of best polynomial approximants to a function $f \in C[-1,1]$ and the analyticity of the function f. Walsh [180] studied the relationship between analyticity and the behavior of the zeros of maximally convergent polynomials. He showed that every boundary point of the domain E_{R-1} determined by the maximal Green line L_{R-1} in which f is analytic, is a limit point of zeros of maximally convergent polynomials. Hence, Walsh could generalize Jentzsch's result about the zeros of partial sums of power series to the larger class of maximally convergent polynomials.

Since partial sums are least-square approximants to the corresponding series, it is natural to ask whether other sequences of approximating polynomials possess the Jentzsch–Szegő property. Walsh's investigation of maximally convergent polynomials was a step in this direction. M. Fekete and J.L. Walsh [59] analyzed the asymptotic behavior of zeros of certain sequences of extremal polynomials, such as Chebyshev polynomials associated with a compact set E.

The situation for continuous functions on a compact set E having a singularity on the boundary shows a peculiar flavor. P. Borwein [41] investigated the case $E = [-1,1]$. First, he showed that f is analytic on $[-1,1]$ if all zeros of the best polynomial approximants remain outside the ellipse E_σ for some $\sigma > 0$. Second, he showed that the rate at which the zeros of the best approximants approach the interval $[-1,1]$, affects the number of derivatives of the function being approximated.

In [37] Blatt and Saff proved that every point of the boundary of E is a limit point of zeros of best approximants if f is continuous on E, analytic in E°, but not analytic at every boundary point of E and if f does not vanish identically on any component of E.

In Theorem 1.2 these results were sharpened to the weak convergence of the zero distributions to the equilibrium measure of E, at least for a subsequence, a result due to Blatt, Saff, Simkani [38]. That weak convergence does not hold for the whole sequence of best approximants was proved by Grothmann and Saff [80] and generalized by Ivanov, Saff, and Totik [88] (Theorem 1.3). Moreover, Saff and Totik [155] have shown that the zero distributions $\nu^*_{p_n}$ of the best polynomial approximants can have every possible limit distribution for subsequences $\Lambda \in \mathbb{N}$.

The discrepancy estimate of Theorem 1.4 was proved in [20]. The scheme of its proof is essentially borrowed from Blatt, Grothmann [31].

The proof of Theorem 2.3 uses ideas of Ivanov, Saff, Totik [88], where the analogous statement is proved for the case of the unit disk.

Theorem 2.4 is presented in accordance with Grothmann and Saff [80].

Theorem 3.1 is due to R. Grothmann [79], as is Theorem 3.3 [76].

The results of Section 6.4 are given in accordance with [20]. Lemma 4.6 is due to F.G. Abdullaev [1].

The last section contains results of Papamichael, Saff, and Gong [137] as well as results of [20].

7
Applications of Discrepancy Theorems

From the numerous systems of points and associated monic polynomials to which discrepancy theorems can be applied we choose Fekete points for compact sets E of $\mathbb{C}$. It is known that the counting measures for Fekete point sets converge to the equilibrium distribution of E. Furthermore, if E is a Jordan curve or arc, then this weak*-convergence can be estimated by discrepancy bounds. For analytic Jordan curves Pommerenke [144, 145] has proved sharp asymptotic estimates, which can be found in Section 7.2.

In Section 7.3 we consider the distribution of extreme points in polynomial Chebyshev approximation on admissible sets E. For the special case $E = [-1, 1]$ the best approximation is characterized by alternants that show, at least for a subsequence, the typical discrepancy estimate $O((\log n)^2/n)$ with respect to the equilibrium measure $\mu_{[-1,1]}$ as $n \to \infty$.

For L^p-approximation the characteristic point sets of the optimal error curves are sign changes, or interpolation points of best approximants to the given function f. The distribution of such sign changes of the optimal error curve is investigated in Section 7.5.

Section 7.6 deals with the problem of approximating functions on the whole real line, a problem first investigated by G. Freud in the 1970s. We restrict ourselves to the investigation of the zero distribution of extremal polynomials. Besides discrepancy estimates with respect to the associated extremal measures, lower and upper bounds for the largest zeros of the extremal polynomials are derived.

Another important topic in this chapter is the problem of determining the distribution of zeros of orthogonal polynomials on real intervals and on the unit circle (Section 7.7). It turns out that the importance of the

7.1 Distribution of Fekete Points

For any compact set $E \in \mathbb{C}$ we denote by $\mathcal{F}_n(E)$, $n \in \mathbb{N}$ and $n \geq 2$, an nth Fekete point set of E.

Theorem 1.1. *Let E be a compact set with connected and regular complement. Then for any sequence of nth Fekete point sets $\mathcal{F}_n(E)$ we have*

$$\nu(\mathcal{F}_n(E)) \xrightarrow{*} \mu_E \quad as \ n \to \infty. \tag{1.1}$$

Proof. Let $\mathcal{F}_n(E) = \{z_{1,n}, \ldots, z_{n,n}\}$. Observing that $z_{j,n} \in \partial E$ and since

$$\lim_{n \to \infty} \|q_n\|_E^{1/n} = \operatorname{cap} E$$

(cf. Lemma 1.1.14), for the nth Fekete polynomial

$$q_n(z) := \prod_{k=1}^n (z - z_{k,n})$$

we can apply Theorem 2.1.7 to get (1.1). $\square$

For a quasiconformal arc or curve L the deviation between $\nu(\mathcal{F}_n(L))$ and μ_L, measured in terms of discrepancy, can be estimated in terms of n.

Theorem 1.2. *Let L be a quasiconformal curve or arc. Then for any integer $n \geq 2$,*

$$D[\mu_L - \nu(\mathcal{F}_n(L))] \leq c \frac{(\log n)^2}{n}, \tag{1.2}$$

where the constant $c > 0$ depends only on L.

Proof. By virtue of Lemma 1.1.13 and Theorem 1.3.3 the nth Fekete polynomial q_n satisfies the conditions of Theorem 3.1.8 with $C_n = c_1 n \log n$, $n \geq 2$. From this, (1.2) follows immediately. $\square$

For a quasiconformal curve L and the domain $\Omega := \operatorname{ext} L$ we introduce as before the conformal mapping $\Phi : \Omega \to \Delta$ with standard normalization at ∞, the inverse mapping $\Psi := \Phi^{-1}$, and the level lines

$$L_r := \{\zeta : |\Phi(\zeta)| = 1 + r\}, \quad r > 0.$$

7.1 Distribution of Fekete Points

Theorem 1.3. *Let the points $z_k = z_{k,n} = \Psi(e^{i\theta_k})$, $\theta_1 < \theta_2 < \cdots < \theta_n < \theta_{n+1} := \theta_1 + 2\pi$, form a Fekete point set on the K-quasiconformal curve L, $K \geq 1$. Then*

$$\frac{c_1}{n} \leq \theta_{j+1} - \theta_j \leq \frac{c_2}{n} \quad (j=1,\ldots,n) \tag{1.3}$$

with some positive constants c_1, c_2 depending only on K.

Since the functions Φ and Ψ can be extended to quasiconformal homeomorphisms of the complex plane onto itself (cf. Theorem B.2.1), in what follows we shall often use Theorem 1.2.7 (sometimes without special reference to it) for suitable, each time naturally defined, triplets of points in $\overline{\Omega}$ and $\overline{\Delta}$.

In the proof of the above theorem we shall use the properties of the Dzjadyk kernel $K_{m'}(\zeta, z) := K_{0,1,1,m}(\zeta, z)$ (see Appendix C), which is a polynomial with respect to z of degree $m' := 3(m-1) - 1$.

Lemma 1.4. *There exist constants $c > 1$ and $k \in \mathbb{N}$ depending only on the coefficient of quasiconformality K and such that for $m > 2c$, $2 \geq |w| \geq (1 - c/m)^{-1}$, $\zeta := \Psi(w)$, and $z \in L$,*

$$\frac{1}{2|\zeta - z|} \leq |K_{m'}(\zeta, z)| \leq \frac{3}{2|\zeta - z|}. \tag{1.4}$$

Proof. Let $\widetilde{\zeta} = \widetilde{\zeta}(m) := \Psi(\Phi(\zeta)(1 + 1/m))$. From Theorem C.2.4 we see that

$$\left|\frac{1}{\zeta - z} - K_{m'}(\zeta, z)\right| \leq \frac{c_1}{|\zeta - z|} \left|\frac{\widetilde{\zeta} - \zeta}{\widetilde{\zeta} - z}\right| \leq \frac{c_2}{|\zeta - z|} \left|\frac{w}{(|w| - 1)m}\right|^{1/K^2},$$

from which (1.4) immediately follows if we choose $c := c_2^{K^2} 2^{K^2+1}$. $\square$

Proof of Theorem 1.3. To see that $\theta_{j+1} - \theta_j$ is appropriately bounded from below, we consider the fundamental polynomials

$$h_j(z) = h_j(z_j, z) := \prod_{k \neq j} \frac{z - z_k}{z_j - z_k},$$

associated with the system of points $\{z_j\}_1^n$. Since $\|h_j\|_L \leq 1$, by the Bernstein–Walsh lemma (cf. Lemma 1.1.11) we obtain for ζ with $|\zeta - z_j| \leq \text{dist}(z_j, L_{1/n}) =: d_j$ the estimate

$$|h_j(\zeta)| \leq \left(1 + \frac{1}{n}\right)^n < e.$$

Consequently, if $|\zeta - z_j| \leq d_j/2$, then

$$|h'_j(\zeta)| \leq \frac{1}{2\pi} \int_{|\zeta - \xi| = d_j/2} \frac{|h_j(\xi)|}{|\xi - \zeta|^2} |d\xi| \leq \frac{2e}{d_j}.$$

We claim that
$$|z_{j+1} - z_j| \succeq d_j \asymp |z_j - \widetilde{z}_j|, \tag{1.5}$$
where $\widetilde{z}_j := \Psi[(1+1/n)\Phi(z_j)]$.

Indeed, we have to verify this inequality only if $|z_{j+1} - z_j| \leq d_j/2$. But in this case we obtain
$$1 = |h_j(z_{j+1}) - h_j(z_j)| \leq \int_{[z_j, z_{j+1}]} |h'_j(z)|\,|dz| \preceq \frac{|z_{j+1} - z_j|}{d_j},$$
and by (1.5) via Theorem 1.2.7 we get the left-hand side of (1.3).

In order to prove the right-hand inequality in (1.3) we choose ζ such that
$$|w| = \left(1 - \frac{c}{m}\right)^{-1},$$
where c is the constant from Lemma 1.4. Further, let $s \in \mathbb{N}$, $s \geq 2K^2$, be fixed. Define the function
$$p_{s,m}(z) = p_{s,m}(\zeta, z) := K_{m'}(\zeta, z)^s (\zeta - \zeta_L)^s,$$
where $\zeta_L := \Psi(w/|w|)$. Then $p_{s,m}(z)$ is a polynomial in z of degree at most $3s(m-1)$, and, by Lemma 1.4,
$$|p_{s,m}(\zeta_L)| \geq 2^{-s}. \tag{1.6}$$

In addition, for any $z \in L$ and $\tau := \Phi(z)$ we have by Theorem 1.2.7
$$|p_{s,m}(z)| \preceq \left|\frac{\zeta - \zeta_L}{\zeta - z}\right|^s \preceq (m|w - \tau|)^{-s/K^2} \preceq (m|w-\tau|)^{-2}. \tag{1.7}$$

Set $\delta = \delta_j := \frac{1}{2}(\theta_{j+1} - \theta_j)$, $\theta_0 = \theta_{0,j} := \frac{1}{2}(\theta_{j+1} + \theta_j)$. We rename the points $\{\theta_j\}_1^n$ by $\{\theta'_j\}_1^\nu$ and $\{\theta''_j\}_1^{n-\nu}$ in such a way that
$$\theta_0 < \theta'_1 < \cdots < \theta'_\nu \leq \pi + \theta_0,$$
$$\theta_0 - \pi < \theta''_{n-\nu} < \cdots < \theta''_1 < \theta_0.$$

By the left-hand-side inequality in (1.3) we obtain
$$\theta'_j - \theta_0 \geq c_1 j/n + \delta, \qquad j = 1, \ldots, \nu,$$
$$\theta_0 - \theta''_j \geq c_1 j/n + \delta, \qquad j = 1, \ldots, n - \nu.$$

Set $m := \left[\frac{n-1}{3s}\right]$, $w := (1 + c_2/m)e^{i\theta_0}$, $\zeta := \Psi(w)$, where the constants $c_2 > 0$ and $s \in \mathbb{N}$ are chosen such that for the polynomial $p(z) := p_{s,m}(\zeta, z) \in \mathbb{P}_{n-1}$ the relations (1.6) and (1.7) are fulfilled.

Since by Lagrange's interpolation formula

$$p(z) = \sum_{j=1}^{\nu} p(\zeta_j') \, h(z_j', z) + \sum_{j=1}^{n-\nu} p(z_j'') \, h(z_j'', z),$$

where $z_j' := \Psi(e^{i\theta_j'})$, $z_j'' := \Psi(e^{i\theta_j''})$, the quantity δ can easily be estimated from above by the following reasoning:

$$2^{-s} \leq |p(\Psi(e^{i\theta_0}))| \leq \sum_{j=1}^{\nu} |p(z_j')| + \sum_{j=1}^{n-\nu} |p(z_j'')|$$

$$\preceq n^{-2} \left(\sum_{j=1}^{\nu} \left(\delta + \frac{c_1 j}{n}\right)^{-2} + \sum_{j=1}^{n-\nu} \left(\delta + \frac{c_1 j}{n}\right)^{-2} \right)$$

$$\leq 2 \sum_{j=1}^{\infty} (\delta n + c_1 j)^{-2} \leq 2 \int_0^{\infty} \frac{dx}{(\delta n + c_1 x)^2} = \frac{2}{c_1 \delta n}.$$

$\square$

7.2 Fekete Points for Domains with Analytic Boundary

In this section we will show how the results of the previous one can be supplemented and even improved if we have much more information about E, i.e., if $E = \overline{G}$ and G is a Jordan domain bounded by an arbitrary analytic curve L.

Let E be as above, and let $\mathcal{F}_n(E) = \{\omega_{n\nu}\}_{\nu=0}^{n-1}$, where

$$\omega_{n\nu} = \Psi(e^{i\theta_{n\nu}}), \qquad \theta_{n0} < \theta_{n1} < \cdots < \theta_{nn} := \theta_{n0} + 2\pi, \tag{2.1}$$

be an nth Fekete point set corresponding to E. According to Theorems 1.2 and 1.3,

$$\theta_{n\nu} = \theta_{n0} + \frac{2\pi\nu}{n} + O\left(\frac{(\log n)^2}{n}\right), \qquad \nu = 1, \ldots, n, \tag{2.2}$$

$$|\theta_{n\nu} - \theta_{nt}| \asymp \frac{|\nu - t|}{n}, \qquad \nu, t = 0, \ldots, n-1. \tag{2.3}$$

On the unit circle, the nth Fekete point sets are given by the nth roots of unity modulo a rotation. One might expect that the nth order Fekete point sets of a Jordan curve L are approximately the conformal images of suitably rotated sets of the nth roots of unity. However, the precise structure of these sets is more complicated. Pommerenke [144, 145] discovered that for curves different from a circle, a certain nonconstant shift function χ will enter the picture.

Theorem 2.1. *Let E be bounded by an analytic curve. Then the Fekete points (2.1) satisfy*

$$\theta_{n\nu} = \alpha_n + \frac{2\pi\nu}{n} + \frac{1}{n}\chi\left(\frac{2\pi\nu}{n} + \alpha_n\right) + \tau_{n\nu}, \qquad \nu = 1,\ldots,n, \qquad (2.4)$$

where $\chi(\theta) := \operatorname{Im}\{\chi^*(e^{-i\theta})\}$, $\chi^*(z)$ is analytic on $\overline{\mathbb{D}}$,

$$\frac{1}{n}\sum_{\nu=1}^{n} \tau_{n\nu}^2 \le \frac{c_1}{n^4}, \qquad (2.5)$$

$$|\tau_{n\nu}| \le \frac{c_2}{n^2}\sqrt{\log n}, \qquad \nu = 1,\ldots,n, \qquad (2.6)$$

with some constants $c_1, c_2 > 0$ depending only on E.

It will be seen below that the function χ satisfies the following integral equation:

$$\chi(\theta) = \frac{1}{\pi}\int_0^{2\pi} \operatorname{Re}\left[\frac{e^{it}\Psi'(e^{it})}{\Psi(e^{it})-\Psi(e^{i\theta})} - \frac{e^{it}}{e^{it}-e^{i\theta}}\right]\chi(t)\,dt - \arg\Psi'\left(e^{i\theta}\right). \qquad (2.7)$$

The method of proof of Theorem 2.1 is based on the following idea: First one introduces explicitly known comparison points such that the Vandermonde determinant with respect to these points is nearly maximal. Then it is shown that the Fekete points (for which this determinant takes a maximum value) must be close to the comparison points.

We begin with some auxiliary facts. For simplicity, let $\operatorname{cap} E = 1$. We recall that the Faber polynomials $\Phi_k(w) \in \mathbb{M}_k$ are defined by the relation

$$\frac{\Psi'(\zeta)}{\Psi(\zeta) - w} = \sum_{k=0}^{\infty} \frac{\Phi_k(w)}{\zeta^{k+1}}. \qquad (2.8)$$

We have already seen (for details, see Section 1.3) that

$$\Phi_k(\Psi(z)) = z^k + k\sum_{l=1}^{\infty} a_{kl}\,z^{-l} \qquad (2.9)$$

with $a_{kl} = a_{lk}$, and that for $z, \zeta \in \Delta\setminus\{\infty\}$,

$$\log\frac{\Psi(\zeta) - \Psi(z)}{\zeta - z} = -\sum_{k=1}^{\infty}\sum_{l=1}^{\infty} a_{kl}\,\zeta^{-k}z^{-l}. \qquad (2.10)$$

The numbers a_{kl} are the Grunsky coefficients, and for all complex numbers v_k they satisfy the inequality (cf. Theorem 1.3.1)

$$\sum_{k=1}^{\infty} k \left| \sum_{l=1}^{\infty} a_{kl} v_l \right|^2 \leq \sum_{k=1}^{\infty} \frac{|v_k|^2}{k}. \quad (2.11)$$

From (2.10) and the symmetry of the Grunsky coefficients we have

$$\log \Psi'(z) = - \sum_{k=1}^{\infty} c_k z^{-k}, \quad c_k := \sum_{j=1}^{k-1} a_{j,k-j} = \frac{2}{k} \sum_{j=1}^{k-1} j \, a_{j,k-j}. \quad (2.12)$$

Since L is analytic, there exists ρ, $0 < \rho < 1$, such that $\Psi(z)$ is analytic and univalent in $\{z : |z| > \rho\}$.

Lemma 2.2. *For $k, l \in \mathbb{N}$,*

$$|a_{kl}| \leq \rho^{k+l}, \quad |c_k| \leq 4 \rho^k. \quad (2.13)$$

Moreover, for $v_k \in \mathbb{C}$,

$$\sum_{k=1}^{\infty} k \rho^{-2k} \left| \sum_{l=1}^{\infty} a_{kl} v_l \right|^2 \leq \sum_{k=1}^{\infty} \frac{\rho^{2k}}{k} |v_k|^2. \quad (2.14)$$

Proof. The dilation

$$\Psi^*(z) := \rho^{-1} \Psi(\rho z) = z + \cdots$$

of the function Ψ is analytic and univalent in Δ. Let a_{kl}^* and c_k^* be defined by (2.9) and (2.12) (for Ψ^* instead of Ψ). Then

$$a_{kl} = \rho^{k+l} a_{kl}^*, \quad c_k = \rho^k c_k^*.$$

Since by (2.11), written for Ψ^* with $v_k := \delta_{kl}$ (the Kronecker symbol), $|a_{kl}^*| \leq (k \, l)^{-1/2} \leq 1$, we obtain the first part of (2.13) and

$$|c_k| \leq \rho^k \sum_{j=1}^{k-1} |a_{j,k-j}^*| \leq \rho^k \sum_{j=1}^{k-1} (j(k-j))^{-1/2} \leq 4 \rho^k.$$

From (2.11) we obtain

$$\sum_{k=1}^{\infty} k \rho^{-2k} \left| \sum_{l=1}^{\infty} a_{kl} v_l \right|^2 = \sum_{k=1}^{\infty} k \left| \sum_{l=1}^{\infty} a_{kl}^* \rho^l v_l \right|^2 \leq \sum_{k=1}^{\infty} \frac{\rho^{2k}}{k} |v_k|^2,$$

and (2.14) is proved. $\square$

Lemma 2.3. *The infinite system of equalities*

$$s_k + k \sum_{l=1}^{\infty} a_{kl} \overline{s_l} = \frac{k}{2} c_k, \quad k \in \mathbb{N}, \tag{2.15}$$

has a unique solution $\{s_k\}_{k \in \mathbb{N}}$ *satisfying*

$$\sum_{k=1}^{\infty} \frac{\rho^{-k}}{k} |s_k|^2 < \infty. \tag{2.16}$$

Proof. Set $v_{0k} := 0$ and successively

$$v_{n+1,k} := -k \sum_{l=1}^{\infty} a_{kl} \overline{v}_{nl} + \frac{k}{2} c_k.$$

By (2.14)

$$\sum_{k=1}^{\infty} \frac{\rho^{-2k}}{k} |v_{n+1,k} - v_{nk}|^2 \le \sum_{k=1}^{\infty} \frac{\rho^{2k}}{k} |v_{nk} - v_{n-1,k}|^2$$

$$\le \rho^4 \sum_{k=1}^{\infty} \frac{\rho^{-2k}}{k} |v_{nk} - v_{n-1,k}|^2 \le \cdots \le \frac{\rho^{4n}}{4} \sum_{k=1}^{\infty} k \rho^{2k} |c_k|^2.$$

Hence, taking into account (2.13) we have

$$\sum_{k=1}^{\infty} \frac{\rho^{-2k}}{k} |v_{n+1,k} - v_{nk}|^2 \le 4 \rho^{4n} (1-\rho)^{-2}.$$

This inequality provides the existence of

$$s_k := \lim_{n \to \infty} v_{nk}, \quad k \in \mathbb{N},$$

which satisfy (2.15) and (2.16).

If $\{s_k^*\}_{k=1}^{\infty}$ is another solution of (2.15), then by (2.14)

$$\sum_{k=1}^{\infty} \frac{|s_k - s_k^*|^2}{k} = \sum_{k=1}^{\infty} k \left| \sum_{l=1}^{\infty} a_{kl} (\overline{s_l} - \overline{s_l^*}) \right|^2 \le \rho^2 \sum_{k=1}^{\infty} \frac{|s_k - s_k^*|^2}{k}.$$

Hence, $s_k = s_k^*$, which proves the uniqueness of the solution. $\square$

Next, we introduce the functions

$$\chi^*(z) := 2 \sum_{k=1}^{\infty} \frac{s_k}{k} z^k,$$

$$\chi(\theta) := \operatorname{Im}\{\chi^*(e^{-i\theta})\} = i \sum_{k=1}^{\infty} \left(\frac{\overline{s_k}}{k} e^{ik\theta} - \frac{s_k}{k} e^{-ik\theta} \right).$$

From (2.16) it follows that $|s_k| \leq \sqrt{k}\, \rho^{k/2}$ for k large enough. Therefore, χ^* is analytic on $\overline{\mathbb{D}}$. A straightforward calculation shows that $\chi(\theta)$ satisfies the integral equation (2.7).

Example 2.4. For $0 < d < 1$ consider the function

$$\Psi(z) = \Psi_d(z) := z + \frac{d}{z}, \quad z \in \Delta,$$

and suppose that the closed domain $E = E_d$ is bounded by the ellipse $\partial E = \Psi(\partial D)$ with semiaxes of length $1+d$ and $1-d$. Since for $\zeta, z \in \Delta$,

$$\log \frac{\Psi(\zeta) - \Psi(z)}{\zeta - z} = \log\left(1 - \frac{d}{\zeta z}\right),$$

we obtain for the Grunsky coefficients

$$a_{kk} = \frac{d^k}{k}, \quad a_{kl} = 0, \quad k \neq l,$$

and for the coefficients from (2.12)

$$c_{2k} = \frac{d^k}{k}, \quad c_{2k+1} = 0.$$

It is simple to see that the solution of (2.15) (which is unique) is given by the formula

$$s_{2k} = \frac{d^k}{2(1+d^{2k})}, \quad s_{2k+1} = 0.$$

Thus, for this example

$$\chi(\theta) = -\sum_{k=1}^{\infty} \frac{d^k}{k(1+d^{2k})} \sin 2k\theta.$$

To describe the distribution of the points $\{\theta_{n\nu}\}$, we introduce the families of sequences

$$t_{n\nu} := \alpha_n + \frac{2\pi\nu}{n} + \frac{1}{n}\chi\left(\frac{2\pi\nu}{n} + \alpha_n\right), \quad \nu = 1, \ldots, n, \tag{2.17}$$

of comparison points. Then, if we take (2.4) as definition for $\tau_{n\nu}$,

$$\sum_{\nu=1}^{n} \tau_{n\nu} = \sum_{\nu=1}^{n}\left(\theta_{n\nu} - \frac{2\pi\nu}{n}\right) - n\,\alpha_n - \frac{1}{n}\sum_{\nu=1}^{n}\chi\left(\alpha_n + \frac{2\pi\nu}{n}\right).$$

Since the last term is uniformly bounded, we can choose α_n such that

$$\sum_{\nu=1}^{n} \tau_{n\nu} = 0. \tag{2.18}$$

234   7. Applications of Discrepancy Theorems

We may also assume that $n_0 \geq 3$ is so large that

$$|\tau_{n\nu}| < \varepsilon, \quad \nu = 1, \ldots, n, \quad n > n_0, \qquad (2.19)$$

where the choice of $\varepsilon = \varepsilon(E)$ will be fixed later. Equation (2.2) shows that this is possible.

Lemma 2.5. *For $n \to \infty$ we have uniformly in $k \in \mathbb{N}$,*

$$\sum_{\nu=1}^{n} e^{ikt_{n\nu}} = s_k + k^2 O\left(\frac{1}{n}\right), \qquad (2.20)$$

and uniformly in $\mu = 1, \ldots, n$,

$$\operatorname{Im}\left\{\sum_{\substack{\nu=1 \\ \nu \neq \mu}}^{n} e^{it_{n\mu}} \Psi'(e^{it_{n\mu}})/(\Psi(e^{it_{n\mu}}) - \Psi(e^{it_{n\nu}}))\right\} = O\left(\frac{1}{n}\right). \qquad (2.21)$$

Proof. In order to simplify the notation, we set for $\nu = 0, 1, \ldots, n$,

$$t_\nu := t_{n\nu}, \quad t_{\nu \pm n} := t_{n\nu \pm 2\pi}, \quad \beta_\nu = \beta_{n\nu} := \alpha_n + \frac{2\pi\nu}{n}. \qquad (2.22)$$

For $k \geq n$ (2.20) is trivial. Thus, we assume $0 < k < n$. By virtue of (2.17) and by definition of the function χ,

$$\sum_{\nu=1}^{n} e^{ikt_\nu} = \sum_{\nu=1}^{n} e^{ik(\alpha_n + 2\pi\nu/n)}(e^{ik\chi(2\pi\nu/n + \alpha_n)/n} - 1)$$

$$= \sum_{\nu=1}^{n} e^{ik(\alpha_n + 2\pi\nu/n)}\left(\frac{ik}{n}\chi\left(\frac{2\pi\nu}{n} + \alpha_n\right) + k^2 O\left(\frac{1}{n^2}\right)\right)$$

$$= \frac{ik}{n}\sum_{\nu=1}^{n} e^{ik(\alpha_n + 2\pi\nu/n)} \chi(2\pi\nu/n + \alpha_n) + k^2 O\left(\frac{1}{n}\right)$$

$$= \frac{ik}{2\pi}\int_{\alpha_n}^{2\pi+\alpha_n} e^{ikt}\chi(t)\,dt + k^2 O\left(\frac{1}{n}\right) = s_k + k^2 O\left(\frac{1}{n}\right).$$

To prove (2.21) we first note that for $t \neq \theta$,

$$\lim_{r \to 1+0} \operatorname{Im}\left\{\sum_{k=0}^{\infty} e^{ikt} e^{-ik\theta} r^{-k}\right\} = \operatorname{Im}\left\{\frac{1}{1 - e^{i(t-\theta)}}\right\} = \frac{1}{2}\cot\frac{t-\theta}{2}.$$

By (2.8) and (2.9) we have for $t \neq \theta$,

$$\operatorname{Im}\left\{\frac{e^{i\theta}\Psi'(e^{i\theta})}{\Psi(e^{i\theta}) - \Psi(e^{it})}\right\} = \frac{1}{2}\cot\frac{t-\theta}{2} + \operatorname{Im}\left\{\sum_{k=1}^{\infty}\sum_{l=1}^{\infty} k\, a_{kl}\, e^{-ilt} e^{-ik\theta}\right\}.$$

7.2 Fekete Points for Domains with Analytic Boundary

Consequently, for $\mu = 1, \ldots, n$,

$$\operatorname{Im}\left\{\sum_{\substack{\nu=1 \\ \nu \neq \mu}}^{n} e^{it_\mu} \Psi'(e^{it_\mu})/(\Psi(e^{it_\mu}) - \Psi(e^{it_\nu}))\right\}$$

$$= \frac{1}{2}\sum_{\substack{\nu=1 \\ \nu \neq \mu}}^{n} \cot \frac{t_\nu - t_\mu}{2} + \operatorname{Im}\left\{\sum_{\substack{\nu=1 \\ \nu \neq \mu}}^{n}\sum_{k=1}^{\infty}\sum_{l=1}^{\infty} k\, a_{kl}\, e^{-ilt_\nu - ikt_\mu}\right\}. \quad (2.23)$$

Further,

$$\frac{1}{2}\sum_{\substack{\nu=1 \\ \nu \neq \mu}}^{n} \cot \frac{t_\nu - t_\mu}{2} = \frac{1}{4}{\sum_{\nu}}'\left(\cot \frac{t_{\mu+\nu} - t_\mu}{2} + \cot \frac{t_{\mu-\nu} - t_\mu}{2}\right)$$

$$= \frac{1}{4}{\sum_{\nu}}' \frac{\sin \frac{1}{2}(t_{\mu+\nu} + t_{\mu-\nu} - 2t_\mu)}{\sin \frac{1}{2}(t_{\mu+\nu} - t_\mu)\sin \frac{1}{2}(t_{\mu-\nu} - t_\mu)}, \quad (2.24)$$

where ${\sum_{\nu}}'$ means that the summation is taken over all $\nu \neq 0$ with

$$-\left[\frac{n}{2}\right] \leq \nu \leq \left[\frac{n}{2}\right] - \frac{1}{2}(1 + (-1)^n).$$

By virtue of (2.17) and (2.22),

$$t_{\mu \pm \nu} - t_\mu = \pm\frac{2\pi\nu}{n} + \frac{\chi\left(\frac{2\pi(\mu \pm \nu)}{n} + \alpha_n\right) - \chi\left(\frac{2\pi\mu}{n} + \alpha_n\right)}{n}$$

$$= \pm\frac{2\pi\nu}{n}\left(1 + O\left(\frac{1}{n}\right)\right),$$

$$t_{\mu+\nu} + t_{\mu-\nu} - 2t_\mu = \frac{1}{n}\left(\chi\left(\frac{2\pi(\mu+\nu)}{n} + \alpha_n\right) + \chi\left(\frac{2\pi(\mu-\nu)}{n} + \alpha_n\right)\right.$$
$$\left. - 2\chi\left(\frac{2\pi\mu}{n} + \alpha_n\right)\right) = \frac{\nu^2}{n^3}O(1).$$

From (2.24) we obtain

$$\frac{1}{2}\sum_{\substack{\nu=1 \\ \nu \neq \mu}}^{n} \cot \frac{t_\nu - t_\mu}{2}$$

$$= -\frac{1}{8n}{\sum_{\nu}}' \frac{\chi(\beta_\mu + 2\pi\nu/n) + \chi(\beta_\mu - 2\pi\nu/n) - 2\chi(\beta_\mu)}{\sin^2 \pi\nu/n} + O\left(\frac{1}{n}\right)$$

$$= -\frac{1}{16\pi}\int_{-\pi}^{\pi} \frac{\chi(\beta_\mu + \theta) + \chi(\beta_\mu - \theta) - 2\chi(\beta_\mu)}{\sin^2 \theta/2}\, d\theta + O\left(\frac{1}{n}\right).$$

236   7. Applications of Discrepancy Theorems

The definition of the function χ and the elementary relation,

$$\int_{-\pi}^{\pi} \frac{\sin^2 k\theta/2}{\sin^2 \theta/2} \, d\theta = 2\pi k, \qquad k \in \mathbb{N},$$

(cf. (C.1.4)) yield

$$\frac{1}{2} \sum_{\substack{\nu=1 \\ \nu \neq \mu}}^{\infty} \cot \frac{t_\nu - t_\mu}{2} = \operatorname{Im}\left\{ \sum_{k=1}^{\infty} \frac{1}{2\pi} \frac{s_k}{k} e^{-ik\beta_\mu} \int_{-\pi}^{\pi} \frac{\sin^2 k\theta/2}{\sin^2 \theta/2} \, d\theta \right\} + O\left(\frac{1}{n}\right)$$

$$= \operatorname{Im}\left\{ \sum_{k=1}^{\infty} s_k \, e^{-ik\beta_\mu} \right\} + O\left(\frac{1}{n}\right)$$

$$= \operatorname{Im}\left\{ \sum_{k=1}^{\infty} s_k \, e^{-ikt_\mu} \right\} + O\left(\frac{1}{n}\right). \tag{2.25}$$

The last conclusion in (2.25) follows from (2.17), (2.22), and (2.16).
    Next, by virtue of (2.13) and (2.20),

$$\sum_{\substack{\nu=1 \\ \nu \neq \mu}}^{n} \sum_{k=1}^{\infty} \sum_{l=1}^{\infty} k \, a_{kl} \, e^{-ikt_\mu} e^{-ilt_\nu}$$

$$= \sum_{k=1}^{\infty} \sum_{l=1}^{\infty} k \, a_{kl} \left(\sum_{\nu=1}^{n} e^{-ilt_\nu} \right) e^{-ikt_\mu} - \sum_{k=1}^{\infty} \sum_{l=1}^{\infty} k \, a_{kl} \, e^{-i(k+l)t_\mu}$$

$$= \sum_{k=1}^{\infty} \left(k \sum_{l=1}^{\infty} a_{kl} \, \overline{s_l} - \sum_{j=1}^{k-1} j \, a_{j, k-j} \right) e^{-ikt_\mu} + O\left(\frac{1}{n}\right).$$

Hence, (2.21) follows from (2.25), (2.15), and (2.12).   □

Lemma 2.6. *Let* $\sum_{\mu=1}^{n} z_\mu = 0, n \geq 3$, *and*

$$a^2 := \sum_{\mu=1}^{n} \sum_{\substack{\nu=1 \\ \nu \neq \mu}}^{n} \frac{|z_\mu - z_\nu|^2}{4 \sin^2(\pi(\mu-\nu)/n)}.$$

Then

$$\sum_{\mu=1}^{n} |z_\mu|^2 \leq \frac{2a^2}{n}, \tag{2.26}$$

$$|z_\nu| \leq \frac{2a}{n} \sqrt{\log n}, \qquad \nu = 1, \ldots, n. \tag{2.27}$$

7.2 Fekete Points for Domains with Analytic Boundary

Proof. Consider the function

$$h(t) := \sum_{m=0}^{n-1} \gamma_m e^{imt}, \qquad 0 \leq t \leq 2\pi,$$

satisfying the interpolation conditions

$$h\left(\frac{2\pi\nu}{n}\right) = z_\nu, \qquad \nu = 1, \ldots, n.$$

Since

$$\sum_{\nu=1}^{n} z_\nu = \sum_{\nu=1}^{n} h\left(\frac{2\pi\nu}{n}\right) = n\gamma_0,$$

we obtain $\gamma_0 = 0$. Some computations show that

$$a^2 = \sum_{\mu=1}^{n} \sum_{\substack{\nu=1 \\ \nu \neq \mu}}^{n} \left|\frac{h(2\pi\mu/n) - h(2\pi\nu/n)}{e^{2\pi i\mu/n} - e^{2\pi i\nu/n}}\right|^2 = n \sum_{m=1}^{n-1} m(n-m)|\gamma_m|^2.$$

Furthermore,

$$\sum_{\nu=1}^{n} |z_\nu|^2 = n \sum_{m=1}^{n-1} |\gamma_m|^2 \leq \frac{n}{n-1} \sum_{m=1}^{n-1} m(n-m)|\gamma_m|^2 = \frac{a^2}{n-1} \leq \frac{2a^2}{n}.$$

To complete the proof of the lemma we have for $\nu = 1, \ldots, n$,

$$|z_\nu| = \left|h\left(\frac{2\pi\nu}{n}\right)\right| \leq \sum_{m=1}^{n-1} |\gamma_m|$$

$$\leq \left(\sum_{m=1}^{n-1} m(n-m)|\gamma_m|^2\right)^{1/2} \left(\sum_{m=1}^{n-1} \frac{1}{m(n-m)}\right)^{1/2}$$

$$\leq \frac{a}{\sqrt{n}} \left(\frac{2}{n} \sum_{m=1}^{n-1} \frac{1}{m}\right)^{1/2} \leq \frac{2a}{n} \sqrt{\log n}.$$

Hence (2.27) is true. □

Now we are in the position to prove Theorem 2.1.

Proof of Theorem 2.1. In what follows we will omit the index n and use an abbreviated notation for the summation over μ, ν, j (from 1 to n) and k, l

238   7. Applications of Discrepancy Theorems

(from 1 to ∞). From (2.10) it follows that

$$F(u) = F(u_1, \ldots, u_n) := \log \prod_{\mu \neq \nu} \prod |\Psi(e^{iu_\mu}) - \Psi(e^{iu_\nu})|$$

$$= \sum_{\mu \neq \nu} \sum \left(\log |e^{iu_\mu} - e^{iu_\nu}| + \mathrm{Re}\left\{ \log \frac{\Psi(e^{iu_\mu}) - \Psi(e^{iu_\nu})}{e^{iu_\mu} - e^{iu_\nu}} \right\} \right)$$

$$= \sum_{\mu \neq \nu} \sum \left(\log \left|2 \sin \frac{u_\mu - u_\nu}{2}\right| - \mathrm{Re}\left\{ \sum_k \sum_l a_{kl}\, e^{-iku_\mu - ilu_\nu} \right\} \right). \tag{2.28}$$

By Taylor's formula and (2.4),

$$F(\theta) = F(t) + \sum_\nu \tau_\nu F'_\nu(t) + \frac{1}{2} \sum_\mu \sum_\nu \tau_\mu \tau_\nu F''_{\mu\nu}(u), \tag{2.29}$$

where

$$\theta = (\theta_1, \ldots, \theta_n),\ t = (t_1, \ldots, t_n),\ u = (u_1, \ldots, u_n),$$
$$u_\nu = t_\nu + \lambda \tau_\nu,\ 0 \leq \lambda = \lambda_n \leq 1, \tag{2.30}$$
$$F'_\nu(t) := \frac{\partial F(t)}{\partial t_\nu},\ F''_{\mu\nu}(t) := \frac{\partial^2 F(t)}{\partial t_\mu \partial t_\nu}.$$

Since

$$F'_\nu(t_1, \ldots, t_n) = 2 \sum_{\substack{\mu \\ \mu \neq \nu}} \mathrm{Im} \frac{e^{it_\nu} \Psi'(e^{it_\nu})}{(\Psi(e^{it_\mu}) - \Psi(e^{it_\nu}))},$$

we get by (2.21)

$$|F'_\nu(t_1, \ldots, t_n)| \leq \frac{c_1}{n},\quad \nu = 1, \ldots, n. \tag{2.31}$$

The constant c_1 in (2.31) and the following constants $c_2, \ldots$ depend on E only. According to the maximal property of Fekete points,

$$F(t_1, \ldots, t_n) \leq F(\theta_1, \ldots, \theta_n).$$

Therefore, by (2.29) and (2.31),

$$-\sum_\mu \sum_\nu \tau_\mu \tau_\nu F''_{\mu\nu}(u_1, \ldots, u_n) \leq \frac{2c_1}{n} \sum_\nu |\tau_\nu|. \tag{2.32}$$

Next, (2.28) and (2.23) yield

$$F''_{\mu\nu}(u) = \frac{1}{2 \sin^2((u_\mu - u_\nu)/2)} + 2\, \mathrm{Re}\left\{ \sum_k \sum_l kl\, a_{kl}\, e^{-iku_\mu - ilu_\nu} \right\},$$

7.2 Fekete Points for Domains with Analytic Boundary

for $\mu \neq \nu$ and,

$$F''_{\mu\mu}(u) = \sum_{\substack{j \\ j \neq \mu}} \left(-\frac{1}{2\sin^2((u_\mu - u_j)/2)} + 2\operatorname{Re}\left\{\sum_k \sum_l k^2 a_{kl}\, e^{-iku_\mu - ilu_j}\right\}\right).$$

Consequently, we obtain

$$-\sum_\mu \sum_\nu \tau_\mu \tau_\nu F''_{\mu\nu}(u) =$$

$$\sum_{\mu \neq \nu} \sum \frac{(\tau_\mu - \tau_\nu)^2}{4\sin^2((u_\mu - u_\nu)/2)}$$

$$- \operatorname{Re}\left\{2\sum_{\mu \neq \nu}\sum \sum_k \sum_l a_{kl}\, e^{-iku_\mu - ilu_\nu}\left(kl\,\tau_\mu \tau_\nu + k^2 \tau_\mu^2\right)\right\}. \quad (2.33)$$

To investigate the second term on the right-hand side of (2.33) we write it in the form $\operatorname{Re}\{A + B\}$, where

$$A := 2\sum_k \sum_l a_{kl} \sum_\mu \Bigl(k^2 \sum_{\nu \neq \mu} e^{-ilu_\nu} - kl\, e^{-ilu_\mu}\Bigr) e^{-iku_\mu} \tau_\mu^2,$$

$$B := 2\sum_k \sum_l a_{kl}\left(k\sum_\mu e^{-iku_\mu}\tau_\mu\right)\left(l\sum_\nu e^{-ilu_\nu}\tau_\nu\right).$$

The inequalities (2.13) yield

$$|A| \leq 2\sum_k \sum_l \rho^{k+l}\left(k^2 \left|\sum_\nu e^{ilu_\nu}\right| + k(k+l)\right)\sum_\mu \tau_\mu^2.$$

By (2.30), $|u_\nu - t_\nu| \leq |\tau_\nu|$. Using (2.19) and (2.20) we get

$$\left|\sum_\nu e^{ilu_\nu}\right| \leq \left|\sum_\nu e^{ilt_\nu}\right| + l\sum_\nu |\tau_\nu| \leq c_2\, l^2 + ln\varepsilon, \quad (2.34)$$

and, consequently,

$$|A| \leq c_3\,(1 + n\varepsilon)\sum_\mu \tau_\mu^2. \quad (2.35)$$

To estimate $|B|$ we use (2.14) and the Cauchy–Schwarz inequality to write

$$|B| = 2 \sum_k \left(k^{1/2} \rho^k \sum_\mu e^{-iku_\mu} \tau_\mu \right) \left(k^{1/2} \rho^{-k} \sum_l l\, a_{kl} \sum_\nu e^{-ilu_\nu} \tau_\nu \right)$$

$$\leq 2 \left(\sum_k k\rho^{2k} \left| \sum_\mu e^{-iku_\mu} \tau_\mu \right|^2 \right)^{1/2}$$

$$\left(\sum_k k\rho^{-2k} \left(\sum_l l\, a_{kl} \sum_\nu e^{-ilu_\nu} \tau_\nu \right)^2 \right)^{1/2}$$

$$\leq 2 \sum_k k\rho^{2k} \left| \sum_\mu e^{-iku_\mu} \tau_\mu \right|^2$$

$$= \sum_\mu \sum_\nu \sum_k k\rho^{2k} e^{ik(u_\nu - u_\mu)} \left(\tau_\nu^2 + \tau_\mu^2 - (\tau_\mu - \tau_\nu)^2 \right)$$

$$\leq \left| \sum_{\mu \neq \nu} \frac{\rho^2 e^{i(u_\nu - u_\mu)}}{(1 - \rho^2 e^{i(u_\nu - u_\mu)})^2} (\tau_\nu - \tau_\mu)^2 \right| + 2 \sum_k k\rho^{2k} \left| \sum_\nu e^{iku_\nu} \right| \sum_\mu \tau_\mu^2.$$

Taking into account (2.34) we obtain

$$|B| \leq \sum_{\mu \neq \nu} \frac{\rho^2 (\tau_\nu - \tau_\mu)^2}{(1 - \rho^2)^2 + 4\rho^2 \sin^2((u_\nu - u_\mu)/2)} + c_4 (1 + n\varepsilon) \sum_\mu \tau_\mu^2$$

$$\leq \sum_{\mu \neq \nu} \frac{1 - (1 - \rho^2)^2/5}{4 \sin^2((u_\mu - u_\nu)/2)} (\tau_\nu - \tau_\mu)^2 + c_4 (1 + n\varepsilon) \sum_\mu \tau_\mu^2.$$

Inserting the last inequality and (2.35) into (2.33) we obtain

$$\frac{(1 - \rho^2)^2}{20} \sum_{\mu \neq \nu} \frac{(\tau_\mu - \tau_\nu)^2}{\sin^2((u_\mu - u_\nu)/2)} - c_5 (1 + n\varepsilon) \sum_\mu \tau_\mu^2$$

$$\leq -\sum_\mu \sum_\nu F''_{\mu\nu}(u) \tau_\mu \tau_\nu. \qquad (2.36)$$

Set
$$\sigma^2 := \sum_\nu \tau_\nu^2.$$

Then
$$\sum_\nu |\tau_\nu| \leq \sqrt{n}\, \sigma.$$

Further, by (2.30),
$$u_\mu - u_\nu = (1 - \lambda)(t_\mu - t_\nu) + \lambda(\theta_\mu - \theta_\nu)$$

with some $0 \leq \lambda \leq 1$. From (2.3) and (2.17) we conclude that

$$|u_\mu - u_\nu| \leq c_6 \frac{|\mu - \nu|}{n},$$

and, consequently,

$$\left|\sin \frac{u_\mu - u_\nu}{2}\right| \leq c_7 \sin \frac{\pi |\mu - \nu|}{n}.$$

Hence, by (2.36) and (2.32),

$$\sum\sum_{\mu \neq \nu} \frac{(\tau_\mu - \tau_\nu)^2}{4\sin^2(\pi(\mu - \nu)/n)} \leq \frac{c_8 \sigma}{\sqrt{n}} + c_9 (1 + n\varepsilon) \sigma^2. \qquad (2.37)$$

According to (2.26) and (2.18),

$$\sigma^2 \leq 2 c_8 \sigma n^{-3/2} + 2 c_9 (n^{-1} + \varepsilon) \sigma^2$$

for $n > n_0 \geq 3$. If we choose $\varepsilon := (4c_9)^{-1}$, then for sufficiently large $n > n_1 \geq n_0$,

$$\sigma^2 \leq c_{10} \sigma n^{-3/2},$$

which corresponds to (2.5).

Next, from (2.37) we derive

$$\sum\sum_{\mu \neq \nu} \frac{(\tau_\mu - \tau_\nu)^2}{4\sin^2(\pi(\mu - \nu)/n)} \leq \frac{c_{11}}{n^2}.$$

Inequality (2.6) follows from the estimate (2.27). □

We complete this section with the remark that for E bounded by an analytic curve we have, according to Theorem 2.1,

$$\frac{1}{n} \leq D[\mu_E - \nu(\mathcal{F}_n(E))] \leq \frac{c_{12}}{n}$$

(cf. Theorem 1.2).

7.3 Extreme Points in Uniform Polynomial Approximation

Let $E \subset \mathbb{C}$ be admissible, i.e., E is a compact set of positive capacity such that its complement $\Omega := \overline{\mathbb{C}} \backslash E$ is connected. As before, denote by $A(E)$ the set of functions continuous on E and analytic in the interior of E, endowed with the supremum norm $\|\cdot\|_E$. For $f \in A(E)$ let

$$E_n(f, E) := \inf_{p \in \mathbb{P}_n} \|f - p\|_E, \qquad n \in \mathbb{N},$$

and let $p_n^*(f)$ be the best uniform approximant to f out of $\mathbb{P}_n$, i.e.,

$$\|f - p_n^*(f)\|_E = E_n(f, E).$$

We are interested in the properties of the *set of extreme points*

$$A_n(f) := \{z \in E : |f(z) - p_n^*(f)(z)| = E_n(f, E)\}.$$

It is well known that $A_n(f)$ consists of at least $n+2$ points. However, $A_n(f)$ can actually coincide with ∂E for every n. As before, denote by $\mathcal{F}_n(K)$ any nth Fekete point set of a compact set $K \subset \mathbb{C}$.

One of the main results of this section is following theorem.

Theorem 3.1. *Let $E \subset \mathbb{C}$ be admissible, $f \in A(E)$. Then for some subsequence $\Lambda \subset \mathbb{N}$,*

$$\nu\big(\mathcal{F}_{n+1}(A_{n-1}(f))\big) \xrightarrow{*} \mu_E \quad \text{as } n \in \Lambda,\ n \to \infty. \tag{3.1}$$

Before giving the proof we want to discuss some auxiliary facts. For a compact set $K \subset \mathbb{C}$, we consider the Chebyshev number

$$t_n(K) := \|T_n\|_K = \inf_{p \in \mathbb{P}_{n-1}} \{\|p\|_K : p \in \mathbb{M}_n\},$$

where $T_n(z)$, $n \in \mathbb{N}$, is the nth Chebyshev polynomial of the set K (cf. Section 1.1.9). The proof of Theorem 3.1 is based on simple lemmas.

Lemma 3.2. *For any compact set K containing at least $n+1$ points,*

$$t_n(\mathcal{F}_{n+1}(K)) \geq \frac{1}{n+1} t_n(K).$$

Proof. Let $\mathcal{F}_{n+1}(K) = \{z_0, z_1, z_2, \ldots, z_n\}$ and let $T_n(z)$ be the nth Chebyshev polynomial of K. It follows from Lagrange's interpolation formula that

$$T_n(z) = \sum_{k=0}^{n} T_n(z_k) L_k(z),$$

where

$$L_k(z) = \prod_{i \neq k} \frac{z - z_i}{z_k - z_i}.$$

The extremal property of the Fekete point set $\mathcal{F}_{n+1}(K)$ implies that

$$|L_k(z)| \leq 1, \quad z \in K.$$

Hence,

$$t_n(K) = \max_{z \in K} |T_n(z)| \leq (n+1) \max_{0 \leq k \leq n} |T_n(z_k)| \leq (n+1) t_n(\mathcal{F}_{n+1}(K)).$$

$\square$

Lemma 3.3. Let $\{a_n\}_{n=1}^{\infty}$ be a decreasing sequence of positive real numbers with limit 0. Then there exists a subsequence $\Lambda \subset \mathbb{N}$ such that
$$\frac{a_n + a_{n+1}}{a_n - a_{n+1}} \leq n^2 \quad \text{for } n \in \Lambda, \quad n \to \infty.$$

Proof. Since $\lim\limits_{m \to \infty} a_m = 0$ and
$$a_{m+1} = a_1 \prod_{n=1}^{m} \frac{a_{n+1}}{a_n} = a_1 \prod_{n=1}^{m} \left(1 - \frac{a_n - a_{n+1}}{a_n}\right),$$
we conclude that the series
$$\sum_{n=1}^{\infty} \frac{a_n - a_{n+1}}{a_n} \quad \text{as well as} \quad \sum_{n=1}^{\infty} \frac{a_n - a_{n+1}}{a_n + a_{n+1}}$$
cannot converge. Therefore, there exists $\Lambda \subset \mathbb{N}$ such that $a_n \neq a_{n+1}$ for $n \in \Lambda$ and
$$\frac{a_n + a_{n+1}}{a_n - a_{n+1}} \leq n^2, \quad n \in \Lambda.$$
$\square$

Lemma 3.4. Suppose $f \in A(E)$. Then there exists a subsequence $\Lambda \subset \mathbb{N}$ with the property
$$t_n(E) \leq n^2 \, t_{n-1}(A_{n-1}(f)), \quad n \in \Lambda.$$

Proof. Let $p_n^*(f) - a_n z^n \in \mathbb{P}_{n-1}$ and let us denote by $T_n^*(z)$ the Chebyshev polynomial of degree n on the set $A_{n-1}(f)$; i.e.,
$$t_n(A_{n-1}(f)) = \|T_n^*\|_{A_{n-1}(f)}.$$
For abbreviation set $E_n := E_n(f, E)$. Since $E_{n-1}(f, A_{n-1}(f)) = E_{n-1}$, we obtain
$$E_{n-1} \leq \|f - p_n^*(f) - a_n T_n^*\|_{A_{n-1}(f)} \leq E_n + |a_n| \, t_n(A_{n-1}(f)).$$
On the other hand,
$$|a_n| \, t_n(E) \leq \|p_n^*(f) - p_{n-1}^*(f)\| \leq E_{n-1} + E_n.$$
Hence, if $E_n \neq E_{n-1}$, then
$$t_n(E) \leq \frac{E_{n-1} + E_n}{E_{n-1} - E_n} \, t_n(A_{n-1}(f)).$$
Since $\lim_{n \to \infty} E_n = 0$, Lemma 3.3 yields
$$t_n(E) \leq n^2 \, t_{n-1}(A_{n-1}(f)).$$
$\square$

244   7. Applications of Discrepancy Theorems

Combining Lemma 3.2 and Lemma 3.4 we see that

$$t_n(\mathcal{F}_{n+1}(A_{n-1}(f))) \geq (n+1)^{-3} t_n(E), \qquad n \in \Lambda. \tag{3.2}$$

Let $\mathcal{F}_{n+1}(A_{n-1}(f)) = \{z_{0,n}, \ldots, z_{n,n}\}$. Since this is an $(n+1)$-point set, we can easily compute the Chebyshev number

$$\tilde{t}_n := t_n(\mathcal{F}_{n+1}(A_{n-1}(f))).$$

In fact, $\tilde{t}_n$ is determined (cf. Meinardus [121, §6, pp. 72 ff.]) by

$$\tilde{t}_n = \left|\sum_{k=0}^n \alpha_k p(z_{k,n})\right| \bigg/ \sum_{k=0}^n |\alpha_k|,$$

where p is any polynomial in $\mathbb{M}_n$, $\alpha_k = \omega'_{n+1}(z_{k,n})$ for $0 \leq k \leq n$, and

$$\omega_{n+1}(x) = \prod_{k=0}^n (x - z_{k,n}). \tag{3.3}$$

Hence, choosing $p(z) = \prod_{k=1}^n (z - z_{k,n})$ we obtain

$$\tilde{t}_n = \left(\sum_{k=0}^n |\alpha_k|\right)^{-1}.$$

Then (3.2) shows that

$$|\omega'_{n+1}(z_{k,n})| \geq (n+1)^{-3} t_n(E) \tag{3.4}$$

for every $0 \leq k \leq n$ and $n \in \Lambda$.

Proof of Theorem 3.1. Let $\mathcal{F}_{n+1}(E) = \{y_{0,n+1}, \ldots, y_{n,n+1}\}$,

$$q_{n+1}(z) = q_{n+1}(E, z) := \prod_{k=0}^n (z - y_{k,n+1}).$$

By the Lagrange interpolation formula, we have for the polynomials ω_{n+1} of (3.3),

$$q_{n+1}(z) - \omega_{n+1}(z)$$
$$= \sum_{k=0}^n (q_{n+1}(z_{k,n}) - \omega_{n+1}(z_{k,n})) \frac{\omega_{n+1}(z)}{\omega'_{n+1}(z_{k,n})(z - z_{k,n})}$$

(note that $q_{n+1} - \omega_{n+1} \in \mathbb{P}_n$). Since $\omega_{n+1}(z_{k,n}) = 0$ for every k and

$$|q_{n+1}(z_{k,n})| \leq \|q_{n+1}\|_E =: M_{n+1},$$

7.3 Extreme Points in Uniform Polynomial Approximation

we see that for the n's satisfying (3.4), i.e., for $n \in \Lambda$,

$$|q_{n+1}(z) - \omega_{n+1}(z)| \leq (n+1)^3 \frac{M_{n+1}}{t_n(E)} |\omega_{n+1}(z)| \sum_{k=0}^{n} \frac{1}{|z - z_{k,n}|},$$

$$\left|\frac{q_{n+1}(z)}{\omega_{n+1}(z)}\right| \leq 1 + (n+1)^3 \frac{M_{n+1}}{t_n(E)} \sum_{k=0}^{n} \frac{1}{|z - z_{k,n}|},$$

which implies that for $n \in \Lambda$ and every closed subset K of $\Omega = \overline{\mathbb{C}} \setminus E$,

$$\frac{1}{n+1} \log \left|\frac{q_{n+1}(z)}{\omega_{n+1}(z)}\right| \leq c(K) \, \varepsilon_{n+1}, \quad z \in K, \tag{3.5}$$

where

$$\varepsilon_{n+1} := \frac{\log(n+1)}{n+1} + \frac{1}{n+1} \log\left(1 + \frac{M_{n+1}}{t_n(E)}\right).$$

By (1.1.31),

$$\lim_{n \to \infty} M_n^{1/n} = \operatorname{cap} E,$$

which, together with Lemma 1.1.14, implies that $\varepsilon_n \to 0$ as $n \to \infty$. It is known from (1.1.32) that for $z \in \mathbb{C} \setminus E$,

$$\lim_{n \to \infty} \frac{1}{n+1} \log |q_{n+1}(z)| = G(z) + \log \operatorname{cap} E, \tag{3.6}$$

where $G(z)$ is the Green function for Ω with pole at infinity.

Since all the zeros of ω_{n+1} and q_{n+1} lie in E, the functions

$$u_{n+1}(z) := \frac{1}{n+1} \log \left|\frac{q_{n+1}(z)}{\omega_{n+1}(z)}\right|, \quad n \in \Lambda,$$

are harmonic in Ω, vanish at ∞, and have the upper bound $c(M) \, \varepsilon_{n+1}$ on every closed subset M of Ω (cf. (3.5)). Thus, we can apply Harnack's inequality [132, p. 115] to the function $-u_{n+1}(z) + c(M) \varepsilon_{n+1}$ and get that uniformly on every closed subset K' of the interior of K,

$$|u_{n+1}(z)| \leq c_1 \, \varepsilon_{n+1}, \quad c_1 = c_1(K') \quad (n \in \Lambda).$$

Making use of (3.6), we finally obtain that uniformly on compact subsets of $\mathbb{C} \setminus E$,

$$\lim_{\substack{n \to \infty \\ n \in \Lambda}} \frac{1}{n+1} \log |\omega_{n+1}(z)| = G(z) + \log \operatorname{cap} E,$$

which can be written via potentials in the equivalent form:

$$\lim_{\substack{n \to \infty \\ n \in \Lambda}} U^{\tau_n}(z) = U^{\mu_E}(z), \quad z \in \mathbb{C} \setminus E,$$

where
$$\tau_n := \nu(\mathcal{F}_{n+1}(A_{n-1}(f))).$$

Thus, application of Theorem 1.1.10 and Lemma 1.1.8 yields (3.1). □

It is natural to ask whether equidistribution is achieved for *all* large n (instead of *some* large n), and, if this is not true in general, whether it can be established at least when f is nice, say analytic. Below, we construct a counterexample that implies that (3.1) need not be true for $\Lambda = \mathbb{N}$, even for entire functions.

Theorem 3.5. *Let E be admissible. For each $z_0 \in \partial E$ there exists an entire function f such that for some $\delta > 0$ and some infinite sequence $\Lambda \subset \mathbb{N}$,*
$$|z_0 - z| \geq \delta, \qquad z \in A_n(f), \quad n \in \Lambda. \tag{3.7}$$

To establish Theorem 3.5 we require some auxiliary results.

Lemma 3.6. *For a given $z_0 \in E$ consider a sequence of polynomials*
$$P_{2n}(z) := (z - z_0)^n p_n(z), \qquad p_n \in \mathbb{P}_n, \quad n \in \mathbb{N} \tag{3.8}$$
such that $\|P_{2n}\|_E \leq 1$ for all $n \in \mathbb{N}$. Then there exists a $\delta > 0$ such that $P_{2n}(z)$ tends to zero uniformly on $D(z_0, \delta) = \{z : |z - z_0| < \delta\}$.

Proof. Let $E_\delta := E \setminus D_0(\delta)$. It can easily be seen that cap $E_\delta > 0$ for $0 < \delta < \delta_0$. Furthermore, we have $|\delta_0^n p_n(z)| \leq 1$ for all $z \in E_{\delta_0}$. Applying the Bernstein–Walsh lemma (cf. Lemma 1.1.11), we can find a constant $M_0 > 0$ (independent of n) such that
$$|p_n(z)| \leq M_0^n \qquad \text{for } z \in \overline{D(z_0, \delta))}.$$

Thus, for every $0 < \delta < \delta_0$ and $z \in D_0(\delta)$ we have
$$|P_{2n}(z)| \leq \delta^n M_0^n, \qquad n \in \mathbb{N}.$$

Finally, choosing $\delta < 1/M_0$, we can ensure that $P_{2n}(z)$ tends to 0 uniformly on $D(z_0, \delta)$. □

The next statement is a straightforward application of a result of Krein (cf. Theorem 1.5.2).

Lemma 3.7. *Let $z_0 \in E$. Then there exists a polynomial $P_{2n}(z)(\not\equiv 0)$ of the form (3.8) such that 0 is its best uniform approximant on E with respect to $\mathbb{P}_{n-1}$.*

We can now proof Theorem 3.5.

Proof of Theorem 3.5. Set $n_0 := 0$, $n_{k+1} := 2n_k + 1$, $q_k(z) := (z - z_0)^{n_k} p_{n_k}(z)$, where q_k is chosen in such a way that $\|q_k\|_E = 1$ and 0 is its best approximating element with respect to $\mathbb{P}_{n_k-1}$ (cf. Lemma 3.7).

Next, let $\{a_{k,0}\}_{k=0}^{\infty}$ be a sequence of positive numbers converging monotonically to 0 and such that

$$\sum_{k=0}^{\infty} a_{k,0} M_k < \infty, \tag{3.9}$$

where $M_k := \max_{|z_k| \leq k} |q_k(z)|$, $k = 0, 1, 2, \ldots$. The last condition ensures that functions of type

$$f = \sum_{k=0}^{\infty} c_k \, q_k, \tag{3.10}$$

where $0 < c_k \leq a_{k,0}$, are entire functions.

Now we define the coefficients c_k inductively. Assume that $c_0, c_1, \ldots, c_s$ are fixed as well as a sequence $\{a_{k,s}\}_{k=s}^{\infty}$, where $a_{k,s} \leq a_{k,s-1}$ for $k = s, s+1, \ldots$. Let

$$f_j := \sum_{k=0}^{j} c_k \, q_k, \qquad j = s, s-1.$$

Then f_{s-1} is the best approximant of f with respect to $\mathbb{P}_{2n_{s-1}}$, and

$$\|f_s - f_{s-1}\|_E = c_s \|q_s\|_E = c_s. \tag{3.11}$$

Because of Lemma 3.6 there exists $\delta > 0$ such that $|q_m(z)| < \frac{1}{2}$ for $z \in D(z_0, \delta)$ and all $m \geq m_0$. Hence

$$|(f_s - f_{s-1})(2)| < c_s/2 \qquad \text{for } z \in D(z_0, \delta) \text{ and } m \geq m_0. \tag{3.12}$$

Let $p_m^*(g)$ denote the best uniform approximation of $g \in C(E)$ with respect to $\mathbb{P}_m$, $m = 0, 1, 2, \ldots$. We know from Theorem 1.5.1 that the operator $p_m^*(g)$ is continuous on $C(E)$. Therefore, there exists a sequence

$$\{a_{k,s+1}\}_{k=s+1}^{\infty} \qquad (0 < a_{k,s+1} \leq a_{k,s})$$

such that the subset

$$V_{s+1} := \left\{ g : g = f_s + \sum_{k=s+1}^{\infty} b_k \, q_k, \, 0 < b_k \leq a_{k,s+1} \right\}$$

of entire functions has the following property, which is a consequence of (3.11) and (3.12): All best approximants $p_{2n_{s-1}}^*(g)$ ($g \in V_{s+1}$) satisfy

$$A_{2n_{s-1}}(g) \cap D(z_0, \delta) = \emptyset.$$

Then we define $c_{s+1} := a_{k,s+1}$. Summarizing, the function

$$f = \sum_{k=0}^{\infty} c_k q_k$$

is entire, and $A_n(f) \cap D(z_0, \delta) = \emptyset$ for a subsequence $n \in \Lambda \subset \mathbb{N}$. □

Next, we give an application of Theorem 4.1.6 to the distribution of extreme points in best complex polynomial approximation.

Theorem 3.8. *Let L be a quasiconformal curve or arc and let $E = \overline{\text{int } L}$ if L is a curve, and $E = L$ when L is an arc. Let $f \in A(E)$. Then there exist infinitely many integers n satisfying*

$$D[\nu(\mathcal{F}_{n+1}(A_{n-1}(f))) - \mu_E] \leq c \frac{(\log n)^2}{\sqrt{n}}. \tag{3.13}$$

Proof. Let $\mathcal{F}_{n+1}(A_{n-1}(f)) = \{z_{0,n}, \ldots, z_{n,n}\}$ and $\omega_{n+1}(z)$ be defined as in (3.3). Then estimate (3.4) and Lemma 1.1.12 yield for infinitely many integers n,

$$|\omega'_{n+1}(z_{j,n})| \geq \frac{(\text{cap } E)^n}{(n+1)^3}, \qquad 0 \leq j \leq n. \tag{3.14}$$

Thus, (3.13) follows immediately from Theorem 4.1.6. □

Note that for piecewise Dini-smooth arcs and some domains bounded by a piecewise Dini-smooth curve estimate, (3.13) can even be slightly improved (up to the logarithmic term) if we take into account inequality (3.14) and apply Theorem 4.1.5.

7.4 Alternation Points in Chebyshev Approximation

Let $C(I)$ denote the class of continuous real-valued functions on $I = [-1, 1]$ endowed with the Chebyshev norm

$$\|g\|_\infty := \max_{-1 \leq x \leq 1} |g(x)|, \quad g \in C(I).$$

Let $f \in C(I)$. Then there exists a unique best uniform polynomial $B_{n,\infty} := B_{n,\infty}(f) \in \mathbb{P}_n$ such that

$$E_{n,\infty} := E_{n,\infty}(f) := \inf_{P_n \in \mathbb{P}_n} \|f - P_n\|_\infty = \|f - B_{n,\infty}\|_\infty.$$

It is well known that $B_{n,\infty}$ is uniquely determined by an *alternant* X_n consisting of $n+2$ points

$$-1 \leq x_0^{(n)} < x_1^{(n)} < \cdots < x_{n+1}^{(n)} \leq 1,$$

7.4 Alternation Points in Chebyshev Approximation

where
$$(f - B_{n,\infty})\left(x_i^{(n)}\right) = \delta(-1)^i E_{n,\infty}(f) \tag{4.1}$$

with $\delta = +1$ or $\delta = -1$ is fixed.

Let $\mathcal{A}_n(f)$ denote the collection of all alternants of length $n+2$ and let $\nu(X_n)$ denote the normalized counting measure of $X_n \in \mathcal{A}_n$. The next theorem provides a discrepancy estimate $D[\nu(X_n) - \mu]$, where μ is again the equilibrium distribution of I.

Theorem 4.1. *There exists an absolute constant $c > 0$ (independent of $f \in C(I)$) and a subsequence $\Lambda \subset \mathbb{N}$ such that*

$$D[\nu(X_n) - \mu] \leq c \frac{(\log n)^2}{n}, \quad n \in \Lambda,$$

for all $X_n \in \mathcal{A}_n(f)$.

The proof uses some techniques that we also need later on. Hence, it is worthwhile to formulate these results as separate lemmas.

Lemma 4.2. *Let k, n be positive integers and $P_{n+k} \in \mathbb{P}_{n+k}$ a polynomial such that there exist $n+2$ points x_i, $-1 \leq x_0 < x_1 < \cdots < x_{n+1} \leq 1$, with*

$$P_{n+k}(x_i) P_{n+k}(x_{i+1}) < 0, \quad 0 \leq i \leq n,$$

and
$$|P_{n+k}(x_i)| \geq 1, \quad 0 \leq i \leq n+1.$$

Set
$$\omega_{n+2} := \prod_{i=0}^{n+1}(x - x_i) \in \mathbb{M}_{n+2}.$$

Then
$$|\omega'_{n+2}(x_i)| \geq \left[\binom{2n+k}{2n+1} \frac{n+k}{n+1} E_{n,\infty}(P_{n+k})\right]^{-1} \frac{1}{2^n}, \quad 0 \leq i \leq n+1. \tag{4.2}$$

Proof. Since the $P_{n+k}(x_i)$ alternate in sign, the well-known de la Vallée Poussin Theorem yields

$$E_{n,\infty}(P_{n+k}) \geq E_{n,\infty,X}(P_{n+k}) \geq 1, \tag{4.3}$$

where
$$E_{n,\infty,X}(P_{n+k}) = \min_{P \in \mathbb{P}_n} \max_{x \in X}|(P_{n+k} - P)(x)|$$

denotes the minimal error on the discrete set,

$$X = \{x_0, \ldots, x_{n+1}\},$$

while $E_{n,\infty,X}(P_{n+k})$ is determined (cf. Meinardus [121, §6, pp. 72 ff.]) by

$$E_{n,\infty,X}(P_{n+k}) = \left|\sum_{j=0}^{n+1} \alpha_j (P_{n+k} - P)(x_j)\right| \bigg/ \sum_{j=0}^{n+1} |\alpha_j|, \qquad (4.4)$$

where one can choose for P any polynomial in $\mathbb{P}_n$ and

$$\alpha_i = \prod_{\substack{j=0 \\ j \neq i}}^{n+1} (x_i - x_j)^{-1}, \quad 0 \leq i \leq n+1. \qquad (4.5)$$

Now consider the polynomial $q_0 \in \mathbb{P}_n$ that interpolates $R_{n+k} := P_{n+k} - B_{n,\infty}(P_{n+k})$ at the points x_j, $1 \leq j \leq n+1$. Then

$$R_{n+k}(x) - q_0(x) = \frac{R_{n+k}^{(n+1)}(\xi)}{(n+1)!} \prod_{j=1}^{n+1}(x - x_j), \qquad (4.6)$$

where $\xi = \xi(x) \in (-1, 1)$. A result of Duffin and Schaefer (cf. [150, p. 119]) shows that

$$\left|R_{n+k}^{(n+1)}(x)\right| \leq \frac{(n+k)^2((n+k)^2 - 1^2)\cdots((n+k)^2 - n^2)}{1 \cdot 3 \cdot 5 \cdots (2n+1)} E_{n,\infty}(P_{n+k})$$

for all $x \in I$. Hence,

$$\left|R_{n+k}^{(n+1)}(x)\right| \leq 2^n n! \frac{(n+k+n)\cdots(n+k-n)}{1 \cdot 2 \cdots (2n+1)} (n+k) E_{n,\infty}(P_{n+k})$$

$$= 2^n n! \binom{2n+k}{2n+1} (n+k) E_{n,\infty}(P_{n+k}). \qquad (4.7)$$

Inserting $P = q_0 + B_{n,\infty}(P_{n,k})$ in (4.4) we obtain with (4.3)

$$\sum_{j=0}^{n+1} |\alpha_j| \leq |\alpha_0 (R_{n+k} - q_0)(x_0)|,$$

and it follows, using (4.6) and (4.7), that

$$\frac{1}{|\omega'_{n+2}(x_j)|} = |\alpha_j| \leq E_{n,\infty}(P_{n+k}) \binom{2n+k}{2n+1} \frac{n+k}{n+1} 2^n$$

for $j = 0, 1, \cdots, n+1$. □

Lemma 4.3. *Let $n \geq 2$ and $-1 \leq x_1 < x_2 < \cdots < x_n \leq 1$. Moreover, let $P_n \in \mathbb{M}_n$ with*

$$P_n(x_i) P_n(x_{i+1}) < 0 \quad \text{for } i = 1, \ldots, n-1.$$

7.4 Alternation Points in Chebyshev Approximation

Then there exists a constant $c > 0$ such that

$$\|\omega_n\|_I \leq c\, n^3 \|P_n\|_I,$$

where $\omega_n(x) = \prod_{i=1}^{n}(x - x_i)$.

Proof. Let ξ_i, $1 \leq i \leq n-1$, be the zeros of P_n that separate the points x_i; i.e.,

$$x_1 < \xi_1 < x_2 < \cdots < x_{n-1} < \xi_{n-1} < x_n.$$

Set $x_0 := 0$ and $x_{n+1} := 1$. Fix $x \in I$ and choose k such that $0 \leq k \leq n+1$ and $x_k \leq x \leq x_{k+1}$. Then

$$|\omega_n(x)| \leq 2 \prod_{\substack{j=1 \\ j \neq k}}^{n-1} |x - \xi_j|$$

and therefore

$$|\omega_n(x)| \leq 2\, n^3 \|P_n\|_I \quad \text{if } |x - \xi_k| \geq n^{-3}.$$

Summarizing,

$$|\omega_n(x)| \leq 2\, n^3 \|P_n\|_I, \quad x \in A,$$

where

$$A = \left\{ x \in I : \min_{1 \leq j \leq n-1} |x - \xi_j| \geq n^{-3} \right\}.$$

Since $\operatorname{mes}(A) > 2 - n^2$, we get by Remez's inequality (cf. Theorem 1.5.4)

$$\|\omega_n\|_I \leq c\, n^3 \|P_n\|_I,$$

where

$$c \leq 2\, T_n\left(\frac{2 + 1/n^2}{2 - 1/n^2}\right) \leq 2 \left(\frac{\sqrt{2} + 1/n}{\sqrt{2} - 1/n}\right)^n < 2\, e^5.$$

$\square$

Proof of Theorem 4.1. First we note that

$$\|B_{n+1,\infty} - B_{n,\infty}\|_I \leq E_{n,\infty} + E_{n+1,\infty}. \tag{4.8}$$

Let $B_{n,\infty}(z) = a_n z^n + \cdots \in \mathbb{P}_n$. Then

$$B_{n+1,\infty} - B_{n,\infty} = \frac{a_{n+1}}{2^n} T_{n+1} + q_n,$$

where T_{n+1} is the classical Chebyshev polynomial of degree $n+1$, normalized by $\|T_{n+1}\|_I = 1$ and $q_n \in \mathbb{P}_n$. Hence,

$$\frac{|a_{n+1}|}{2^n} \|T_{n+1}\|_I \geq \|f - B_{n,\infty} - q_n\|_I - \|f - B_{n+1,\infty}\|_I$$
$$\geq E_{n,\infty} - E_{n+1,\infty},$$

or

$$|a_{n+1}| \geq 2^n (E_{n,\infty} - E_{n+1,\infty}). \tag{4.9}$$

Since $\lim_{n \to \infty} E_{n,\infty}(f) = 0$, according to Lemma 3.3, there exists a subsequence $\Lambda \subset \mathbb{N}$ such that $E_{n,\infty} \neq E_{n+1,\infty}$ and

$$\frac{E_{n,\infty} + E_{n+1,\infty}}{E_{n,\infty} - E_{n+1,\infty}} \leq n^2, \quad n \in \Lambda. \tag{4.10}$$

Hence, $a_{n+1} \neq 0$, and the polynomial

$$P_{n+1} := \frac{B_{n+1,\infty} - B_{n,\infty}}{a_{n+1}}, \quad n \in \Lambda,$$

is monic, and by (4.8)–(4.10),

$$\|P_{n+1}\|_I \leq n^2 \left(\frac{1}{2}\right)^n, \quad n \in \Lambda. \tag{4.11}$$

Let $X_n = \left\{x_i^{(n)}\right\}_{i=0}^{n+1}$ be an alternant of $f - B_{n,\infty}$ (cf. 4.1). Then we obtain

$$\delta(-1)^i P_{n+1}\left(x_i^{(n)}\right) \geq \frac{E_{n,\infty} - E_{n+1,\infty}}{|a_{n+1}|} \geq \frac{1}{n^2} \frac{1}{2^n}, \quad 0 \leq i \leq n+1,$$

for $n \in \Lambda$. Let $\omega_{n+2}(x) = \prod_{i=0}^{n+1} \left(x - x_i^{(n)}\right)$. Then by Lemma 4.2, combined with properties of Chebyshev polynomials,

$$\left|\omega'_{n+2}\left(x_i^{(n)}\right)\right| \geq \frac{1}{n^2 \, 2^n \, E_{n,\infty}(P_n)} \frac{1}{2^n} = \frac{1}{n^2} \frac{1}{2^n}. \tag{4.12}$$

By Lemma 4.3 and (4.11) we get

$$\|\omega_{n+2}\|_I \leq c n^5 \left(\frac{1}{2}\right)^n \quad \text{for } n \in \Lambda, \tag{4.13}$$

with some absolute constant c. Combining (4.12) and (4.13) with Theorem 3.1.8 we obtain the discrepancy estimate of the theorem. $\square$

In Theorem 4.1 the discrepancy estimate is in general true only for a subsequence. This follows already from Theorem 3.5. On the other hand, it is interesting to get some insight into so-called *bad* subsequences $\{n_j\}_{j=1}^\infty$ where ν_{n_j} does not converge to μ in the weak* sense as $j \to \infty$.

7.4 Alternation Points in Chebyshev Approximation

Theorem 4.4. *If $\{n_j\}_{j=1}^\infty$ is a subsequence such that the equilibrium measure is not a limit point of $\{\nu_{n_j}\}_{j=1}^\infty$ in the weak* topology, then*

$$\limsup_{j \to \infty} \frac{n_{j+1}}{n_j} > 1.$$

Proof. We may assume that f is not a polynomial. Contrary to the theorem, let us assume that the subsequence $\{n_j\}_{j=1}^\infty$ satisfies $\lim_{j \to \infty} n_{j+1}/n_j = 1$. Again Lemma 3.3 provides a subsequence $\{n_{i_k}\}_{k=1}^\infty$ such that

$$\frac{E_{n_{i_k},\infty} + E_{n_{i_k+1},\infty}}{E_{n_{i_k},\infty} - E_{n_{i_k+1},\infty}} \leq i_k^2 \leq n_{i_k}^2.$$

Let

$$P_{n_{j+1}} = \frac{B_{n_{j+1},\infty} - B_{n_j,\infty}}{E_{n_j,\infty} - E_{n_{j+1},\infty}} \in \mathbb{P}_{n_{j+1}}.$$

Then

$$\left| P_{n_{j+1}}\left(x_i^{(n_j)}\right) \right| \geq 1, \quad 0 \leq i \leq n_j + 1,$$

and

$$P_{n_{j+1}}\left(x_i^{(n_j)}\right) P_{n_{j+1}}\left(x_{i+1}^{(n_j)}\right) < 0, \quad 0 \leq i \leq n_j.$$

Again set

$$w_{n_j+2}(x) = \prod_{i=0}^{n_j+1} \left(x - x_i^{(n_j)}\right).$$

Then by Lemma 4.2

$$\left| w'_{n_j+2}\left(x_i^{(n_j)}\right) \right| \geq \alpha_j \frac{1}{2^n},$$

where

$$\alpha_j^{-1} = \binom{n_j + n_{j+1}}{2n_j + 1} \frac{n_{j+1}}{n_j + 1} E_{n_j,\infty}(P_{n_{j+1}}).$$

Theorem 4.1.5 now yields

$$D\left[\nu_{n_j} - \mu\right] \leq c \sqrt{\frac{\log(n_j \alpha_j)}{n_j}}.$$

Since

$$n_j \alpha_j \leq c \left(\frac{n_j + n_{j+1}}{2n_j}\right)^3 E_{n_j,\infty}(P_{n_{j+1}}),$$

we obtain

$$D[\nu_{n_j} - \mu] \leq c \sqrt{\frac{1}{n} \log \|P_{n_{j+1}}\|_I + \log\left(\frac{n_j + n_{j+1}}{2n_j}\right)}. \tag{4.14}$$

Now,
$$\|P_{n_{j+1}}\|_I \le \frac{E_{n_j,\infty} + E_{n_{j+1},\infty}}{E_{n_j,\infty} - E_{n_{j+1},\infty}} \le n_{i_k}^2$$

for $j = i_k$, $k = 1, 2, \ldots$. By Stirling's formula
$$\log\binom{n_j + n_{j+1}}{2n_j} = o(n_j) \quad \text{as } j \to \infty.$$

Hence, (4.14) yields
$$D[\nu_n - \mu] \le c_1 \sqrt{\frac{\log n}{n}}, \quad n = n_{i_k}, \, k = 1, 2, \ldots,$$

and therefore $\nu_{n_{i_k}} \xrightarrow{*} \mu$ as $k \to \infty$, which contradicts the fact that $\{\nu_{n_j}\}_{j=1}^\infty$ does not have the equilibrium measure μ as a limit point. $\square$

7.5 Sign Changes in L^p-Approximation

Let τ be a positive, finite Borel measure on $I = [-1, 1]$ with infinite support. For $1 \le p < \infty$, let $L_\tau^p(I)$ be the class of all τ-measurable functions f on I such that $|f|^p$ is integrable with respect to τ. We denote by

$$\|f\|_{p,\tau} := \left(\int |f|^p \, d\tau\right)^{1/p}$$

the weighted L^p-norm on $L_\tau^p(I)$.

Given $f \in L_\tau^p(I)$ and $n \in \mathbb{N}$, we set $B_{n,p} = B_{n,p}(f)$ for a best L_τ^p-approximant of f on I with respect to $\mathbb{P}_n$, i.e.,

$$\|f - B_{n,p}\|_{p,\tau} = E_{n,p} = E_{n,p}(f) := \min_{P_n \in \mathbb{P}_n} \|f - P_n\|_{p,\tau}.$$

For $p > 1$ the best approximant $B_{n,p}$ is unique. For $p = 1$ uniqueness does not hold in general.

In this section we are interested in the distribution of sign changes of the optimal error functions $f - B_{n,p}$. To prepare the results we need some auxiliary considerations.

In the following, we assume that τ has a *Radon–Nikodym derivative* $d\tau(x) = w(x)\, dx$ with respect to the Lebesgue measure. Moreover, if w is positive a.e. on I and $\int_I w(x)\, dx = 1$, we define

$$\varphi(w, \varepsilon) := \inf\left\{\int_A w\, dx : A \subset I, \, \mu(A) \ge \varepsilon\right\}, \quad 0 < \varepsilon \le 1,$$

where μ again denotes the equilibrium distribution on I. Then $\varphi(w,\varepsilon)$ is a continuous strictly increasing function on $[0,1]$ with $\varphi(w,0) = 0$ and $\varphi(w,1) = 1$. Therefore, for every $n \geq 0$ there exists a unique number $\varepsilon_n(w) \in (0,1]$ satisfying

$$\varphi(w,\varepsilon) = e^{-n\varepsilon}. \tag{5.1}$$

Then $\varepsilon_n(w) \to 0$ as $n \to \infty$ and

$$\varepsilon_n(w) \geq \frac{1}{2} \frac{\log n}{n}, \quad n = 1, 2, \ldots. \tag{5.2}$$

Indeed, let us assume that $\varepsilon_n(w) < \log n/(2n)$ for some integer $n \geq 2$. Set $\varepsilon_n := \varepsilon_n(w)$ and let N denote the maximal number of pairwise disjoint open intervals $I_j \subset [-1,1]$, $j = 1, \ldots, N$, satisfying $\mu(I_j) \geq \varepsilon_n$. Then $N \geq \varepsilon_n^{-1} - 1$ and therefore

$$1 = \int_{-1}^{1} w(x)\,dx \geq \sum_{j=1}^{N} \int_{I_j} w(x)\,dx \geq N\,\varphi(w,\varepsilon_n) = N\,e^{-n\varepsilon_n}$$

$$\geq \left(\frac{1}{\varepsilon_n} - 1\right) e^{-n\varepsilon_n} > \left(2\frac{n}{\log n} - 1\right) n^{-1/2} \geq \frac{n^{1/2}}{\log n} > 1,$$

a contradiction.

Now it is possible to prove a *Nikol'skiĭ-type inequality*:

Lemma 5.1. *Let $\tau \in \mathcal{M}(I)$ have a Radon–Nikodym derivative $d\tau(x) = w(x)\,dx$ and $w > 0$ a.e. in I. Then for every $P_n \in \mathbb{P}_n$,*

$$\|P_n\|_I \leq e^{cn\varepsilon_n(w)} \|P_n\|_{p,\tau}, \tag{5.3}$$

where c is a constant, depending only on w and p.

Proof. If $n = 0$, then $\|P_n\|_I = \|P_n\|_{p,\tau}$ and (5.3) is true for any $c > 0$. Let $n \geq 1$, $P_n \in \mathbb{P}_n$, and $0 < \gamma \leq 1$. Set

$$A(\gamma) = \{x \in I : |P_n(x)| \geq \gamma \|P_n\|_I\}$$

and $S(\gamma) = \mu(A(\gamma))$. Evidently, $S(\gamma)$ is a continuous function of $\gamma \in [0,1]$, strictly decreasing in $[0,1]$ from 1 to 0. Since $\varepsilon_n(w) \in (0,1]$, we can choose $\gamma = \gamma_n$ such that $S(\gamma_n) = \varepsilon_n(w)$. Define

$$g_n := \frac{1}{\gamma_n \|P_n\|_I} P_n \in \mathbb{P}_n.$$

Then

$$\mu(\{x \in I : g_n(x) \leq 1\}) = \mu(I \setminus A(\gamma_n)) = 1 - \varepsilon_n(w).$$

Next, we distinguish two cases.

Case 1: $0 < \varepsilon_n(w) \leq \frac{1}{2}$.

Then the inequality of Remez yields

$$\gamma_n^{-1} \leq \|g_n\|_I \leq e^{c^* n \varepsilon_n(w)}$$

with some absolute constant $c^* > 0$, and consequently,

$$\|P_n\|_{p,\tau} \geq \left(\int_{A(\gamma_n)} |P_n(x)|^p w(x)\, dx\right)^{1/p} \geq \gamma_n \|P_n\|_I \left(\int_{A(\gamma_n)} w(x)\, dx\right)^{1/p}$$

$$\geq \gamma_n \|P_n\|_I \left(\varphi(w, \varepsilon_n(w))\right)^{1/p} \geq e^{-c^* n \varepsilon_n(w)} \|P_n\|_I \, e^{-n \varepsilon_n(w)/p}.$$

Hence,

$$\|P_n\|_I \leq e^{(c^* + 1/p) n \varepsilon_n(w)} \|P_n\|_{p,\tau}.$$

Case 2: $\frac{1}{2} \leq \varepsilon_n(w) \leq 1$.

Since $\lim_{n\to\infty} \varepsilon_n(w) = 0$, there exists $n_0 = n_0(w) \in \mathbb{N}$ such that $n \leq n_0$.
Set

$$c_0 = 2 \max_{P \in \mathbb{P}_{n_0}} \log \frac{\|P_n\|_I}{\|P_n\|_{p,\tau}}.$$

Then $c_0 > 0$ and

$$\|P\|_I \leq e^{c_0/2} \|P\|_{p,\tau} \leq e^{c_0 n \varepsilon_n(w)} \|P\|_{p,\tau} \quad \text{for all } P \in \mathbb{P}_n,\, n \leq n_0.$$

$\square$

To state the results, we need the sign function sgn g of an extended real-valued function $g : I \to \overline{\mathbb{R}}$. Set

$$\operatorname{sgn} g(x) := \begin{cases} 1 & \text{if } g(x) > 0, \\ 0 & \text{if } g(x) = 0, \\ -1 & \text{if } g(x) < 0. \end{cases} \tag{5.4}$$

Then $\operatorname{sgn} g : I \to \mathbb{R}$ is the function defined by $(\operatorname{sgn} g)(x) := \operatorname{sgn} g(x)$.

7.5.1 The Case $1 < p < \infty$

In this case, it is well known (cf. [163, Theorem 1.11]) that the linear functional

$$L(g) = \int g\, |f - B_{n,p}|^{p-1} \operatorname{sgn}(f - B_{n,p})\, d\tau \tag{5.5}$$

vanishes for $g \in \mathbb{P}_n$. Hence, the error function $f - B_{n,p}$ must have at least $n+1$ sign changes in $[-1, 1]$. But in general, this is practically useless information if we assume only that $f \in L^p(I)$, because we can modify, for example, the given function f at a countable number of points without changing the best-approximating polynomial. Hence, let us assume in the following that $f \in C(I)$ if $1 < p < \infty$.

7.5 Sign Changes in L^p-Approximation

Definition 5.2. Let $g \in C(I)$. Then
$$X : x_1 < x_2 < \cdots < x_m$$
is an *alternating sequence of length m for g* if $X \subset I$ and $g(x_i)\, g(x_{i+1}) < 0$ for $i = 1, \ldots, m-1$.

Let $\mathcal{A}_{n,p}$ denote the collection of all alternating sequences X_n of length $n+2$ for $f - B_{n,p}$ and let $\nu(X_n)$ be the normalized counting measure of X_n.

Theorem 5.3. *Let $\tau \in \mathcal{M}(I)$ and let $d\tau(x) = w(x)\,dx$, $w(x) > 0$ a.e. in I and $\varepsilon_n(w)$ defined by (5.1). If $f \in C(I)$ is not a polynomial, then there exists a subsequence $\Lambda \subset \mathbb{N}$ and alternating sequences*
$$X_n : x_0^{(n)} < x_1^{(n)} < \cdots < x_{n+1}^{(n)}$$
of $f - B_{n,p}$ such that
$$D[\nu(X_n) - \mu] \le c\sqrt{\varepsilon_n(w)}, \quad n \in \Lambda,$$
where μ is the equilibrium measure of I and c is a constant, depending only on w and p.

The proof will be based on the following idea.
Let us consider the linear functional
$$L(g) := \frac{1}{\|f - B_{n,p}\|_p^{p-1}} \int g\,|f - B_{n,p}|^{p-1}\,\text{sgn}(f - B_{n,p})\,d\tau.$$

Then $L(P_n) = 0$ for all $P_n \in \mathbb{P}_n$. Moreover, $L(f - B_{n,p}) = \|f - B_{n,p}\|_p = E_{n,p}$, and by Hölder's inequality
$$\|L\|_p = \sup\{|L(g)| : \|g\|_p = 1\} = 1. \tag{5.6}$$

Next, for any $k \ge 1$,
$$L(B_{n+k,p} - B_{n,p}) = L(f - B_{n,p}) - L(f - B_{n+k,p})$$
$$\ge \|f - B_{n,p}\|_p - \|f - B_{n+k,p}\|_p = E_{n,p} - E_{n+k,p}.$$

Since $L(P_n) = 0$ for all $P_n \in \mathbb{P}_n$, we obtain
$$L(B_{n+k,p}) = L(B_{n+k,p} - P_n) \ge E_{n,p} - E_{n+k,p}. \tag{5.7}$$

By inequality (5.7) we shall show that a polynomial $P_n^* \in \mathbb{P}_n$ can be found with the following properties: There exist $n+2$ points $-1 \le x_0 < x_1 < \cdots < x_{n+2} \le 1$ such that both $f - B_{n,p}$ and $B_{n+k,p} - P_n^*$ have alternating signs at the points x_i and
$$|(B_{n+k,p} - P_n^*)(x_i)| \ge E_{n,p} - E_{n+k,p}$$

258   7. Applications of Discrepancy Theorems

for $0 \leq i \leq n+1$ (Lemma 5.4 and Lemma 5.5). Hence, $B_{n+k,p} - P_n^*$ is an oscillating polynomial to which Lemma 4.2 can be applied. This leads to a distribution estimate for the points x_i that are points where the error function $f - B_{n,p}$ alternates.

Now, the crucial key is the following minimization problem: Let $E_{n,p} \neq E_{n+k,p}$, $P_n \in \mathbb{P}_n$, and set

$$\lambda_n(P_n) := \max_{-1 \leq x \leq 1} (B_{n+k,p} - P_n)(x) \, \text{sgn}\, (f - B_{n,p})(x).$$

Define the minimization problem $(\mathcal{O}_{n,p})$ by

$$\lambda_n^* := \inf\{\lambda_n(P_n) : P_n \in \mathbb{P}_n\}.$$

Lemma 5.4. *There exists $P_n^* \in \mathbb{P}_n$ such that $\lambda_n(P_n^*) = \lambda_n^*$ and*

$$\lambda_n^* \geq E_{n,p} - E_{n+k,p} > 0.$$

Proof. Let $\{P_{n,\nu}\}_{\nu=1}^\infty$ be a sequence such that $P_{n,\nu} \in \mathbb{P}_n$ and further $\lambda_n^* = \lim_{\nu \to \infty} \lambda_{n,\nu}$, where $\lambda_{n,\nu} := \lambda_n(P_{n,\nu})$. Because of (5.7), we have $\lambda_{n,\nu} > 0$. Next, we assert that the sequence $\{P_{n,\nu}\}_{\nu=1}^\infty$ is bounded: To see this, note that there exist $n+2$ points

$$-1 \leq x_0 < x_1 < \cdots < x_{n+1} \leq 1$$

such that $(f - B_{n,p})(x_j) \neq 0$ and

$$\text{sgn}\,(f - B_{n,p})(x_j) = (-1)^j \delta, \quad 0 \leq j \leq n+1,$$

where $\delta \in \{-1, +1\}$ is fixed. Hence, there exists a constant $\varepsilon > 0$ such that

$$\text{sgn}\,(f - B_{n,p})(x) = (-1)^j \delta \quad \text{for} \quad x \in (x_j - \varepsilon, x_j + \varepsilon) \cap I.$$

In each interval $I_j = [x_j - \varepsilon, x_j + \varepsilon] \cap I$ we fix a point $\xi_{j,\nu}$ such that

$$\|B_{n+k,p} - P_{n,\nu}\|_{j,\infty} := \max_{x \in I_j} |(B_{n+k,p} - P_{n,\nu})(x)| = |(B_{n+k,p} - P_{n,\nu})(\xi_{j,\nu})|.$$

First, let $\nu_1, \nu_2, \ldots$ denote all indices such that there exists a point $\xi_{i(\nu_m),\nu_m}$ with

$$\text{sgn}(B_{n+k,p} - P_{n,\nu_m})(\xi_{i(\nu_m),\nu_m}) = (-1)^{i(\nu_m)} \delta.$$

Then

$$\lambda_{n,\nu_m} \geq |(B_{n+k,p} - P_{n,\nu_m})(\xi_{i(\nu_m),\nu_m})| = \|B_{n+k,p} - P_{n,\nu_m}\|_{i(\nu_m),\infty}. \quad (5.8)$$

Next, let $\mu_1, \mu_2, \ldots$ denote all indices such that for each $j = 0, \ldots, n+1$,

$$\text{sgn}(B_{n+k,p} - P_{n,\mu_m})(\xi_{j,\mu_m}) = -(-1)^j \delta.$$

Then $B_{n+k,p} - P_{n,\mu_m}$ has at least $n+1$ zeros

$$y_1 < y_2 < \cdots < y_{n+1}$$

in $(-1,1)$. Hence, by the Lagrange remainder formula,

$$\|B_{n+k,p} - P_{n,\mu_m}\|_\infty \leq \sup_{x\in[-1,1]} \left| \frac{B_{n+k,p}^{(n+1)}(\xi(x))}{(n+1)!} \prod_{i=1}^{n+1}(x-y_i) \right|$$
$$\leq \frac{2^{n+1}}{(n+1)!} \left\|B_{n+k,p}^{(n+1)}\right\|_\infty. \tag{5.9}$$

Summarizing the results (5.8) and (5.9) about the two subsequences, we obtain

$$\|B_{n+k,p} - P_{n,\nu}\|_\infty \leq c \max\left(\lambda_{n,\nu}, \left\|B_{n+k,p}^{(n+1)}\right\|_\infty\right)$$

for all $\nu = 1, 2, \ldots$, where $c > 0$ is some fixed constant. Therefore, the sequence $\{P_{n,\nu}\}_{\nu=1}^\infty$ is bounded.

Hence, there exists a subsequence of $\{P_{n,\nu}\}_{\nu=1}^\infty$ that converges to some polynomial $P_n^* \in \mathbb{P}_n$. Then

$$\operatorname{sgn}(f - B_{n,p})(x)(B_{n+k,p} - P_n^*)(x) \leq \lambda_n^*$$

for all $x \in I$ and $\lambda_n^* = \lambda_n(P_n^*)$. Consequently, P_n^* is optimal for $(\mathcal{O}_{n,p})$, and taking (5.6) into account, (5.7) yields

$$\lambda_n^* \geq E_{n,p} - E_{n+k,p}.$$

$\square$

Using standard arguments the next lemma can be proved easily.

Lemma 5.5. *Let $E_{n+k,p} \neq E_{n,p}$. Then there exists in I an alternating sequence*

$$X_n: x_0^{(n)} < x_1^{(n)} < \cdots < x_{n+1}^{(n)}$$

of length $n+2$ for the error function $f - B_{n,p}$ and

$$(B_{n+k,p} - P_n^*)(x_i) \operatorname{sgn}(f - B_{n,p})(x_i) = \lambda_n^*, \quad 0 \leq i \leq n+1.$$

An intermediate step for proving Theorem 5.3 is the following lemma.

Lemma 5.6. *Let τ and f satisfy the conditions of Theorem 5.3, and let k, n be positive integers such that $E_{n,p} \neq E_{n+k,p}$. Then the alternating sequence X_n of length $n+2$ for $f-B_{n,p}$, determined by Lemma 5.5, satisfies*

$$D[\nu(X_n) - \mu] \leq c\sqrt{\varepsilon_{n+k}(w) + \frac{1}{n}\left(\log\frac{E_{n,p} + E_{n+k,p}}{E_{n,p} - E_{n+k,p}} + \log\left(\frac{2n+k}{2n}\right)\right)},$$

where c is a constant, depending only on w and p.

Proof. Set
$$P_{n+k} := \frac{B_{n+k,p} - P_n^*}{E_{n,p} - E_{n+k,p}}$$
and let X_n be the alternating sequence for $f - B_{n,p}$ of Lemma 5.5. Then
$$\left|P_{n+k}\left(x_i^{(n)}\right)\right| \geq 1, \quad 0 \leq i \leq n+1,$$
and
$$P_{n+k}\left(x_i^{(n)}\right) P_{n+k}\left(x_{i+1}^{(n)}\right) < 0, \quad 0 \leq i \leq n.$$
Hence, for $P_{n+k} \in \mathbb{P}_{n+k}$ all conditions of Lemma 4.2 are satisfied, and we obtain
$$\left|\omega'_{n+2}\left(x_i^{(n)}\right)\right| \geq \frac{1}{\beta_{n,k} E_{n,\infty}(P_{n+k})} \frac{1}{2^n}, \quad 0 \leq i \leq n+1,$$
with
$$\beta_{n,k} = \binom{2n+k}{2n+1} \frac{n+k}{n+1}.$$
Since
$$n\beta_{n,k} \leq c \binom{2n+k}{2n}^3$$
with some absolute constant c, Theorem 4.1.5 yields that for any $q_n \in \mathbb{P}_n$,
$$D\left[\nu(X_n) - \mu\right] \leq c \sqrt{\frac{1}{n}\left(\log \|P_{n+k} - q_n\|_\infty + \log \binom{2n+k}{2n}\right)}, \quad (5.10)$$
where c is an absolute constant. If we choose
$$q_n := \frac{B_{n,p} - P_n^*}{E_{n,p} - E_{n+k,p}},$$
then
$$P_{n+k} - q_n = \frac{B_{n+k,p} - B_{n,p}}{E_{n,p} - E_{n+k,p}},$$
and together with the Nikol'skiĭ inequality of Lemma 5.1 it follows that
$$\frac{1}{n}\log \|P_{n+k} - q_n\|_\infty \leq c \left(\varepsilon_{n+k}(w) + \frac{1}{n}\log \frac{E_{n,p} + E_{n+k,p}}{E_{n,p} - E_{n+k,p}}\right),$$
where c depends on w and p. Inserting this in (5.10), we get the desired estimate of Lemma 5.6. □

Now we are in position to prove Theorem 5.3.

7.5 Sign Changes in L^p-Approximation

Proof of Theorem 5.3. Because of Lemma 3.3 there exists a subsequence $\Lambda \subset \mathbb{N}$ such that

$$\frac{E_{n,p} + E_{n+1,p}}{E_{n,p} - E_{n+1,p}} \leq n^2, \quad n \in \Lambda.$$

Thus, the discrepancy estimate of Theorem 5.3 is a direct consequence of Lemma 5.6 if we bear in mind (5.2) and $\varepsilon_n(w) \geq \varepsilon_{n+1}(w)$. $\square$

As a corollary of Lemma 5.6 and the proof of Theorem 5.3 we discuss the zeros of polynomials of minimal L^p_τ-norm. Let $T_{n,p} \in \mathbb{P}_n$ denote the monic polynomial satisfying the equation

$$\|T_{n,p}\|_{p,\tau} = \min_{P \in \mathbb{P}_{n-1}} \|x^n - P(x)\|_{p,\tau}.$$

Then $T_{n,p}$ has n simple zeros in $(-1,1)$. Let $\nu_{T_{n,p}}$ denote the zero counting measure of $T_{n,p}$.

Corollary 5.7. *Let τ be as in Theorem 5.3. Then there exists a constant $c > 0$, depending on w and p, such that*

$$D[\nu_{T_{n,p}} - \mu] \leq c\sqrt{\varepsilon_n(w)}, \quad n = 0, 1, \ldots.$$

Proof. Let $n \geq 1$ and define $f(x) = x^n$. Then there exists an alternating sequence X_{n-1} of length $n+1$ such that, according to Lemma 5.6,

$$D[\nu(X_{n-1}) - \mu] \leq c\sqrt{\varepsilon_n(w)}, \quad n = 1, 2, \ldots.$$

Since the zeros of $T_{n,p}$ are separated by the points of X_{n-1}, the corollary is proved. $\square$

For special weights, the discrepancy estimate of the zero distribution of $T_{n,p}$ can be strengthened.

Corollary 5.8. *Let $-1 = t_1 < \cdots < t_m = 1$ be fixed points, $a_1, \ldots, a_m > -1$ fixed numbers, and w a weight function that satisfies*

$$w(x) \geq c_1 \prod_{i=1}^{m} |x - t_i|^{a_i} \tag{5.11}$$

with $c_1 > 0$, $d\tau(x) = w(x)\,dx$, and $\tau \in \mathcal{M}(I)$. Then there exists a constant $c > 0$ such that

$$D[\nu(T_{n,p}) - \mu] \leq c\frac{(\log n)^2}{n}, \quad n = 2, 3, \ldots, \tag{5.12}$$

where c depends only on w and p.

Proof. Analogous to the proof of Lemma 5.6 there exists an alternating sequence $X_{n-1} = \{x_i^{(n-1)}\}_{i=0}^{n}$ of length $n+1$ such that

$$\left|\omega'_{n+1}\left(x_i^{(n-1)}\right)\right| \geq \frac{\|T_{n,p}\|_{p,\tau}}{E_{n-1,\infty}(x^n)} \frac{1}{2^{n-1}}, \quad 0 \leq i \leq n,$$

with $\omega_{n+1}(x) = \prod_{i=0}^{n}\left(x - x_i^{(n-1)}\right)$. Again Nikol'skiĭ's inequality yields

$$E_{n-1,\infty}(x^n) \leq \|T_{n,p}\|_\infty \leq e^{c\,n\,\varepsilon_n(w)} \|T_{n,p}\|_{p,\tau},$$

and therefore

$$\left|\omega'_{n+1}\left(x_i^{(n-1)}\right)\right| \geq e^{-c\,n\,\varepsilon_n(w)} \frac{1}{2^{n-1}}, \quad 0 \leq i \leq n.$$

Since the zeros of $T_{n,p}$ are separated by the points $x_i^{(n-1)}$, we obtain with Lemma 4.3 that

$$\|\omega_{n+1}\|_I \leq c_1\, n^3\, \|T_{n,p}\|_\infty \leq c_1\, n^3\, e^{c\,n\,\varepsilon_n(w)} \frac{1}{2^{n-1}}.$$

Observing that $\varepsilon_n(w) = O(\log n/n)$ as $n \to \infty$, Theorem 3.1.8 yields

$$D[\nu(X_{n-1}) - \mu] \leq c\,\frac{(\log n)^2}{n}, \quad n = 2, 3, \ldots,$$

where c is a constant depending on w and p. This implies (5.12). $\square$

Typical examples of weight functions satisfying (5.11) are generalized Jacobi weights (cf. [131]).

As in the case of Chebyshev approximation on an interval, the question arises as to whether the discrepancy estimate can hold for every $n \in \mathbb{N}$. This is not true analogous to Theorem 3.5 for uniform approximation.

Theorem 5.9. *Let $1 < p < \infty$, $\tau \in \mathcal{M}(I)$ with Radon–Nikodym derivative $d\tau(x) = w(x)\,dx$ with respect to the Lebesgue measure. Then there exists an entire function f and an interval $[a,b] \subset I$ such that $f - B_{n,p}(f)$ has no zeros in $[a,b]$ for some infinite subset $\Lambda \subset \mathbb{N}$.*

For the proof, we use an idea of G.G. Lorentz [116]. We construct a sequence $\{n_k\}_{k=0}^{\infty} \subset \mathbb{N}$ and polynomials $p_{n_k+1} \in \mathbb{P}_{n_k+1}$ with no zeros in a fixed interval $[\alpha, \beta] \subset I$ such that 0 is their best L_τ^p-approximation. Then setting $f = \sum_{k=0}^{\infty} c_k\, p_{n_k}$ with $c_k \to 0$ sufficiently fast we can assume that the error functions

$$f - B_{n_k,p}(f)$$

are close enough to p_{n_k+1} so that they do not vanish in some subinterval $[a,b] \subset [\alpha,\beta]$. The next lemma will provide the appropriate polynomials.

Lemma 5.10. *Let τ be as in Theorem 5.9. Then there exists a constant $c_{p,w} > 0$, depending on p and w, and polynomials $q_{2n+1} \in \mathbb{P}_{2n+1}$ such that $B_{n,p}(q_{2n+1}) = 0$ and q_{2n+1} has no zeros in the interval $(-1, -1+c_{p,w})$, $n \in \mathbb{N}$.*

Proof. Consider the $(n+2)$-dimensional subspace
$$U_{n+2} = \{g(x) = (x+1)^n p_{n+1}(x) : p_{n+1} \in \mathbb{P}_{n+1}\}.$$
By a theorem of Krein (cf. Theorem 1.5.2) there exists an element
$$q_{2n+1}(x) = (x+1)^n p^*_{n+1}(x) \in U_{n+2}$$
having 0 as its best L^p_τ-approximant in the $(n+1)$-dimensional subspace $\mathbb{P}_n$. We may assume that $\|q_{2n+1}\|_I = 1$. Then, $|p^*_{n+1}(x)| \leq 1$ for $x \in [0,1]$. Set $h_{n+1}(x) := p^*_{n+1}((x+1)/2) \in \mathbb{P}_{n+1}$. Then $\|h_{n+1}\|_I \leq 1$, and therefore
$$\|p^*_{n+1}\|_I \leq \|h_{n+1}\|_{[-3,1]} \leq \|T_{n+1}\|_{[-3,1]} \leq 6^n,$$
where T_{n+1} is the Chebyshev polynomial. Let $-1 < x_1 < 1$ denote the smallest zero of p^*_{n+1} and set
$$\widetilde{p}_n(x) := \frac{p^*_{n+1}(x)}{x - x_1} \in \mathbb{P}_n.$$
By (5.5),
$$\int \widetilde{p}_n |q_{2n+1}|^{p-1} \operatorname{sgn} q_{2n+1} \, d\tau = 0,$$
and therefore
$$-\int_{-1}^{x_1} \frac{|p^*_{n+1}|}{|x-x_1|} |q_{2n+1}|^{p-1} \, d\tau + \int_{x_1}^{1} \frac{|p^*_{n+1}|}{|x-x_1|} |q_{2n+1}|^{p-1} \, d\tau = 0.$$
Thus
$$\int \frac{|p^*_{n+1}|}{|x-x_1|} |q_{2n+1}|^{p-1} \, d\tau = 2 \int_{-1}^{x_1} \frac{|p^*_{n+1}|}{|x-x_1|} |q_{2n+1}|^{p-1} \, d\tau. \tag{5.13}$$
Using Markov's inequality we obtain for the integral on the right hand side
$$\int_{-1}^{x_1} \frac{|p^*_{n+1}|}{|x-x_1|} |q_{2n+1}|^{p-1} \, d\tau$$
$$\leq (1+x_1)^{n(p-1)} \int_{-1}^{1} \left| \frac{p^*_{n+1}(x) - p^*_{n+1}(x_1)}{x - x_1} \right| |p^*_{n+1}|^{p-1} \, d\tau$$
$$\leq (1+x_1)^{n(p-1)} (n+1)^2 \|p^*_{n+1}\|_I^p$$
$$\leq (n+1)^2 6^{np} (1+x_1)^{n(p-1)}$$
$$\leq e^{c^*_p n}(1+x_1)^{n(p-1)}$$

264   7. Applications of Discrepancy Theorems

with some constant $c_p^* > 0$, depending only on p.

On the other hand, Nikol'skiĭ's inequality (5.3) and $\varepsilon_n(w) \geq \frac{1}{2}\log n/n$ (cf. (5.2)) yields

$$\int \frac{|p_{n+1}^*|}{|x-x_1|} |q_{2n+1}|^{p-1}\, d\tau \geq \frac{1}{2} \int |p_{n+1}^*| |q_{2n+1}|^{p-1}\, d\tau$$

$$= \frac{1}{2} \int \frac{|q_{2n+1}|^p}{(x+1)^n}\, d\tau \geq \frac{1}{2^{n+1}} \|q_{2n+1}\|_{p,\tau}$$

$$\geq \frac{1}{2^{n+1}} e^{-c\, n\, \varepsilon_n(w)} \geq e^{-c_1 n}$$

with some constant $c_1 > 0$, depending on w and p.

Substituting the last two inequalities in (5.13) we obtain

$$1 + x_1 \geq c_{p,w}$$

with some constant $c_{p,w}$, depending only on p and w. □

Proof of Theorem 5.9. The function of the theorem will be given by a function of type

$$f = \sum_{k=0}^{\infty} c_k\, g_k, \quad c_k > 0, \tag{5.14}$$

where $g_k = q_{4^k+1} \in \mathbb{P}_{4^k+1}$ are polynomials as in Lemma 5.10 and moreover, g_k is positive in the interval $(-1, -1+c_p)$.

First, let us determine positive coefficients $\{a_{k,0}\}_{k=0}^{\infty}$ such that this series converges monotonically to 0 and

$$\sum_{k=0}^{\infty} a_{k,0}\, M_k < \infty, \tag{5.15}$$

where $M_k = \max_{|z| \leq k} |g_k(z)|$, $k = 0, 1, \ldots$. The last condition ensures that functions of type (5.14) are entire functions if $0 < c_k \leq a_{k,0}$.

Now we define the coefficients c_k inductively. Assume that $c_0, c_1, \ldots, c_s$ are already fixed as well as a sequence

$$\{a_{k,s}\}_{k=s}^{\infty} \quad \text{where } a_{k,s} \leq a_{k,s-1} \text{ for } k = s, s+1, \ldots .$$

Let

$$f_j := \sum_{k=0}^{j} c_k\, g_k, \quad j = s, s-1.$$

Then f_{s-1} is the best L_τ^p-approximant in $\mathbb{P}_{4^{s-1}+1}$ for the function f_s. Hence f_{s-1} is positive in $(-1, -1+c_p)$. Since the L_τ^p-norm is strictly convex, the

7.5 Sign Changes in L^p-Approximation

best approximant is unique and the best approximating operator $B_{n,p}(f)$ is continuous in $L^p_\tau(I)$. Therefore, there exists a sequence

$$\{a_{k,s+1}\}_{k=s+1}^\infty \quad (a_{k,s+1} \leq a_{k,s})$$

of positive real numbers such that the subset

$$V_{s+1} := \left\{ g : g = f_s + \sum_{k=s+1}^\infty b_k\, g_k,\ 0 < b_k \leq a_{k,s+1} \right\}$$

of entire functions has the following property: If $g \in V_{s+1}$ and $n = 4^{s-1}+1$, then all best approximants $B_{n,p}(g)$ have no zeros in $[a,b]$, where

$$a = -1 + c_p/4, \quad b = -1 + 3c_p/4.$$

Then we define

$$c_{s+1} := a_{k,s+1},$$

and finally, the function

$$f = \sum_{k=0}^\infty c_k\, g_k$$

has all properties of the theorem. $\square$

Next, we consider the problem of obtaining, in analogy to the Chebyshev case, some knowledge about possible bad subsequences $\{n_j\}$.

Theorem 5.11. *Let τ be as in Theorem 5.3, $\alpha > 1$ fixed, $Y_n \in \mathcal{A}_n$ such that*

$$D[\nu(Y_n) - \mu] \leq \alpha \inf_{X \in \mathcal{A}_n} D[\nu(X) - \mu], \quad n = 1, 2, \ldots .$$

If $\{n_j\}_{j=1}^\infty \subset \mathbb{N}$ such that μ is not a limit point of $\{\nu(Y_{n_j})\}_{j=1}^\infty$ in the weak topology, then*

$$\limsup_{j \to \infty} n_{j+1}/n_j > 1.$$

Proof. Contrary to the assertion, let us assume that $\limsup_{j \to \infty} n_{j+1}/n_j = 1$. According to the proof of Theorem 4.4 there exists a subsequence $\{n_{i_k}\}_{k=1}^\infty$ such that

$$\frac{E_{n_{i_k},p} + E_{n_{i_k}+1,p}}{E_{n_{i_k},p} - E_{n_{i_k}+1,p}} \leq n_{i_k}^2.$$

Stirling's formula yields

$$\log\left(\frac{n_j + n_{j+1}}{2n_j}\right) = o(n_j) \quad \text{as } j \to \infty.$$

Let X_n be the alternating sequence of Lemma 5.5 ($n = 1, 2, \ldots$). Then Lemma 5.6 yields

$$D[\nu(X_n) - \mu] \leq c\sqrt{\varepsilon_n(w)} \quad \text{for } n = n_{i_k},\ k = 1, 2, \ldots.$$

Hence

$$D[\nu(X_{n_{i_k}}) - \mu] \to 0 \quad \text{as } k \to \infty,$$

and $\{\nu(Y_{n_j})\}_{j=1}^\infty$ has μ as limit point. $\qquad\square$

7.5.2 The Case $p = 1$

Let $f \in L^1_\tau(I)$ and let $B_{n,1} := B_{n,1}(f) \in \mathbb{P}_n$ be a best-approximating polynomial with respect to $\mathbb{P}_n$. Such best approximations are characterized in such a way that the linear functional

$$L(g) = \int g\,\mathrm{sgn}(f - B_{n,1})\,d\tau, \quad g \in L^1_\tau(I), \tag{5.16}$$

satisfies

$$|L(P_n)| \leq \int_{Z(f - B_{n,1})} |P_n|\,d\tau \quad \text{for } P_n \in \mathbb{P}_n, \tag{5.17}$$

where for $g \in L^1_\tau(I)$,

$$Z(g) := \{x : g(x) = 0\}. \tag{5.18}$$

If $\tau(Z(f - B_{n,1})) = 0$, then (5.17) implies that the error function $f - B_{n,1}$ should change its sign at least $n + 1$ times in $[-1, 1]$.

For formulating the results, we need a generalized notion of sign change.

Definition 5.12. *The function $f \in L^1_\tau(I)$ does not change its sign at x_0, $x_0 \in I$, if $f > 0$ (or $f < 0$) a.e. in some neighborhood U of x_0. In the first case, x_0 is called $(+)$-point of f, while otherwise, x_0 is a $(-)$-point of f.*

All points of $I = [-1, 1]$ at which the function f does not have a fixed sign in the above sense are called *sign changes* of f.

Definition 5.13. *$\varphi \in L^1_\tau(I)$ is called a sign function if $\varphi^2 = 1$ on I. A sign function φ is orthogonal to $\mathbb{P}_n$ (written $\varphi \perp \mathbb{P}_n$) if*

$$\int \varphi P_n\,d\tau = 0 \quad \text{for every } P_n \in \mathbb{P}_n.$$

Two sign functions φ and ψ are called *equivalent* if $\varphi = \psi$ or $\varphi = -\psi$ a.e. in I.

7.5 Sign Changes in L^p-Approximation

We note again that in general, the best-approximating polynomial is not unique. But for $f \in C(I)$, uniqueness holds due to the Jackson–Krein theorem (cf. [45, Corollary 25]).

Lemma 5.14. *Let φ be a sign function, $\varphi \perp \mathbb{P}_n$, and assume φ has exactly $n+1$ sign changes at the points*

$$(y_0 = -1 <) \; y_1 < y_2 < \cdots < y_{n+1} \; (< 1 = y_{n+2})$$

in $(-1, 1)$. If the sign function ψ is not equivalent to φ and $\psi \perp \mathbb{P}_n$, then ψ has a sign change in each interval (y_i, y_{i+1}), $0 \leq i \leq n+1$.

Proof. We may assume that $\varphi(x) > 0$ for $x > y_{n+1}$. Assume now that ψ does not have a sign change in (y_j, y_{j+1}) for some $0 \leq j \leq n+1$. We may restrict ourselves to the case that $\psi = \varphi$ a.e. in (y_j, y_{j+1}).

We define $\eta := \varphi - \psi$ and obtain

$$\eta \varphi = \varphi^2 - \psi \varphi \geq 0 \quad \text{and} \quad \int \eta P_n \, d\tau = 0 \quad \text{for all } P_n \in \mathbb{P}_n. \tag{5.19}$$

If $1 \leq j \leq n$, we consider

$$P_n^*(x) = \prod_{\substack{i=1 \\ i \neq j, j+1}}^{n+1} (x - y_i) \in \mathbb{P}_{n-1}.$$

Then $\operatorname{sgn} P_n^*(x) = \operatorname{sgn} \varphi(x)$ for $x \notin [y_j, y_{j+1}]$. Since $\eta = 0$ a.e. in (y_j, y_{j+1}), the property (5.19) yields

$$\int_{-1}^{y_j} \eta P_n^* \, d\tau = 0 = \int_{y_{j+1}}^{1} \eta P_n^* \, d\tau \,,$$

and therefore $\eta = 0$ a.e. in I; i.e., $\varphi = \psi$ a.e. in I.

If $j = n+1$, then we define

$$P_n^*(x) = -\prod_{i=1}^{n}(x - y_i) \in \mathbb{P}_n$$

and again obtain a contradiction. For $j = 0$, the polynomial

$$P_n^*(x) = \prod_{i=2}^{n+1}(x - y_i)$$

has the appropriate behavior. $\square$

Lemma 5.14 can be applied to the sign function induced by the signs of the Chebyshev polynomial $U_{n+1} \in \mathbb{P}_{n+1}$ of the second kind. Let $x = \cos \theta$. Then

$$U_{n+1}(x) := \frac{\sin(n+2)\theta}{\sin \theta} = 2^{n+1} x^{n+1} + \ldots \in \mathbb{P}_{n+1}. \tag{5.20}$$

It is well known that $2^{-n-1}U_{n+1}(x)$ is the minimal error function if the function $f(x) = x^{n+1}$ is approximated in $L^1_\tau(I)$ with respect to $\mathbb{P}_n$, where τ is the Lebesgue measure on I; i.e.,

$$\min_{P_n \in \mathbb{P}_n} \int_{-1}^{1} |x^{n+1} - P_n(x)|\, dx = \frac{1}{2^{n+1}} \int_{-1}^{1} |U_{n+1}(x)|\, dx.$$

Using the characterization (5.17), we see that

$$\varphi(x) := \begin{cases} \operatorname{sgn} U_{n+1}(x) & \text{for } x \text{ where } U_{n+1}(x) \neq 0, \\ 1 & \text{elsewhere}, \end{cases} \qquad (5.21)$$

is a sign function and $\varphi \perp \mathbb{P}_n$ for the Lebesgue measure.

Theorem 5.15. *Let $L^1(I)$ be the collection of all real-valued functions that are integrable on I in the sense of Lebesgue. For given $f \in L^1(I)$, let $B_{n,1}(f)$ denote a best L^1-approximation with respect to $\mathbb{P}_n$. Then there exists a set Y_{n+1} of $n+1$ sign changes of $f - B_{n,1}(f)$ such that*

$$D[\nu(Y_{n+1}) - \mu] \leq \frac{2}{n+2}, \qquad (5.22)$$

where $\nu(Y_{n+1})$ is the normalized counting measure of the point set Y_{n+1} and μ is the equilibrium distribution on I.

Proof. From the definition (5.20) we conclude that U_{n+1} has zeros at the points

$$x_k = \cos \theta_k = \cos \frac{k\pi}{n+2}, \qquad k = 1, \ldots, n+1.$$

Therefore,

$$\mu[x_k, x_{k-1}] = \frac{1}{n+2} \quad \text{for } k = 1, 2, \ldots, n$$

and

$$\mu[-1, x_{n+1}] = \mu[x_1, 1] = \frac{1}{n+2}.$$

In this case there exists a sign function $\psi \perp \mathbb{P}_n$ such that $\psi(x) = \operatorname{sgn}(f - B_{n,1}(f))(x)$ for all $x \in I \setminus Z_n(f - B_{n,1}(f))$ (cf. [139, Lemma 2, p. 648]). Using Lemma 5.14 with φ as in (5.21), we get the estimate (5.22). $\square$

We denote by $U_{n+1,\tau}$ the *Chebyshev polynomial of second kind of degree $n+1$ with respect to τ*; i.e., $U_{n+1,\tau} \in \mathbb{P}_{n+1}$ is the monic polynomial such that

$$\int |U_{n+1,\tau}|\, d\tau = \min_{p_n \in \mathbb{P}_n} \int |x_{n+1} - p_n(x)|\, d\tau(x). \qquad (5.23)$$

The polynomial $U_{n+1,\tau}$ is unique and characterized by exactly $n+1$ sign changes (or zeros) at points

$$-1 < y_1^{(n)} < y_2^{(n)} < \cdots < y_{n+1}^{(n)} < 1. \tag{5.24}$$

In this case, we denote by $\nu_{n,\tau}$ the normalized counting measure that associates the mass $1/(n+1)$ with each point $y_i^{(n)}$, $1 \leq i \leq n+1$.

Theorem 5.16. *Let $B_{n,1}(f)$ denote a best L^1_τ-approximation to $f \in L^1_\tau(f)$ with respect to $\mathbb{P}_n$ and let $\tau \in \mathcal{M}(I)$ have a Radon–Nikodym derivative $d\tau(x) = w(x)\,dx$ with $w > 0$ a.e. on I. Then there exists a set Y_{n+1} of $n+1$ sign changes of $f - B_{n,1}(f)$ such that the normalized counting measure $\nu(Y_{n+1})$ of the point set Y_{n+1} satisfies*

$$D[\nu(Y_{n+1}) - \mu] \leq 2\, D[\nu_{n,\tau} - \mu] \leq c\,\sqrt{\varepsilon_n(w)} \tag{5.25}$$

for all $n = 0, 1, 2, \ldots$, where c is a constant, depending only on w.

Proof. Addressing the polynomial $U_{n+1,\tau}(x)$ instead of $U_{n+1}(x)$, we define

$$\varphi(x) := \begin{cases} \operatorname{sgn} U_{n+1,\tau}(x) & \text{for } x \text{ with } U_{n+1,\tau}(x) \neq 0, \\ 1 & \text{elsewhere.} \end{cases}$$

Again ψ is defined as in the proof of Lemma 5.14. Then with the same arguments we get the left inequality in (5.25).

In what follows we are going to show that

$$D[\nu_{n,\tau} - \mu] \leq c\,\sqrt{\varepsilon_n(w)} \tag{5.26}$$

for all n. First, let us present some auxiliary results. We consider the linear functional

$$L(g) := \int g\,(\operatorname{sgn} U_{n+1,\tau})\,d\tau$$

for $g \in L^1_\tau(I)$. The characterization of best L^1_τ-approximation yields

$$L(P_n) = 0 \quad \text{for all } P_n \in \mathbb{P}_n. \tag{5.27}$$

Moreover,

$$L(U_{n+1,\tau}) = \|U_{n+1,\tau}\|_{1,\tau} = E_{n,1}(x^{n+1})$$

and

$$\|L\|_{1,\tau} = \sup\{L(g) : \|g\|_{1,\tau} = 1\} = 1.$$

We use for abbreviation

$$E_n := E_{n,1}(x^{n+1}) = \|U_{n+1,\tau}\|_{1,\tau}.$$

Then (5.27) yields

$$L(P_{n+1}) = E_n \quad \text{for all } P_{n+1} \in \mathbb{M}_{n+1}. \tag{5.28}$$

Now let us consider the following minimization problem: For $P_{n+1} \in \mathbb{M}_{n+1}$ let
$$\lambda_n(P_{n+1}) := \sup_{-1 \leq x \leq 1} P_{n+1}(x)\,\mathrm{sgn}\,U_{n+1,\tau}(x).$$

Then we define the minimization problem $(\mathcal{O}_{n,1})$ by
$$\lambda_n^* := \inf\{\lambda_n(P_{n+1}) : P_{n+1} \in \mathbb{M}_{n+1}\}.$$

Concerning the existence of an optimal solution of $(\mathcal{O}_{n,1})$ we have the following lemma.

Lemma 5.17. *There exists $P_{n+1}^* \in \mathbb{M}_{n+1}$ such that $\lambda(P_{n+1}^*) = \lambda_n^*$ and $\lambda_n^* \geq E_n$.*

We remark that
$$\lambda_n(P_{n+1}) = \max_{0 \leq j \leq n+1} \max_{x \in [y_j^{(n)}, y_{j+1}^{(n)}]} (-1)^{n+1-j} P_{n+1}(x)$$

with $y_0^{(n)} := -1$ and $y_{n+2}^{(n)} = 1$.

The optimal solution P_{n+1}^* of $(\mathcal{O}_{n,1})$ is characterized by the following result.

Lemma 5.18. *P_{n+1}^* is optimal for the problem $(\mathcal{O}_{n,1})$ if and only if there exist $n+2$ points*
$$-1 \leq x_0^{(n)} < x_1^{(n)} < \cdots < x_{n+1}^{(n)} \leq 1$$

such that
$$(-1)^{n+1-j} P_{n+1}^*(x_i^{(n)}) = \lambda_n^* = \max_{x \in [y_j^{(n)}, y_{j+1}^{(n)}]} (-1)^{n+1-j} P_{n+1}^*(x)$$

and, for all $0 \leq i \leq n+1$,
$$x_i^{(n)} \in \left[y_i^{(n)}, y_{i+1}^{(n)}\right].$$

Moreover, the optimal solution of $(\mathcal{O}_{n,1})$ is unique.

The proof of Lemma 5.17 and Lemma 5.18 will be postponed. Lemma 5.18 yields
$$\left|P_{n+1}^*\left(x_i^{(n)}\right)\right| = \lambda_n^*, \quad 0 \leq i \leq n+1,$$

and
$$P_{n+1}^*\left(x_i^{(n)}\right) P_{n+1}^*\left(x_{i+1}^{(n)}\right) < 0, \quad 0 \leq i \leq n.$$

Since $E_{n,\infty}(P_{n+1}^*) = E_{n,\infty}(U_{n+1,\tau})$, we get by Theorem 4.1.5, after taking into account Lemma 4.2,

$$D[\mu_n - \mu] \leq c_1 \sqrt{\frac{1}{n}(\log E_{n,\infty}(U_{n+1,\tau}) - \log \lambda_n^* + \log(2n+1))}$$

with some constant $c_1 > 0$. Recall that μ_n is the normalized counting measure of the point set $\left\{x_i^{(n)}\right\}_{i=0}^{n+1}$, which appears in Lemma 5.18.
Nikol'skiĭ's inequality and Lemma 5.17 yield

$$\log E_{n,\infty}(U_{n+1,\tau}) \leq c_2\, n\, \varepsilon_n(w) + \log E_n \leq c_2\, n\varepsilon_n(w) + \log \lambda_n^*,$$

and observing (5.2), we get

$$D[\mu_n - \mu] \leq c\sqrt{\varepsilon_n(w)}.$$

By Lemma 5.18 the points $x_i^{(n)}$ are points with alternating sign with respect to $U_{n+1,w}$. Hence (5.26) is proved. □

It remains to prove Lemma 5.17 and Lemma 5.18.

Proof of Lemma 5.17 and Lemma 5.18. First, we solve the problem $(\mathcal{O}_{n,1})$ among all $p \in \mathbb{P}_{n+1}$ that satisfy $p(2) = 1$. Let us denote this modified problem by $(\widetilde{\mathcal{O}}_{n,1})$. Let $\mathcal{P}$ denote the set of all polynomials p in $\mathbb{P}_{n+1}$ with

$$\max_{0\leq j\leq n+1} \max_{x \in [y_j^{(n)}, y_{j+1}^{(n)}]} (-1)^{n+1-j} p(x) \leq 1$$

and $p(2) = 1$. It is clear that $\mathcal{P}$ is not empty. Further, since

$$|p(y_j^{(n)})| \leq 1, \quad j = 1, \ldots, n+1$$

and

$$|p(2)| = 1$$

for all $p \in \mathcal{S}$, the set $\mathcal{S}$ is bounded. Hence, $(\widetilde{\mathcal{O}}_{n,1})$ has a solution, say p_{n+1}. By standard arguments from the theory of semi-infinite optimization, we get $n+2$ characteristic points in the interval $[-1, 1]$, which satisfy for all $0 \leq j \leq n+1$,

$$(-1)^{n+1-j} p_{n+1}(x_j^{(n)}) = h_n = \max_{x\in[y_j^{(n)}, y_{j+1}^{(n)}]} (-1)^{n+1-j} p_{n+1}(x)$$

with some $h_n > 0$, and

$$x_j^{(n)} \in [y_j^{(n)}, y_{j+1}^{(n)}].$$

Let α_{n+1} be the leading coefficient of p_{n+1}. For $p_{n+1}(2) = 1$, α_{n+1} is positive. Then

$$P_{n+1}^* = \frac{1}{\alpha_{n+1}} p_{n+1} \in \mathbb{M}_{n+1}$$

is the optimal solution of $(\mathcal{O}_{n,1})$ and

$$\lambda_n^* = \lambda(P_{n+1}^*) = h_n/\alpha_{n+1}.$$

By a simple zero counting argument, P_{n+1}^* is the unique solution of the minimization problem $(\mathcal{O}_n)$.

From (5.28) we immediately get

$$\lambda_n(P_{n+1}^*) = \lambda_n^* \geq E_{n,1}\left(x^{n+1}\right) = E_n.$$

This completes the proof of both lemmas. □

It is remarkable that, contrary to the case $1 < p < \infty$, the whole sequence $\{\nu(Y_n)\}_{n=1}^{\infty}$ converges in the weak* sense to the equilibrium measure. Moreover, the discrepancy estimate of the preceding theorem can be considerably improved for special weights.

Theorem 5.19. *Let $\tau \in \mathcal{M}(I)$ with Radon–Nikodym derivative $d\tau(x) = w(x)\,dx$ with respect to the Lebesgue measure and w as in Corollary 5.8. If $B_{n,1}(f)$ is a best L_τ^1-approximation of $f \in L_\tau^1(I)$ with respect to $\mathbb{P}_n$, then there exists a point set Y_{n+1} of $n+1$ sign changes of $f - B_{n,1}(f)$ such that*

$$D[\nu(Y_{n+1} - \mu] \leq c\,\frac{(\log n)^2}{n}, \quad n = 2, 3, \ldots, \tag{5.29}$$

where c is a constant, depending only on w.

Proof. Let P_{n+1}^* be the optimal solution of Problem $(\mathcal{O}_{n,1})$ and further let $\left\{x_i^{(n)}\right\}_{i=0}^{n+1}$ be the alternating sequence of Lemma 5.18. Set

$$Q_{n+2}(x) := \prod_{i=0}^{n+1}\left(x - x_i^{(n)}\right) \in \mathbb{M}_{n+2}.$$

Then Lemma 4.2 yields

$$\left|Q'_{n+2}\left(x_i^{(n)}\right)\right| \geq \lambda_n^*, \quad 0 \leq i \leq n+1.$$

Next,

$$\frac{1}{2^n} = \|T_{n+1}\|_\infty \leq \|U_{n+1,\tau}\|_\infty \leq e^{c_2\,n\,\varepsilon_n(w)}\,E_n,$$

where T_{n+1} denotes the Chebyshev polynomial of first kind and of degree $n+1$ and c_2 is a constant, depending on w. For weight functions of type (5.11) it is known that $\varepsilon_n(w) = O(\log n/n)$ as $n \to \infty$. Since $\lambda_n^* \geq E_n$, we therefore obtain

$$|Q'_{n+2}(x_i^{(n)})| \geq \frac{1}{n^\kappa}\,\frac{1}{2^n}, \quad i = 0, \ldots, n+1, \tag{5.30}$$

7.5 Sign Changes in L^p-Approximation

where $\kappa > 0$ is a constant, independent of n, but depending on w.

To estimate the Chebyshev norm of P_{n+1}^* on $I = [-1, 1]$, fix $\zeta \in I$ such that
$$|P_{n+1}^*(\zeta)| = \|P_{n+1}^*\|_I.$$

Then Markov's inequality yields
$$|P_{n+1}^*(x)| \geq \frac{1}{2}\|P_{n+1}^*\|_I, \quad x \in \left[\zeta - \frac{1}{2n^2}, \zeta + \frac{1}{2n^2}\right] \cap I.$$

Introduce, for n sufficiently large, the sets
$$A_n := I \setminus B_n, \quad B_n := \bigcup_{i=1}^m \left[t_i - \frac{1}{4n^2}, t_i + \frac{1}{4n^2}\right],$$

where t_i are the points in the definition (5.11) of w. Then there exists an interval $J_n \subset A_n$, $|J_n| = 1/4n^2$, such that
$$|P_{n+1}^*(x)| \geq \frac{1}{2}\|P_{n+1}^*\|_I, \quad x \in J_n. \tag{5.31}$$

Using the definitions of the sets A_n and B_n, we conclude that there exists a constant $\kappa_1 > 0$, depending on w, such that
$$|w(x)| \geq \frac{1}{n^{\kappa_1}}, \quad x \in J_n. \tag{5.32}$$

We recall that we are looking for an upper estimate for the Chebyshev norm $\|P_{n+1}^*\|_I$. If $\|P_{n+1}^*\|_I \leq 2\lambda_n^*$, then the theorem of de la Vallée Poussin yields
$$\|P_{n+1}^*\|_\infty = 2\lambda_n^* \leq 2\|T_{n+1}\|_\infty = 2^{-n+1}. \tag{5.33}$$

Consider the case where $\|P_{n+1}^*\|_\infty > 2\lambda_n^*$. Then we can conclude that
$$(\operatorname{sgn} U_{n+1,w}(x))P_{n+1}^*(x) > 0, \quad x \in J_n.$$

Next, we notice that
$$\lambda_n^* = \lambda_n(P_{n+1}^*) \geq \int_{-1}^{1} (\operatorname{sgn} U_{n+1,w}(x))P_{n+1}^*(x)\, w(x)\, dx.$$

Furthermore,
$$L(P_{n+1}^*) = \int_{I \setminus J_n} (\operatorname{sgn} U_{n+1,w}(x))P_{n+1}^*(x)\, w(x)\, dx - \int_{J_n} |P_{n+1}^*(x)|\, w(x)\, dx.$$

Because of (5.28), we have $L(P^*_{n+1}) > 0$. Thus

$$\int_{J_n} |P^*_{n+1}(x)| w(x) \, dx < \int_{I \setminus J_n} (\text{sgn } U_{n+1,w}(x)) P^*_{n+1}(x) w(x) \, dx \le \lambda^*_n.$$

Combining (5.31) and (5.32) with the last inequality, we obtain

$$\|P^*_{n+1}\|_\infty \le 8 n^{\kappa_1+2} \lambda^*_n \le 8 n^{\kappa_1+2} \lambda_n(T_{n+1}) \le n^{\kappa_1+2} \frac{1}{2^{n-3}}, \qquad (5.34)$$

which is also true if $\|P^*_{n+1}\|_I \le 2\lambda^*_n$ according to (5.33).

According to Lemma 4.3 we obtain for Q_{n+2},

$$\|Q_{n+2}\|_I \le c n^{\kappa_1+5} \frac{1}{2^n}, \quad n = 1, 2, \dots \,. \qquad (5.35)$$

Hence, if we combine (5.30) and (5.35) with Theorem 3.1.8, we obtain

$$D[\mu_n - \mu] \le c \frac{(\log n)^2}{n} \quad \text{for all } n,$$

where μ_n is the normalized counting measure of the point set $\left\{x_i^{(n)}\right\}_{i=0}^{n+1}$. Since the zeros of $U_{n+1,w}$ are separated by these points, the theorem is proved. □

7.6 Extremal Polynomials on the Real Line

The problem of approximating functions defined on the whole real line by weighted polynomials can be traced back to S.N. Bernstein. In the 1970s, G. Freud started an investigation of the degree of approximation by weighted polynomials, under not necessarily the weakest conditions possible on the weight function, but nevertheless including many interesting weight functions. He considered the weight functions of the form $w_Q(x) = \exp(-Q(x))$ and obtained many important results under various conditions on Q.

We shall always assume in this section that the function $Q : \mathbb{R} \to \mathbb{R}$ satisfies the following condition.

Condition A: Q is even, the derivative $Q'(x)$ exists in $(0, \infty)$ and is nonnegative, $x Q'(x)$ is increasing on $(0, \infty)$, and $\lim_{x \to \infty} x Q'(x) = \infty$. Moreover, we assume that for some $r > 1$ the functions $Q'(c \cos \theta)$ are in $L^r[0, \pi]$ for all $c > 0$.

7.6 Extremal Polynomials on the Real Line

If f is a Lebesgue measurable function and $A \subset \mathbb{R}$ is Lebesgue measurable, we write

$$\|f\|_{p,A} := \begin{cases} \left(\int_A |f(t)|^p \, dt\right)^{1/p}, & 0 < p < \infty, \\ \operatorname*{ess\,sup}_{t \in A} |f(t)|, & p = \infty. \end{cases} \tag{6.1}$$

If $A = \mathbb{R}$, we omit the mention of A in the notation; thus, $\|f\|_\infty = \|f\|_{\infty,\mathbb{R}}$, and generally $\|f\|_p = \|f\|_{p,\mathbb{R}}$.

For $1 \leq p \leq \infty$ and every integer $n \geq 1$ we define

$$E_{n,p,Q} := \inf_{P \in \mathbb{P}_{n-1}} \|w_Q(x)(x^n - P(x))\|_p. \tag{6.2}$$

Since $x\,Q'(x)$ is increasing on $(0, \infty)$, all norms $\|w_Q P\|_p$ are finite for every polynomial P, $1 \leq p \leq \infty$. Moreover, there exists a unique polynomial

$$T_{n,p,Q}(x) := x^n + \cdots \in \mathbb{M}_n \tag{6.3}$$

such that

$$E_{n,p,Q} = \|w_Q T_{n,p,Q}\|_p. \tag{6.4}$$

$T_{n,p,Q}$ is called the nth L^p-extremal polynomial. We note that $T_{n,2,Q}$ is the monic orthogonal polynomial of degree n with respect to w_Q^2.

Now we are interested in the distribution of the zeros of $T_{n,p,Q}$. Therefore, we need essentially the theory of weighted approximation on $\mathbb{R}$.

One of the most important developments in this theory is summarized in the following theorem. An elementary and self-contained proof can be found in [124].

Theorem 6.1. *Let $w_Q(x) = \exp(-Q(x))$ and let Q satisfy condition A. Then for each $y > 0$, the equation*

$$\frac{2}{\pi} \int_0^1 \frac{a\,t\,Q'(a\,t)}{\sqrt{1-t^2}} \, dt = y \tag{6.5}$$

has a unique solution $a = a_y$. We define for $x \in (-1, 1)$,

$$\mu_n^*(x) := \frac{1}{\pi\sqrt{1-x^2}} - \frac{1}{n\pi^2} \frac{1}{\sqrt{1-x^2}} \operatorname{PV} \int_{-1}^1 \frac{a_n\,Q'(a_n t)\sqrt{1-t^2}}{x-t} \, dt, \tag{6.6}$$

where the symbol PV means that the integral is taken in the principal value sense. Define $\mu_n^*(x) = 0$ for $|x| \geq 1$ and

$$\lambda_n^*(B) := \int_B \mu_n^*(t) \, dt$$

7. Applications of Discrepancy Theorems

for any Borel set $B \subset \mathbb{R}$. Then the measure λ_n^* with density μ_n^* has the following properties:

(a)
$$\mu_n^*(x) > 0 \quad \text{for } x \in (-1,1), \quad \text{and} \quad \int_{-1}^{1} \mu_n^*(t)\, dt = 1.$$

(b) With
$$F_n^* := \log 2 + \frac{2}{n\pi} \int_0^1 \frac{Q(a_n t)}{\sqrt{1-t^2}}\, dt, \tag{6.7}$$

we have
$$U^{\lambda_n^*}(x) = \int_{-1}^{1} \log \frac{1}{|x-t|} \mu_n^*(t)\, dt = -\frac{Q(a_n x)}{n} + F_n^*, \quad x \in [-1,1],$$

and
$$U^{\lambda_n^*}(x) = \int_{-1}^{1} \log \frac{1}{|x-t|} \mu_n^*(t)\, dt > -\frac{Q(a_n x)}{n} + F_n^*, \quad x \in \mathbb{R} \setminus [-1,1].$$

The number a_y is called *Mhaskar–Rakhmanov–Saff* or *MRS number* of Q.

We observe that the conditions of Theorem 6.1 are satisfied by the prototypical Freud weights $w_Q(x) = \exp(-|x|^\alpha)$ for all $\alpha > 0$. In this case it can be shown (cf. [125]) that the formulas for a_n, F_n^*, and μ_n^* can be substantially simplified. With
$$\beta_\alpha := \left\{ \frac{2^{\alpha-2}\Gamma(\alpha/2)}{\Gamma(\alpha)} \right\}^{1/\alpha},$$

we have
$$a_n = \beta_\alpha n^{1/\alpha}, \quad F_n^* = \log(1/2) - 1/\alpha, \quad n = 1, 2, \ldots,$$

and
$$\mu_n^*(t) := \frac{\alpha}{\pi} \int_{|t|}^{1} \frac{y^{\alpha-1}}{\sqrt{y^2 - t^2}}\, dy, \quad |t| \le 1.$$

In particular, F_n^* and μ_n^* are independent of n. In this case, the function $w_Q(a_n x)$ has the form $W_\alpha(x)^n$, where $W_\alpha(x) := \exp(-|\beta_\alpha x|^\alpha)$.

It is convenient to transfer the measure with density μ_n^* back to $[-a_n, a_n]$. Define $\mu_n^*(x) := 0$ for $|x| \geq 1$ and

$$\lambda_n(B) := \frac{1}{a_n} \int_B \mu_n^*(t/a_n)\, dt, \quad n = 1, 2, \ldots, \tag{6.8}$$

for all Borel sets $B \subset \mathbb{R}$. Then $\mathrm{supp}(\lambda_n) = [-a_n, a_n]$, $\lambda_n \in \mathcal{M}([-a_n, a_n])$ and $U^{\lambda_n}(z) = U^{\lambda_n^*}(z/a_n) - \log a_n$. Let

$$F_n := F_n^* - \log a_n. \tag{6.9}$$

Then

$$U^{\lambda_n}(x) = -\frac{Q(x)}{n} + F_n, \quad x \in [-a_n, a_n], \tag{6.10}$$

and

$$U^{\lambda_n}(x) > -\frac{Q(x)}{n} + F_n, \quad x \in \mathbb{R} \setminus [-a_n, a_n]. \tag{6.11}$$

Hence, the measure λ_n is the *extremal* measure associated with the weight function

$$w_n(x) = \exp(-Q(x)/n),$$

and F_n is the *modified Robin constant* for w_n (cf. [156, Theorem I.1.3]).

Corollary 6.2. *Let $P \in \mathbb{M}_n$. Then*

$$\|w_Q P\|_{\infty, [-a_n, a_n]} \geq e^{-n F_n}. \tag{6.12}$$

For any $P \in \mathbb{P}_n$ and $z \in \mathbb{C}$,

$$\log |P(z)| \leq -n\, U^{\lambda_n}(z) + n F_n + \log \|w_Q P\|_{\infty, [-a_n, a_n]}.$$

In particular,

$$|w_Q(x) P(x)| < \|w_Q P\|_{\infty, [-a_n, a_n]}, \quad x \in \mathbb{R} \setminus [-a_n, a_n]. \tag{6.13}$$

Proof. Let $M = \|w_Q P\|_{\infty, [-a_n, a_n]}$. If $P \in \mathbb{M}_n$, we obtain with (6.10) that

$$\frac{1}{n} \log |P(x)| \leq \frac{Q(x)}{n} + \frac{\log M}{n} \leq -U^{\lambda_n}(x) + F_n + \frac{\log M}{n},$$

for $x \in [-a_n, a_n]$. Hence,

$$U^{\lambda_n - \nu(P)}(x) \leq F_n + \frac{\log M}{n}, \quad x \in [-a_n, a_n].$$

Since $U^{\lambda_n - \nu(P)}$ is harmonic in $\mathbb{C} \setminus [-a_n, a_n]$ and 0 at infinity, the maximum principle for harmonic functions yields

$$F_n + \frac{\log M}{n} \geq 0,$$

and inequality (6.12) is proved.

Let $P \in \mathbb{P}_n$. Then we define

$$h(z) := \log |P(z)| + n\, U^{\lambda_n}(z)$$

and obtain, as above,

$$h(x) \leq n\, F_n + \log M, \quad x \in [-a_n, a_n].$$

The function h is subharmonic in $\overline{\mathbb{C}} \setminus [-a_n, a_n]$ and bounded from above. Hence, the maximum principle again yields that for $z \in \mathbb{C} \setminus [-a_n, a_n]$,

$$h(z) < n\, F_n + \log M,$$

or

$$\log |P(z)| < -n\, U^{\lambda_n}(z) + n\, F_n + \log M.$$

In particular,

$$|w_Q(x)\, P(x)| < M \quad \text{for } x \in \mathbb{R} \setminus [-a_n, a_n].$$

$\square$

The inequality (6.13) is called the *infinite-finite-range inequality for the uniform norm* $\|w_Q P\|_\infty$.

Let σ be a signed Borel measure on $\mathbb{R}$ with compact support and let

$$D_\mathbb{R}[\sigma] := \sup |\sigma(J)|,$$

where the supremum is taken over all finite intervals $J \subset \mathbb{R}$. Then the distribution of the zeros of the extremal polynomials $T_{n,p,Q}$ can be determined via the discrepancy $D_\mathbb{R}[\nu_{T_{n,p,Q}} - \lambda_n]$, where λ_n is defined by (6.8).

Theorem 6.3. *Let Q satisfy Condition A. Moreover, suppose that there exist $B \geq 1$, $c > 0$ and $x_0 \geq 0$ such that*

$$2x\, Q'(2x) \leq c\, (x\, Q'(x))^B, \quad x > x_0; \tag{6.14}$$

furthermore, suppose that there exists $\rho < 1$ such that

$$x^\rho\, Q'(x) \text{ is increasing in } (x_0, \infty). \tag{6.15}$$

Then for $1 < p \leq \infty$,

$$D_\mathbb{R}[\nu_{T_{n,p,Q}} - \lambda_n] \leq c_1 \frac{(\log n)^2}{n}, \quad n \geq \max\left(2, \frac{1}{p-1}\right),$$

where c_1 is a constant independent of n and p.

7.6 Extremal Polynomials on the Real Line

In the following, the symbols $c, c_1, c_2, \ldots$ will denote positive constants depending only on the weight function w_Q. Their values may be different at different occurrences, even within the same formula.

To get in position to prove this theorem, we need some tools from weighted approximation. First, we need *infinite-finite-range inequalities for weighted L^p-norms*. Namely, if $1 \leq p < \infty$ and Q satisfies Condition A, then for $P \in \mathbb{P}_n$

$$\|w_Q P\|_p \leq 2 \|w_Q P\|_{p,[-a_{n+2}, a_{n+2}]}. \tag{6.16}$$

This is a consequence of Theorem 6.2.4 in [124].

Moreover, we need some crude Nikol'skiĭ-type inequalities.

Lemma 6.4. *Let Q satisfy $\lim_{x \to \infty} x Q'(x) = \infty$ (Condition A is not required). Then there exists a constant $c > 0$ depending only on Q such that for $n \geq 1$, $P \in \mathbb{P}_n$, and $1 \leq p < \infty$,*

$$\|w_Q P\|_\infty \leq c n^2 \|w_Q P\|_p.$$

Proof. It follows from Markov's inequality that

$$\|P\|_{\infty, J} \leq 8 |J|^{-1} n^2 \|P\|_{1, J}$$

for any interval J of length $|J|$. Then an application of Hölder's inequality yields

$$\|P\|_{\infty, J} \leq 8 n^2 |J|^{-1/p} \|P\|_{p, J}.$$

Let $h(x) := w_Q(x)$. Then

$$h'(x) = -Q'(x) w_Q(x), \quad x > 0.$$

Since $\lim_{x \to \infty} x Q'(x) = \infty$, there exists $x_0 > 0$ such that h is strictly decreasing for $x \geq x_0$. Next, let $\zeta \in \mathbb{R}$ such that

$$\|w_Q P\|_\infty = |w_Q(\zeta) P(\zeta)|.$$

Suppose first that $\zeta \geq x_0 + 2$. Then

$$\begin{aligned}
\|w_Q P\|_p &\geq \|w_Q P\|_{p, [\zeta-2, \zeta]} \geq w_Q(\zeta) \|P\|_{p, [\zeta-2, \zeta]} \\
&\geq c_1 w_Q(\zeta) n^{-2} \|P\|_{\infty, [\zeta-2, \zeta]} \\
&\geq c_1 n^{-2} |w_Q(\zeta) P(\zeta)| \\
&= c_1 n^{-2} \|w_Q P\|_\infty,
\end{aligned}$$

where c_1 is a constant independent of p. If $0 \leq \zeta \leq x_0 + 2$, we obtain with some constant $c_2 > 0$,

$$\begin{aligned}
\|w_Q P\|_p &\geq \|w_Q P\|_{p, [0, x_0+2]} \geq c_2 n^{-2} \|w_Q P\|_{\infty, [0, x_0+2]} \\
&= c_2 n^{-2} \|w_Q P\|_\infty,
\end{aligned}$$

and the lemma is proved. $\square$

280   7. Applications of Discrepancy Theorems

Lemma 6.5. *All zeros of the polynomials $T_{n,p,Q}$ are located in the interval $[-3\,a_{n+2}, 3\,a_{n+2}]$, $1 \le p \le \infty$.*

Proof. First we note that all zeros of $T_{n,p,Q}$ are real. Now let us recall that according to (6.16),
$$\|w_Q P\|_p \le 2\,\|w_Q P\|_{p,[-a_{n+2}, a_{n+2}]}$$
for $n \ge 1$ and $P \in \mathbb{P}_n$. Let x_n denote the largest positive zero of $T_{n,p,Q}$. If $x_n > 3\,a_{n+2}$, then we consider the rational function
$$r(x) := \frac{x - a_{n+2}}{x - x_n}$$
and observe that for $x \in [-a_{n+2}, a_{n+2}]$ we obtain
$$|r(x)| \le |r(-a_{n+2})| < \frac{1}{2}.$$
Hence, with $S(x) = r(x)\,T_{n,p,Q}(x)$ we see that S is a monic polynomial of degree n and
$$\|w_Q S\|_p \le 2\|w_Q S\|_{p,[-a_{n+2}, a_{n+2}]} < \|w_Q T_{n,p,Q}\|_p,$$
which is a contradiction to the extremal property of $T_{n,p,Q}$. Since the smallest zero of $T_{n,p,Q}$ is $-x_n$, the lemma is proved. $\square$

Next, we prove some results about the behavior of the numbers a_n.

Lemma 6.6. *The MRS numbers a_n have the following properties:*
(i) $n \le a_n\,Q'(a_n)$ for $n \ge 1$.

(ii) If Q satisfies (6.14), then there exists $c_1 > 0$ such that
$$a_n\,Q'(a_n) \le c_1\,n^B, \quad n \ge 1.$$

(iii) If Q satisfies (6.14) and (6.15), then there exists $c_2 > 0$ such that
$$a_n \le c_2\,n^{B/(1-\rho)} \quad \text{for } n \ge 1.$$

Proof. In view of (6.5) and (6.14) we obtain
$$a_n\,Q'(a_n) \ge n = \frac{2}{\pi} \int_0^1 \frac{a_n\,t\,Q'(a_n\,t)}{\sqrt{1-t^2}}\,dt$$
$$\ge \frac{2}{\pi} \frac{a_n}{2} Q'\left(\frac{a_n}{2}\right) \int_{1/2}^1 \frac{dt}{\sqrt{1-t^2}} = \frac{1}{3} \frac{a_n}{2} Q'\left(\frac{a_n}{2}\right)$$
$$\ge c\,(a_n\,Q'(a_n))^{1/B}$$

if n is sufficiently large. Therefore, (i) and (ii) hold.

Let n be so large that $a_n \geq x_0 + 1$. Then

$$c_1 n^B \geq a_n Q'(a_n) \geq a_n^{1-\rho} a_n^{\rho} Q'(a_n) \geq a_n^{1-\rho} (x_0 + 1)^{\rho} Q'(x_0 + 1).$$

Hence $a_n \leq c_2 n^{B/(1-\rho)}$ for $n \geq 1$ in the case (iii). $\square$

For the proof of Theorem 6.3 we use in addition the method of *discretizing a logarithmic* potential.

Lemma 6.7. *Let $I = [-1, 1]$ and let $\nu \in \mathcal{M}(I)$. Suppose there exist K and $\varepsilon > 0$ such that*

$$\int_{|x-t| \leq n^{-K}} |\log|x - t|| \, d\nu(t) \leq \varepsilon \frac{\log n}{n}, \quad x \in I, \quad (6.17)$$

where n is a positive integer. Then there exists a monic polynomial $P_{n-1} \in \mathbb{M}_{n-1}$ with zeros in I such that

$$\frac{1}{n} \log |P_{n-1}(x)| \leq -U^{\nu}(x) + (K + \varepsilon) \frac{\log n}{n}, \quad x \in \mathbb{R}, \quad (6.18)$$

and, for $z \in \mathbb{C} \setminus I$,

$$\frac{1}{n} \log |P_{n-1}(z)| \geq -U^{\nu}(z) - \frac{2}{n} \log \left(\frac{1 + |z|}{d(z, I)} \right). \quad (6.19)$$

Proof. Let

$$-1 = y_{0,n} < y_{1,n} < \cdots < y_{n-1,n} < y_{n,n} = 1$$

be chosen such that

$$\int_{y_{j,n}}^{y_{j+1,n}} d\nu(t) = \frac{1}{n}, \quad j = 0, 1, 2, \ldots, n - 1.$$

Let

$$P_{n-1}(x) := \prod_{j=1}^{n-1} (x - y_{j,n}).$$

Then (6.18) is true (for a proof see [156, Theorem VI.4.1]).

Next, let $z \in \mathbb{C}$ and $\operatorname{Im} z \neq 0$. We prove (6.19) for $\operatorname{Re} z \in [y_{j,n}, y_{j+1,n}]$, $1 \leq j \leq n - 1$. The proof of (6.19) for other values of z is similar, but simpler, and hence will be omitted. Geometric considerations show that $\log |z - t|$ is decreasing for $t < \operatorname{Re} z$ and increasing for $t > \operatorname{Re} z$ ($t \in \mathbb{R}$).

Hence

$$\int_{-1}^{1} \log|z-t|\,d\nu(t) = \sum_{\substack{k=0 \\ k\neq j}}^{n-1} \int_{y_{k,n}}^{y_{k+1,n}} \log|z-t|\,d\nu(t) + \int_{y_{j,n}}^{y_{j+1,n}} \log|z-t|\,d\nu(t)$$

$$\leq \frac{1}{n}\sum_{\substack{k=0 \\ k\neq j,j+1}}^{n} \log|z-y_{k,n}| + \frac{1}{n}\log(1+|z|)$$

$$\leq \frac{1}{n}\log|P_{n-1}(z)| + \frac{2}{n}\log(1+|z|)$$
$$-\frac{1}{n}\log|z-y_{j,n}| - \frac{1}{n}\log|z-y_{j+1,n}|$$

$$\leq \frac{1}{n}\log|P_{n-1}(z)| + \frac{2}{n}\log(1+|z|) - \frac{2}{n}\log d(z,I).$$

Consequently,

$$\frac{1}{n}\log|P_{n-1}(z)| \geq -U^{\nu}(z) - \frac{2}{n}\log(1+|z|) + \frac{2}{n}\log d(z,I)$$
$$= -U^{\nu}(z) - \frac{2}{n}\log\left(\frac{1+|z|}{d(z,I)}\right).$$

□

To apply this lemma to $\nu = \lambda_n^*$ we have to prove property (6.17) for λ_n^*. Moreover, for the application of the discrepancy theorems we need a comparison between λ_n^* and the equilibrium distribution of I.

Lemma 6.8. *Let Q satisfy the conditions of Theorem 6.3 and let μ_I be the equilibrium measure of $I = [-1,1]$. Then there exist positive constants c, B^*, and $0 < \gamma < 1$, independent of n, such that*

$$\lambda_n^*(J) \leq c n^{B^*} \mu_I(J)^{\gamma} \tag{6.20}$$

for any subinterval J of I and

$$\int_{|x-t|\leq n^{-2(B^*+1)}} |\log|x-t|| \, |\mu_n^*(t)|\,dt \leq \frac{c}{n}, \quad x \in I. \tag{6.21}$$

Proof. Let
$$R_n(\theta) := \frac{1}{n} Q(a_n \cos\theta), \quad \theta \in [-\pi, \pi].$$

Since $R'_n \in L^r[-\pi, \pi]$, $r > 1$, the conjugate function

$$\widetilde{R}'_n(\theta) = \frac{1}{2\pi} \mathrm{PV} \int_{-\pi}^{\pi} R'(\varphi) \cot\left(\frac{\theta - \varphi}{2}\right) d\varphi$$

7.6 Extremal Polynomials on the Real Line

exists for almost all $\theta \in [-\pi, \pi]$ and belongs to $L^r[-\pi, \pi]$ as well. If we define
$$\mu^\circ(\theta) := \mu_n^*(\cos\theta)|\sin\theta|, \quad \theta \in [-\pi, \pi],$$
then we want to show first that
$$\mu^\circ = \frac{1}{\pi} + \frac{1}{\pi}\widetilde{R}_n'. \tag{6.22}$$

Since
$$\frac{-\sin\varphi}{\cos\varphi - \cos\theta} = \frac{\sin\left(\frac{\varphi+\theta}{2} + \frac{\varphi-\theta}{2}\right)}{2\sin\frac{\varphi+\theta}{2}\sin\frac{\varphi-\theta}{2}}$$
$$= \frac{1}{2}\left\{\cot\frac{\varphi-\theta}{2} + \cot\frac{\varphi+\theta}{2}\right\}$$

and R_n' is an odd function, we obtain
$$\widetilde{R}_n'(\theta) = \frac{1}{\pi} \text{PV} \int_0^\pi \frac{R_n'(\varphi)\sin\varphi}{\cos\theta - \cos\varphi}\, d\varphi$$
$$= \frac{1}{n\pi} \text{PV} \int_{-1}^1 \frac{a_n Q'(a_n t)\sqrt{1-t^2}}{t-x}\, dt,$$

where $x = \cos\theta$. Then (6.22) follows from (6.6).

Note that
$$R_n'(\theta) = \frac{d}{d\theta} R_n(\theta) = \frac{a_n}{n}\sin\theta\, Q'(a_n\cos\theta),$$
where $Q'(x) = dQ(x)/dx$. Using the transformation $t = a_n\cos\theta$ and the symmetry of Q we obtain
$$\int_{-\pi}^\pi |R_n'(\theta)|^r\, d\theta \leq \frac{4\, a_n^{r-1}}{n^r} \int_0^{a_n} |Q'(t)|^r\, dt. \tag{6.23}$$

Since $Q'(c\cos\theta) \in L^r[-\pi,\pi]$ for any $c > 0$, we obtain
$$\int_{-\pi}^\pi |Q'(a_n\cos\theta)|^r\, d\theta = 4\int_0^{\pi/2} |Q'(a_n\cos\theta)|^r\, d\theta$$
$$= 4\int_0^{a_n} |Q'(t)|^r \frac{dt}{\sqrt{a_n^2 - t^2}}$$
$$\geq \frac{4}{a_n}\int_0^{a_n} |Q'(t)|^r\, dt.$$

Hence, the integral on the right-hand side of (6.23) exists.

Let n be large enough that $a_n > x_0$. For $t \in (x_0, a_n]$ we obtain by (6.15)

$$Q'(t) \le a_n^\rho \, Q'(a_n) \, t^{-\rho}.$$

In the definition of ρ by (6.15) we may assume that $\rho > 1/r$. Then,

$$\int_{x_0}^{a_n} |Q'(t)|^r \, dt \le \frac{1}{1 - \rho r} (Q'(a_n))^r \, a_n$$

and by (6.23) and Lemma 6.6,

$$\int_{-\pi}^{\pi} |R_n'(\theta)|^r \, d\theta \le c \left(\frac{a_n^{r-1}}{n^r} + \left(\frac{a_n \, Q'(a_n)}{n} \right)^r \right) \le c \, n^{B/(1-\rho)-1}.$$

Set $B^* := B/(1-\rho) - 1$. The theorem of M. Riesz yields

$$\int_{-\pi}^{\pi} \left| \mu^\circ(\theta) - \frac{1}{\pi} \right|^r d\theta \le \frac{1}{\pi} \int_{-\pi}^{\pi} |\widetilde{R}_n'(\theta)|^r \, d\theta \le c \int_{-\pi}^{\pi} |R_n'(\theta)|^r \, d\theta \le c_4 \, n^{B^* r}. \tag{6.24}$$

Hence, for any interval J of I we have with $J' := \{\theta \in [0, \pi] : \cos \theta \in J\}$,

$$\lambda_n^*(J) = \int_{J'} \mu^\circ(\theta) \, d\theta \le \int_{J'} \left| \mu^\circ(\theta) - \frac{1}{\pi} \right| d\theta + \mu_I(J).$$

Then Hölder's inequality leads to

$$\int_J \left| \mu^\circ(\theta) - \frac{1}{\pi} \right| d\theta \le \left\{ \int_J \left| \mu^\circ(\theta) - \frac{1}{\pi} \right|^r d\theta \right\}^{1/r} \left\{ \int_J d\theta \right\}^{1-1/r}$$

$$\le c n^{B^*} \mu_I(J)^{1-1/r},$$

where μ_I is the equilibrium measure of $I = [-1, 1]$. Therefore,

$$\lambda_n^*(J) \le c n^{B^*} \mu_I(J)^{1-1/r},$$

which proves (6.20).

Let $x = \cos \theta$ with $0 \le \theta \le \pi$ and let us define

$$\mathcal{I}_n := \int_{|x-t| \le n^{-2(B^*+1)}} |\log |x - t|| \, \mu_n^*(t) \, dt.$$

7.6 Extremal Polynomials on the Real Line

Let $|x - t| \leq n^{-2(B^*+1)}$. Then we choose φ in such a way that $t = \cos\varphi$ and $|\varphi - \theta| \leq \pi/2$ for $n \geq 2$. Then there exists $c_1 > 0$ such that

$$|\cos\varphi - \cos\theta| \geq c_1^2 |\varphi - \theta|^2.$$

Let $B_n := \{\varphi \in \mathbb{R} : |\theta - \varphi| \leq \kappa_n\}$, where $\kappa_n := n^{-(B^*+1)}/c_1$. Then

$$\mathcal{I}_n \leq \int_{B_n} |\log|c_1(\varphi - \theta)|^2| \mu^\circ(\varphi) \, d\varphi$$

$$\leq \frac{2}{\pi} \int_{B_n} |\log|c_1(\varphi - \theta)|| \, d\varphi + \frac{2}{\pi} \int_{B_n} |\log|c_1(\varphi - \theta)|| \widetilde{R}'_n(\varphi)| \, d\varphi.$$

The first integral on the right-hand side can be estimated by c_2/n. Using Hölder's inequality again we obtain for the second integral

$$\int_{B_n} |\log|c_1(\varphi - \theta)|| \widetilde{R}'_n(\varphi)| \, d\varphi$$

$$\leq \left\{ \int_{B_n} |\log|c_1(\varphi - \theta)||^{r/(r-1)} \, d\varphi \right\}^{1-1/r} \left\{ \int_{B_n} |\widetilde{R}'_n(\varphi)|^r \, d\varphi \right\}^{1/r}$$

$$\leq c n^{B^*} \left\{ \int_0^{n^{-(B^*+1)}} |\log\varphi|^{r/(r-1)} \, d\varphi \right\}^{1-1/r},$$

where r is the number of Condition A such that $Q'(c\cos\theta) \in L^r[0, \pi]$ and where we have used (6.24). Now, there exists $c_2 > 0$ such that

$$|\log\varphi|^{r/(r-1)} \leq \frac{c_2}{2\sqrt{\varphi}}, \qquad 0 \leq \varphi \leq 1,$$

and consequently,

$$\int_0^{n^{-(B^*+1)}} |\log\varphi|^{r/(r-1)} \, d\varphi \leq c_2 \, n^{-(B^*+1)/2} \leq \frac{c_2}{n}.$$

Hence,

$$\mathcal{I}_n \leq \frac{c}{n},$$

where c is an absolute constant, independent of n. □

If we apply this lemma and Lemma 6.7 to the measure λ_n, we obtain the following.

Corollary 6.9. *Let Q satisfy the conditions of Theorem 6.3. Then for $n \geq 2$ there exists a monic polynomial $P_{n-1} \in \mathbb{M}_{n-1}$ with zeros in $J_{a_n} = [-a_n, a_n]$ such that*

$$\frac{1}{n}\log|P_{n-1}(x)| \leq -U^{\lambda_n}(x) - \frac{\log a_n}{n} + c\frac{\log n}{n}, \quad x \in \mathbb{R},$$

and, for $z \in \mathbb{C} \setminus J_{a_n}$,

$$\frac{1}{n}\log|P_{n-1}(z)| \geq -U^{\lambda_n}(z) - \frac{\log a_n}{n} - \frac{c}{n}\log\left(\frac{a_n + |z|}{d(z, J_{a_n})}\right),$$

where c is a positive constant independent of n.

Proof of Theorem 6.3. Because of Corollary 6.9 there exists a monic polynomial $P_{n-1} \in \mathbb{M}_{n-1}$ such that

$$\|w_Q P_{n-1}\|_\infty \leq a_n^{-1} n^c e^{-n F_n}. \tag{6.25}$$

Consequently, by the infinite-finite-range inequality (6.13),

$$\frac{1}{n}\log\|w_Q T_{n,\infty,Q}\|_\infty \leq \frac{1}{n}\log\|w_Q x P_{n-1}\|_\infty$$
$$\leq \frac{1}{n}\log\|w_Q x P_{n-1}\|_{\infty,[-a_n,a_n]} \leq c\frac{\log n}{n} - F_n.$$

If $1 \leq p < \infty$, then the infinite-finite-range inequality (6.16) leads to

$$\|w_Q x P_{n-1}\|_p \leq 2\|w_Q x P_{n-1}\|_{p,[-a_{n+2}, a_{n+2}]}.$$

Therefore,

$$\begin{aligned}\|w_Q T_{n,p,Q}\|_p &\leq \|w_Q x P_{n-1}\|_p \leq 2^{1+1/p} a_{n+2}^{1/p} \|w_Q x P_{n-1}\|_\infty \\ &\leq 2^{1+1/p} a_{n+2}^{1/p} n^c e^{-n F_n}.\end{aligned} \tag{6.26}$$

Now Nikol'skiĭ's inequality (Lemma 6.4) and Lemma 6.6 yield

$$\frac{1}{n}\log\|w_Q T_{n,p,Q}\|_\infty \leq c\frac{\log n}{n} - F_n. \tag{6.27}$$

But this inequality also includes the case $p = \infty$.

Hence, (6.27) yields for $x \in \mathbb{R}$,

$$U^{\lambda_n}(x) - U^{\nu_{T_{n,p,Q}}}(x) \leq c\frac{\log n}{n}.$$

Since the right-hand side is harmonic in $\mathbb{C} \setminus \mathbb{R}$ and 0 at infinity, we finally have

$$U^{\lambda_n - \nu_{T_{n,p,Q}}}(z) \leq c\frac{\log n}{n}, \quad z \in \mathbb{C}, \tag{6.28}$$

7.6 Extremal Polynomials on the Real Line

where $c > 0$ depends only on Q.

To get lower bounds for this logarithmic potential, let n be fixed. Then we define

$$P_{n,p} := E_{n,p,Q}^{-1} T_{n,p,Q}, \qquad (6.29)$$

and, for $1 < p < \infty$,

$$W_{n,p} := |P_{n,p}|^{p-2} w_Q^p. \qquad (6.30)$$

For $1 < p < \infty$ all moments of $W_{n,p}$ are finite, and hence there exists a system of orthonormal polynomials

$$\widehat{p}_m(x) = \gamma_m^{(p)} x^m + \cdots = \gamma_m^{(p)} \prod_{k=1}^{n} (x - x_{k,m}) \in \mathbb{P}_m \qquad (6.31)$$

such that

$$\int \widehat{p}_m(t)\,\widehat{p}_k(t)\,W_{n,p}(t)\,dt = \begin{cases} 1, & \text{if } k = m, \\ 0, & \text{if } k \neq m. \end{cases} \qquad (6.32)$$

Note that

$$\int P_{n,p}^2\,W_{n,p}(t)\,dt = 1,$$

and the characterization of the extremal polynomials $T_{n,p,Q}$ shows that for all $P \in \mathbb{P}_{n-1}$,

$$\int |T_{n,p,Q}(t)|^{p-1}\,\mathrm{sgn}\,T_{n,p,Q}(t)\,P(t)\,w_Q^p(t)\,dt = 0.$$

By some simple calculations we obtain

$$\int P_{n,p}(t)\,P(t)\,W_{n,p}(t)\,dt = 0.$$

Therefore, $P_{n,p} = \widehat{p}_n$. Hence, by Lemma 6.5 all zeros of $\widehat{p}_n$ are located in $[-3\,a_{n+2}, 3\,a_{n+2}]$.

The next lemma provides some estimates on $\gamma_{n-1}^{(p)}/\gamma_n^{(p)}$.

Lemma 6.10. *Let Q satisfy Condition A and let $1 < p < \infty$. Then there exists a constant $c_1 > 0$, independent of p, such that*

$$\frac{\gamma_{n-1}^{(p)}}{\gamma_n^{(p)}} \leq c_1\,a_{n+5}. \qquad (6.33)$$

The proof of this lemma will be postponed.

288   7. Applications of Discrepancy Theorems

Let P_{n-1} again be the monic polynomial of Corollary 6.9. Then as in the proof of Lemma 7.1 we obtain by Lagrange interpolation that

$$P_{n-1}(z) = \frac{\gamma_{n-1}^{(p)}}{\gamma_n^{(p)}} \widehat{p}_n(z) \sum_{k=1}^n \widehat{\lambda}_{k,n} \frac{\widehat{p}_{n-1}(x_{k,n}) P_{n-1}(x_{k,n})}{z - x_{k,n}},$$

where $\widehat{\lambda}_{k,n}$ are the Christoffel numbers and $x_{k,n}$ are the zeros of $\widehat{p}_n$, $k = 1, \ldots, n$.

Using Lemma 6.6 and Lemma 6.10, the Cauchy–Schwarz inequality, and the Gauss quadrature formula we get for $z \in \mathbb{C}\setminus J_n$, $J_n := [-3\,a_{n+2}, 3\,a_{n+2}]$,

$$\left|\frac{P_{n-1}(z)}{\widehat{p}_n(z)}\right| \leq \frac{c_1\,a_{n+5}}{d(z, J_n)} \sum_{k=1}^n \widehat{\lambda}_{k,n} |\widehat{p}_{n-1}(x_{k,n})|\, |P_{n-1}(x_{k,n})|$$

$$\leq \frac{c_1\,a_{n+5}}{d(z, J_n)} \left(\sum_{k=1}^n \widehat{\lambda}_{k,n} \widehat{p}_{n-1}^2(x_{k,n})\right)^{1/2} \left(\sum_{k=1}^n \widehat{\lambda}_{k,n} P_{n-1}^2(x_{k,n})\right)^{1/2}$$

$$\leq \frac{c_1\,a_{n+5}}{d(z, J_n)} \left(\int P_{n-1}^2(t)\, W_{n,p}(t)\, dt\right)^{1/2}, \qquad (6.34)$$

where $d(z, J_n)$ denotes the distance from z to the interval J_n.

Now we distinguish the cases $p \geq 2$ and $1 < p < 2$.

The case $p \geq 2$.

Using Hölder's inequality with $p/2$ and $p/(p-2)$ we get with (6.32),

$$\left(\int P_{n-1}^2(t)\, W_{n,p}(t)\, dt\right)^{1/2} = \left(\int P_{n-1}^2(t)\, w_Q^2(t)\, |P_{n,p}(t)\, w_Q(t)|^{p-2}\, dt\right)^{1/2}$$

$$\leq \|w_Q\, P_{n-1}\|_p.$$

Inserting this inequality in (6.34) we obtain by Lemma 6.6, analogously to (6.26),

$$\left|\frac{P_{n-1}(z)}{\widehat{p}_n(z)}\right| \leq \frac{c_2}{d(z, J_n)}\, n^c\, e^{-n F_n},$$

or

$$\frac{1}{n} \log \left|\frac{P_{n-1}(z)}{\widehat{p}_n(z)}\right| \leq -F_n + c\,\frac{\log n}{n} + \frac{1}{n} \log \frac{1}{d(z, J_n)} \qquad (6.35)$$

for $n \geq 2$. Now,

$$U^{\nu_{T_{n,p,Q}}} = \frac{1}{n} \log \left|\frac{1}{\widehat{p}_n(z)}\right| + \frac{1}{n} \log \gamma_n = \frac{1}{n} \log \left|\frac{1}{\widehat{p}_n(z)}\right| - \frac{1}{n} \log E_{n,p,Q},$$

7.6 Extremal Polynomials on the Real Line

and Nikol'skiĭ's inequality and Corollary 6.2 yield

$$E_{n,p,Q} = \|w_Q T_{n,p,Q}\|_p \geq \frac{c}{n^2} \|w_Q T_{n,p,Q}\|_\infty$$

$$= \frac{c}{n^2} \|w_Q T_{n,p,Q}\|_{[-a_n, a_n]} \geq \frac{c}{n^2} e^{-n F_n}.$$

Therefore, we obtain from (6.35) and Corollary 6.9,

$$U^{\nu T_{n,p,Q} - \lambda_n}(z) \leq c \frac{\log n}{n} + \frac{c}{n} \log\left(\frac{a_n + |z|}{d(z, J_{a_n})}\right) + \frac{1}{n} \log \frac{1}{d(z, J_n)} \quad (6.36)$$

for all $z \in \mathbb{C} \setminus J_n$, where c is a constant independent of n and p.

Let $p \to \infty$. Then the last inequality is also true for $p = \infty$, since the constants c_1 and c_2 are independent of p and ∞, and $T_{n,p,Q}$ converges uniformly to $T_{n,\infty,Q}$ as $p \to \infty$.

The case $1 < p < 2$.

To estimate the integral in (6.34) we use Lagrange interpolation, namely

$$P_{n-1}(z) = \widehat{p}_n(z) \sum_{k=1}^n \frac{P_{n-1}(x_{k,n})}{\widehat{p}_n'(x_{k,n})(z - x_{k,n})}, \quad (6.37)$$

and therefore

$$\left|\frac{P_{n-1}(z)}{\widehat{p}_n(z)}\right|^{2-p} \leq \sum_{k=1}^n \left|\frac{P_{n-1}(x_{k,n})}{\widehat{p}_n'(x_{k,n})}\right|^{2-p} \frac{1}{|z - x_{k,n}|^{2-p}}$$

$$\leq M_n^{2-p} \sum_{k=1}^n \frac{1}{|z - x_{k,n}|^{2-p}},$$

where

$$M_n = \max_{1 \leq k \leq n} \left|\frac{P_{n-1}(x_{k,n})}{\widehat{p}_n'(x_{k,n})}\right|.$$

Hence,

$$\int P_{n-1}^2(t) W_{n,p}(t) \, dt = \int |P_{n-1}(t) w_Q(t)|^p \left|\frac{P_{n-1}(t)}{\widehat{p}_n}\right|^{2-p} dt$$

$$\leq M_n^{2-p} \sum_{k=1}^n \int \frac{|P_{n-1}(t) w_Q(t)|^p}{|t - x_{k,n}|^{2-p}} dt.$$

We subdivide the integrals on the right-hand side into

$$\int \frac{|P_{n-1}(t) w_Q(t)|^p}{|t - x_{k,n}|^{2-p}} dt = \int_{|t-x_{k,n}|\leq a_n} + \int_{|t-x_{k,n}|\geq a_n} =: \mathcal{I}_{1,k} + \mathcal{I}_{2,k}.$$

Then (6.25) and Lemma 6.6 yield

$$\mathcal{I}_{1,k} \leq \|w_Q P_{n-1}\|_\infty^p \int_{|t-x_{k,n}|\leq a_n} \frac{dt}{|t-x_{k,n}|^{2-p}} \leq c_1 \left(n^c e^{-nF_n}\right)^p \frac{1}{p-1},$$

and

$$\mathcal{I}_{2,k} \leq \frac{1}{a_n^{2-p}} \int |P_{n-1}(t) w_Q(t)|^p dt \leq c_2 \left(n^c e^{-nF_n}\right)^p.$$

Therefore,

$$\int P_{n-1}^2 W_{n,p}(t) dt \leq c_3 M_n^{2-p} \left(n^c e^{-nF_n}\right)^p \frac{1}{p-1}. \tag{6.38}$$

Since $\widehat{\lambda}_{k,n}$ are the Christoffel numbers,

$$\widehat{\lambda}_{k,n} = \frac{\gamma_n^{(p)}}{\gamma_{n-1}^{(p)}} \frac{1}{\widehat{p}_{n-1}(x_{k,n})\, \widehat{p}_n{}'(x_{k,n})},$$

we get by Schwarz's inequality and Gauss quadrature that

$$\sum_{k=1}^n \left|\frac{P_{n-1}(x_{k,n})}{\widehat{p}_n(x_{k,n})}\right|^2 = \sum_{k=1}^n \left|\frac{\gamma_{n-1}^{(p)}}{\gamma_n^{(p)}} \widehat{\lambda}_{k,n} \widehat{p}_{n-1}(x_{k,n}) P_{n-1}(x_{k,n})\right|^2$$

$$\leq c_4 a_{n+5}^2 \int P_{n-1}^2(t) W_{n,p}(t) dt.$$

Therefore,

$$M_n^2 \leq c_4 a_{n+5}^2 \int P_{n-1}^2(t) W_{n,p}(t) dt \leq c_5 M_n^{2-p} \left(n^c e^{-nF_n}\right)^p \frac{1}{p-1},$$

or

$$M_n \leq c_6 n^c e^{-nF_n} \frac{1}{p-1}.$$

Then (6.37) yields

$$\left|\frac{P_{n-1}(z)}{\widehat{p}_n(z)}\right| \leq \frac{c_6}{\mathrm{dist}(z, J_n)} n^c e^{-nF_n} \frac{1}{p-1}.$$

We therefore obtain from (6.35) and Corollary 6.9 that for $1 < p \leq 2$

$$U^{\nu T_{n,p,Q} - \lambda_n}(z) \leq c \frac{\log n}{n} + \frac{c}{n} \log\left(\frac{a_n + |z|}{d(z, J_{a_n})}\right)$$

$$+ \frac{1}{n} \log \frac{1}{d(z, J_n)} + \frac{1}{n} \log \frac{1}{p-1} \tag{6.39}$$

Next, we want to apply the discrepancy estimate of Theorem 3.1.5 to the interval J_n. According to Remark 3.1.6 there exists a constant $K \geq 1$, independent of n, such that the estimate (3.1.19) holds.

Let us consider (6.20), i.e.,

$$\lambda_n^*(J) \leq c n^{B^*} \mu_I(J)^\gamma$$

for any subinterval J of I. Then we have to take $\beta = \gamma$ in the inequality (3.1.19).

Let $G(z)$ be Green's function for $\mathbb{C} \setminus J_n$ and let

$$L_\delta = \{z \in \mathbb{C} : G(z) = \log(1 + \delta)\}$$

with $\delta := n^{-\kappa}$ and $\kappa := 2 K^2 (B^* + 1) / \gamma$. Then (6.20) yields for any subinterval $J^* \subset J_n$,

$$\lambda_n(J^*) = \lambda_n^*(J) \leq c n^{B^*} \mu_I(J)^\gamma = c n^{B^*} \mu_{J_n}(J^*)^\gamma,$$

where c is an absolute constant, $J := \{z : z = x/(3 a_{n+2}), x \in J^*\}$, and μ_{J_n} denotes the equilibrium measure of J_n. Next, we note that

$$\mathrm{dist}(z, J_n) \leq c_1 a_n n^{-2\kappa}, \quad z \in L_\delta,$$

with some absolute constant c_1. Combine (6.28), (6.36), and (6.39) to obtain

$$\left| U^{\lambda_n - \nu_{T_n,p,Q}}(z) \right| \leq c_2 \frac{\log n}{n} + \frac{1}{n} \log \max \left(2, \frac{1}{p-1} \right), \quad z \in L_\delta. \quad (6.40)$$

Now Theorem 3.1.5 leads to

$$D[\nu_{T_n,p,Q} - \lambda_n] \leq c_3 \frac{(\log n)^2}{n}$$

with some absolute constant c_3 for all $n \geq \max(2, 1/(p-1))$. $\square$

The proof of Theorem 6.3 will be completed by appending the proof of Lemma 6.10.

Proof of Lemma 6.10. Using the Cauchy–Schwarz inequality in the equation

$$\frac{\gamma_{n-1}^{(p)}}{\gamma_n^{(p)}} = \int x \widehat{p}_{n-1}(x) \widehat{p}_n(x) W_{n,p}(x) \, dx$$

and the orthogonality relation (6.32), we obtain

$$\frac{\gamma_{n-1}^{(p)}}{\gamma_n^{(p)}} \leq \left(\int x^2 |P_{n,p}(x)|^p w_Q^p(x) \, dx \right)^{1/2}. \quad (6.41)$$

First, let $2 < p < \infty$. Then by Hölder's inequality with $p/2$ and $p/(p-2)$ we obtain

$$\frac{\gamma_{n-1}^{(p)}}{\gamma_n^{(p)}} \leq \left\{\int |x|^p |P_{n,p}(x)|^p \, w_Q^p(x) \, dx\right\}^{\frac{1}{p}} \left\{\int |P_{n,p}(x)|^p \, w_Q^p(x) \, dx\right\}^{\frac{p-2}{2p}}$$

$$= \|w_Q \, x \, P_{n,p}\|_p,$$

which is also true for $p = 2$. Then the infinite-finite-range inequality (6.16) yields

$$\frac{\gamma_{n-1}^{(p)}}{\gamma_n^{(p)}} \leq \|w_Q \, x \, P_{n,p}\|_p \leq 2 \, \|w_Q \, x \, P_{n,p}\|_{p,[-a_{n+3},a_{n+3}]}$$

$$\leq 2 \, a_{n+3} \, \|w_Q \, P_{n,p}\|_p = 2 \, a_{n+3}.$$

Let $1 < p < 2$. Since $x \, Q'(x)$ is increasing to ∞ as $x \to \infty$ and $x \, Q'(x) \to 0$ as $x \to 0$, there exists a constant $C > 1$ and points $x_0 = x_0(p)$ such that $x_0(p) \leq C$ for $1 < p < \infty$ and

$$x_0 \, Q'(x_0) = \frac{2}{p}$$

as well as

$$x \, Q'(x) > \frac{2}{p}, \quad x > x_0.$$

Now we define an even function $\widehat{Q}(x)$ where

$$\widehat{Q}(x) := \begin{cases} Q(x_0) - \dfrac{2}{p} \log x_0, & x \leq x_0, \\ Q(x) - \dfrac{2}{p} \log x, & x \geq x_0. \end{cases}$$

Then $\widehat{Q}(x)$ satisfies Condition A. Next, we choose n_0 such that

$$\frac{2}{\pi} \int_0^{C/a_n} \frac{a_n \, t \, Q'(a_n \, t)}{\sqrt{1-t^2}} \, dt < 1, \quad n \geq n_0.$$

Hence, for $n \geq n_0$,

$$\frac{2}{\pi} \int_0^1 \frac{a_n \, t \, \widehat{Q}'(a_n \, t)}{\sqrt{1-t^2}} \, dt > n - 3,$$

or

$$\widehat{a}_n \leq a_{n+3}, \quad n \geq n_0,$$

where $\widehat{a}_n$ denote the MRS numbers associated with $\widehat{Q}$. Then the infinite-finite-range inequality implies

$$\left(\frac{\gamma_{n-1}^{(p)}}{\gamma_n^{(p)}}\right)^2 \le \int |P_{n,p}(x)|^p\, x^2\, w_Q^p(x)\, dx$$

$$\le \int_{-C}^{C} |P_{n,p}(x)|^p\, x^2\, w_Q^p(x)\, dx + \int |P_{n,p}(x)|^p\, w_{\widehat{Q}}^p(x)\, dx$$

$$\le 2\, \|w_{\widehat{Q}}\, P_{n,p}\|_{p,[-\widehat{a}_{n+2},\widehat{a}_{n+2}]}^p + \|x^{2/p}\, w_Q\, P_{n,p}\|_{p,[-C,C]}^p$$

$$\le c_1\, \widehat{a}_{n+2}^2\, \|w_Q\, P_{n,p}\|_p^p = c_1\, \widehat{a}_{n+2}^2 \le c_2\, a_{n+5}^2. \qquad \square$$

Investigating the behavior of μ_n^* we are now able to prove asymptotic inequalities for the largest zero of the extremal polynomial $T_{n,p,Q}$.

Theorem 6.11. *Let Q satisfy the conditions of Theorem 6.3 and let us denote the zeros of $T_{n,p,Q}$ by*

$$-\infty < x_{n,n}^{(p)} < x_{n-1,n}^{(p)} < \cdots < x_{1,n}^{(p)} < \infty. \qquad (6.42)$$

For any fixed positive integer j, there exists a constant $c = c(j) > 0$, independent of p, such that

$$\left|\frac{x_{j,n}^{(p)}}{a_n} - 1\right| \le c\left(\frac{\log n}{n^{1/B}}\right)^{2/3} \qquad (6.43)$$

for all $1 < p \le \infty$ and every $n \ge \max(2, 1/(p-1))$.

For any $0 < \kappa < 1$ there exists $c_1 = c_1(\kappa)$, independent of p and n, such that

$$\frac{x_{k,n}^{(p)}}{a_n} - \frac{x_{k+1,n}^{(p)}}{a_n} \le c_1\,\frac{\log n}{n}, \qquad \kappa n \le k \le (1-\kappa)n, \qquad (6.44)$$

for all $1 < p \le \infty$ and every $n \ge \max(2, 1/(p-1))$.

Proof. First, we prove that there exist constants $c_1 > 0$ and $n_0 \in \mathbb{N}$ such that

$$\mu_n^*(x) \ge c_1\, n^{1/B-1}\sqrt{1-x^2}, \qquad x \in [-1,1], \qquad (6.45)$$

for all $n \ge n_0$. To this end, we use a representation of μ_n^* [124, Theorem 6.16, p. 137], that is different from (6.6), namely

$$\mu_n^*(x) = \frac{2}{n\pi}\sqrt{1-x^2}\, \mathrm{PV}\int_0^1 \frac{a_n\, t\, Q'(a_n\, t) - a_n\, x\, Q'(a_n\, x)}{(t^2-x^2)\sqrt{1-t^2}}\, dt.$$

294   7. Applications of Discrepancy Theorems

First, we note that for $x_0 < a_n t \leq a_n x$ we obtain from (6.15):

$$a_n x\, Q'(a_n x) - a_n x\, Q'(a_n t)$$
$$= (a_n x^\rho\, Q'(a_n x) - (a_n t)^\rho\, Q'(a_n t))\,(a_n x)^{1-\rho}$$
$$+ a_n t\, Q'(a_n t)\left(\left(\frac{x}{t}\right)^{1-\rho} - 1\right)$$
$$\geq a_n t\, Q'(a_n t)\left(\left(\frac{x}{t}\right)^{1-\rho} - 1\right) \tag{6.46}$$

Now we consider two cases:

(a) $\frac{3}{4} \leq x \leq 1$. Let $\frac{1}{4} \leq t \leq \frac{1}{2}$. Then the monotonicity of $x\, Q'(x)$ and (6.14) implies together with (6.46) and Lemma 6.6 that there exists $n_1 \in \mathbb{N}$ with

$$a_n x\, Q'(a_n x) - a_n t\, Q'(a_n t) \geq \frac{3}{4} a_n Q'\left(\frac{3}{4} a_n\right) - \frac{1}{2} a_n Q'\left(\frac{1}{2} a_n\right)$$
$$\geq \frac{a_n}{2} Q'\left(\frac{a_n}{2}\right)\left(\left(\frac{3}{2}\right)^{1-\rho} - 1\right)$$
$$\geq \left(\frac{1}{c} a_n Q'(a_n)\right)^{1/B}\left(\left(\frac{3}{2}\right)^{1-\rho} - 1\right)$$
$$\geq C_1\, n^{1/B},$$

for all $n \geq n_1$, where $C_1 := c^{-1/B}\left((3/2)^{1-\rho} - 1\right)$.

Hence,

$$\frac{\mu_n^*(x)}{\sqrt{1-x^2}} \geq C_2\, n^{1/B-1}\int_{1/4}^{1/''}\frac{dt}{\sqrt{1-t^2}} \geq C_3\, n^{1/B-1},\quad n \geq n_1.$$

where C_2 and C_3 are absolute constants, independent of n.

(b) $0 \leq x \leq \frac{3}{4}$. Let $\frac{7}{8} \leq t \leq 1$. The same reasoning as above yields that there exists $n_2 \in \mathbb{N}$ such that

$$a_n t\, Q'(a_n t) - a_n x\, Q'(a_n x) \geq \frac{7}{8} a_n Q'\left(\frac{7}{8} a_n\right) - \frac{3}{4} a_n Q'\left(\frac{3}{4} a_n\right)$$
$$\geq \frac{3}{4} a_n Q'\left(\frac{3}{4} a_n\right)\left(\left(\frac{7}{6}\right)^{1-\rho} - 1\right)$$
$$\geq \left(\frac{1}{c}\frac{3}{2} a_n Q'\left(\frac{3}{2} a_n\right)\right)^{1/B}\left(\left(\frac{7}{6}\right)^{1-\rho} - 1\right)$$
$$\geq C_4\, n^{1/B},$$

for all $n \geq n_2$, where $C_4 := c^{-1/B}\left((7/6)^{1-\rho} - 1\right)$. Hence again,

$$\frac{\mu_n^*(x)}{\sqrt{1-x^2}} \geq C_5 \, n^{1/B-1} \int_{7/8}^{1} \frac{dt}{\sqrt{1-t^2}} \geq C_6 \, n^{1/B-1},$$

for all $n \geq n_2$. Summarizing, (6.45) is proved.

Now let

$$\kappa_n := \left(\alpha \frac{\log n}{n^{1/B}}\right)^{2/3}, \quad \alpha > 0, \tag{6.47}$$

and let us assume that $\kappa_n \leq \frac{1}{2}$. Define

$$J_n = [1 - \kappa_n, 1 + \kappa_n].$$

Then

$$\mu_n^*(J_n) = \int_{1-\kappa_n}^{1} \mu_n^*(x) \, dx \geq c_1 \, n^{1/B-1} \int_{1-\kappa_n}^{1} \sqrt{1-x^2} \, dx$$

$$\geq c_2 \, n^{1/B-1} \kappa_n^{3/2}$$

$$= c_2 \, \alpha \, \frac{\log n}{n},$$

where c_2 is an absolute constant. Next we apply Theorem 3.2.2 on the interval $[-3, 3]$ to the signed measure $\sigma = \mu_n^* - \nu_n$, where ν_n is the measure that associates the mass $1/n$ with each point $\zeta_k := x_{k,n}^{(p)}/a_n$, $1 \leq k \leq n$. Because of (6.40) we have with some constant c_3 (independent of n and p)

$$|U^\sigma(z)| \leq c_3 \frac{\log n}{n}$$

for $n \geq 2$ and all $z = x + iy$ with $|y| \geq n^{-3}$. If at most j of the points $\{\zeta_k\}_{k=1}^n$ are located in $[1 - 2\kappa_n, 1 + 2\kappa_n]$, then the inequality (2.4) of Theorem 3.2.2 yields

$$\mu_n^*(J_n) \leq c_4 \frac{\log n}{n}$$

with some absolute constant $c_4 = c_4(j)$ independent of n and p. This leads to a contradiction if $\alpha > c_4/c_2$ in the definition (6.47) of κ_n. Hence, (6.43) is proved. The proof of (6.44) follows the same lines. $\square$

Finally, we want to apply Theorem 6.3 to some well investigated classes of weights.

Definition 6.12. $w_Q(x) = \exp(-Q(x))$ is called *Freud-type weight function* if $Q(x)$ is even and continuous in $\mathbb{R}$, Q is twice continuously differentiable on $(0, \infty)$, $Q'(x)$ is positive in $(0, \infty)$, and for some $C_1, C_2 > 0$,

$$C_1 \leq \frac{(x\,Q'(x))'}{Q'(x)} \leq C_2, \quad x \in (0, \infty).$$

Lemma 6.13. *If w_Q is a Freud-type weight function, then Q satisfies Condition A, (6.14) with $B = 1$ and (6.15).*

Proof. To get Condition A, we have only to prove that $x\,Q'(x)$ is increasing on $(0, \infty)$ with $\lim_{x \to \infty} x\,Q'(x) = \infty$. This follows by the mean value theorem, namely

$$y\,Q'(y) - x\,Q'(x) = (\xi\,Q'(\xi))'(y - x) \leq C_2\,Q'(\xi)(y - x),$$

where $x < \xi < y$.

Next,

$$\frac{2x\,Q'(2x)}{x\,Q'(x)} = \exp\int_x^{2x} \frac{(t\,Q'(t))'}{t\,Q'(t)}\,dt \leq \exp\left(C_2 \int_x^{2x} \frac{1}{t}\,dt\right) = 2^{C_2},$$

or

$$2x\,Q'(2x) \leq 2^{C_2} x\,Q'(x), \quad x > 0,$$

i.e., (6.14) holds with $B = 1$ and $x_0 = 0$.

Moreover, the mean value theorem yields

$$y\,Q'(y) - x\,Q'(x) = (\xi\,Q'(\xi))'(y - x) \geq C_1\,Q'(\xi)(y - x)$$

$$\geq C_1 \frac{x\,Q'(x)}{\xi}(y - x)$$

$$\geq \frac{1}{2} C_1\,Q'(x)(y - x),$$

where $0 < x < \xi < y \leq 2x$.

We may assume that $C_1 < 2$. Hence,

$$(x\,Q'(x))' = x\,Q''(x) + Q'(x) \geq \frac{C_1}{2} Q'(x),$$

or

$$x\,Q''(x) + (1 - C_1/2)\,Q'(x) \geq 0.$$

Set $\rho := (1 - C_1/2)$. Then this is equivalent to $(x^\rho\,Q'(x))' \geq 0$ for $x > 0$. □

Corollary 6.14. *Let w_Q be a Freud-type weight function. Then for $1 < p \leq \infty$ and $n \geq \max\left(2, \frac{1}{p-1}\right)$,*

$$D_{\mathbb{R}}[\nu_{T_{n,p,Q}} - \lambda_n] \leq c \frac{(\log n)^2}{n},$$

where c is a constant independent of n and p.

Let $\left\{x_{j,n}^{(p)}\right\}_{j=1}^{n}$ denote the zeros of $T_{n,p,Q}$ according to (6.42) and let j be a fixed integer. Then there exists $c_1 = c_1(j) > 0$ such that

$$\left|\frac{x_{j,n}^{(p)}}{a_n} - 1\right| \leq c_1 \left(\frac{\log n}{n}\right)^{2/3} \tag{6.48}$$

for all $1 < p \leq \infty$ and every $n \geq \max\left(2, \frac{1}{p-1}\right)$.

For any $0 < \kappa < 1$ there exists $c_2 = c_2(\kappa)$, independent of n and p, such that

$$\frac{x_{k,n}^{(p)}}{a_n} - \frac{x_{k+1,n}^{(p)}}{a_n} \leq c_2 \frac{\log n}{n}, \quad \kappa n \leq k \leq (1-\kappa)n \tag{6.49}$$

for all $1 < p \leq \infty$ and every $n \geq \max\left(2, \frac{1}{p-1}\right)$.

The proof of Corollary 6.14 follows directly from the preceeding lemma and Theorem 6.3 and Theorem 6.11.

7.7 Zeros of Orthogonal Polynomials

7.7.1 Orthogonal Polynomials on Real Intervals

Let τ be a positive, finite Borel measure on $I = [-1, 1]$ with infinite support. Then the system of orthonormal polynomials

$$p_n(\tau; x) := \gamma_n(\tau) x^n + \cdots + \gamma_0(\tau) \in \mathbb{P}_n, \quad \gamma_n(\tau) > 0,$$

is uniquely defined by the normalization

$$\int p_n p_m \, d\tau = \begin{cases} 1, & n = m, \\ 0, & n \neq m. \end{cases}$$

Since τ is fixed, we often use for abbreviation p_n instead of $p_n(\tau; \cdot)$ and γ_n for $\gamma_n(\tau)$. For each $n \geq 1$, the polynomial p_n has n simple zeros in $(-1, 1)$. We denote by $\nu_n = \nu_n(\tau)$ the zero counting measure of $p_n = p_n(\tau; \cdot)$.

If $p \in \mathbb{P}_n$ is an arbitrary real polynomial with highest coefficient $\gamma_n(\tau)$, then

$$p = p_n + \beta_{n-1} p_{n-1} + \beta_{n-2} p_{n-2} + \cdots + \beta_0 p_0, \quad \beta_i \in \mathbb{R},$$

and therefore
$$\int p^2 \, d\tau = 1 + \sum_{\nu=0}^{n-1} \beta_\nu^2 \geq 1.$$

Hence, we have obtained the so-called *minimum property* of orthogonal polynomials: For each $n \geq 1$,
$$\gamma_n^{-2}(\tau) = \min_{p \in \mathbb{P}_{n-1}} \int |x^n - p(x)|^2 \, d\tau(x), \tag{7.1}$$

and the minimum is attained if and only if
$$x^n - p(x) = \gamma_n^{-1}(\tau) \, p_n(\tau; x).$$

An important expression in the theory of orthogonal polynomials is the Christoffel–Darboux kernel
$$K_{n+1}(\tau; x, u) = \sum_{k=0}^{n} p_k(\tau; x) \, p_k(\tau; u),$$

which reproduces polynomials $P_n \in \mathbb{P}_n$, i.e.,
$$P_n(x) = \int P_n(u) K_{n+1}(\tau; x, u) \, d\tau(u). \tag{7.2}$$

In particular,
$$K_{n+1}(\tau; x, u) = \int K_{n+1}(\tau; x, u)^2 \, d\tau(u). \tag{7.3}$$

If the Radon–Nikodym derivative $\tau' = w$ of τ with respect to the Lebesgue measure exists, we write $d\tau = w(x) \, dx$. Then τ belongs to the *Szegő class* **S** if
$$\int_{-1}^{1} \frac{|\log w(x)|}{\sqrt{1 - x^2}} \, dx$$

exists in the Lebesgue sense.

To apply discrepancy theorems for the normalized zero counting measure $\nu_n(\tau)$ we need lower and upper bounds for the leading coefficient $\gamma_n(\tau)$ and the polynomial $|p_n(\tau; x)|$ itself. Remember that the inverse of the Joukowski map
$$w = \Phi(z) = z + \sqrt{z^2 - 1}$$

maps $\overline{\mathbb{C}} \setminus I$ onto the exterior Δ of the unit circle, where in $\sqrt{z^2 - 1}$ we choose the branch such that $\sqrt{z^2 - 1}/z$ is 1 at ∞. For any $\alpha \leq 1$, let $\mathcal{E}_\alpha$ be the ellipse with foci at ± 1 and major axis $\alpha + 1/\alpha$. Then $\mathcal{E}_\alpha$ is just a level line of the function $|\Phi(z)|$.

7.7 Zeros of Orthogonal Polynomials

Lower bounds for $\gamma_n(\tau)$. There is a simple way to show that the highest coefficient satisfies

$$\gamma_n(\tau) \geq 2^{n-1}\gamma_0(\tau). \tag{7.4}$$

Indeed, let

$$T_n(x) = 2^{n-1}x^n + \cdots$$

denote the nth Chebyshev polynomial. Then $\|T_n\|_I = 1$, and we have by the minimum property

$$1 \leq \left(\frac{\gamma_n(\tau)}{2^{n-1}}\right)^2 \int T_n^2 \, d\tau \leq \left(\frac{\gamma_n(\tau)}{2^{n-1}\gamma_0(\tau)}\right)^2,$$

which implies (7.4).

Upper bounds for $\gamma_n(\tau)$. Let $d\tau(x) = w(x)\,dx$ and let τ be in the Szegő class **S**. Write

$$p_n(\tau; x) = \gamma_n(\tau) \prod_{k=1}^{n}(x - x_{k,n}),$$

where $x_{k,n}$ denote the zeros of the orthogonal polynomial p_n. Then (7.1) yields

$$\log\left(\gamma_n(\tau)^{-2}\right) = \log \int \prod_{k=1}^{n} |x - x_{k,n}|^2 w(x)\,dx,$$

and by Jensen's inequality,

$$\log\left(\gamma_n(\tau)^{-2}\right) \geq \int \log\left\{\prod_{k=1}^{n} |x - x_{k,n}|^2 w(x) \pi \sqrt{1-x^2}\right\} d\mu(x),$$

where μ is the equilibrium measure of $[-1, 1]$. Since

$$U^\mu(z) = \int \log \frac{1}{|x-z|} d\mu(x) = -\log \operatorname{cap} I = \log 2$$

for any $z \in [-1, 1]$, we obtain

$$\log\left(\gamma_n(\tau)^{-2}\right) \geq (2n+1)\log\frac{1}{2} + \log \pi + \int \log w(x)\,d\mu(x),$$

or

$$\gamma_n(\tau) \leq c\, 2^n \quad \text{for } \tau \in \mathbf{S} \tag{7.5}$$

with

$$c = \left(\frac{2}{\pi}\right)^{1/2} \exp\left(-\frac{1}{2}\int \log w(x)\,d\mu(x)\right)$$

300   7. Applications of Discrepancy Theorems

Note that the last integral exists, since $\tau \in \mathbf{S}$.

Upper bounds for $|p_n(\tau;x)|$. Suppose that $S = \mathrm{supp}(\tau)$ consists of a finite number of compact, nondegenerate, disjoint intervals. Moreover, let $\tau'(x) = w(x) \geq \kappa > 0$ for $x \in S$. Then

$$\|p_n\|_S = O(n), \quad n \to \infty. \tag{7.6}$$

This can be proved by Markov's inequality: Let $S =: \bigcup_{i=1}^{r}[a_i, b_i]$, where $a_i < b_i$ ($1 \leq i \leq r$). Fix $x_0 \in S$ with

$$\|p_n\|_S = |p_n(x_0)|.$$

Then $x_0 \in [a_j, b_j]$ for some j, $1 \leq j \leq r$. According to Markov's inequality,

$$|p'_n(x)| \leq \frac{2\|p_n\|_S}{b_j - a_j} n^2, \quad x \in [a_j, b_j].$$

Let $\delta := mn^{-2}$, where $m = \min_{1 \leq i \leq r}(b_i - a_i)/4$. Then at least one of the intervals $[x_0 - \delta, x_0]$ or $[x_0, x_0 + \delta]$, say the latter, lies in $[a_j, b_j]$ for all sufficiently large n. Therefore,

$$|p_n(x)| \geq \frac{\|p_n\|_S}{2}, \quad x \in [x_0, x_0 + \delta],$$

and

$$1 = \int_{-1}^{1} p_n^2 \, d\tau \geq \int_{x_0}^{x_0+\delta} p_n^2 \, d\tau \geq (\kappa\delta/4)\|p_n\|_S^2,$$

or

$$\|p_n\|_S \leq \frac{2n}{\sqrt{m\kappa}}.$$

Lower bounds for $|p_n(\tau;x)|$. There exist lower bounds for $|p_n(\tau;x)|$ outside the interval $[-1,1]$, a result that can be derived using an idea of G. Freud [60, Theorem 7.1, p. 120]. We want to formulate this result in terms of the logarithmic potential $U^{\nu_n(\tau)-\mu}(z)$ and the inverse Joukowski map $\Phi(z)$ from $\mathbb{C} \setminus I$ onto the exterior of the unit circle.

Lemma 7.1. *Let τ be a positive, finite Borel measure in I with infinite support. Then, for all $z \in \mathbb{C} \setminus I$,*

$$U^{\nu_n(\tau)-\mu}(z) \leq \frac{1}{n}\log\left(\frac{1}{\gamma_0(\tau)}\frac{\gamma_{n-1}(\tau)}{2^{n-1}}\frac{|\Phi(z)|^2}{d(z,I)(|\Phi(z)|-1)}\right), \tag{7.7}$$

where $d(z,I)$ denotes the distance of z to I.

7.7 Zeros of Orthogonal Polynomials

Proof. Let T_{n-1} be the Chebyshev polynomial of degree $n-1$ normalized to $\|T_{n-1}\|_I = 1$ and let us denote by $x_{k,n}$, $k = 1, \ldots, n$, the zeros of p_n. By Lagrange's interpolation formula we obtain

$$T_{n-1}(z) = \sum_{k=1}^{n} T_{n-1}(x_{k,n}) L_{k,n}(z), \quad L_{k,n}(z) := \frac{p_n(z)}{(z - x_{k,n}) p_n'(x_{k,n})}.$$

Let

$$\lambda_{k,n} := \int L_{k,n}(x) \, d\tau(x)$$

be the Christoffel numbers, i.e., the weights of the Gauss quadrature formula with respect to τ. Then

$$\lambda_{k,n} = \frac{\gamma_n(\tau)}{\gamma_{n-1}(\tau)} \frac{1}{p_{n-1}(x_{k,n}) p_n'(x_{k,n})}$$

(see [173, p. 48]) and

$$T_{n-1}(z) = \frac{\gamma_{n-1}(\tau)}{\gamma_n(\tau)} p_n(z) \sum_{k=1}^{n} \lambda_{k,n} \frac{p_{n-1}(x_{k,n}) T_{n-1}(x_{k,n})}{z - x_{k,n}}.$$

Using the Cauchy–Schwarz inequality and the Gauss quadrature formula we have for $z \in \mathbb{C} \setminus I$,

$$\left| \frac{T_{n-1}(z)}{p_n(z)} \right| \leq \frac{\gamma_{n-1}(\tau)}{\gamma_n(\tau)} \frac{1}{d(z, I)} \sum_{k=1}^{n} \lambda_{k,n} |p_{n-1}(x_{k,n})| \, |T_{n-1}(x_{k,n})|$$

$$\leq \frac{\gamma_{n-1}(\tau)}{\gamma_n(\tau)} \frac{1}{d(z, I)} \left(\sum_{k=1}^{n} \lambda_{k,n} p_{n-1}^2(x_{k,n}) \right)^{1/2} \left(\sum_{k=1}^{n} \lambda_{k,n} T_{n-1}^2(x_{k,n}) \right)^{1/2}$$

$$= \frac{\gamma_{n-1}(\tau)}{\gamma_n(\tau)} \frac{1}{d(z, I)} \left(\int T_{n-1}^2 \, d\tau \right)^{1/2}$$

$$\leq \frac{\gamma_{n-1}(\tau)}{\gamma_n(\tau)} \frac{1}{d(z, I)} \left(\int d\tau \right)^{1/2} = \frac{\gamma_{n-1}(\tau)}{\gamma_0(\tau) \gamma_n(\tau)} \frac{1}{d(z, I)}.$$

Therefore, with $\nu_n = \nu_n(\tau)$,

$$U^{\nu_n - \mu}(z) = \frac{1}{n} \log \left(\frac{\gamma_n(\tau)}{|p_n(z)|} \frac{|\Phi(z)|^n}{2^n} \right)$$

$$\leq \frac{1}{n} \log \left(\frac{1}{\gamma_0(\tau)} \frac{\gamma_{n-1}(\tau)}{2^n} \frac{|\Phi(z)|^n}{d(z, I) |T_{n-1}(z)|} \right).$$

Now, with $\Psi = \Phi^{-1}$ for $|w| > 1$,

$$T_{n-1}(\Psi(w)) = \frac{1}{2} \left(w^{n-1} + w^{-n+1} \right)$$

(cf. [60, p. 120]), and thus for $n \geq 2$,

$$\frac{|T_{n-1}(\Psi(w))|}{|w|^n} \geq \frac{1}{2|w|}\left(1 - |w|^{-2(n-1)}\right) \geq \frac{1}{2|w|}\left(1 - \frac{1}{|w|}\right).$$

Hence, (7.7) holds. □

Theorem 7.2. *Let τ be a positive, finite Borel measure on I with infinite support. Assume that the leading coefficients $\gamma_n(\tau)$ of the orthogonal polynomials $p_n(\tau;x)$ satisfy*

$$\gamma_n(\tau) \leq n^\alpha 2^n, \quad n = 2, 3, \ldots,$$

where α is a positive constant. Then

$$D[\nu_n(\tau) - \mu] \leq c\sqrt{\frac{\log n}{n}}, \quad n = 2, 3, \ldots,$$

where c depends only on α.

Proof. Set

$$q_n(x) = \frac{1}{\gamma_n(\tau)} p_n(\tau;x), \quad n = 2, 3, \ldots.$$

Then $q_n \in \mathbb{M}_n$, and because of Lemma 7.1 there exists a constant $c_1 > 0$, independent of n, such that

$$U^{\nu_n(\tau)-\mu}(z) \leq c_1 \frac{\log n}{n}$$

for all $z \in \mathcal{E}_{1+\sigma}$, where

$$\mathcal{E}_{1+\sigma} = \left\{z \in \mathbb{C} : |\Phi(z)| = 1 + \sigma, \ \sigma = (n^{-1}\log n)^{1/2}\right\}.$$

Following Section 2.2.2 we see that the modules of continuity satisfy formula (2.2.29) with $\alpha = 1$. Hence Theorem 2.3.1 yields

$$D[\nu_n(\tau) - \mu] \leq c_2 \left(\frac{\log n}{n\sigma} + \sigma\right) \leq 2c_2\sqrt{\frac{\log n}{n}}.$$

□

Because of the upper bound (7.5), Theorem 7.2 applies to weights of the Szegő class. Another interesting example is the zero distribution of Pollaczek polynomials, which we shall discuss later, in connection with the next theorem.

Theorem 7.3. *Let τ be a positive, finite Borel measure with infinite support. Moreover, let $S = \operatorname{supp}(\tau)$ consist of a finite number of compact intervals. Then there exists a constant $c > 0$, depending only on $\operatorname{supp}(\tau)$, such that*

$$D[\nu_n(\tau) - \mu_S] \leq c \frac{\log n}{n} \left[\log n + \log(1 + \|p_n\|_S) \right].$$

Proof. The zeros $x_{i,n}$, $1 \leq i \leq n$, of the orthogonal polynomial p_n are all simple and contained in $[-1, 1]$. The Christoffel–Darboux formula (cf. [173, p. 43]) yields

$$\sum_{k=0}^{n-1} p_k^2(x_{i,n}) = \frac{\gamma_{n-1}}{\gamma_n} p_n'(x_{i,n}) p_{n-1}(x_{i,n}),$$

and therefore, using the fact that $\gamma_{n-1} \leq \gamma_n$ [60, p. 45], we have

$$2\gamma_0 |p_{n-1}(x_{i,n})| = 2|p_0(x_{i,n}) p_{n-1}(x_{i,n})| \leq p_0^2(x_{i,n}) + p_{n-1}^2(x_{i,n})$$
$$\leq \frac{\gamma_{n-1}}{\gamma_n} p_n'(x_{i,n}) p_{n-1}(x_{i,n}) \leq |p_n'(x_{i,n}) p_{n-1}(x_{i,n})|.$$

Hence,

$$|\gamma_n^{-1} p_n'(x_{i,n})| \geq 2 \frac{\gamma_0}{\gamma_n}.$$

On the other hand, let $T_{n,S} \in \mathbb{M}_n$ be the Chebyshev polynomial of degree n on the compact set S. Then the minimum property of $T_{n,S}$ yields

$$\gamma_n^{-1} \|p_n\|_S \geq \|T_{n,S}\| \geq (\operatorname{cap} S)^n,$$

and therefore for $1 \leq i \leq n$,

$$|\gamma_n^{-1} p_n'(x_{i,n})| \geq \frac{2\gamma_0}{\|p_n\|_S} (\operatorname{cap} S)^n. \tag{7.8}$$

Moreover, the minimal property of the orthonormal polynomials p_n leads to

$$1 = \int p_n^2 \, d\tau \leq \gamma_n^2 \int T_{n,S}^2 \, d\tau \leq \left(\frac{\gamma_n}{\gamma_0} \right)^2 \|T_{n,S}\|_S.$$

According to a result of Widom [182, Theorem 11.5], there exists $c > 0$ such that

$$\|T_{n,S}\|_S \leq c (\operatorname{cap} S)^n$$

and therefore

$$\gamma_n^{-1} \|p_n\|_S \leq \frac{c}{\gamma_0} \|p_n\|_S (\operatorname{cap} S)^n. \tag{7.9}$$

Hence, the discrepancy estimate follows from Theorem 3.1.8, taking into account (7.8) and (7.9). □

In the case that $\tau' \geq \kappa > 0$ on S, then under the conditions of Theorem 7.3 we obtain with (7.6) that

$$D[\nu_n(\tau) - \mu_S] \leq c \frac{(\log n)^2}{n}.$$

For generalized Jacobi weights τ it is known that $\|p_n(\tau;\cdot)\|_I = O(n^\kappa)$, where κ is a constant depending on τ (cf. Nevai [130, p. 170]). Hence, in this case,

$$D[\nu_n(\tau) - \mu] \leq c \frac{(\log n)^2}{n},$$

where μ is the equilibrium distribution of I.

If we apply Theorem 7.3 to the zeros of orthogonal *Pollaczek polynomials* (cf. [173, Appendix, p. 393]), then we get the estimate

$$D[\nu_n(\tau) - \mu] \leq c \frac{\log n}{\sqrt{n}}, \quad n = 2, 3, \ldots, \quad (7.10)$$

which is weaker than the estimate $O\left(\sqrt{\log n/n}\right)$ obtained by Theorem 7.2.

We recall that the Pollaczek polynomials $P_n(x;a,b)$ are orthonormal polynomials with respect to $d\tau(x) = w(x;a,b)\,dx$, where a and b are real parameters, $a > |b|$, and the weight function $w(x;a,b)$ is defined for $x = \cos\theta$, $0 < \theta < \pi$, by

$$w(x;a,b) := e^{(2\theta-\pi)h(\theta)}[\cosh(\pi h(\theta))]^{-1} \quad (7.11)$$

with

$$h(\theta) = \frac{a\cos\theta + b}{2\sin\theta}.$$

The highest coefficient in

$$P_n(x;a,b) = \gamma_n x^n + \cdots$$

satisfies $\gamma_n = O(n^{a/2} 2^n)$ as $n \to \infty$ (cf. [121, Appendix] or Lemma 7.6). Moreover, the Pollaczek weight functions satisfy for all $-1 < x < 1$ the inequality

$$w(x;a,b) \geq c_1 g(x),$$

where $g(x) := \exp\{-c_2/\sqrt{1-x^2}\}$ with positive constants $c_1 = c_1(a,b)$ and $c_2 = c_2(a,b)$.

If $\cos\theta_{j,n}$ are the zeros of the Pollaczek polynomials, where

$$0 < \theta_{1,n} < \theta_{2,n} < \cdots < \theta_{n,n} < \pi,$$

then for any fixed value of j,

$$\lim_{n\to\infty} n^{1/2} \theta_{j,n} = (a+b)^{1/2}. \quad (7.12)$$

Hence, the estimate (7.10) is optimal up to the $\log n$ term. On the other hand, we will show with Theorem 3.2.2 that the estimate for the distribution of the zeros of $P_n(x; a, b)$ can be sharpened considerably inside the interval I.

Motivated by the example of Pollaczek weights we consider the following generalization.

Definition 7.4. A weight function W is a *generalized Pollaczek* (or *GP*) *weight function*, if there exists an algebraic polynomial P the zeros of which are all on $[-1, 1]$ and simple such that

$$W(x) = \exp\left(-\frac{\sqrt{1-x^2}}{|P(x)|}\right), \quad -1 \leq x \leq 1.$$

Let $z_1, z_2, \ldots, z_m$ be the zeros of $P(x)$, i.e., $-1 \leq z_1 \leq z_2 \leq \ldots \leq z_m \leq 1$, and let $z(W) := \{z_i\}_{i=1}^m$ be the zero set of $P(x)$. Define the intervals

$$O_n(W; z_i) := \left\{y \in I : \mu([z_i, y]) < \frac{1}{\sqrt{n}}\right\}$$

for $i = 1, \ldots, m$. Moreover, we need

$$I_n(W) := I \setminus \bigcup_{i=1}^m O_n(W; z_i). \tag{7.13}$$

In the sequel, the symbols $c, c_1, \ldots$ will denote positive constants depending only on the weight functions and intervals involved. Moreover, for abbreviation we set $I_n := I_n(W)$ in the proofs.

The following theorem gives local discrepancy bounds for the measures $\nu_n(\tau) - \mu$ for a large class of mass distributions τ related to GP weight functions.

Theorem 7.5. *Let τ be a positive Borel measure on I, W a GP weight function, and let the Radon Nikodym derivative τ' satisfy $\tau'(x) \geq W(x)$ a.e. in $[-1, 1]$. Then*

$$D[\nu_n(\tau) - \mu] \leq c\sqrt{\frac{\log n}{n}}, \tag{7.14}$$

where c is a positive constant, independent of n. Moreover,

$$D_\Gamma[\nu_n(\tau) - \mu] \leq c_1 \frac{(\log n)^2}{n} \tag{7.15}$$

for every closed interval $\Gamma \subset I \setminus Z(W)$, where $c_1 = c_1(\Gamma)$ is a constant, depending only on the subinterval Γ.

306   7. Applications of Discrepancy Theorems

We remark that in the case where $\tau'(x)$ is the Pollaczek weight $w(x;a,b)$, then Theorem 7.2 together with (7.12) shows that

$$\frac{c}{\sqrt{n}} \leq D[\nu_n(\tau) - \mu] \leq c_1 \sqrt{\frac{\log n}{n}}.$$

Hence, the local discrepancy estimate (7.15) is seen to be a substantial improvement.

Before proving Theorem 7.5 we want to consider some basic facts about GP weight functions. In the following, we will use $\gamma_n(W)$, $p_n(W;x)$, and $K_{n+1}(W;x,t)$ instead of $\gamma_n(\tau)$, $p_n(\tau;x)$, $K_{n+1}(\tau;x,t)$ if $d\tau(x) = W(x)\,dx$.

Lemma 7.6. *Let τ be a positive Borel measure and W the corresponding GP weight function as in Theorem 7.5. Then there exist positive constants c and c_1 such that*

$$c_1 2^n \leq \gamma_n(\tau) \leq \gamma_n(W) \leq n^c\, 2^n, \quad n = 2, 3 \ldots. \tag{7.16}$$

Proof. Since

$$\gamma_n^{-2}(\tau) = \min_{p \in \mathbb{P}_{n-1}} \int (x^n - p(x))^2\, d\tau(x),$$

we have

$$1 = \int p_n(\tau;x)^2\, d\tau(x) \leq \int \left(\frac{\gamma_n}{2^{n-1}} T_n(x)\right)^2 d\tau(x)$$

$$\leq \left(\frac{\gamma_n}{2^{n-1}}\right)^2 \int d\tau,$$

where $T_n(x)$ is the nth Chebyshev polynomial. This proves the first inequality in (7.16). Furthermore,

$$\gamma_n^{-2}(\tau) \geq \min_{p \in \mathbb{P}_{n-1}} \int (x^n - p(x))^2\, W(x)\, dx = \gamma_n^{-2}(W)$$

Therefore, it is sufficient to prove the third inequality in (7.16).

The polynomial $p_n(W;x)$ has n simple zeros

$$-1 < x_{n,n} < x_{n-1,n} < \cdots < x_{1,n} < 1,$$

and we have

$$p_n(W;x) = \gamma_n(W) \prod_{k=1}^{n} (x - x_{k,n}).$$

Therefore,

$$1 = \gamma_n^2(W) \int \prod_{k=1}^{n} (x - x_{k,n})^2 W(x)\, dx,$$

or
$$\gamma_n^{-2}(W) \geq \pi \int_{I_{n^2}} \prod_{k=1}^{n}(x - x_{k,n})^2 \, W(x) \sqrt{1-x^2} \, d\mu(x).$$

For abbreviation, define $\mu_n := \mu/\mu(I_{n^2})$. Then $\mu_n(I_{n^2}) = 1$, and by Jensen's inequality,

$$\log \gamma_n^{-2}(W) \geq \log \pi + \log \mu(I_{n^2})$$
$$+ \int_{I_{n^2}} \log \left\{ \prod_{k=1}^{n} |x - x_{k,n}|^2 \, W(x) \sqrt{1-x^2} \right\} d\mu_n(x)$$
$$= \log \pi + \log \mu(I_{n^2}) - \frac{1}{\mu(I_{n^2})} \left\{ \int_{I_{n^2}} \frac{\sqrt{1-x^2} \, d\mu(x)}{|P(x)|} \right.$$
$$- \frac{1}{2} \int_{I_{n^2}} \log|1-x| \, d\mu(x) - \frac{1}{2} \int_{I_{n^2}} \log|1+x| \, d\mu(x)$$
$$\left. - 2 \sum_{k=1}^{n} \int_{I_{n^2}} \log|x_{k,n} - x| \, d\mu(x) \right\}.$$

We observe that
$$\mu(I \setminus I_{n^2}) = |1 - \mu(I_{n^2})| \leq \frac{c}{n}.$$

Therefore, for $\xi \in I$,
$$\int_{I_{n^2}} \log|\xi - x| \, d\mu(x) = \int_{I} \log|\xi - x| \, d\mu(x) - \int_{I \setminus I_{n^2}} \log|\xi - x| \, d\mu(x)$$
$$\geq \log \frac{1}{2} - \mu(I \setminus I_{n^2}) \log 2 \geq \log \frac{1}{2} - \frac{c}{n}.$$

Finally, since all the zeros of P are simple, it is easy to deduce that
$$\int_{I_{n^2}} \frac{dx}{|P(x)|} \leq c \log n.$$

Summarizing, we obtain that
$$\log \gamma_n^{-2}(W) \geq (2n+1) \log \frac{1}{2} - c_3 - c_4 \log n,$$

with absolute constants c_3 and c_4. Hence, (7.16) is proved. □

Lemma 7.7. *Let W be a GP weight function. Then there exists $n_0 > 0$ such that for $n \geq n_0$,*

$$|p_n(W;x)|^2 \leq K_{n+1}(W;x,x) \leq \begin{cases} \exp(c\sqrt{n}) & \text{for } x \in I, \\ cn^3 W(x)^{-1} & \text{for } x \in I_n(W), \end{cases} \quad (7.17)$$

where c is an absolute constant.

Proof. We define for any $x \in I_n$ the intervals

$$l(x,n) := \left[x - \frac{1}{n^2}, x + \frac{1}{n^2}\right].$$

Then there exist $n_0 > 0$ and $c_0 > 0$ such that $\text{dist}(l(x,n), Z(W)) \geq c_0/n$ for all $n \geq n_0$. Consider the function

$$h(x) := \frac{\sqrt{1-x^2}}{|P(x)|}.$$

Then the mean value theorem yields for $x, y \in l(x,n)$,

$$|h(x) - h(y)| \leq c_1,$$

where $c_1 > 0$ is independent of n. Hence,

$$\frac{1}{c_2} \leq \frac{W(y)}{W(x)} \leq c_2 \quad (7.18)$$

for all $x, y \in l(x,n) \cap I$.

First, let us assume $|x| \leq 1 - 1/n^2$. We write $\eta := 1/n^2$. Let v be the function defined by $v(u) := 1/(\pi\sqrt{\eta^2 - (x-u)^2})$ if $u \in l(x,n)$, and equal to 0 if $u \notin l(x,n)$. We have

$$p_n(v; x + \eta\cos\theta) = \begin{cases} \sqrt{2}\cos(n\theta), & \text{if } n \neq 0, \\ 1, & \text{if } n = 0, \end{cases}$$

and hence (cf. [124, example 1.1.1])

$$K_{n+3}(v;y,y) \leq 2n + 5, \quad y \in l(x,n). \quad (7.19)$$

Now using (7.2), Schwarz's inequality, and (7.3), we get for $t \in l(x,n)$,

$$(\eta^2 - (x-t)^2)^2 K_{n+1}(W;x,t)^2$$

$$= \left(\int (\eta^2 - (x-u)^2) K_{n+1}(W;x,u) K_{n+3}(v;t,u) v(u)\, du\right)^2$$

$$\leq \left(\int K_{n+1}(W;x,u)^2 W(u)\, du\right)$$

$$\times \left(\int (\eta^2 - (x-u)^2)^{3/2} K_{n+3}(v;t,u)^2 \frac{v(u)}{W(u)}\, du\right)$$

$$\leq K_{n+1}(W;x,x)\, \eta^3 \int K_{n+3}(v;t,u)^2 \frac{v(u)}{W(u)}\, du. \quad (7.20)$$

Using (7.18) and (7.19), we obtain

$$\int K_{n+3}(v;t,u)^2 \frac{v(u)}{W(u)} du \leq \frac{c}{W(x)} K_{n+3}(v;t,t) \leq \frac{cn}{W(x)}.$$

Therefore, (7.20) yields for $t \in l(x,n)$:

$$(\eta^2 - (x-t)^2)^2 K_{n+1}(W;x,t)^2 \leq cn\eta^3 K_{n+1}(W;x,x)W(x)^{-1}.$$

We use this estimate with $t = x$, and simplify to arrive at (7.17) when $x \in I_n(W)$ and $|x| \leq 1 - \eta$. If $x \in I_n(W)$, then (7.17) follows for the remaining points $1 \geq |x| > 1 - \eta$ from Bernstein's inequality.

Again using Bernstein's inequality we get from the lower inequality in (7.17) the upper one. □

Proof of Theorem 7.5. Since (7.14) is a direct consequence of Theorem 7.2 and (7.16), it remains to prove (7.15). There exists $n_0 > 0$ such that the intervals $O_n(W; z_i)$ are pairwise disjoint. Because of Lemma 7.7 the potential $U^{\mu-\nu_n(\tau)}(x)$ satisfies

$$U^{\mu-\nu_n(\tau)}(x) \leq \begin{cases} \dfrac{c}{\sqrt{n}}, & x \in I, \\ \dfrac{c}{n}\left(\log n - \sum_{j=1}^{m} \dfrac{\sqrt{1-x^2}}{|x - z_i|}\right), & x \in I_n. \end{cases} \quad (7.21)$$

Let Ψ be the inverse mapping of Φ, continuously extended to $|t| \geq 1$. Then

$$u_n(\zeta) = U^{\mu-\nu_n(\tau)}(\Psi(1/\zeta))$$

is harmonic in $|\zeta| < 1$. Moreover, $u_n(e^{i\theta})$ is in $L^p[-\pi,\pi]$, $p > 1$, and by (7.21),

$$u_n(e^{i\theta}) \leq \frac{c_1}{n}\left(\log n + \sum_{j=1}^{m}\left(\frac{1}{|\theta - \theta_j| + 1/\sqrt{n}} + \frac{1}{|\theta + \theta_j| + 1/\sqrt{n}}\right)\right) \quad (7.22)$$

for all $\theta \in [-\pi, \pi]$, where we have used

$$z_j = \Psi(e^{i\theta_j}) = \Psi(e^{-i\theta_j}), \quad j = 1, \ldots, m. \quad (7.23)$$

Let $\Gamma = [a,b]$ and $\kappa > 0$. Then we define

$$\widetilde{B}_\kappa = \{\zeta = 1/t : t = \Phi(z), z = x + iy \text{ with } a - \kappa \leq x \leq b + \kappa, y \neq 0\}$$

and

$$E_\kappa = \{\zeta : |\zeta| < 1, |\zeta - e^{\pm i\theta_j}| \geq 2\kappa, j = 1, \ldots, m\}.$$

The constant κ can be chosen in such a way that $0 < \kappa < \pi/2$, $\widetilde{B}_\kappa \subset E_\kappa$. By Poisson's formula,

$$u_n(\zeta) = \frac{1-|\zeta|^2}{2\pi} \int_{-\pi}^{\pi} \frac{u_n(e^{i\theta})}{|\zeta - e^{i\theta}|^2} d\theta,$$

and we note that

$$1 = \frac{1-|\zeta|^2}{2\pi} \int_{-\pi}^{\pi} \frac{d\theta}{|\zeta - e^{i\theta}|^2}. \tag{7.24}$$

Let j be fixed and let

$$g(\theta) = \frac{1}{|\theta - \theta_j| + 1/\sqrt{n}}.$$

Then we consider for $A_1 := [\theta_j - \kappa, \theta_j + \kappa]$ and $A_2 = [\theta_j + \kappa, \theta_j - \kappa + 2\pi]$ the integrals

$$\mathcal{I}_k = \frac{1-|\zeta|^2}{2\pi} \int_{A_k} \frac{g(\theta)\, d\theta}{|\zeta - e^{i\theta}|^2}, \quad k = 1, 2.$$

By (7.24) we have $\mathcal{I}_2 \leq c_2$. For $|\zeta - e^{i\theta_j}| \geq 2\kappa$ we obtain

$$|\zeta - e^{i\theta}| \geq |\zeta - e^{i\theta_j}| - |e^{i\theta_j} - e^{i\theta}| \geq \kappa, \quad \theta \in A_1,$$

and therefore

$$\mathcal{I}_1 \leq c_3 \int_0^\kappa \frac{d\theta}{\theta + 1/\sqrt{n}} \leq c_4 \log n.$$

Using such estimates for $j = 1, \ldots, m$, and also for $-\theta_j$ instead of θ_j, we finally get by (7.22)

$$u_n(\zeta) \leq c_3 \frac{\log n}{n} \tag{7.25}$$

for all $\zeta \in E_\kappa$. Next, we choose $\delta = 1/n^2$ and combine Lemma 7.1 with (7.25). Thus

$$|U^{\mu-\nu_n(\tau)}(z)| \leq c \frac{\log n}{n}, \quad z \in B(\delta, [a,b]),$$

where $c = c(\kappa, \tau)$ is constant and

$$B(\delta, [a,b]) := \{z = x + iy \,:\, a - \delta \leq x \leq b + \delta,\, |y| \geq \delta^3\}.$$

Hence, Theorem 3.2.2 yields the discrepancy estimate (7.15), taking into account that $\gamma = \frac{1}{2}$. □

Finally, we want to consider the distance between two consecutive zeros of the orthogonal polynomial p_n. Though the results can be formulated for generalized Pollaczek polynomials, we restrict ourselves for simplicity to the classical Pollaczek polynomials associated with the weight function $w(x; a, b)$ in (7.11).

Let

$$-1 < x_{n,n} < x_{n-1,n} < \cdots < x_{1,n} < 1 \qquad (7.26)$$

be the zeros of the Pollaczek polynomials $P_n(x; a, b)$ and let $x_{j,n} = \cos\theta_{j,n}$, $0 < \theta_{j,n} < \pi$.

Theorem 7.8. *There exists a constant $c_1 = c_1(a, b) > 0$ such that*

$$0 < \theta_{1,n} \leq \frac{c_1}{\sqrt{n}}, \qquad \pi > \theta_{n,n} \geq \pi - \frac{c_1}{\sqrt{n}} \qquad (7.27)$$

for all $n = 1, 2, \ldots$.

For any $0 < \kappa < 1$ there exists a constant $c_2 = c_2(\kappa; a, b)$ such that

$$\theta_{j+1,n} - \theta_{j,n} < c_2 \frac{\log n}{n}, \qquad n = 2, 3, \ldots, \qquad (7.28)$$

for all j with $\kappa n \leq j \leq (1-\kappa)n$.

Proof. The highest coefficients γ_n of $P_n(x; a, b)$ satisfy $\gamma_n \leq n^{c_1} 2^n$, $n = 2, 3, \ldots$, with some constant $c_1 > 0$ (Lemma 7.6). Let ν_n be the zero counting measure of $P_n(x; a, b)$. Then, by (7.21),

$$U^{\mu-\nu_n}(z) \leq \frac{c_2}{\sqrt{n}}, \qquad z \in \mathbb{C}. \qquad (7.29)$$

Let $J_n = [-1, -1 + \alpha/n]$, where $\alpha \geq 1$ will be chosen appropriately later. Lemma 7.1 yields

$$U^{\nu_n-\mu}(z) \leq c_3 \frac{\log n}{n}, \qquad z \in \mathcal{E}_{1+1/n}, \qquad (7.30)$$

where the constant c_3 is independent of α. Because of (7.29) and (7.30),

$$\varepsilon := \varepsilon_{\mu-\nu_n}(n^{-1}, J_n) = \max_{z \in B(n^{-1}, J_n)} |U^{\mu-\nu_n}(z)| \leq \frac{c_4}{\sqrt{n}}$$

with some $c_4 = c_4(c_2, c_3)$ and

$$B(n^{-1}, J_n) := \{z = x + iy : -1 - n^{-1} \leq x \leq 1 + (\alpha+1)n^{-1}, \ |y| \geq n^{-2}\}.$$

Using the discrepancy estimate of Theorem 3.2.2 with $\delta = 1/n$ we obtain

$$|(\mu - \nu_n)(J_n)| \leq \frac{4c_4}{\sqrt{n}}(2 + \log(1+\alpha)) + 16\frac{\log n}{n} + \frac{2}{\pi\sqrt{n}},$$

keeping in mind that

$$\frac{\sqrt{2}}{\pi}|J|^{1/2} \leq \mu(J_n) \leq \frac{2}{\pi}|J|^{1/2}$$

for any subinterval J of I. Since

$$\log(1+\alpha) \leq \alpha^{1/3} \quad \text{for } \alpha \geq 27,$$

we conclude that there exists a constant $\alpha_0 > 1$ with

$$|(\mu - \nu_n)(J_n)| \leq \frac{1}{3}\sqrt{\frac{\alpha_0}{n}}$$

for $J_n = [-1, -1 + \alpha_0/n]$. Since

$$\mu(J_n) \geq \frac{\sqrt{2}}{\pi}|J_n|^{1/2} > \frac{1}{3}\sqrt{\frac{\alpha_0}{n}},$$

it follows that there exists a zero of $P_n(x; a, b)$ in the interval J_n. The same is true for the interval $[1 - \alpha_0/n, 1]$, and (7.27) is proved.

To prove (7.28) we note first that because of (7.14), there exists an interval $\Gamma \subset (-1, 1)$ such that $x_{j,n} \in \Gamma$ for all j with $\kappa n \leq j \leq (1 - \kappa)n$. From the proof of Theorem 7.5 we know that

$$\varepsilon_{\mu - \nu_n}(n^{-1}, \Gamma) \leq c_5 \frac{\log n}{n},$$

where $c_5 = c_5(\Gamma) > 0$ is independent of n. The estimate (3.2.4) of Theorem 3.2.2 leads to the inequality (7.28). □

Estimates of type (7.28) were proved for certain weights of the Szegő class by Krawtchouk [103], Erdős–Turán [53], G. Szegő [173, Section 6.11], and P. Nevai [130, Theorem 5, Section 9].

7.7.2 Orthogonal Polynomials on the Unit Circle

Next, we study orthogonal polynomials on the unit circle. Let τ be a positive, finite Borel measure on $\mathbb{T}$ whose support is an infinite set. Then there exists an infinite sequence of polynomials

$$\omega_n(z) := \omega_n(\tau, z) := \kappa_n z^n + \cdots \in \mathbb{P}_n, \quad \kappa_n > 0,$$

such that

$$\int \omega_n \omega_m \, d\tau = \delta_{n,m}, \quad n, m = 0, 1, 2, \ldots.$$

For $p(z) = c_0 + c_1 z + \cdots + c_n z^n \in \mathbb{P}_n$ we define the "reciprocal" polynomial

$$p^*(z) := z^n \overline{p(1/\bar{z})} = \overline{c_n} + \overline{c_{n-1}} z + \cdots + \overline{c_0} z^n.$$

If the zeros of p are $z_1, \ldots, z_n$, those of p_n^* are $z_1^*, \ldots, z_n^*$, where $z_i^* = 1/\overline{z_i}$ is the point obtained from z_i by reflection with respect to the unit circle $\mathbb{T}$. Hereby, the zeros are counted with their multiplicity and $0^* = \infty$, $\infty^* = 0$; ∞ as a zero of order k means that the coefficients of the k highest powers vanish.

Next, we consider the modified *Christoffel–Darboux formula*

$$\sum_{\nu=o}^{n-1} \overline{\omega_\nu(\xi)}\, \omega_\nu(z) = \frac{\overline{\omega_n^*(\xi)}\,\omega_n^*(z) - \overline{\omega_n(\xi)}\,\omega_n(z)}{1 - \overline{\xi} z} \qquad (7.31)$$

and the recurrence relation

$$\kappa_n\, \omega_{n+1}^*(z) = \kappa_{n+1}\, \omega_n^*(z) + \overline{\omega_{n+1}(0)}\, z\, \omega_n(z) \qquad (7.32)$$

(cf. [69, pp. 6–8], or [173, Theorem 11.4.2, p. 293]).

Inserting $\xi = z$ in (7.31) we obtain

$$|\omega_n^*(z)|^2 = (1 - |z|^2) \sum_{\nu=0}^{n-1} |\omega_\nu(z)|^2 + |\omega_n(z)|^2$$

$$\geq (1 - |z|^2)\, |\omega_o(z)|^2 = (1 - |z|^2)\, \kappa_0^2. \qquad (7.33)$$

Hence, $\omega_n^*(z) \neq 0$ for all $|z| < 1$. Moreover, let $z = e^{i\theta}$. Then

$$|\omega_n^*(re^{i\theta})|^2 \geq (1 - r)\, \kappa_0^2,$$

and therefore, ω_n^* has no zero on the unit circle $\mathbb{T}$. That is, all zeros of ω_n^* lie in Δ; i.e., all zeros of ω_n lie in $\mathbb{D}$. Let

$$z_{1,n},\ z_{2,n},\ \ldots,\ z_{n,n}$$

denote the zeros of ω_n. We define the sequence $\widehat{\nu}_n$ of measures on the unit circle to be the *balayage measure* associated with these zeros as follows. Let

$$z_{k,n} =: r_{k,n} \exp(i\, t_{k,n}), \qquad k = 1, \ldots, n;\ n = 1, 2, \ldots\ .$$

With the Poisson kernel

$$P(r, \theta) = \frac{1 - r^2}{1 - 2r\cos\theta + r^2}$$

we define, for any Borel measurable subset B of the unit circle,

$$\widehat{\nu}_n(B) := \frac{1}{2\pi n} \sum_{k=1}^{n} \int_B P(r_{k,n}, t - t_{k,n})\, dt. \qquad (7.34)$$

We observe that if f is any function continuous on $|z| \leq 1$ and harmonic on $|z| < 1$, then

$$\int f\, d\widehat{\nu}_n = \frac{1}{n} \sum_{k=1}^{n} f(z_{k,n}).$$

In particular,
$$U^{\nu_n}(z) = U^{\widehat{\nu}_n}(z) = \frac{1}{n} \log \frac{\kappa_n}{\omega_n(z)}, \quad |z| > 1. \tag{7.35}$$

Equation (7.35) persists on the unit circle as well.

Lemma 7.9. *Let τ be a positive, finite Borel measure on $\mathbb{T}$ with infinite support. Then*
$$U^{\nu_n - \mu_{\mathbb{T}}}(z) \leq \frac{1}{n} \log \left(\frac{\kappa_n}{\kappa_0} \frac{|z|}{|z|-1} \right), \quad z \in \Delta.$$

Proof. The inequality (7.33) yields
$$|\omega_n^*(z)|^2 \geq (1 - |z|) \kappa_0^2 \quad \text{for } |z| < 1,$$
or
$$|\omega_n(z)|^2 \geq |z|^{2n} \left(1 - \frac{1}{|z|}\right) \kappa_0^2 \quad \text{for } |z| \geq 1,$$
so that
$$\frac{1}{n} \log |\omega_n(z)| \geq \log |z| + \frac{1}{n} \log \left(\frac{|z|-1}{|z|} \kappa_0 \right),$$
or
$$U^{\nu_n - \mu_{\mathbb{T}}}(z) \leq \frac{1}{n} \log \left(\frac{\kappa_n}{\kappa_0} \frac{|z|}{|z|-1} \right), \quad z > 1.$$
□

Theorem 7.10. *Let τ be a positive, finite Borel measure on $\mathbb{T}$ with infinite support and let $\widehat{\nu}_n$ be the balayage measure of ν_n on $\mathbb{T}$. Then there exists a constant $c > 0$ such that*
$$D[\widehat{\nu}_n - \mu_{\mathbb{T}}] \leq c \left(\frac{\log(n \kappa_n / \kappa_0)}{n} \right)^{1/2}$$
for $n = 2, 3, \ldots$.

Proof. Since $\kappa_{n+1}^2 - \kappa_n^2 = |\omega_{n+1}(0)|^2$ (cf. [69, p. 7]) we note that $\kappa_n / \kappa_0 \geq 1$. Because of Lemma 7.9, there exists a constant $c_1 > 0$ such that for $n = 2, \ldots,$
$$U^{\widehat{\nu}_n - \mu_{\mathbb{T}}}(z) = U^{\nu_n - \mu_{\mathbb{T}}}(z) \leq c_1 \frac{1}{n} \log(n \kappa_n / \kappa_0), \quad |z| = 1 + \frac{\delta}{2},$$
where $\delta = \frac{1}{n} \log(n \kappa_n / \kappa_0)$. Then Theorem 4.1.1 yields
$$D[\widehat{\nu}_n - \mu_{\mathbb{T}}] \leq c \left(\frac{\log(n \kappa_n / \kappa_0)}{n} \right)^{1/2}, \quad n = 2, 3, \ldots,$$

with some absolute constant c. □

If τ belongs to the Szegő class on $\mathbb{T}$, i.e., the integral

$$\int_{\mathbb{T}} |\log \tau'(\theta)|\, d\theta$$

exists in the Lebesgue sense, then the leading coefficients κ_n are bounded (cf. [69, formula (1.27), p. 14]). Hence, the above theorem yields

$$D[\widehat{\nu}_n - \mu_{\mathbb{T}}] \leq c \sqrt{\frac{\log n}{n}}, \quad n = 2, \ldots, \tag{7.36}$$

which reflects the same behavior as for the Szegő class on the real interval $[-1, 1]$ (cf. remarks on Theorem 7.2).

Theorem 7.11. *With the measures $\widehat{\nu}_n$ and $\mu_{\mathbb{T}}$ defined as above, we have*

$$D[\widehat{\nu}_n - \mu_{\mathbb{T}}] \leq c \frac{\log n}{n} \left\{ \log\left(n \frac{\kappa_n}{\kappa_0}\right) + \sum_{k=1}^{n} \kappa_k^{-1}|\omega_k(0)| \right\}, \tag{7.37}$$

where c is an absolute constant.

Proof. Introducing the monic polynomials

$$\Phi_n(z) := \kappa_n^{-1} \omega_n(z) \in \mathbb{M}_n,$$

we obtain from (7.32) the recurrence relation

$$\Phi_{n+1}^*(z) = \Phi_n^* - a_n z\, \Phi_n(z), \tag{7.38}$$

where $a_n := -\overline{\Phi_{n+1}(0)} = -\overline{\omega_{n+1}(0)}/\kappa_{n+1}$. Hence,

$$\Phi_n^*(z) = \prod_{k=0}^{n-1}\left(1 - a_k z \frac{\Phi_k(z)}{\Phi_k^*(z)}\right), \quad n = 1, 2, \ldots. \tag{7.39}$$

Let

$$z_{1,k} = z_{2,k} = \cdots = z_{m_k,k} = 0, \quad z_{i,k} \neq 0 \text{ for } m_k + 1 \leq i \leq k.$$

Then

$$\frac{\Phi_k(z)}{\Phi_k^*(z)} = z^{m_k} \prod_{\nu=m_k+1}^{k} \frac{\overline{z_{\nu,k}}\,(z - z_{\nu,k})}{\overline{z_{\nu,k}}\, z - 1}.$$

Each function $(z - z_{\nu,k})/(\overline{z_{\nu,k}} z - 1)$ is a Möbius transformation that maps $\mathbb{D}$ onto $\mathbb{D}$, and therefore

$$\left|\frac{\Phi_k(z)}{\Phi_k^*(z)}\right| < 1 \quad \text{for } z \in \mathbb{D}.$$

Then, by (7.39) we obtain for $z \in \mathbb{T}$,

$$|\Phi_n^*(z)| \le \prod_{k=0}^{n-1}(1+|a_k|) \le \exp\left(\sum_{k=0}^{n-1}|a_k|\right)$$

and

$$|\Phi_n(z)| \le \exp\left(\sum_{k=1}^{n}|\Phi_k(0)|\right).$$

Next the Bernstein–Walsh lemma (Lemma 1.1.11) yields

$$U^{\mu_\mathbb{T}-\nu_n}(z) = U^{\mu_\mathbb{T}-\widehat{\nu}_n}(z) \le \frac{1}{n}\sum_{k=1}^{n}|\Phi_k(0)|, \quad z \in \mathbb{C}. \tag{7.40}$$

Together with Lemma 7.9, we have obtained the bound

$$|U^{\mu_\mathbb{T}-\widehat{\nu}_n}(z)| \le \frac{c}{n}\left(\log\frac{\kappa_n}{\kappa_0}\frac{|z|}{|z|-1} + \sum_{k=1}^{n}\frac{|\omega_k(0)|}{\kappa_k}\right), \quad z \in \Delta.$$

We apply Theorem 3.1.2 and choose for the level curve

$$L_\delta = \{z = |z| = 1+\delta\}$$

the parameter $\delta = 1/n^2$. This yields the estimate (7.37). □

In the case that

$$\alpha := \limsup_{n\to\infty}|a_n| = \limsup_{n\to\infty}\frac{|\omega_n(0)|}{\kappa_n} < 1,$$

then the relation

$$\kappa_n^2 = \kappa_0^2 \prod_{k=0}^{n-1}(1-|a_k|^2)^{-1}$$

(cf. [69, formula 8.6, p. 156]) can be used to express (7.37) in terms of the reflection coefficients a_n only. Indeed, let $\alpha < 1$. Then there exists $k_0 \in \mathbb{N}$ such that

$$\frac{1}{1-|a_k|} \le 1 + \frac{2}{1-\alpha}|a_k| \quad \text{for } k \ge k_0.$$

Consequently, there exists $c > 0$ such that

$$\left(\frac{\kappa_n}{\kappa_0}\right)^2 \le c\prod_{k=0}^{n-1}\left(1+\frac{2}{1-\alpha}|a_k|\right), \quad n = 0, 1, \ldots,$$

which is $O\left(\exp\left(\sum_{k=0}^{n-1}|a_k|\right)\right)$ as $n \to \infty$. Hence, if $\alpha < 1$, we obtain

$$D[\widehat{\nu}_n - \mu_\mathbb{T}] \le c\frac{\log n}{n}\left\{\sum_{k=1}^{n}\frac{|\omega_k(0)|}{\kappa_k} + \log n\right\}, \tag{7.41}$$

In the case where $\tau' > 0$ a.e., Rakhmanov [148] and Máté, Nevai, and Totik [120] have proved that $\kappa_n^{-1}|\omega_n(0)| \to 0$ as $n \to \infty$. Thus the estimate (7.41) is applicable in this important case.

If $\log \tau'$ is integrable (i.e., τ belongs to the Szegő class on $\mathbb{T}$), then (7.41) leads only to

$$D[\widehat{\nu}_n - \mu_{\mathbb{T}}] \leq c \frac{\log n}{\sqrt{n}},$$

since the condition "τ belongs to the Szegő class on $\mathbb{T}$" is equivalent to $\sum_{k=0}^{\infty} |a_k|^2 < \infty$ (cf. [69, Theorem 8.2, p. 159]). But, according to (7.36) we already have obtained for the Szegő class the better estimate

$$D[\widehat{\nu}_n - \mu_{\mathbb{T}}] \leq c \sqrt{\frac{\log n}{n}}.$$

In the "Jacobi case," where $d\tau = |\sin(\theta/2)|^{2p} d\theta$, $p > 0$, defines the measure τ it is known ([131]) that $|a_n| = p/(n+p)$, so that (7.41) yields

$$D[\widehat{\nu}_n - \mu_{\mathbb{T}}] \leq c \frac{(\log n)^2}{n}. \tag{7.42}$$

If $\tau' \geq m > 0$, then $|\omega_n(z)| \leq c\sqrt{n}$ for all $z \in \overline{\mathbb{D}}$ and all n (cf. [69, Theorem 3.5, p. 40]), and $\{\kappa_n\}$ is bounded. Hence, the inequality in (7.40) can be replaced by

$$U^{\mu_{\mathbb{T}} - \widehat{\nu}_n}(z) \leq c \frac{\log n}{n},$$

and therefore, together with Lemma 7.9, this yieds

$$|U^{\mu_{\mathbb{T}} - \widehat{\nu}_n}(z)| \leq \frac{c}{n} \log \frac{n|z|}{|z|-1} \quad \text{for } z \in \Delta.$$

Using Theorem 3.1.2 on L_δ, $\delta = 1/n^2$, we get again (7.42).

7.8 Historical Comments

The assertion of Theorem 1.3 for some subclasses of domains with quasiconformal boundary (namely, bounded by a smooth or convex curve) was proved in [102, 100]. We follow the proof given in [19].

The results of Section 2 are presented in accordance with Pommerenke [144, 145]. For a generalization of Theorem 2.1 to some smooth curves and related questions, see [97, 98, 99]. To see how Theorem 2.1 can be generalized to other systems of extremal points we quote the papers of Menke (see, for example, [122]) and recent work of M. Götz [70].

Theorem 3.1 is proved by Blatt, Saff, and Totik in [39], and we follow the reasoning presented in that paper.

We note that the first result corresponding to Theorem 3.8 (even with $\log n$ instead of $(\log n)^2$ on the right-hand side of (3.13)) for curves $L \in C^{1+}$ was proved in [39].

Potential and discrepancy estimates for weighted Fekete and Tsuji points on quasiconformal curves and arcs were obtained by M. Götz and E.B. Saff [74]. Moreover, M. Götz [73] proved such discrepancy estimates for weighted extremal points on surfaces in $\mathbb{R}^d$, $d \geq 3$.

The investigation of the distribution of alternation points in polynomial Chebyshev approximation was initiated by Kadec [92]. He proved that there exists a subsequence $\Lambda \subset \mathbb{N}$ such that for any $\varepsilon > 0$,

$$D[\nu(X_n) - \mu] = O\left(n^{-1/2+\varepsilon}\right) \quad \text{for } n \in \Lambda, n \to \infty,$$

where X_n is an alternant of length $n + 2$ and μ is the equilibrium measure of $I = [-1, 1]$. This result was sharpened by Blatt and Lorentz [33] to

$$D[\nu(X_n) - \mu] = O((\log n/n)^{1/2}) \quad \text{for } n \in \Lambda, n \to \infty.$$

The sharper estimate of Theorem 4.1 was proved in [28].

W.H.J. Fuchs [61] extended the work of Kadec by studying the limit distribution of alternation points for real polynomial approximation on a finite number of compact real intervals. His discrepancy estimates of order $O\left(n^{-1/5}\right)$ were improved to $O\left(\log n/n\right)^{1/2})$ by Blatt and Grothmann [31].

G.G. Lorentz [116] showed that in general, Kadec's result does not hold for all subsequences.

Theorem 3.5 is a generalization of this result of Lorentz and was proved by Króo and Saff in [106]. The corresponding results for $1 < p < \infty$, contained in Theorem 5.9, were proved by Króo and Swetits [107], where Lemma 5.10 can be found, too.

Sign changes in L^p-approximation were investigated for $p = 1$ by Króo and Peherstorfer [105], for $p = 2$ by Saff and Shekhtman [153], and for $1 < p < \infty$ by Króo and Swetits [107]. All these results were concerned only with density properties of the sign changes. Discrepancy estimates as in Theorem 5.3 and Lemma 5.14 were proved for $p = 2$ by Blatt [30], for $1 < p < \infty$ in [29], and for $p = 1$ by Blatt, Grothmann, and Kovacheva [32]. For $1 < p \leq \infty$, the lacunary property of so-called *bad sequences* was proved by A. Króo [104].

Corollary 5.8 is contained in [35].

The discrepancy result of Theorem 6.3 was obtained in [36] by Blatt and Mhaskar for weight functions $w_Q(x)$ satisfying (6.14) with $B = 1$. In the case of Freud-type weight functions the estimate (6.48) was proved by Lubinsky and Saff [118] for $2 \leq p \leq \infty$.

The distribution of the zeros of orthogonal polynomials was considered by Krawtchouk [103], Erdős–Turán [53, 54], Szegő [173], and G. Freud [60]. In [57] Erdős and Turán obtained for special weights the discrepancy estimate of Theorem 7.2. Moreover, in the same paper (p. 111) they announced that P. Erdős succeeded in improving the error term to $O((\log n)^2/n)$ "using a much more difficult argument". That means that P. Erdős was aware of discrepancy results of type (7.15) for special weights. But we could not find a proof of this result of P. Erdős.

The discrepancy estimates of Theorem 7.3 and Theorem 7.5 for the zeros of orthogonal polynomials were proved in [34] and [21].

8
Special Topics

In this chapter we consider the approximation of a piecewise analytic function on touching domains by polynomials and the convergence of Bieberbach polynomials in domains with quasiconformal boundary. We have already used some of the results of this chapter in previous sections.

We separate this material from the main text because its study needs some special methods that are not typical for our previous considerations. At the same time these results essentially complete our knowledge about the behavior of some special polynomials.

8.1 Polynomial Approximation to a Piecewise Analytic Function on Touching Domains

If K is the unit interval $[-1,1]$, then a celebrated result of de la Vallée Poussin tells us that the best uniform approximation $E_n(f,K)$ to the function $f(x) := |x|$ on K by polynomials of degree at most $n \in \mathbb{N}$ satisfies

$$E_n(f,K) \leq \frac{c}{n}, \tag{1.1}$$

where $c = \text{const.} > 0$.

Let $\overline{G}_1 \supset [-1,0]$ and $\overline{G}_2 \supset [0,1]$ be two closed bounded Jordan domains such that $\overline{G}_1 \cap \overline{G}_2 = \{0\}$. The function $|x|$, $-1 \leq x \leq 1$, can be extended to the continuum

$$K := \overline{G}_1 \cup \overline{G}_2$$

in the following way:

$$f(z) := \begin{cases} -z, & \text{if } z \in \overline{G}_1, \\ z, & \text{if } z \in \overline{G}_2. \end{cases}$$

In connection with the distribution of the zeros of certain "near-best" approximating polynomials (see Section 4.2) it is natural to raise the following question: Does (1.1) remain true if the interval $[-1,1]$ is replaced by some continuum K consisting of two touching domains as was described above? (Note that in Example 6.2.2, $K = E_2$ was a parabolic region). We shall say that a continuum K (consisting of two touching domains) has the *de la Vallée Poussin property*, or more briefly, has the *VP-property*, if (1.1) remains true with some constant $c = c(K)$.

The purpose of this paragraph is to give an exhaustive geometric description of continua with the VP-property.

Let $\Omega := \overline{\mathbb{C}} \setminus K$ be the (simply connected) complement of K with respect to the extended complex plane $\overline{\mathbb{C}}$, and let $L := \partial\Omega = \partial K$ be their common boundary. Let $w = \Phi(z)$ map Ω conformally and univalently onto the exterior $\Delta := \overline{\mathbb{C}} \setminus \overline{\mathbb{D}}$ of the unit disk $\mathbb{D}$ with Φ normalized by the conditions $\Phi(\infty) = \infty$ and $\Phi'(\infty) > 0$.

Abusing notation, we sometimes use the symbol Φ both for the conformal mapping of Ω and for its extension to the Carathéodory prime end compactification of the domain Ω (see, for example, [142, Chapter 9]). Note that each point of $L \setminus \{0\}$ is an impression of exactly one boundary prime end; only the origin will be the impression of two boundary prime ends, which we denote by Z_1 and Z_2, respectively.

For definiteness we assume that the prime end Z_1 can be defined by some null-chain of crosscuts of Ω lying in $\{z : \operatorname{Im} z > 0\}$.

Let $l \subset \Omega$, $\infty \in l$, be an arbitrary fixed crosscut of Ω dividing Ω into two subdomains Ω_1 and Ω_2 such that for Ω_j, $j = 1, 2$, the prime end Z_j is *adjacent*, i.e., Z_j can be defined by one and the same null-chain of crosscuts of Ω and its subdomain Ω_j.

For $j = 1, 2$ and $u > 0$ set $\Psi := \Phi^{-1}$,

$$L_u := \{z : |\Phi(z)| = 1 + u\},$$
$$d_{u,j} := \operatorname{dist}(0, L_u \cap \Omega_j),$$
$$d_u := \max_{j=1,2} d_{u,j}.$$

We denote by $c, c_1, \ldots$ positive constants and by ε sufficiently small positive constants (each time different, in general) that either are absolute or depend on parameters not essential for the arguments; otherwise, such a dependency will be indicated.

Let $Z = Z_j$ with some fixed $j = 1, 2$, and let $\zeta_1, \zeta_2 \in \Omega$. We will say that an arc or curve $\gamma \subset \Omega$ separates the prime end Z and the point ζ_1 from ζ_2

8.1 Polynomial Approximation to a Piecewise Analytic Function

and ∞ if it divides Ω into two domains, one of which contains ζ_2 and ∞, while for the other one containing ζ_1 the prime end Z is adjacent.

Denote by $\Gamma(Z, \zeta_1, \zeta_2)$ the family of all arcs and curves separating in Ω the prime end Z and the point ζ_1 from ζ_2 and ∞.

Comparing the level curves L_u for the case of an arbitrary K and the interval $[-1, 1]$ we see that

$$d_{u,j} \geq c_1 u \qquad j = 1, 2, \, u > 0. \tag{1.2}$$

Theorem 1.1. *For any continuum K as above,*

$$E_n(f, K) \geq c\, d_{1/n}, \qquad n \in \mathbb{N}, \tag{1.3}$$

with some constant $c = c(K) > 0$.

Proof. Without loss of generality, we may assume that

$$E_n := E_n(f, K) \leq d := d_{1/n} = d_{1/n,1}.$$

Set

$$\gamma_n := \{z = d\, e^{i\theta} : 0 \leq \theta \leq \pi\},$$
$$\nu_n := [-d, d], \quad z_n := -\varepsilon\, i\, d,$$

where the choice of the sufficiently small fixed constant $0 < \varepsilon < \tfrac{1}{2}$ will be specified later on.

Under a suitable orientation of the arcs γ_n and ν_n, we have

$$I := \int_{\nu_n} \frac{f(\zeta)}{(\zeta - z_n)^2}\, d\zeta = \int_{\nu_n} \frac{f(\zeta) - p_n^*(\zeta)}{(\zeta - z_n)^2}\, d\zeta + \int_{\gamma_n} \frac{p_n^*(\zeta)}{(\zeta - z_n)^2}\, d\zeta, \tag{1.4}$$

where $p_n^* = p_n^*(f, K)$ is the polynomial in $\mathbb{P}_n$ of best uniform approximation to f on K.

The left-hand side of (1.4) can be estimated from below as follows:

$$I = \log \frac{z_n^2 - d^2}{z_n^2} - \frac{2d^2}{d^2 - z_n^2} \geq 2\log\frac{1}{\varepsilon} - 2. \tag{1.5}$$

Next, we estimate the right-hand side of (1.4) from above. First, we have

$$\left| \int_{\nu_n} \frac{f(\zeta) - p_n^*(\zeta)}{(\zeta - z_n)^2}\, d\zeta \right| \leq E_n \int_{\nu_n} \frac{d\zeta}{|\zeta - z_n|^2} \leq \frac{c_1 E_n}{\varepsilon\, d}. \tag{1.6}$$

In order to estimate the last integral in (1.4), we note that for $z \in K$ the following inequality holds:

$$|p_n^*(z)| \leq |p_n^*(z) - f(z)| + |f(z)| \leq d\left(1 + \frac{|z|}{d}\right).$$

Consider the function $\varphi_n(z)$ that maps $\overline{\mathbb{C}}\setminus\nu_n$ conformally and univalently onto Δ with the normalization $\varphi_n(\infty) = \infty$, $\varphi_n'(\infty) > 0$.

It is easy to see that for $z \in \mathbb{C}\setminus\nu_n$,

$$c_2\left(1 + \frac{|z|}{d}\right) \leq |\varphi_n(z)| \leq c_3\left(1 + \frac{|z|}{d}\right)$$

(cf. Lemma C.3.1). Furthermore,

$$|p_n^*(z)| \leq c_4 d, \qquad z \in \gamma_n.$$

Indeed, if $z \in \gamma_n \cap K$, then the above inequality is valid with $c_4 = 2$. If $z \in \gamma_n \cap \Omega$, then we have to apply the maximum principle to the function

$$\frac{p_n^*(z)}{\varphi_n(z)\,\Phi(z)^n}$$

in Ω.

Now, since

$$\left| \int_{\gamma_n} \frac{p_n^*(\zeta)}{(\zeta - z_n)^2} d\zeta \right| \leq c_4 \pi, \tag{1.7}$$

we obtain by (1.4)–(1.7) for sufficiently small (but fixed) ε that

$$\frac{c_1 E_n}{\varepsilon\, d} \geq 2\log\frac{1}{\varepsilon} - 2 - c_4\pi \geq \log\frac{1}{\varepsilon}.$$

Consequently,

$$E_n \geq c_5 d, \quad c_5 = c_5(\varepsilon) > 0.$$

$\square$

Let $g_\varepsilon(x)$, $0 \leq x \leq \varepsilon$, be an arbitrary continuous monotone function with the property $0 = g_\varepsilon(0) \leq g_\varepsilon(\varepsilon) < \pi/2$. Set

$$E(g_\varepsilon) := \{z = re^{i\theta} : 0 \leq r \leq \varepsilon,\ |\theta| \leq g_\varepsilon(r) \text{ or } |\theta - \pi| \leq g_\varepsilon(r)\},$$
$$D(r) := \{z : |z| < r\}, \qquad r > 0.$$

Theorem 1.2. *Let K satisfy*

$$d_{1/n} \leq \frac{c}{n}, \qquad n \in \mathbb{N}. \tag{1.8}$$

Then there exist an ε and a function g_ε such that

$$\overline{D(\varepsilon)} \cap K \subset E(g_\varepsilon), \tag{1.9}$$

and for all z with $0 < |z| < \varepsilon/2$, $\operatorname{Re} z = 0$, the inequality

$$|z| \leq c_1 (|\Phi(z)| - 1) \tag{1.10}$$

holds with some constant $c_1 = c_1(K) > 0$.

8.1 Polynomial Approximation to a Piecewise Analytic Function

Proof. We shall carry out the proof of (1.9) by contradiction. Denote by $\gamma_j(r)$, $j = 1, 2$, $0 < r < 1$, any crosscut of Ω lying on the circle $\{z : |z| = r\}$ and separating the prime end Z_j from ∞.

Assume that the assertion of Theorem 1.2 concerning (1.9) is false. This means that for some continuum K satisfying (1.8) and for some $0 < \alpha < \pi/2$ there exists a sequence of numbers $1 = r_0 > r_1 > r_2 > \cdots$ such that

$$\lim_{k \to \infty} r_k = 0,$$

and for some fixed value of j (for definiteness, say $j = 1$),

$$|\gamma_1(r_k)| \leq (\pi - 2\alpha) r_k, \qquad k \in \mathbb{N}.$$

We may suppose without loss of generality that $r_k \leq e^{\pi/2} r_{k-1}$ and that $\gamma_1(r_k)$ separates Z_1 from $\gamma_1(r_{k-1})$.

Let n be large enough and let $z_n \in L_{1/n} \cap \Omega_1$ be any point with the property $|z_n| = d := d_{1/n,1}$. Denote by $l_n \subset \{z : |z| = 2d\}$ any crosscut of Ω that separates z_n from ∞. The arcs $\gamma_1(1)$ and l_n are the sides of some quadrilateral $Q_n \subset \Omega$ whose other two sides are part of the boundary L. Let Γ_n be the family of all crosscuts of Q_n separating the sides $\gamma_1(1)$ and l_n. Denote by Γ_n^* the family of all arcs $\gamma \subset Q_n$ joining the sides $\gamma_1(1)$ and l_n (see Figure 8.1).

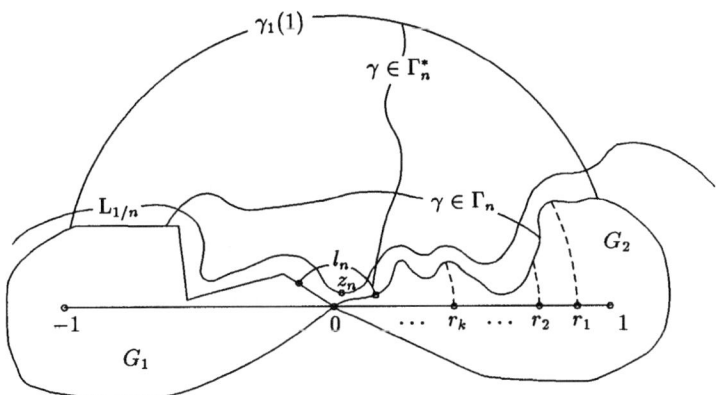

Fig. 8.1

It is known that the modules of these families satisfy the condition

$$m(\Gamma_n) \, m(\Gamma_n^*) = 1 \tag{1.11}$$

(cf. (B.1.3)).

Lemma A.1.15 yields

$$m(\Gamma_n) = m(\Phi(\Gamma_n)) \leq \frac{1}{\pi} \log n + c_1. \tag{1.12}$$

Our next task is to estimate the quantity $m(\Gamma_n^*)$ from above.

Choose the positive integer k_n so that

$$\gamma_1(r_{k_n}) \subset Q_n, \quad \gamma_1(r_{k_n+1}) \not\subset Q_n.$$

By our assumption,

$$\lim_{n \to \infty} k_n = \infty. \tag{1.13}$$

Now, for $k \in \mathbb{N}$ set

$$S_k := \begin{cases} \{z = re^{i\theta} : 0 \leq \theta \leq \alpha/2,\ e^{-\alpha/2} \leq r/r_k \leq e^{\alpha/2}\}, \\ \qquad \text{if } \gamma_1(r_k) \cap \{z = r_k e^{i\theta} : \pi - \alpha \leq \theta \leq \pi\} \neq \emptyset, \\ \{z = re^{i\theta} : \pi - \alpha/2 \leq \theta \leq \pi,\ e^{-\alpha/2} \leq \frac{r}{r_k} \leq e^{\alpha/2}\}, \quad \text{elsewhere}. \end{cases}$$

Further, consider the admissible function

$$\rho(z) := \begin{cases} \dfrac{1}{|z|}, & \text{if } 2d \leq |z| \leq 1,\ \operatorname{Im} z \geq 0,\ z \notin \bigcup_{0 < 2k < k_n} S_{2k}, \\ 0, & \text{elsewhere}. \end{cases}$$

Notice that

$$\int_\gamma \rho(z)\,|dz| \geq \log \frac{1}{2d}, \quad \gamma \in \Gamma_n^*. \tag{1.14}$$

Indeed, denote by γ_k, $1 \leq k \leq k_n + 1$, any subarc of $\gamma \in \Gamma_n^*$ lying in the annulus

$$\begin{cases} \{z : r_{k-1} < |z| < r_k\}, & \text{if } k \leq k_n, \\ \{z : r_{k_n} < |z| < 2d\}, & \text{if } k = k_n + 1, \end{cases}$$

and joining its boundary components.

For any arc γ_k we have

$$\int_{\gamma_k} \rho(z)\,|dz| \geq \begin{cases} \log \dfrac{r_{k-1}}{r_k}, & \text{if } k \leq k_n, \\ \log \dfrac{r_{k_n}}{2d}, & \text{if } k = k_n + 1. \end{cases} \tag{1.15}$$

In fact, if $\gamma_k \cap \bigcup_{0 < 2k < k_n} S_{2k} = \emptyset$, then (1.15) follows from the inequality

$$\int_{\gamma_k} \rho(z)\,|dz| \geq \left| \int_{\gamma_k} \frac{dz}{z} \right|.$$

The situation is more complicated when γ_k intersects S_{2m} (for an appropriate m).

We describe how (1.15) can be proved if $k \leq k_n$ is even; the other cases can be handled similarly.

There exist two subarcs γ'_k and γ''_k of γ_k with the following properties: γ'_k joins $\gamma_1(r_{k-1})$ to ∂S_k, and γ''_k joins ∂S_k to $\gamma_1(r_k)$. Therefore,

$$\int_{\gamma_k} \rho(z)\,|dz| \geq \left|\int_{\gamma'_k} \frac{dz}{z}\right| + \left|\int_{\gamma''_k} \frac{dz}{z}\right| \geq \frac{\alpha}{2} + \log\left(e^{-\alpha/2}\frac{r_{k-1}}{r_k}\right) = \log\frac{r_{k-1}}{r_k}.$$

Inequality (1.14) is an immediate consequence of estimate (1.15) and the obvious relation

$$\int_\gamma \rho(z)\,|dz| \geq \sum_{k=1}^{k_n+1} \int_{\gamma_k} \rho(z)\,|dz|.$$

On the other hand, the ρ-area of $\mathbb{C}$ is (cf. Appendix A)

$$A(\rho) = \int_\mathbb{C} \rho^2(z)\,dm(z) = \pi \log\frac{1}{2d} - a_n,$$

where

$$a_n := \sum_{0<2k<k_n} \int_{S_{2k}} \frac{dm(z)}{|z|^2} = \sum_{0<2k<k_n} \frac{\alpha^2}{2} \geq c_2 k_n.$$

Recalling the definition of the module of a family of curves, we have

$$m(\Gamma^*_n) \leq \frac{\pi \log(1/(2d)) - c_2 k_n}{(\log(1/(2d)))^2}.$$

Combining the last estimate with (1.11), (1.12) and our assumption (1.8) we obtain

$$\left(\log\frac{1}{2d}\right)^2 \leq \left(\pi\log\frac{1}{2d} - c_2 k_n\right)\left(\frac{1}{\pi}\log\frac{1}{2d} + c_3\right)$$

$$\leq \left(\log\frac{1}{2d}\right)^2 + \left(\pi c_3 - \frac{c_2}{\pi} k_n\right)\log\frac{1}{2d}.$$

Thus,

$$k_n \leq \frac{\pi^2 c_3}{c_2} = c_4.$$

The last relation, however, contradicts (1.13), and the proof of (1.9) is complete.

We begin the proof of (1.10) with the remark that (1.8) and an elementary argument involving Lemma 1.2.3 imply

$$d_{u,j} \preceq u, \qquad 0 < u < 1. \tag{1.16}$$

Set

$$J_j := \left(0, (-1)^{j+1} \frac{\varepsilon}{2} i\right), \quad J'_j := \Phi(J_j), \quad w_j := \Phi(Z_j).$$

We claim that for any $w \in J'_j$,

$$|w - w_j| \preceq |w| - 1. \tag{1.17}$$

Indeed, let Γ_1 be the family of all crosscuts of Ω separating the point $z := \Psi(w)$ from the prime end Z_j and ∞. An elementary computation shows that

$$m(\Gamma_1) \le c_5,$$
$$m(\Phi(\Gamma_1)) \ge \frac{1}{2\pi} \log \frac{|w - w_j|}{|w| - 1},$$

which, in view of the conformal invariance of the module of a family of curves (cf. (1.2.11)), yields (1.17).

Let $z \in J_j$ be an arbitrary point. As before, set $w := \Phi(z)$, $u := |w| - 1$. Denote by z^* any point in $L_u \cap \Omega_j$ with

$$|z^*| = d_{u,j} =: d.$$

For our purposes it is enough to establish the estimate

$$|z| \preceq d. \tag{1.18}$$

Denote by Γ_2 the family of all crosscuts of Ω separating z^* from ∞.
By inequality (A.1.9) we have

$$m(\Phi(\Gamma_2)) \le 2 + \frac{1}{\pi} \log \frac{2|w^*|}{|w^*| - 1} \le 2 + \frac{1}{\pi} \log \frac{4}{u} \le c_6 + \frac{1}{\pi} \log \frac{1}{d}, \tag{1.19}$$

where $w^* := \Phi(z^*)$.

The circle $\{\zeta : |\zeta| = r\}$, $d < r < 1$, contains one crosscut, or finitely many crosscuts of Ω separating z^* from ∞. Let $\gamma(r)$ denote the one that is chosen such that it necessarily intersects $[0, (-1)^{j+1}i]$ if such a choice is possible.

Let r^* be the infimum of all r with the property that $\gamma(r)$ intersects $[0, (-1)^{j+1}i]$.
Notice that

$$r^* \preceq d. \tag{1.20}$$

8.1 Polynomial Approximation to a Piecewise Analytic Function

Indeed, if $r^* > d$, then (1.20) follows from (1.19) and a direct consequence of the integrated version of composition laws for modules, namely from the inequality

$$m(\Gamma_2) \geq \frac{2}{\pi} \log \frac{r^*}{d} + \frac{1}{\pi} \log \frac{1}{r^*} = \frac{1}{\pi} \log \frac{r^*}{d} + \frac{1}{\pi} \log \frac{1}{d}.$$

It suffices to prove (1.18) only for the case $|z| > r^*$.

Set $\Gamma_3 := \Gamma(Z_j, z^*, z)$. Then we find with (1.17) that

$$m(\Phi(\Gamma_3)) \leq c_7.$$

On the other hand,

$$m(\Gamma_3) \geq \frac{1}{\pi} \log \frac{|z|}{r^*} \geq \frac{1}{\pi} \log \frac{|z|}{c_8 d}.$$

Comparing the last two inequalities we get (1.18). □

Theorem 1.3. *If K satisfies (1.9) and (1.10), then K possesses the VP-property.*

Proof. By Cauchy's formula,

$$f(z) = f_1(z) + f_2(z), \qquad z \in K \setminus \{0\},$$

where

$$f_1(z) := -\frac{1}{\pi i} \int_{[-i\varepsilon, i\varepsilon]} \frac{\zeta}{\zeta - z} d\zeta,$$

and the function $f_2(z)$ can be analytically continued to some fixed neighborhood of K (depending on ε).

Thus, our objective will be the proof of the estimate

$$E_n(f_1, K) \preceq n^{-1}, \tag{1.21}$$

for n large enough.

We begin with the problem of approximation of the Cauchy kernel $(\zeta - z)^{-1}$ by polynomial kernels of the form

$$K_n(\zeta, z) = \sum_{j=0}^{n} a_j(\zeta) z^j, \qquad n \in \mathbb{N}. \tag{1.22}$$

We claim that for each positive integer n there exists a polynomial kernel of the form (1.22) and constants $c_j = c_j(K)$, $j = 1, 2$, satisfying the inequality

$$\left| \frac{1}{\zeta - z} - K_n(\zeta, z) \right| \leq c_1 n^{-2} |\zeta|^{-3} \tag{1.23}$$

for $z \in K$ and $\zeta \in [-\varepsilon i, \varepsilon i]$, $|\zeta| \geq c_2/n$.

Indeed, for $\zeta \in \Omega$, $z \in K$, and $m \in \mathbb{N}$ consider the *Dzjadyk polynomial kernel* $K_{0,1,2,m}(\zeta, z)$ (see Appendix C). It is known that the degree of $K_{0,1,2,m}(\zeta, z)$ in the variable z is equal to $10m - 9$. Set $m := [n/10]$,

$$K_n(\zeta, z) := K_{0,1,2,m}(\zeta, z).$$

According to Theorem C.2.4,

$$\left| \frac{1}{\zeta - z} - K_n(\zeta, z) \right| \preceq \frac{1}{|\zeta - z|} \left| \frac{\widetilde{\zeta} - \zeta}{\widetilde{\zeta} - z} \right|^2, \tag{1.24}$$

where $\widetilde{\zeta} := \Psi\left((1 + 1/m)\Phi(\zeta)\right)$. By virtue of (1.10), there exists a constant c_2 such that for $\zeta \in [-\varepsilon i, \varepsilon i]$ with $|\zeta| \geq c_2/n$ we have

$$|\Phi(\zeta)| - 1 \geq \frac{64}{m}.$$

Thus, for $m \geq 64$ by Lemma 1.2.3,

$$|\widetilde{\zeta} - \zeta| \leq \frac{1}{2} d\left(\widetilde{\zeta}, L\right).$$

Consequently,

$$\frac{1}{2}|\zeta - z| \leq |\widetilde{\zeta} - z| \leq \frac{3}{2}|\zeta - z|, \quad z \in K,$$

and again by Lemma 1.2.3,

$$\left| \frac{\widetilde{\zeta} - \zeta}{\widetilde{\zeta}} \right| \preceq \frac{1/m}{|\Phi(\zeta)| - 1} \preceq \frac{1/m}{|\zeta|}.$$

Recalling (1.24), we get (1.23).

Returning to the proof of (1.21) we consider the polynomial

$$p_n(z) := -\frac{1}{\pi i} \int_{[-\varepsilon i, -c_2 i/n] \cup [c_2 i/n, \varepsilon i]} \zeta K_n(\zeta, z) \, d\zeta,$$

where K_n and c_2 are the polynomial kernel and the respective constant from (1.23).

Finally,

$$|f_1(z) - p_n(z)| \leq \frac{1}{\pi} \left| \int_{[-\varepsilon i, -c_2 i/n] \cup [c_2 i/n, \varepsilon i]} \zeta \left(\frac{1}{\zeta - z} - K_n(\zeta, z) \right) d\zeta \right|$$

$$+ \frac{1}{\pi} \left| \int_{[-c_2 i/n, c_2 i/n]} \frac{\zeta}{\zeta - z} d\zeta \right| \leq \frac{c_3}{n^2} \int_{c_2/n}^{\varepsilon} \frac{dx}{x^2} + \frac{c_4}{n} \preceq n^{-1}.$$

$\square$

8.1 Polynomial Approximation to a Piecewise Analytic Function

As a simple direct consequence of Theorems 1.1–1.3, we obtain the following statement.

Theorem 1.4. *The following three assertions are equivalent:*

(i) K *has the VP-property.*

(ii) K *satisfies (1.8).*

(iii) K *satisfies (1.9) and (1.10).*

It might be interesting to investigate the connection between the VP-property of K and the structure of the function g_ε in relation (1.9).

The next assertion is the first step in this direction.

Theorem 1.5. *Let K satisfy (1.9) with some function g_ε such that*

$$\int_0^\varepsilon \frac{g_\varepsilon(x)}{x}\, dx < \infty. \tag{1.25}$$

Then K has the VP-property.

Proof. Without loss of generality we assume that

$$K \cap \overline{D(\varepsilon)} = E(g_\varepsilon).$$

According to Theorem 1.4 it is enough to establish the validity of (1.10). Hence, by virtue of Theorem A.1.17, we can restrict ourselves to the proof of the estimate

$$m(\Gamma_4) \leq \frac{1}{\pi} \log \frac{1}{\delta} + c_1, \tag{1.26}$$

where $0 < \delta < \varepsilon\, e^{-\pi}$, $\Gamma_4 := \Gamma(Z_1, \delta i, e^{-\pi}\varepsilon i)$.

In order to do this we introduce the admissible function

$$\rho(z) := \begin{cases} (|z|\,(\pi - 2\,g_\varepsilon(e^\pi|z|)))^{-1}, & \text{if } z \in \Omega,\ \mathrm{Im}\, z > 0, \\ & e^{-\pi}\delta \leq |z| < e^{-\pi}\varepsilon, \\ (|z|\,(\pi - 2\,g_\varepsilon(\varepsilon)))^{-1}, & \text{if } z \in \Omega,\ \mathrm{Im}\, z > 0, \\ & e^{-\pi}\varepsilon \leq |z| \leq \varepsilon, \\ 0, & \text{elsewhere in } \mathbb{C}, \end{cases}$$

and note that

$$\int_\gamma \rho(z)\,|dz| \geq 1, \qquad \gamma \in \Gamma_4. \tag{1.27}$$

Indeed, for $\gamma \in \Gamma_4$ set

$$r_{\max} := \sup_{z\in\gamma} |z|, \quad r_{\min} := \inf_{z\in\gamma} |z|.$$

If $r_{\max} \geq e^\pi r_{\min}$, then there exists an arc $\gamma_1 \subset \gamma$ contained in the annulus $\{z : re^{-\pi} < |z| < r\}$ with some $r \in (\delta, e^{-\pi}\varepsilon)$ whose endpoints connect the two boundary components of the annulus. Therefore,

$$\int_\gamma \rho(z)\,|dz| \geq \left|\int_{\gamma_1} \rho(z)\,dz\right| \geq 1.$$

If $r_{\max} < e^\pi r_{\min}$, then the monotony of $g_\varepsilon(r)$ gives

$$\int_\gamma \rho(z)\,|dz| \geq \frac{1}{\pi - 2\,g_\varepsilon(r_{\max})} \left|\int_\gamma \frac{dz}{z}\right| \geq 1.$$

Set $c_2 := \pi - 2\,g_\varepsilon(\varepsilon)$. From the L-definition of the module of a curve family we obtain

$$m(\Gamma_4) \leq \int_{\mathbb{C}} \rho^2(z)\,dm(z)$$

$$= \int_{e^{-\pi}\delta}^{e^{-\pi}\varepsilon} \frac{\pi - 2\,g_\varepsilon(r)}{(\pi - 2\,g_\varepsilon(e^\pi r))^2}\,\frac{dr}{r} + \int_{e^{-\pi}\varepsilon}^{\varepsilon} \frac{\pi - 2\,g_\varepsilon(r)}{(\pi - 2\,g_\varepsilon(\varepsilon))^2}\,\frac{dr}{r}$$

$$\leq \frac{1}{\pi}\int_{e^{-\pi}\delta}^{e^{-\pi}\varepsilon} \frac{dr}{r} + \frac{1}{\pi}\int_{e^{-\pi}\delta}^{e^{-\pi}\varepsilon} \frac{4\pi\,g_\varepsilon(e^\pi r)}{(\pi - 2\,g_\varepsilon(e^\pi r))^2}\,\frac{dr}{r} + \frac{\pi^2}{c_2^2}$$

$$\leq \frac{1}{\pi}\log\frac{\varepsilon}{\delta} + \frac{4}{c_2^2}\int_\delta^\varepsilon g_\varepsilon(r)\,\frac{dr}{r} + \frac{\pi^2}{c_2^2} \leq \frac{1}{\pi}\log\frac{1}{\delta} + c_1.$$

□

Note that the result of Example 6.2.2 concerning the approximation of $f(x) = |x|$ on a parabolic region is an immediate consequence of Theorem 1.5.

The sharpness of the last theorem can be shown from the following result. Set

$$J(g_\varepsilon) := \left\{ z = re^{i\theta} : 0 \leq r \leq \varepsilon,\ \left|\theta - \frac{\pi}{2}\right| = \frac{\pi}{2} - g_\varepsilon(r) \right\}.$$

Theorem 1.6. Suppose $K \supset J(g_\varepsilon)$ with some function g_ε satisfying

$$\int_0^\varepsilon \frac{g_\varepsilon(x)}{x}\,dx = \infty. \tag{1.28}$$

8.1 Polynomial Approximation to a Piecewise Analytic Function

Then

$$\lim_{n\to\infty} n\, E_n(f,K) = \infty \, ; \qquad (1.29)$$

that is, K does not have the VP-property.

Proof. We can assume that $K = J = J(g_\varepsilon)$ is the arc defined by the function g_ε satisfying (1.28). Note that in this case, K and the interval $[-1,1]$ have only one common point: the origin.

Let Z_1 be the boundary prime end of Ω that can be defined by the null-chain of crosscuts

$$\left\{ \left\{ z = \frac{1}{m} e^{i\theta} : g_\varepsilon\left(\frac{1}{m}\right) < \theta < \pi - g_\varepsilon\left(\frac{1}{m}\right) \right\}, \, m \in \mathbb{N}, \, m > \frac{1}{\varepsilon} \right\}.$$

Set $\tilde{z}_1 = \tilde{z}_1(n) := \Psi((1+1/n)\,\Phi(Z_1))$.

By repeating word for word the proof of Theorem 1.1 we have

$$E_n(f,K) \succeq \delta \quad (\delta := |\tilde{z}_1|) \qquad (1.30)$$

(for details, see [17, p. 152] or [12]). Further, consider the family of curves

$$\Gamma_5 := \Gamma(Z_1, \tilde{z}, \varepsilon\, i)\,.$$

For its module (when n is large enough) we have

$$m(\Gamma_5) \geq \int_\delta^\varepsilon \frac{dr}{(\pi - 2g_\varepsilon(r))\,r} \geq \frac{1}{\pi} \int_\delta^\varepsilon \frac{dr}{r} + \frac{2}{\pi} \int_\delta^\varepsilon \frac{g_\varepsilon(r)}{r}\,dr\,.$$

Therefore, by Theorem A.1.17,

$$\delta \succeq \frac{1}{n} \exp\left(c_1 \int_\delta^\varepsilon \frac{g_\varepsilon(r)}{r}\,dr \right). \qquad (1.31)$$

Thus, relation (1.29) follows immediately from (1.28), (1.30), and (1.31). □

The last theorems raise the following question: Is it possible to characterize K with the VP-property in terms similar to relations (1.9) and (1.25)?

The next assertion gives a negative answer to this question.

Theorem 1.7. *There exists a continuum K_0 satisfying the following two conditions:*

(i) *K_0 has the VP-property.*

(ii) *For any function g_ε such that (1.9) is valid, relation (1.28) holds.*

334   8. Special Topics

Proof. Consider the function
$$h(x) := \frac{1}{\log \frac{1}{x}}, \qquad 0 < x < 1,$$
the sequence of numbers
$$r_k := e^{-k\pi}, \qquad k \in \mathbb{N},$$
and the circular arcs
$$l_k := \{z = r_k e^{i\theta} : |\theta| \leq h(r_k)\}, \qquad k \in \mathbb{N}.$$
Let K be symmetric with respect to the origin,
$$K \cap \{\operatorname{Re} z \geq 0\} := [0,1] \cup \left(\bigcup_{k=10}^{\infty} l_k \right),$$
and
$$A := [-1-i, 1-i] \cup [1-i, 1+i] \cup [1+i, -1+i] \cup [-1+i, -1-i].$$

Denote by K_0 the continuum bounded by $L_0 := \Psi(A)$ (see Figure 8.2).

We claim that the continuum K_0 satisfies the properties (i)–(ii) mentioned in Theorem 1.7.

Indeed, by virtue of Theorem 1.4, it is enough to check the validity of (1.10) and (ii) from Theorem 1.7 for K instead of K_0 and $\varepsilon = 1$.

We begin with the remark that if $K \subset E(g_1)$, then
$$\int_0^1 \frac{g_1(x)}{x} dx \geq \sum_{k=10}^{\infty} \int_{r_{k+1}}^{r_k} \frac{g_1(x)}{x} dx \geq \sum_{k=10}^{\infty} h(r_{k+1}) \log \frac{r_k}{r_{k+1}}$$
$$= \sum_{k=10}^{\infty} \frac{1}{k+1} = \infty.$$

As in the proof of Theorem 1.5 we need to prove (1.26) (with $\varepsilon = 1$). In order to do so, we consider the admissible functions
$$\rho_1(z) := \begin{cases} \dfrac{1}{\pi|z|}, & \text{if } \operatorname{Im} z \geq 0,\ \delta e^{-\pi} \leq |z| \leq 1, \\ 0, & \text{elsewhere,} \end{cases}$$
$$\rho_2(z) := \begin{cases} \dfrac{2}{r_k}, & \text{if } \operatorname{dist}(z, l_k) \leq h(r_k) r_k \text{ for some } k \geq 10, \\ 0, & \text{elsewhere.} \end{cases}$$

The function
$$\rho(z) := \max\{\rho_1(z), \rho_2(z)\}$$

8.1 Polynomial Approximation to a Piecewise Analytic Function   335

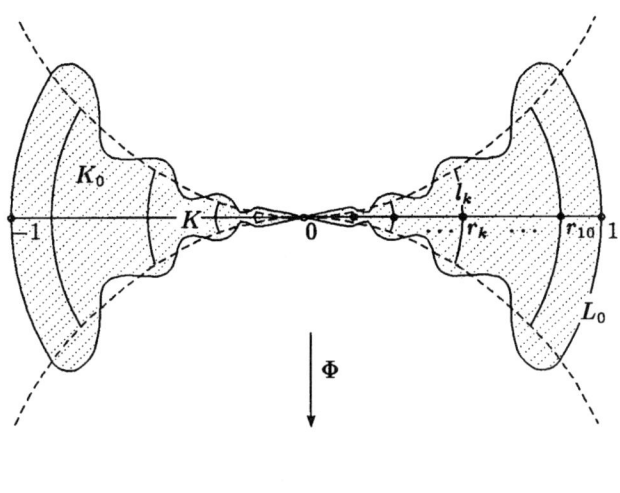

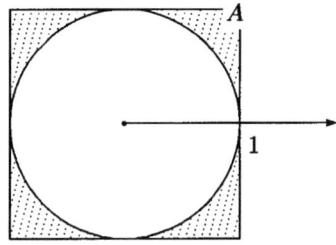

Fig. 8.2

satisfies (1.27).

Indeed, for $\gamma \in \Gamma_4$ let the quantities $r_{\max}$ and $r_{\min}$ be defined as above. If $r_{\max} \geq e^\pi r_{\min}$ or

$$\gamma \cap \left(\bigcup_{k=10}^{\infty} l_k \right) = \emptyset,$$

then the proof of (1.27) is completely the same as in Theorem 1.5.

So suppose now that $r_{\max} < e^\pi r_{\min}$ and for some $k \geq 10$,

$$\gamma \cap l_k \neq \emptyset.$$

Note that this is possible only for a single value of k.

In this case we obtain

$$\int_\gamma \rho(z)\,|dz| \geq \frac{\pi - 2\pi\, h(r_k)}{\pi} + 2\, h(r_k) = 1.$$

At last, we obtain the estimate (1.26) in the following way:

$$m(\Gamma_4) \le \int_{\mathbb{C}} \rho_1^2(z)\,dm(z) + \int_{\mathbb{C}} \rho_2^2(z)\,dm(z)$$

$$\le \frac{1}{\pi}\log\frac{1}{\delta} + 1 + c_1 \sum_{k=10}^{\infty} h^2(r_k)$$

$$= \frac{1}{\pi}\log\frac{1}{\delta} + 1 + c_1 \sum_{k=10}^{\infty} \frac{1}{k^2} \le \frac{1}{\pi}\log\frac{1}{\delta} + c_2.$$

$\square$

8.2 Uniform Convergence of Bieberbach Polynomials in Domains with Quasiconformal Boundary

In this section we give a proof of Theorem 6.5.6, i.e., the inequality

$$\|f - \pi_n\|_{\overline{G}} \le c\,n^{-\gamma}, \qquad n \in \mathbb{N}, \tag{2.1}$$

where the constants $\gamma > 0$ and $c > 0$ are independent of n, G is a domain bounded by an arbitrary quasiconformal curve L, π_n is the nth Bieberbach polynomial associated with G and some fixed point $z_0 \in G$, the function $f = f_{z_0}$ maps G conformally onto the disk $\{w : |w| < r_0 = r_0(z_0, G)\}$ with the normalization $f(z_0) = 0$, $f'(z_0) = 1$.

Let $\Omega := \overline{\mathbb{C}} \setminus \overline{G}$, and suppose that the function Φ maps Ω conformally onto Δ with $\Phi(\infty) = \infty$, $\Phi'(\infty) > 0$. As we know (see Theorem B.2.1), f and Φ can be extended to quasiconformal mappings of the plane onto itself (with infinity as a fixed point). We keep the same notation f and Φ for them.

For $u > 0$ we put

$$L_u := \{\zeta : |\Phi(\zeta)| = 1 + u\}, \quad \Omega_u := (\operatorname{int} L_u)\setminus\overline{G}.$$

A simple consequence of (B.1.9) is the estimate

$$\int_{\Omega_u} |f_{\bar{\zeta}}|^2\,dm \preceq u^{\delta}, \qquad u \preceq 1. \tag{2.2}$$

Indeed, suppose that $K > 1$ is the coefficient of quasiconformality of the mapping f, and $J_f := |f_\zeta|^2 - |f_{\bar{\zeta}}|^2$ its Jacobian. Since

$$\left|\frac{f_{\bar{\zeta}}}{f_\zeta}\right| \le k := \frac{K-1}{K+1} < 1$$

almost everywhere in $\mathbb{C}$,

$$\int_{\Omega_u} |f_{\bar{\zeta}}|^2 \, dm = \int_{\Omega_u} J_f \left(\left|\frac{f_{\zeta}}{f_{\bar{\zeta}}}\right|^2 - 1 \right)^{-1} dm$$

$$\leq (k^{-2} - 1)^{-1} \int_{\Omega_u} J_f \, dm \preceq m(f(\Omega_u))$$

$$= m\left((f \circ \Phi^{-1})(\Phi(\Omega_u))\right) \preceq m(\Phi(\Omega_u))^{\delta} \asymp u^{\delta}.$$

We denote by $L_1^2(G)$ the class of functions $g(z)$ analytic in G and such that $g(z_0) = 0$,

$$\|g\|_{L_1^2(G)}^2 := \int_G |g'|^2 \, dm < +\infty.$$

Lemma 2.1. *For any polynomial $p_n \in \mathbb{P}_n$, $n \in \mathbb{N}$, with $p_n(z_0) = 0$ we have*

$$\|p_n\|_{\overline{G}} \leq c_1 \sqrt{\log n} \, \|p_n\|_{L_1^2(G)}, \tag{2.3}$$

where c_1 is independent of n and p_n.

Proof. Let ψ be the conformal mapping from $\mathbb{D}$ to G with $\psi(z_0) = 0$, $\psi'(z_0) > 0$. We claim that

$$|p_n(z)| \leq \frac{1}{\sqrt{\pi}} \left(\log \frac{1}{1-r} \right)^{1/2} \|p_n\|_{L_1^2(G)}, \qquad z \in G, \tag{2.4}$$

where $r := |\psi^{-1}(z)|$. To see this consider $g(w) := p_n(\psi(w))$ for $w \in \mathbb{D}$. Since $g(0) = 0$, we have $g(w) = \sum_{k=1}^{\infty} b_k w^k$ and

$$\pi \sum_{k=1}^{\infty} k |b_k|^2 = \int_{\mathbb{D}} |g'(w)|^2 \, dm(w) = \int_G |p_n'(z)|^2 \, dm(z) = \|p_n\|_{L_1^2(G)}^2. \tag{2.5}$$

Further, for $w := \varphi(z) \in \mathbb{D}$, $\varphi := \psi^{-1}$, by Schwarz's inequality

$$|g(w)| \leq \sum_{k=1}^{\infty} |b_k| r^k \leq \left(\sum_{k=1}^{\infty} \frac{r^{2k}}{k} \right)^{1/2} \left(\sum_{k=1}^{\infty} k |b_k|^2 \right)^{1/2},$$

and (2.4) follows from (2.5).

For $n \geq 2$ we set

$$L_{1/n}^* := \left\{ z \in G : |\Phi(z)| = 1 - \frac{1}{n} \right\},$$

and denote by $\Phi_{1/n}$ the conformal mapping of ext $L_{1/n}^*$ to Δ with the standard normalization at ∞. By Lemma B.2.2,

$$|\Phi_{1/n}(\zeta)| - 1 \preceq 1/n, \qquad \zeta \in L.$$

338   8. Special Topics

Therefore, according to the Bernstein–Walsh lemma (cf. Lemma 1.1.11)
$$\|p_n\|_L \preceq \|\tilde{p}_n\|_{L^*_{1/n}}. \tag{2.6}$$
Comparing (2.4) and (2.6) with the fact that for $z \in L^*_{1/n}$,
$$\log \frac{1}{1-|\varphi(z)|} = \log \frac{1}{1-|(\varphi \circ \Phi^{-1})(\Phi(z))|} \preceq \log n$$
(cf. Theorem 1.2.7), we get (2.3). □

Next, we apply the Cauchy–Green formula (B.1.14) to write
$$f(z) = \frac{1}{2\pi i} \int_{\partial U} \frac{f(\zeta)}{\zeta - z} d\zeta - \frac{1}{\pi} \int_{U \setminus \overline{G}} \frac{f_{\bar{\zeta}}(\zeta)}{\zeta - z} dm(\zeta), \quad z \in G,$$
where $U (\supset \overline{G})$ is a fixed disk of sufficiently large radius. Using this integral representation, we can construct a sequence of polynomials that converges to f in the norm of the space $L^2_1(G)$, which makes it possible to assert the validity of the following fact, in view of the extremal properties of Bieberbach polynomials.

Lemma 2.2. *For any* $n \in \mathbb{N}$,
$$\|f - \pi_n\|_{L^2_1(G)} \leq c_1 n^{-2\gamma} \tag{2.7}$$
with some positive constants c_1 *and* γ *independent of* n.

Proof. For $z \in G$ we define
$$U_n := U \setminus (\overline{G} \cup \Omega_{n^{-1/2}}),$$
$$I_1(z) := -\frac{1}{\pi} \int_{\Omega_{n^{-1/2}}} \frac{f_{\bar{\zeta}}(\zeta)}{(\zeta - z)^2} dm(\zeta),$$
$$I_2(z) := \frac{1}{2\pi i} \int_{\partial U} \frac{f(\zeta)}{(\zeta - z)^2} d\zeta - \frac{1}{\pi} \int_{U_n} \frac{f_{\bar{\zeta}}(\zeta)}{(\zeta - z)^2} dm(\zeta).$$

Thus, $f'(z) = I_1(z) + I_2(z)$, $z \in G$. By well-known properties of singular integral operators (see [112, Chapter III, §7]) and (2.2),
$$\int_G |I_1(z)|^2 dm(z) \preceq \int_{\Omega_{n^{-1/2}}} |f_{\bar{\zeta}}(\zeta)|^2 dm(\zeta) \preceq n^{-\delta/2}. \tag{2.8}$$

Moreover,
$$|I_1(z_0)| \preceq \int_{\Omega_{n^{-1/2}}} |f_{\bar{\zeta}}| dm \preceq \left(\int_{\Omega_{n^{-1/2}}} |f_{\bar{\zeta}}|^2 dm \right)^{1/2} \preceq n^{-\delta/4}. \tag{2.9}$$

8.2 Uniform Convergence of Bieberbach Polynomials

In order to approximate $I_2(z)$ we first construct the following auxiliary function. Let $m := [n/2]$, $\zeta \in \overline{U}_n$, $z \in G$, and let Φ_m be the appropriate Faber polynomial (see Section 1.3). The function

$$B_{n-1}(\zeta, z) := \frac{(\Phi_m(\zeta) - \Phi_m(z))^2}{(\zeta - z)^2} \frac{1}{\Phi(\zeta)^{2m}}$$

is a polynomial in z of degree at most $n - 1$. Next,

$$B_{n-1}(\zeta, z) - \frac{1}{(\zeta - z)^2}$$
$$= \frac{(\Phi_m(\zeta) - \Phi_m(z) + \Phi(\zeta)^m)(\Phi_m(\zeta) - \Phi(\zeta)^m - \Phi_m(z))}{(\zeta - z)^2 \Phi(\zeta)^{2m}}.$$

Since

$$\Phi_m(\zeta) = \Phi(\zeta)^m + h_m(\zeta),$$

where

$$|h_m(\zeta)| \leq 1 + \sqrt{e} + \sqrt{m \log(m+1)}$$

(see [165, p. 136]), and in addition,

$$|\Phi_m(z)| \preceq n, \qquad z \in \overline{G}$$

(cf. Theorem 1.3.2), we have (taking also into account Lemma 1.2.5)

$$\left| B_{n-1}(\zeta, z) - \frac{1}{(\zeta - z)^2} \right| \preceq e^{-c_1 \sqrt{n}}. \tag{2.10}$$

Further, for $z \in G$ set

$$P'_n(z) := \frac{1}{2\pi i} \int_{\partial U} f(\zeta) B_{n-1}(\zeta, z) \, d\zeta - \frac{1}{\pi} \int_{U_n} f_{\bar{\zeta}}(\zeta) B_{n-1}(\zeta, z) \, dm(\zeta),$$

$$P_n(z) := \int_{z_0}^{z} P'_n(\zeta) \, d\zeta.$$

Taking into account (2.10), we obtain

$$|I_2(z) - P'_n(z)| \preceq e^{-c_1 \sqrt{n}}, \qquad z \in G. \tag{2.11}$$

Consequently, with $p_n(z) := P_n(z) + (1 - P'_n(z_0))(z - z_0)$ we obtain according to (2.8), (2.9), and (2.11),

$$\|f - p_n\|_{L^2_1(G)}^2 \preceq \int_G |I_1|^2 \, dm + \int_G |I_2 - P'_n|^2 \, dm + |1 - P'_n(z_0)|^2 \preceq n^{-\delta/2}.$$

To complete the proof of Lemma 2.2 it remains to make use of the extremal property of Bieberbach polynomials. □

Now we are in position to prove (2.1). Representing

$$f(z) = \pi_n(z) + \sum_{k=0}^{\infty} (\pi_{2^{k+1}n}(z) - \pi_{2^k n}(z)), \qquad z \in G,$$

we have by Lemma 2.2

$$\|\pi_{2^{k+1}n} - \pi_{2^k n}\|_{L_1^2(G)} \preceq (2^k n)^{-2\gamma}$$

for each term in the sum. Now, Lemma 2.1 yields

$$\|\pi_{2^{k+1}n} - \pi_{2^k n}\|_{\overline{G}} \preceq (2^k n)^{-2\gamma}\sqrt{k + \log n} \preceq (2^k n)^{-\gamma}.$$

Summing over k, we finally get (2.1).

8.3 Historical Comments

The results of Section 8.1 are presented in accordance with [16]. The question, does (1.1) remain true when the interval $[-1, 1]$ is replaced by a parabolic region, is due to Grothmann and Saff [80]. The examples of continua with the VP-property can be found in Anderson and Fuchs [8] and Saff and Totik [154]. There is a close connection between the results of this section and the paper of D. Gaier [66].

Various sufficient conditions on the geometry of the domain G in order to guarantee the validity of the relation

$$\lim_{n \to \infty} \|f - \pi_n\|_{\overline{G}} = 0$$

can be found in Keldysh [93], Mergelyan [123], Wu Xue-Mou [179], Suetin [167], and Simonenko [162]. In our proof of (2.1) we follow [11]. We would also like to attract the reader's attention to a series of recent papers by D. Gaier [63, 64, 65] and Andrievskii and Gaier [22].

A
Conformally Invariant Characteristics of Curve Families

This appendix is devoted to some rather well known facts and results from the geometric theory of functions of a complex variable. The presentation of the material has such a form that in most cases the results are accompanied by proofs or references to standard textbooks and monographs.

A.1 Module and Extremal Length of a Curve Family

A.1.1

The notions of module and extremal length of a curve family play a significant role in geometric function theory. We give only the information on the conformal invariants that are necessary for our purposes. For readers interested in a deeper study we recommend the monographs by Ahlfors [3], [4], Ohtsuka [135], Jenkins [90].

An arc or curve γ is called *locally rectifiable* if every compact subarc of γ is rectifiable (for the concept of rectifiability, see [112, p. 122]).

A set Γ the elements of which are arcs and curves will be referred to as a *curve family*.

A nonnegative Borel measurable function (see [112, pp. 111–112]) $\rho(z)$ defined on $\mathbb{C}$ is called an *admissible function*, or *metric* if the ρ-area $A(\rho)$

of $\mathbb{C}$ satisfies
$$A(\rho) := \int_{\mathbb{C}} \rho^2 dm < \infty,$$

where dm is the two-dimensional Lebesgue measure (area) on $\mathbb{C}$.

Let Γ be a curve family, $\mathcal{P}$ the class of all metrics,

$$L_\rho(\Gamma) := \inf_{\gamma \in \Gamma^*} \int_\gamma \rho \, |dz|,$$

where $\Gamma^* \subset \Gamma$ is the subfamily of all locally rectifiable $\gamma \in \Gamma$.

Then the quantity

$$m(\Gamma) := \inf_{\rho \in \mathcal{P}} \frac{A(\rho)}{L_\rho(\Gamma)^2} \tag{1.1}$$

is called the *module* of the family Γ and the quantity $\lambda(\Gamma) := 1/m(\Gamma)$ is the *extremal length* of the family Γ.

Note that the definition of $L_\rho(\Gamma)$ does not take into account arcs and curves that are not locally rectifiable, so that these do not affect the module of Γ.

Let $E \subset \mathbb{C}$ be a Borel set such that all arcs and curves $\gamma \in \Gamma$ lie in E. Then, clearly, the infimum in (1.1) does not change if instead of all admissible functions ρ we consider only those that vanish outside E.

Apart from the basic definition of $m(\Gamma)$ the so-called L-definition and A-definition of a module are useful (see [90]). Let $\mathcal{P}_L(\Gamma)$ be the subclass of metrics ρ from $\mathcal{P}$ such that for all locally rectifiable $\gamma \in \Gamma$,

$$\int_\gamma \rho \, |dz| \geq 1.$$

The metrics $\rho \in \mathcal{P}_L(\Gamma)$ are called *admissible in the L-definition* of $m(\Gamma)$.

Setting

$$m(\Gamma) := \begin{cases} \inf_{\rho \in \mathcal{P}_L(\Gamma)} A(\rho), & \text{if } \mathcal{P}_L(\Gamma) \neq \emptyset, \\ \infty, & \text{if } \mathcal{P}_L(\Gamma) = \emptyset, \end{cases}$$

we get the *L-definition* of a module.

If $\mathcal{P}_A \subset \mathcal{P}$ denotes the subclass of all metrics ρ with $A(\rho) \leq 1$ and if we set

$$m(\Gamma) = \inf_{\rho \in \mathcal{P}_A} L_\rho(\Gamma)^{-2},$$

then we have the *A-definition* of a module.

The two definitions given above are equivalent to the original definition due to the fact that replacing a metric $\rho(z)$ by a metric $K\rho(z)$, where $K > 0$ is an arbitrary constant, does not change the quantity $A(\rho)L_\rho(\Gamma)^{-2}$.

A metric $\rho^*(z) \in \mathcal{P}_L(\Gamma)$ is called *extremal* if $m(\Gamma) = A(\rho^*)$.

A.1.2

The module of a curve family is an invariant under conformal mappings. Clearly, the extremal length will also be an invariant.

Theorem 1.1. *If a curve family Γ is situated in a domain G and $\varphi(z)$ is a conformal mapping of G onto some domain G^*, $\Gamma^* = \varphi(\Gamma)$, then*
$$m(\Gamma) = m(\Gamma^*).$$

Proof. Without loss of generality we may assume that all curves of Γ are locally rectifiable. Assume that $z = \psi(w)$ is the inverse mapping to $w = \varphi(z)$. Between the admissible functions in the z-plane and the admissible functions in the w-plane there is a one-to-one correspondence that is established by the following relations:
$$\rho \to \rho^* := (\rho \circ \psi)\,|\psi'|, \qquad \rho^* \to \rho := (\rho^* \circ \varphi)\,|\varphi'|,$$
where $\rho \in \mathcal{P}_L(\Gamma)$ and $\rho^* \in \mathcal{P}_L(\Gamma^*)$ vanish in the exterior of G and G^*, respectively. Moreover, if $\gamma^* = \varphi(\gamma)$, $\gamma \in \Gamma$, then
$$\int_\gamma \rho(z)|dz| = \int_{\gamma^*} \rho^*(w)|dw|, \qquad A(\rho) = A(\rho^*).$$
Consequently,
$$m(\Gamma) = \inf_{\rho \in \mathcal{P}_L(\Gamma)} A(\rho) = \inf_{\rho^* \in \mathcal{P}_L(\Gamma^*)} A(\rho^*) = m(\Gamma^*).$$
□

A.1.3

Let Γ_1 and Γ_2 be two curve families in $\overline{\mathbb{C}}$. Then $\Gamma_1 \cup \Gamma_2$ denotes the curve family obtained by taking the usual union of the sets Γ_1 and Γ_2, and the relation $\Gamma_1 < \Gamma_2$ indicates that each curve $\gamma_2 \in \Gamma_2$ contains a curve $\gamma_1 \in \Gamma_1$. The next statement is known as the *comparison principle*:

Theorem 1.2. *If $\Gamma_1 < \Gamma_2$, then*
$$m(\Gamma_1) \geq m(\Gamma_2). \tag{1.2}$$
In particular, $\Gamma_1 \subset \Gamma_2$ implies $m(\Gamma_1) \leq m(\Gamma_2)$.

Proof. Since every curve $\gamma_2 \in \Gamma_2$ contains some curve $\gamma_1 \in \Gamma_1$, we have $\mathcal{P}_L(\Gamma_1) \subset \mathcal{P}_L(\Gamma_2)$. According to the L-definition of the module,
$$m(\Gamma_1) = \inf_{\rho \in \mathcal{P}_L(\Gamma_1)} A(\rho) \geq \inf_{\rho \in \mathcal{P}_L(\Gamma_2)} A(\rho) = m(\Gamma_2).$$
□

344   A. Conformally Invariant Characteristics of Curve Families

Theorem 1.3. *For any finite number of families Γ_j, $j = 1, \ldots, n$, the following inequality is satisfied:*

$$m\left(\bigcup_{j=1}^n \Gamma_j\right) \leq \sum_{j=1}^n m(\Gamma_j). \tag{1.3}$$

Proof. The only nontrivial case is that for which all $m(\Gamma_j)$ are finite. Let $\rho_j \in \mathcal{P}_L(\Gamma_j)$, $j = 1, \ldots, n$. Then the metric $\rho(z) := \sup_j \rho_j(z)$ will be admissible in the L-definition of $m(\Gamma)$, where $\Gamma = \bigcup_j \Gamma_j$. Taking in the inequality

$$m(\Gamma) \leq \int_C \rho^2 dm \leq \int_C \sum_{j=1}^n \rho_j^2 dm = \sum_{j=1}^n A(\rho_j)$$

the infimum with respect to each ρ_j, we have

$$m(\Gamma) = m\left(\bigcup_{j=1}^n \Gamma_j\right) \leq \sum_{j=1}^n m(\Gamma_j).$$

□

The next statement presents the so-called *composition laws*:

Theorem 1.4. *Let the families Γ_1 and Γ_2 be situated in disjoint Borel sets E_1 and E_2, respectively, and let Γ be a third family.*

(i) *If $\Gamma > \Gamma_1$ and $\Gamma > \Gamma_2$, then $\lambda(\Gamma) \geq \lambda(\Gamma_1) + \lambda(\Gamma_2)$.*

(ii) *If $\Gamma_1 > \Gamma$ and $\Gamma_2 > \Gamma$, then $m(\Gamma) \geq m(\Gamma_1) + m(\Gamma_2)$.*

(iii) $m(\Gamma_1 \cup \Gamma_2) = m(\Gamma_1) + m(\Gamma_2)$.

Proof. (i). Without loss of generality we can assume that $0 < m(\Gamma_j) < \infty$, $j = 1, 2$. Otherwise, (i) is a simple corollary of Theorem 1.2.

Here we shall use the general definition of a module. Multiplication of $\rho(z)$ by a positive constant does not change the quantity $A(\rho)L_\rho(\Gamma)^{-2}$. Therefore, in the definition of $m(\Gamma_j)$, $j = 1, 2$, it suffices to choose only metrics ρ_j with the properties $L_{\rho_j}(\Gamma_j) = A(\rho_j)$, $j = 1, 2$. The choice of the metric $\rho^*(z) := \max\{\rho_1(z), \rho_2(z)\}$ yields

$$L_{\rho^*}(\Gamma) \geq L_{\rho_1}(\Gamma_1) + L_{\rho_2}(\Gamma_2) = A(\rho_1) + A(\rho_2),$$
$$A(\rho^*) \leq A(\rho_1) + A(\rho_2),$$
$$\lambda(\Gamma) = \sup_\rho \frac{L_\rho^2(\Gamma)}{A(\rho)} \geq \frac{(A(\rho_1) + A(\rho_2))^2}{A(\rho_1) + A(\rho_2)}$$
$$= L_{\rho_1}(\Gamma_1)^2 A(\rho_1)^{-1} + L_{\rho_2}(\Gamma_2)^2 A(\rho_2)^{-1}.$$

This inequality implies (i).

(ii). We can assume that $m(\Gamma) < \infty$. Let the metric $\rho(z)$ be admissible in the L-definition of a module $m(\Gamma)$. Then the metrics

$$\rho_j(z) := \begin{cases} \rho(z), & z \in E_j, \\ 0, & z \notin E_j, \end{cases} \quad j = 1, 2,$$

will be admissible in the L-definition of a module $m(\Gamma_j)$, $j = 1, 2$. Consequently,

$$m(\Gamma_1) + m(\Gamma_2) \leq A(\rho_1) + A(\rho_2) \leq A(\rho),$$

whence because of the arbitrariness of $\rho(z)$, we have (ii).

(iii) follows from (ii) and (1.3). $\square$

A.1.4

The next two statements are devoted to the solution of the uniqueness problem for the extremal metric and its characterization.

Theorem 1.5. *Assume that $m(\Gamma) \neq \infty$ and that $\rho_1(z)$, $\rho_2(z)$ are extremal metrics in the L-definition of $m(\Gamma)$. Then $\rho_1(z) = \rho_2(z)$ almost everywhere in $\mathbb{C}$.*

Proof. Clearly, $\rho_1(z)$ and $\rho_2(z)$ are finite almost everywhere, and the metric $\rho^*(z) := (\rho_1(z) + \rho_2(z))/2$ is admissible in the L-definition of $m(\Gamma)$. Consequently,

$$4\,m(\Gamma) \leq \int_\mathbb{C} (\rho_1 + \rho_2)^2 \, dm \leq \int_\mathbb{C} (\rho_1 + \rho_2)^2 \, dm + \int_\mathbb{C} (\rho_1 - \rho_2)^2 \, dm = 4\,m(\Gamma).$$

Thus, actually, we have only equalities. Therefore, $\rho_1(z) = \rho_2(z)$, almost everywhere. $\square$

If the extremal metric can be found, the problem of evaluating the module of a curve family can be solved completely. *Beurling's criterion* is the most effective tool for searching for such a metric (see, for example, [4, p. 61]):

Theorem 1.6. *The metric ρ_0 will be extremal for the family Γ in the domain G if Γ contains a subfamily Γ_0 with the following properties:*

(i) *ρ_0 is admissible in the L-definition of $m(\Gamma)$ and*

$$\int_\gamma \rho_0 \,|dz| = 1, \quad \gamma \in \Gamma_0.$$

(ii) *For an arbitrary real-valued Borel measurable function $h(z)$ the condition*

$$\int_\gamma h\,|dz| \geq 0, \quad \gamma \in \Gamma_0, \tag{1.4}$$

implies the satisfaction of the inequality

$$\int_G h\rho_0\,dm \geq 0. \tag{1.5}$$

Proof. Suppose $\rho \in \mathcal{P}_L(\Gamma)$. Then, for any $\gamma \in \Gamma_0$,

$$\int_\gamma \rho\,|dz| \geq 1 = \int_\gamma \rho_0\,|dz|,$$

and the inequality (1.4) is valid for the function $h = \rho - \rho_0$. Consequently, the relation

$$\int_G (\rho\rho_0 - \rho_0^2)\,dm \geq 0$$

follows from (1.5). Thus, after application of the Cauchy inequality, we obtain

$$\int_G \rho_0^2\,dm \leq \int_G \rho^2\,dm.$$

Therefore, $\rho_0(z)$ is extremal. □

Beurling's criterion allows us to evaluate the module of a curve family in many cases. The following three examples will be a visual illustration for its efficiency.

Example 1.7. Let us consider the annular sector $G = \{z : r_1 < |z| < r_2,\ 0 < \arg z < \theta\pi\}$, $0 < r_1 < r_2 < \infty$, $0 < \theta \leq 2$. Suppose Γ is the family of arcs joining in G the circular parts of the boundary ∂G. Then

$$m(\Gamma) = \frac{\theta\pi}{\log(r_2/r_1)},$$

and the metric

$$\rho_0(z) := \begin{cases} \left(|z|\log\dfrac{r_2}{r_1}\right)^{-1}, & z \in G, \\ 0, & \text{otherwise,} \end{cases}$$

will be extremal.

A.1 Module and Extremal Length of a Curve Family   347

Indeed, consider
$$\Gamma_0 := \{\gamma_t := \{re^{it} : r_1 < r < r_2\} : 0 < t < \theta\pi\}.$$
Then for each $\gamma \in \Gamma_0$,
$$\int_\gamma \rho_0 \, |dz| = \frac{1}{\log(r_2/r_1)} \int_{r_1}^{r_2} \frac{dr}{r} = 1.$$
On the other hand, for any $\gamma \in \Gamma$,
$$\int_\gamma \rho_0 \, |dz| \geq \frac{1}{\log(r_2/r_1)} \left| \int_\gamma \frac{dz}{z} \right| \geq 1.$$

That is, condition(i) of Theorem 1.6 is fulfilled.

Let us consider an arbitrary function $h(z)$ subject to the condition (1.4). Then
$$\int_G h\, \rho_0 \, dm = \frac{1}{\log(r_2/r_1)} \int_0^{\theta\pi} \int_{r_1}^{r_2} h(re^{it}) \, drdt \geq 0,$$
and the condition (ii) of Theorem 1.6 is also satisfied. Hence, $\rho_0(z)$ is extremal and
$$m(\Gamma) = m(\Gamma_0) = \frac{1}{(\log(r_2/r_1))^2} \int_0^{\theta\pi} \int_{r_1}^{r_2} \frac{drdt}{r} = \frac{\theta\pi}{\log(r_2/r_1)}. \qquad (1.6)$$

Example 1.8. Let G be as in Example 1.7 and let Γ denote the family of arcs joining in G the radial parts of ∂G. Then
$$m(\Gamma) = \frac{1}{\theta\pi} \log \frac{r_2}{r_1}, \qquad (1.7)$$
and the metric
$$\rho_0(z) := \begin{cases} (\theta\pi|z|)^{-1}, & z \in G, \\ 0, & \text{otherwise}, \end{cases}$$
will be extremal.

Example 1.9. Let Γ be the family of curves in the annulus $G = \{z : r_1 < |z| < r_2\}$, $0 < r_1 < r_2 < \infty$, that separate its circular boundary components. Then
$$m(\Gamma) = \frac{1}{2\pi} \log \frac{r_2}{r_1}, \qquad (1.8)$$

348   A. Conformally Invariant Characteristics of Curve Families

and the metric

$$\rho_0(z) := \begin{cases} (2\pi|z|)^{-1}, & z \in G, \\ 0, & \text{otherwise}, \end{cases}$$

will be extremal.

The proof for Examples 1.8 and 1.9 is similar to the proof of Example 1.7.

A.1.5

The calculation of the exact value of the module of a curve family is a rather difficult problem. Its solution is known for a small number of cases only. Some of them were considered in the previous subsection. But upper and lower estimates of a module can always be obtained with the help of standard metric fitting methods and by transition to subfamilies. Below, we give some examples of such kind. All of them, combined with Theorem 1.1, are used in some places of this monograph.

Let $G \subset \overline{\mathbb{C}}$ be a simply connected domain, and let $K_1, K_2 \subset \overline{G}$ be disjoint compact sets (with respect to $\overline{\mathbb{C}}$). We say that an arc or curve $\Gamma \subset G$ *separates* K_1 from K_2 (in G) if γ has nonempty intersection with each arc $l \subset G$ joining K_1 to K_2.

Example 1.10. Let $\Gamma = \Gamma(z)$, $z \in \mathbb{D}$, $z \neq 0$, denote the family of all crosscuts of the unit disk $\mathbb{D}$ that separate 0 from z in $\mathbb{D}$. Then

$$\frac{1}{\pi} \log \frac{1}{1-|z|} \leq m(\Gamma) \leq 2 + \frac{1}{\pi} \log \frac{2}{1-|z|}. \qquad (1.9)$$

Indeed, we may assume $0 < z < 1$. The left-hand side of (1.9) follows from the comparison principle (cf. Theorem 1.2) for Γ and the family Γ_1 consisting of all arcs joining in the annular sector

$$\left\{ \zeta = 1 + re^{i\theta} : 1 - z < r < 1, \; \frac{\pi}{2} < \theta < \frac{3}{2}\pi \right\}$$

its radial boundary parts. That is, by (1.2) and (1.7),

$$m(\Gamma) \geq m(\Gamma_1) = \frac{1}{\pi} \log \frac{1}{1-z}.$$

For the proof of the right-hand inequality we consider the Möbius transformation

$$t = h(\zeta) := \frac{1+\zeta}{1-\zeta}. \qquad (1.10)$$

A.1 Module and Extremal Length of a Curve Family

Since the modulus is a conformal invariant (see Theorem 1.1), we have

$$m(\Gamma) = m(h(\Gamma)).$$

Using $\tau := h(z) > 1$ we define

$$\rho(t) := \begin{cases} (\pi |t|)^{-1}, & \text{if } e^{-\pi} \leq |t| \leq \tau e^{\pi}, \\ 0, & \text{elsewhere.} \end{cases}$$

Some calculation yields

$$\int_\gamma \rho(t) \, |dt| \geq 1, \qquad \gamma \in h(\Gamma). \tag{1.11}$$

In fact, for arbitrary $\gamma \in h(\Gamma)$ we set

$$u_{\max} := \sup_{t \in \gamma} |t|, \qquad u_{\min} := \inf_{t \in \gamma} |t|.$$

If $u_{\min} < e^{-\pi}$, then there exists an arc $\gamma_1 \subset \gamma$ contained in the annulus $\{t : e^{-\pi} < |t| < 1\}$ that joins the two boundary components of the annulus. Therefore,

$$\int_\gamma \rho(t) \, |dt| \geq \int_{\gamma_1} \rho(t) \, |dt| \geq \frac{1}{\pi} \left| \int_{\gamma_1} \frac{dt}{t} \right| \geq 1.$$

A similar argument works if $u_{\max} > \tau e^{\pi}$. Hence, what remains is to consider the case $e^{-\pi} \leq u_{\min} \leq u_{\max} \leq \tau e^{\pi}$. Then

$$\int_\gamma \rho(t) \, |dt| \geq \frac{1}{\pi} \left| \int_\gamma \frac{dt}{t} \right| \geq 1.$$

This completes the proof of (1.11).

Hence,

$$m(\Gamma) = m(h(\Gamma)) \leq \int \rho^2(t) \, dm(t)$$

$$= \frac{1}{\pi} \log \tau + 2 \leq \frac{1}{\pi} \log \frac{2}{1-z} + 2,$$

which is the desired estimate.

Example 1.11. Let J be a subarc of $\mathbb{T}$, and let $\Gamma = \Gamma(J)$ be the family of all crosscuts of $\mathbb{D}$ that separate 0 from J in $\mathbb{D}$. Then

$$\frac{1}{\pi} \log \frac{2}{|J|} \leq m(\Gamma) \leq 2 + \frac{1}{\pi} \log \frac{4}{|J|}. \tag{1.12}$$

In fact, without loss of generality we may assume that $1 \in J$ and that J is symmetric with respect to the real axis.

First, suppose $|J| \geq \pi$. The left-hand side of (1.12) then follows from the nonnegativity of the module. To prove the right-hand-side inequality we note that for the function

$$\rho_1(z) := \begin{cases} 1, & \text{if } z \in \overline{\mathbb{D}}, \\ 0, & \text{elsewhere}, \end{cases}$$

we have

$$\int_\gamma \rho_1(z)\,|dz| \geq 2, \qquad \gamma \in \Gamma.$$

Therefore,

$$m(\Gamma) \leq \frac{\pi}{4} \leq \frac{1}{\pi} \log \frac{4}{|J|} + \frac{\pi}{4} + \frac{1}{\pi} \log \frac{\pi}{2}$$
$$< \frac{1}{\pi} \log \frac{4}{|J|} + 2.$$

If $|J| < \pi$, then the proof of (1.12) is completely the same as the proof of (1.9). That is, for the verification of the left-hand side of (1.12) we have to compare Γ with the family Γ_1 consisting of all arcs that join in the annular sector

$$\left\{ \zeta = 1 + re^{i\theta} : 2\sin\frac{|J|}{4} < r < 1,\ \frac{\pi}{2} < \theta < \frac{3\pi}{2} \right\}$$

its radial boundary parts. For the proof of the right-hand side of (1.12) we need to carry out the auxiliary Möbius transformation (1.10) and consider the metric

$$\rho(t) := \begin{cases} (\pi|t|)^{-1}, & \text{if } e^{-\pi} \leq |t| \leq \dfrac{\sin(|J|/2)}{1 - \cos(|J|/2)}\, e^\pi, \\ 0, & \text{elsewhere}. \end{cases}$$

This leads to the inequality

$$m(\Gamma) \leq 2 + \frac{1}{\pi} \log \frac{\sin(|J|/2)}{1 - \cos(|J|/2)} \leq 2 + \frac{1}{\pi} \log \frac{4}{|J|}.$$

Example 1.12. Let $E := [-1, 1]$, and let $w_1, w_2 \in \Omega := \overline{\mathbb{C}} \setminus [-1, 1]$ satisfy the inequality $|w_1 - 1| < R := \min(|w_2 - 1|, 2)$. Denote by Γ the family of all crosscuts of Ω separating (in Ω) the points 1 and w_1 from w_2 and ∞. Then

$$\frac{1}{2\pi} \log \frac{R}{|w_1 - 1|} \leq m(\Gamma) \leq 2 + \frac{1}{2\pi} \log \frac{R}{|w_1 - 1|}. \tag{1.13}$$

Indeed, let Γ_1 be the family of curves separating in the annulus
$$\{z : |z-1| < r < R\}$$
its boundary components. Applying Theorem 1.2 as well as (1.8) we find that
$$m(\Gamma) \geq m(\Gamma_1) = \frac{1}{2\pi} \log \frac{R}{|w_1 - 1|}.$$
To prove the right-hand side of (1.13), we set
$$\rho(w) := \begin{cases} (2\pi|w-1|)^{-1}, & \text{if } e^{-2\pi}|w_1 - 1| \leq |w-1| \leq e^{2\pi} R, \\ 0, & \text{otherwise.} \end{cases}$$
A simple calculation similar to the proof of (1.11) shows that
$$\int_\gamma \rho(w)\,|dw| \geq 1, \qquad \gamma \in \Gamma.$$
Consequently,
$$m(\Gamma) \leq \int_{\mathbb{C}} \rho^2(w)\,dm(w) = 2 + \frac{1}{2\pi} \log \frac{R}{|w_1 - 1|}.$$

Example 1.13. Let $z_1, z_2 \in \Delta := \{z : |z| > 1\} \cup \{\infty\}$ satisfy the condition $z_1 = |z_2| = 1 + \delta$, $0 < \delta \leq 1$, and let Γ denote the family of all crosscuts of Δ separating (in Δ) z_1 from z_2 and ∞. Then
$$\frac{1}{\pi} \log \frac{2|z_2 - 1|}{3\delta} \leq m(\Gamma) \leq 2 + \frac{1}{\pi} \log \frac{|z_2 - 1|}{\delta}. \tag{1.14}$$

In fact, we consider the mapping $w = w(z) := \frac{1}{2}(z + 1/z)$ of the domain Δ onto the exterior of the segment $[-1, 1]$. Let $w_j := w(z_j)$, $j = 1, 2$. Then
$$|w_1 - 1| = w_1 - 1 = \frac{\delta^2}{2(1+\delta)} < |w_2 - 1| = \frac{|z_2 - 1|^2}{2(1+\delta)} \leq \frac{9}{2}.$$
Combining these inequalities with (1.13), we get (1.14):
$$\frac{1}{\pi} \log \frac{2|z_2 - 1|}{3\delta} = \frac{1}{2\pi} \log \frac{4}{9} \left| \frac{w_2 - 1}{w_1 - 1} \right| \leq m(\Gamma)$$
$$\leq 2 + \frac{1}{2\pi} \log \left| \frac{w_2 - 1}{w_1 - 1} \right| = 2 + \frac{1}{\pi} \log \frac{|z_2 - 1|}{\delta}.$$

Example 1.14. Let $z_1, z_2 \in \Delta$ satisfy $z_1 = |z_2| = 1 + \delta$, $0 < \delta \leq 1$, and let Γ denote the family of all crosscuts of Δ separating (in Δ) the point z_1 from z_2. Then
$$\frac{2}{\pi} \log \frac{|z_2 - 1|}{\delta} - 3 \leq m(\Gamma) \leq 4 + \frac{2}{\pi} \log \frac{|z_2 - 1|}{\delta}. \tag{1.15}$$

Indeed, clearly, $\Gamma = \Gamma_1 \cup \Gamma_2$, where Γ_j, $j=1,2$, denotes the family of all crosscuts of Δ separating (in Δ) z_j from $\{z_1, z_2\} \setminus \{z_j\}$ and ∞. The right-hand side of (1.15) follows immediately from Theorem 1.3 and (1.14).

In order to prove the left-hand side of (1.15) we write $z_2 = (1+\delta)e^{it}$ and assume that $\delta < \frac{1}{4}$, $3\pi\delta < t \leq \pi$ (because in the other case the left-hand side of (1.15) follows from the nonnegativity of a module). Next, we carry out the auxiliary Möbius transformation

$$w = h(z) := \frac{z - e^{it/2}}{z + e^{it/2}},$$

and note that for $w_j = h(z_j)$ we have

$$\overline{w}_2 = w_1 = \frac{1}{|1 + \delta + e^{it/2}|^2} \left(\delta(2+\delta) - 2i(1+\delta)\sin\frac{t}{2} \right) =: u_1 - iv_1.$$

Our assumption yields $v_1 > u_1$. Denote by Γ_j^*, $j = 1, 2$, the family of all arcs joining in the annular sector

$$G_j^* := \{w : \operatorname{Re} w > 0, \ |w - i\operatorname{Im} w_j| < v_1\}$$

the radial parts of its boundary.

Since $\Gamma_1^* \cup \Gamma_2^* \subset h(\Gamma)$, we obtain by virtue of Theorems 1.1, 1.2, 1.4 and (1.7)

$$m(\Gamma) = m(h(\Gamma)) \geq m(\Gamma_1^*) + m(\Gamma_2^*)$$
$$= \frac{2}{\pi} \log \frac{v_1}{u_1} = \frac{2}{\pi} \log \left(\frac{2(1+\delta)}{\delta(2+\delta)} \sin\frac{t}{2} \right)$$
$$\geq \frac{2}{\pi} \log \frac{2t}{3\pi\delta} \geq \frac{2}{\pi} \log \frac{|z_2 - 1|}{\delta} - 3.$$

$\square$

A.1.6

We establish two special results that we will need to state some relations describing the distortion properties of conformal mappings of a simply connected domain onto Δ.

Lemma 1.15. *Let γ_1 and γ_2 be two crosscuts of Δ such that $\gamma_1 \cap \gamma_2 = \emptyset$ and γ_2 separates γ_1 from ∞. Then the module of the family Γ of all crosscuts of Δ separating γ_1 and γ_2 satisfies*

$$m(\Gamma) \leq \frac{1}{\pi} \log \frac{\operatorname{diam} \gamma_2}{\operatorname{diam} \gamma_1} + 2\pi.$$

A.1 Module and Extremal Length of a Curve Family

Proof. Without loss of generality we assume that one of the endpoints of γ_1 coincides with 1. Set $\rho_j := \operatorname{diam} \gamma_j$, $j = 1, 2$, and consider two possibilities.

(i) Let $\rho_1 \geq 2$. Since in this case

$$|\gamma| \geq 2, \qquad \gamma \in \Gamma,$$

the metric

$$\rho(z) := \begin{cases} \frac{1}{2}, & \text{if } 1 \leq |z| \leq 3, \\ 0, & \text{otherwise,} \end{cases}$$

belongs to $\mathcal{P}_L(\Gamma)$. Hence, by the L-definition of a module,

$$m(\Gamma) \leq \int_{\mathbb{C}} \rho^2 \, dm = 2\pi \leq \frac{1}{\pi} \log \frac{\rho_2}{\rho_1} + 2\pi.$$

(ii) Let $\rho_1 < 2$. It is a simple geometrical observation that there exist points $z_1 \in \gamma_1$ and $z_2 \in \gamma_2$ such that

$$|1 - z_j| \leq \rho_j \leq 2|1 - z_j|, \qquad j = 1, 2.$$

Denote by Γ_1 the family of all crosscuts of Δ that separate in Δ the points 1 and z_1 from z_2 and ∞. By the comparison principle (cf. Theorem 1.2),

$$m(\Gamma) \leq m(\Gamma_1).$$

Next, applying the simple transformation $J(z) = \frac{1}{2}(z + 1/z)$, Theorem 1.1, and (1.13), we obtain

$$m(\Gamma_1) = m(J(\Gamma_1)) \leq 2 + \frac{1}{2\pi} \log \left| \frac{J(z_2) - 1}{J(z_1) - 1} \right|$$
$$= 2 + \frac{1}{2\pi} \log \left| \frac{z_2 - 1}{z_1 - 1} \right|^2 \left| \frac{z_1}{z_2} \right| \leq 2\pi + \frac{1}{\pi} \log \frac{\rho_2}{\rho_1}.$$

$\square$

The sharpness of the result above can be seen from the following assertion.

Lemma 1.16. *Under the assumptions of Lemma 1.15 suppose that z_0, $|z_0| = 1$, is a point such that γ_1 separates z_0 from ∞ and the inequalities*

$$c_1 d_j \leq |z - z_0| \leq c_2 d_j, \qquad z \in \gamma_j, \quad j = 1, 2,$$

hold with some $0 < d_1 < d_2 < 2$ and constants $c_2 \geq c_1 > 0$. Then

$$m(\Gamma) \geq \frac{1}{\pi} \log \frac{d_2}{d_1} - c_3, \quad c_3 = c_3(c_1, c_2) > 0. \tag{1.16}$$

354   A. Conformally Invariant Characteristics of Curve Families

Proof. Set
$$\tilde{d}_1 := \sup_{z \in \gamma_1} |z - z_0|, \quad \tilde{d}_2 := \inf_{z \in \gamma_2} |z - z_0|.$$

For the beginning, let $\tilde{d}_2 \leq 2\tilde{d}_1$. For any $\gamma \in \Gamma$ we have $|\gamma| > \tilde{d}_1$. Setting $\rho(z) := \tilde{d}_1^{-1}$ inside the bounded component of $\Delta \setminus \gamma_2$ and 0 otherwise, we obtain $\rho \in \mathcal{P}_L(\Gamma)$ and
$$m(\Gamma) \leq \int_{\mathbb{C}} \rho^2 dm \leq \frac{\pi(c_2 \tilde{d}_2)^2}{\tilde{d}_1^2} \leq 4\pi c_2^2.$$

Therefore, under the above assumptions (1.16) follows from the negativity of a module with sufficiently large c_3.

Now let $2\tilde{d}_1 < \tilde{d}_2 (\leq 2)$. We assume that $z_0 = 1$ and introduce the auxiliary transformation $J(z) := (z + 1/z)/2$, $z \in \Delta$. Next, we compare the family $J(\Gamma)$ with the family Γ_1 of all curves that separate in the annulus
$$\left\{ \frac{\tilde{d}_1^2}{2} < |w - 1| < \frac{\tilde{d}_2^2}{4} \right\}$$

its boundary components. By Theorems 1.1 and 1.2 as well as (1.8),
$$m(\Gamma) = m(J(\Gamma)) \geq m(\Gamma_1) = \frac{1}{2\pi} \log \frac{\tilde{d}_2^2}{2\tilde{d}_1^2} \geq \frac{1}{\pi} \log \frac{c_1 d_2}{\sqrt{2} c_2 d_1}.$$

□

A.1.7

Let Ω be an arbitrary simply connected domain in $\overline{\mathbb{C}}$ whose complement $E := \mathbb{C} \setminus \Omega$ is a continuum. Let $L := \partial\Omega = \partial E$ be the boundary of Ω. We consider the function $w = \Phi(z)$ mapping Ω conformally and univalently onto Δ and normalized by the conditions $\Phi(\infty) = \infty$, $\lim_{z \to \infty} \Phi(z)/z > 0$. Denote by $\widetilde{\Omega}$ the compactification of the domain Ω by prime ends in the sense of Carathéodory (see [146, pp. 271–278]), i.e., elements of $\widetilde{\Omega}$ are prime ends and points of Ω. We denote the elements of $\widetilde{\Omega}$ by capital letters, points of the plane by small letters, except for the evident cases when a point $Z \in \widetilde{\Omega}$ is identified with an inner point of the domain Ω or can be uniquely identified with a boundary point of the domain (for example, in the case of a Jordan boundary). By the Carathéodory theorem (see [146, p. 272]) Φ can be extended to a homeomorphism between $\widetilde{\Omega}$ and $\overline{\Delta}$, for which we preserve the same notation Φ.

Let $Z, \mathfrak{Z} \in \widetilde{\Omega}$. We say that a crosscut γ of Ω or a curve $\gamma \subset \Omega$ separates $Z \in \widetilde{\Omega}$ and $\mathfrak{Z} \in \widetilde{\Omega}$ from ∞ if it divides Ω into two domains, one of which contains ∞ and for the other domain Z and $\mathfrak{Z}$ are adjacent. The term *adjacent* means for $Z \in \widetilde{\Omega} \setminus \Omega$ that in the domain and subdomain the prime ends can be defined by the same null chains of crosscuts. For a point $Z \in \Omega$ "is adjacent" is equivalent to "belongs to." Typical examples of separating crosscuts are shown in Figure 1.1.

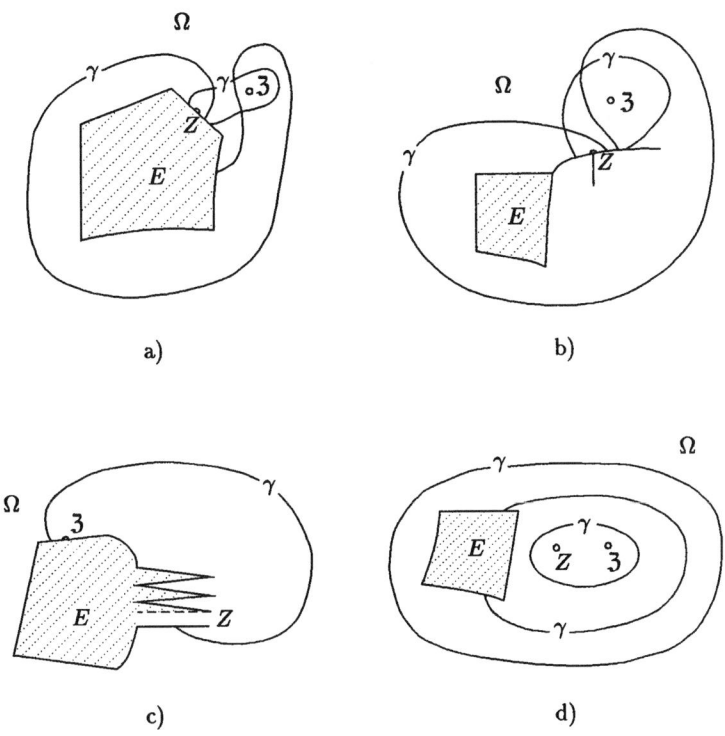

Fig. 1.1

Next we prove a theorem that is essentially due to Belyi [25] (see also [17, p. 34]).

Theorem 1.17. *Let E, Ω, and Φ be as above and let $D_0 \supset E$ be a fixed open disk. Let $Z \in \widetilde{\Omega} \setminus \Omega$, $\mathfrak{Z} \in (\widetilde{\Omega} \setminus \Omega) \cup (D_0 \cap \Omega)$. Then*

$$2e^{-\pi m(\Gamma)} \leq |\Phi(Z) - \Phi(\mathfrak{Z})| \leq c_1 e^{-\pi m(\Gamma)}, \qquad (1.17)$$

where Γ is the family of all crosscuts of Ω that separate Z and $\mathfrak{Z}$ from ∞, and $c_1 > 0$ is independent of Z and $\mathfrak{Z}$.

Proof. Without loss of generality we assume that $\Phi(Z) = 1$. Next we introduce the auxiliary mapping

$$J(\tau) := \frac{1}{2}\left(\tau + \frac{1}{\tau}\right)$$

and the following notation: $w := \Phi(\mathfrak{z}), t := J(w), U := \Phi(\partial D_0), V := J(U)$, $\Gamma_1 := \Phi(\Gamma), \Gamma_2 := J(\Gamma_1)$,

$$r_0 := \max_{\tau \in U} |\tau|, \quad R_0 := \max_{\xi \in V} |1 - \xi|.$$

It is obvious that $r_0 > 0$ and $R_0 > 0$ are independent of Z and $\mathfrak{z}$.

By conformal invariance of a module (cf. Theorem 1.1) we have

$$m(\Gamma) = m(\Gamma_1) = m(\Gamma_2). \tag{1.18}$$

Our next aim is to derive estimates for $m(\Gamma_2)$.

If $|1 - t| \leq 2$, we can use (1.13) (with $w_1 = t, w_2 = \infty$) to write

$$\frac{1}{2\pi} \log \frac{2}{|1-t|} \leq m(\Gamma_2) \leq 2 + \frac{1}{2\pi} \log \frac{2}{|1-t|}. \tag{1.19}$$

If $2 < |1-t| < R_0$, then the left-hand side of (1.19) remains valid because of the nonnegativity of a module. Apart from this, the length of any $\gamma \in \Gamma_2$ is at least 2. Therefore, considering (in the L-definition of $m(\Gamma_2)$) the metric

$$\rho(\xi) := \begin{cases} \max\left\{\frac{1}{2}, \operatorname{dist}\left([-1,1], V\right)^{-1}\right\}, & \text{if } z \in \operatorname{int} V, \\ 0, & \text{otherwise}, \end{cases}$$

we obtain

$$m(\Gamma_2) \leq c_2 \leq \frac{1}{2\pi} \log \frac{R_0}{|1-t|} + c_2 \tag{1.20}$$

with some $c_2 > 0$ independent of Z and $\mathfrak{z}$.

Since

$$|1 - t| = \frac{|1-w|^2}{2|w|},$$

we have, according to (1.18)–(1.20),

$$\frac{1}{\pi} \log \frac{2}{|1-w|} \leq \frac{1}{2\pi} \log \frac{4|w|}{|1-w|^2} \leq m(\Gamma) \leq \frac{1}{2\pi} \log \frac{2R_0|w|}{|1-w|^2} + c_2 + 2$$

$$\leq \frac{1}{\pi} \log \frac{\sqrt{2R_0 r_0}}{|1-w|} + c_2 + 2,$$

which corresponds to (1.17). □

A.2 Reduced Module

A.2.1

In analogy with the reduced module of a simply connected domain (see [6]) we introduce the notion of a reduced module for families of separating curves and arcs.

Let G be a simply connected domain, $\infty \in G$, and let $K \subset G$ be a compact set. Denote by $\Gamma_\infty(K, G)$ the family of curves and crosscuts of G separating K from ∞ in G, i.e., $\gamma \in \Gamma_\infty(K, G)$, if it has nonempty intersection with each arc joining in G the compact set K to ∞. Let $\{D_r\}$ be the family of open disks containing K with center at the origin and of radius $r > 0$. If

$$\Gamma_r(K, G) := \{\gamma \in \Gamma_\infty(K, G) : \gamma \subset D_r\},$$

then the quantity

$$\mu_\infty(K, G) := \lim_{r \to \infty} \left(m(\Gamma_r(K, G)) - \frac{1}{2\pi} \log r \right) \qquad (2.1)$$

is called the *reduced module* of the family $\Gamma_\infty(K, G)$.

The expression $m(\Gamma_r(K, G)) - (\log r)/(2\pi)$ (as a function of r) increases monotonically, and thus the limit in (2.1) exists. In fact, let $r' > r$ and let $\Gamma(r, r')$ be the family of all $\gamma \in \Gamma_\infty(K, G)$ separating the boundary components of the annulus $\{\zeta : r < |\zeta| < r'\}$. In accordance with the main properties of a module (see Theorems 1.2, 1.4 and (1.8)) we have

$$m(\Gamma_{r'}(K, G)) \geq m(\Gamma_r(K, G)) + m(\Gamma(r, r')) = m(\Gamma_r(K, G)) + \frac{1}{2\pi} \log \frac{r'}{r},$$

that is,

$$m(\Gamma_{r'}(K, G)) - \frac{1}{2\pi} \log r' \geq m(\Gamma_r(K, G)) - \frac{1}{2\pi} \log r.$$

A reduced module is not a conformal invariant. However, its behavior under conformal mappings can be described in a simple way.

Theorem 2.1. *Let G_j, $j = 1, 2$, be a simply connected domain, $\infty \in G_j$, $K_j \subset G_j$ a compact set, and $f : G_1 \to G_2$ a conformal mapping normalized by the condition $f(\infty) = \infty$. Then*

$$\mu_\infty(K_1, G_1) = \mu_\infty(K_2, G_2) + \frac{1}{2\pi} \log |f'(\infty)|, \qquad (2.2)$$

where

$$f'(\infty) := \lim_{z \to \infty} \frac{f(z)}{z}.$$

Proof. Using the invariance of a module (cf. Theorem 1.1), for $r > 0$ such that $K \subset D_r$ we obtain

$$m(\Gamma_r) - \frac{1}{2\pi}\log r = m(f(\Gamma_r)) - \frac{1}{2\pi}\log r,$$

where $\Gamma_r := \Gamma_r(K_1, G_1)$.

For sufficiently large r Theorem 1.2 yields

$$m(\Gamma'_{\rho_-}) \leq m(f(\Gamma_r)) \leq m(\Gamma'_{\rho_+}),$$

where $\Gamma'_\rho := \Gamma_\rho(K_2, G_2)$,

$$\rho_- := \min_{|z|=r} |f(z)|, \quad \rho_+ := \max_{|z|=r} |f(z)|.$$

The conformality of $f(z)$ at ∞ implies that

$$\lim_{r \to \infty} \frac{\rho_\pm}{r} = |f'(\infty)|.$$

Thus,

$$\mu_\infty(K_2, G_2) = \lim_{r \to \infty}\left(m(\Gamma'_{\rho_\pm}) - \frac{1}{2\pi}\log \rho_\pm\right)$$

$$= \lim_{r \to \infty}\left(m(\Gamma'_{\rho_\pm}) - \frac{1}{2\pi}\log r\right) - \frac{1}{2\pi}\log|f'(\infty)|$$

$$= \mu_\infty(K_1, G_1) - \frac{1}{2\pi}\log|f'(\infty)|.$$

□

Corollary 2.2. *In (2.1) an arbitrary point $a \in \mathbb{C}$ can be chosen as the center of the disk system $\{D_r\}$.*

Indeed, the change of the center is equivalent to the calculation of the reduced module in the domain that is obtained by the translation $w = z - a$.

Corollary 2.3. *Let $E \subset \mathbb{C}$ be a continuum with connected complement Ω, and in Theorem 2.1 let $G_1 = \Omega$, $G_2 = \Delta$, $K_1 = K$, and $f = \Phi : \Omega \to \Delta$ the conformal mapping with standard normalization at ∞. Then (2.2) takes the form*

$$\mu_\infty(K, \Omega) = \mu_\infty(\Phi(K), \Delta) - \frac{1}{2\pi}\log \operatorname{cap} E. \qquad (2.3)$$

A.2.2

Next, we discuss the following extremal problem.

A.2 Reduced Module

Lemma 2.4. *Let l be a closed arc with endpoints z_1, z_2, and let $[z_1, z_2]$ be the line segment with these endpoints. Then*

$$\mu_\infty(l, \overline{\mathbb{C}}) \leq \mu_\infty([z_1, z_2], \overline{\mathbb{C}}) = \frac{1}{2\pi} \log \frac{4}{|z_1 - z_2|}. \tag{2.4}$$

Proof. Without loss of generality we assume that $z_1 = 0$, $z_2 = \delta > 0$ (cf. Theorem 2.1). Let r be sufficiently large. By Grötzsch's module theorem (see [112, p. 54]),

$$m(\Gamma_r(l, \overline{\mathbb{C}})) \leq m(\Gamma_r([z_1, z_2], \overline{\mathbb{C}})) = \frac{1}{2\pi} \mu\left(\frac{\delta}{r}\right),$$

where $\mu(\rho)$, $0 < \rho < 1$, is the module of the so-called Grötzsch ring domain $D \setminus [0, \rho]$ (see [112, p. 53]). Moreover, the behavior of $\mu(\rho)$ for $\rho \to 0$ is known (see [112, p. 62]):

$$\lim_{\rho \to 0} \left(\mu(\rho) - \log \frac{4}{\rho}\right) = 0.$$

Comparing the last two relations with the definition of a reduced module we get (2.4). □

Corollary 2.5. *Under the assumptions of Corollary 2.3 set $L := \partial E$,*

$$L_u := \{z : |\Phi(z)| = 1 + u\}, \quad u > 0.$$

Then

$$\operatorname{dist}(L, L_u) \geq \frac{u^2}{1 + u} \operatorname{cap} E. \tag{2.5}$$

Besides, the constant $c := \operatorname{cap} E$ is best possible.

Proof. The distance $\operatorname{dist}(L, L_u)$ is realized on some segment $[z_1, z_2]$, $z_1 \in L$, $z_2 \in L_u$. Taking into account (2.3), (2.4), and the obvious fact that the reduced module $\mu_\infty(K, G)$ decreases if G increases, we obtain

$$\frac{1}{2\pi} \log \frac{4}{|z_1 - z_2|} = \mu_\infty([z_1, z_2], \overline{\mathbb{C}}) \leq \mu_\infty([z_1, z_2], \Omega)$$
$$= \mu_\infty(\Phi([z_1, z_2]), \Delta) + \frac{1}{2\pi} \log \frac{1}{c}. \tag{2.6}$$

Let $Z_1 \in \widetilde{\Omega} \setminus \Omega$ be a prime end that is determined by a sequence of points belonging to the interval $[z_1, z_2]$. Set $\Phi(Z_1) =: w_1 = e^{it}$. Consider the auxiliary conformal mapping $J_1 : \Delta \to \overline{\mathbb{C}} \setminus [-1, 1]$ given by the formula

$$\tau = J_1(w) := \frac{1}{2}\left(\frac{w}{e^{it}} + \frac{e^{it}}{w}\right),$$

and introduce $\tau_2 := J_1 \circ \Phi(z_2)$.

By Corollary 2.3 and Lemma 2.4 we have

$$\mu_\infty(\Phi([z_1, z_2]), \Delta) = \mu_\infty(J_1 \circ \Phi([z_1, z_2]), \overline{\mathbb{C}} \setminus [-1, 1]) - \frac{1}{2\pi} \log 2$$

$$\leq \mu_\infty([1, \tau_2], \overline{\mathbb{C}}) - \frac{1}{2\pi} \log 2 \leq \frac{1}{2\pi} \log \frac{4(1+u)}{u^2}.$$

Comparing the last inequality with (2.6) we obtain

$$|z_1 - z_2| \geq \frac{cu^2}{1+u}.$$

Equality in (2.5) occurs, for example, if $E = [-1, 1]$. $\square$

B
Basics in the Theory of Quasiconformal Mappings

Quasiconformal mappings, which play an important role in this monograph, are not the subject of our investigations. They are only a tool for understanding more deeply the different effects that appear in analyzing a conformal mapping of simply connected domains or geometrical properties of such sets as quasicircles and quasiconformal arcs (note that we can introduce this notation merely by means of geometrical terms, without using a quasiconformal mapping). For the same reason we give only the most relevant definitions and facts that are necessary for reading this book. The reader can find more detailed concepts from the theory of quasiconformal mapping in the excellent monographs of Ahlfors [3, 4], Lehto–Virtanen [112], and Lehto [111].

B.1 Quasiconformal Mappings

B.1.1

In the calculus of functions of a complex variable the following formal differential operators (*complex derivatives*) are often used:

$$\partial = \frac{\partial}{\partial z} := \frac{1}{2}\left(\frac{\partial}{\partial x} - i\frac{\partial}{\partial y}\right), \quad \bar{\partial} = \frac{\partial}{\partial \bar{z}} := \frac{1}{2}\left(\frac{\partial}{\partial x} + i\frac{\partial}{\partial y}\right).$$

They are also called the *formal derivatives* with respect to $z = x + iy$ and $\bar{z} = x - iy$. It is also common to use the notation $f_z, f_{\bar{z}}$ for the formal derivatives of f.

362   B. Basics in the Theory of Quasiconformal Mappings

Let f be a homeomorphism of a domain $G \subset \mathbb{C}$ onto a domain $G' \subset \mathbb{C}$ that is *sense-preserving*, i.e., f preserves the orientation of the boundary of every Jordan domain U such that $\overline{U} \subset G$ (see [112, p. 6]). First, we suppose that the partial derivatives f_x and f_y of f are continuous in G, and that its *Jacobian* satisfies

$$J_f := |f_z|^2 - |f_{\bar{z}}|^2 > 0 \quad (\text{in } G).$$

Let $\partial_\alpha f(z)$, $0 \leq \alpha < 2\pi$, denote the *directional derivative* (see [112, pp. 17, 49]):

$$\partial_\alpha f(z) = \lim_{r \to +0} \frac{f(z + re^{i\alpha}) - f(z)}{re^{i\alpha}} = f_z(z) + f_{\bar{z}}(z) e^{-2i\alpha}.$$

Then

$$\max_\alpha |\partial_\alpha f(z)| = |\partial f(z)| + |\overline{\partial} f(z)|,$$
$$\min_\alpha |\partial_\alpha f(z)| = |\partial f(z)| - |\overline{\partial} f(z)|.$$

The *dilatation quotient*

$$D_f := \frac{\max_\alpha |\partial_\alpha f|}{\min_\alpha |\partial_\alpha f|} = \frac{|f_z| + |f_{\bar{z}}|}{|f_z| - |f_{\bar{z}}|}$$

is bounded in every compact subset of G.

If $D_f(z) \leq K$ for all $z \in G$ and some finite $K \geq 1$, we say that f is a *regular K-quasiconformal mapping*. A mapping $f : G \to G'$ is called *regular quasiconformal* if it is regular K-quasiconformal with some K, $1 \leq K < \infty$.

Regular quasiconformal mappings were introduced by Grötzsch [81].

It is evident that a regular 1-quasiconformal mapping is conformal and conversely.

B.1.2

A class of regular K-quasiconformal mappings does not necessarily have the compactness property: Let $\{f_n\}$ be a sequence of regular K-quasiconformal mappings converging locally uniformly in G. It is not necessarily true that the limit function is constant or also K-quasiconformal, even if an appropriate normalization is chosen. Besides, the condition of continuous differentiability is very restrictive. If we use the geometrical definition, we can consider a class of quasiconformal mappings that will be essentially wider. To introduce this concept we need the notions of a quadrilateral and its conformal module.

A configuration that consists of a Jordan domain $Q \subset \overline{\mathbb{C}}$ with four fixed successive points z_1, z_2, z_3, z_4 located on its boundary ∂Q and directed positively with respect to Q is called a *quadrilateral* and denoted by

$Q(z_1, z_2, z_3, z_4)$. The subarcs (z_1, z_2), (z_2, z_3), (z_3, z_4), and (z_4, z_1) of ∂Q (with the corresponding endpoints) are called the sides of the quadrilateral; in addition, the arcs (z_1, z_2) and (z_3, z_4) are called the *a-sides*. Note that quadrilaterals $Q(z_1, z_2, z_3, z_4)$ and $Q(z_2, z_3, z_4, z_1)$ that are situated in the same domain Q are different because their a-sides are different.

The quantity

$$M(Q) = M\left(Q(z_1, z_2, z_3, z_4)\right) := m(\Gamma_a)$$

is called the (conformal) *module* of $Q(z_1, z_2, z_3, z_4)$. Here, Γ_a is the family of all crosscuts of Q that join the a-sides, and $m(\Gamma_a)$ is the module of this family.

Let $w = \varphi(z)$ be a conformal mapping of a Jordan domain Q onto a Jordan domain Q', $w_j = \varphi(z_j)$, $j = 1, \ldots, 4$. Then the quadrilateral $Q'(w_1, w_2, w_3, w_4)$ is called *conformally equivalent* to $Q(z_1, z_2, z_3, z_4)$. By virtue of Theorem A.1.1,

$$m(Q') = m(Q). \tag{1.1}$$

It is easy to verify (see [112, p. 15]) that every quadrilateral $Q(z_1, z_2, z_3, z_4)$ is conformally equivalent to some rectangle $R = R(0, a, a+ib, ib)$, where $R := \{w : 0 < \operatorname{Re} w < a,\, 0 < \operatorname{Im} w < b\}$. Using Beurling's criterion (cf. Theorem A.1.6), we can easily check that the metric

$$\rho_0(w) := \begin{cases} \frac{1}{b}, & \text{if } w \in R, \\ 0, & \text{if } w \notin R, \end{cases}$$

is extremal in the L-definition of $m(\varphi(\Gamma_a))$, and consequently, (1.1) implies

$$M(Q) = M(R) = \frac{a}{b}. \tag{1.2}$$

The same mapping φ is also suited for the quadrilateral $Q(z_2, z_3, z_4, z_1)$, whence, in view of (1.2), we get

$$M\left(Q(z_1, z_2, z_3, z_4)\right) M\left(Q(z_2, z_3, z_4, z_1)\right) = \frac{a}{b} \cdot \frac{b}{a} = 1. \tag{1.3}$$

The quadrilateral $Q^* := Q(z_2, z_3, z_4, z_1)$ is said to be *conjugate* to $Q = Q(z_1, z_2, z_3, z_4)$.

By the definition of the module,

$$M(Q) \leq \frac{A(\rho)}{L_\rho(\Gamma_a)^2} \tag{1.4}$$

for any metric ρ. An important application is obtained if we choose ρ to be the Euclidean metric. Let

$$s_a(Q) := \inf_{\gamma \in \Gamma_a} |\gamma|,$$

and let $m(Q)$ be the area of Q. Then (1.4) with $\rho = 1$ in Q and 0 outside Q gives *Rengel's inequality*

$$M(Q(z_1, z_2, z_3, z_4)) \leq \frac{m(Q)^2}{s_a(Q)}. \tag{1.5}$$

B.1.3

Suppose that f is a sense-preserving homeomorphism of a domain $G \subset \overline{\mathbb{C}}$ onto a domain $G' \subset \overline{\mathbb{C}}$. We consider all kinds of quadrilaterals $Q = Q(z_1, z_2, z_3, z_4)$ such that their closures $\overline{Q}$ lie in G. Let

$$Q' := f(Q), \quad Q' = Q'\left(f(z_1), f(z_2), f(z_3), f(z_4)\right).$$

The quantity

$$K(f, G) := \sup_Q \frac{M(Q')}{M(Q)}$$

is called the *maximal dilatation* of the mapping f in the domain G. Since together with $\overline{Q}$ the conjugated quadrilateral $\overline{Q^*}$ is situated in G and because of (1.3), the maximal dilatation is always at least 1.

We define a *quasiconformal mapping* to be a sense-preserving homeomorphism f whose maximal dilatation $K(f, G)$ is finite. If its maximal dilatation is bounded by a number K, $1 \leq K < \infty$, then f is called a *K-quasiconformal mapping*.

Such a general form of the definition of quasiconformality was first offered by Pfluger [138] and Ahlfors [2] and is called the *geometric definition* of a quasiconformal mapping.

A regular K-quasiconformal mapping is K-quasiconformal with the same K (see [112, p. 18]). The reverse statement is not correct. We can confirm only that for a K-quasiconformal mapping f that is differentiable (i.e. the real and imaginary parts of f are differentiable) at the point $z_0 \in G$ the inequality

$$\max_\alpha |\partial_\alpha f(z_0)| \leq K \min_\alpha |\partial_\alpha f(z_0)| \tag{1.6}$$

holds (see [112, p. 50]). However, a K-quasiconformal mapping is differentiable almost everywhere in G (see [112, p. 164]).

B.1.4

A real-valued continuous function $u(z)$, $z = x + iy$, is said to be *absolutely continuous on lines* (ACL) in a domain G if for every closed rectangle $\{x + iy : a \leq x \leq b;\ c \leq y \leq d\} \subset G$ the function $u(z)$ is absolutely continuous on $[a, b]$ as a function of x for almost all $y \in [c, d]$ and absolutely continuous on $[c, d]$ as a function of y for almost all $x \in [a, b]$.

A complex-valued function f is ACL in G if its real and imaginary parts are ACL in G.

Now we give an *analytic definition* of a quasiconformal mapping: A sense-preserving homeomorphism f of the domain G is said to be a K-*quasiconformal mapping*, $1 \leq K < \infty$, if

(i) f is ACL in G,

(ii) the condition (1.6) holds almost everywhere in G.

The analytic and geometric definitions of quasiconformal mappings are equivalent (see [112, Chapter IV]).

B.1.5

We note the following properties of quasiconformal mappings.

1°. A 1-quasiconformal mapping is conformal and conversely (see [112, p. 28]).

2°. The composition of a K_1-quasiconformal mapping and a K_2-quasiconformal mapping is a $K_1 K_2$-quasiconformal mapping.

3°. The inverse mapping f^{-1} of a K-quasiconformal mapping f is K-quasiconformal.

4°. If a mapping f is K-quasiconformal in a neighborhood of each point $z \in G$, then it is K-quasiconformal in G.

5°. Väisälä's theorem (cf. [112, p. 171]) states that the module of a curve family is a quasi-invariant of quasiconformal mappings: If a curve family Γ is situated in a domain G and $f : G \to G'$ is K-quasiconformal, then

$$K^{-1} m(\Gamma) \leq m(f(\Gamma)) \leq K\, m(\Gamma). \tag{1.7}$$

6°. The Jacobian $J = J_f$ of a quasiconformal mapping f of a domain G satisfies

$$J = |f_z|^2 - |f_{\bar{z}}|^2$$

almost everywhere in G (see [112, p. 130]) and is locally integrable with respect to the two-dimensional Lebesgue measure. Moreover, for every Borel set $E \subset G$ the following formula is valid (see [112, p. 131]):

$$\int_E J\, dm = m(f(E)). \tag{1.8}$$

Hence, taking into account the evident inequality

$$|f_{\bar z}|^2 \le |f_z|^2 \le \frac{J}{1-k^2}, \quad k = \frac{K-1}{K+1},$$

which holds almost everywhere in G, we can ascertain that f_z and $f_{\bar z}$ are locally square-integrable. A more complicated procedure is needed to check that f_z and $f_{\bar z}$ belong locally to L^p, where $p = p(K) > 2$ (see [112, p. 215]).

The last fact yields that a quasiconformal mapping is absolutely continuous with respect to the two-dimensional Lebesgue measure. Moreover, for any compact subset F of $G \setminus (\{\infty\} \cup \{f^{-1}(\infty)\})$ and $\delta < 1 - 2/p(K)$ there exists a constant c such that

$$m(f(E)) \le c(m(E))^\delta \tag{1.9}$$

for every Borel set $E \subset F$ (see [112, p. 216]).

B.1.6

Assume that $\zeta_1, \zeta_2, \zeta_3$ are points in $\mathbb{C}$, and that $\Gamma = \Gamma(\zeta_1, \zeta_2, \zeta_3, \mathbb{C})$ is the family of curves in $\mathbb{C}$ separating ζ_1 and ζ_2 from ζ_3 and ∞. Let

$$m(\zeta_1, \zeta_2, \zeta_3, \mathbb{C}) := m(\Gamma)$$

and consider a K-quasiconformal mapping $F : \overline{\mathbb{C}} \to \overline{\mathbb{C}}$ with the normalization $F(\infty) = \infty$. Then by (1.7) for any $\zeta_1, \zeta_2, \zeta_3 \in \mathbb{C}$, the inequality

$$K^{-1} \le \frac{m(\zeta_1, \zeta_2, \zeta_3, \mathbb{C})}{m(F(\zeta_1), F(\zeta_2), F(\zeta_3), \mathbb{C})} \le K \tag{1.10}$$

holds.

Proof of Theorem 1.2.7. (i). Suppose

$$|\zeta_1 - \zeta_2| \le c_1 |\zeta_1 - \zeta_3|.$$

If $|w_1 - w_2| > |w_1 - w_3|$, then $\Gamma(w_1, w_3, w_2, \mathbb{C})$ contains the family of curves separating the boundary components of the annulus

$$\{w : |w_1 - w_3| < |w_1 - w| < |w_1 - w_2|\},$$

and by Theorem A.1.2 and (A.1.8),

$$m(w_1, w_3, w_2, \mathbb{C}) \ge \frac{1}{2\pi} \log \left| \frac{w_1 - w_2}{w_1 - w_3} \right|.$$

On the other hand,

$$m(\zeta_1, \zeta_3, \zeta_2, \mathbb{C}) \le c_5, \quad c_5 = c_5(c_1) > 0. \tag{1.11}$$

In fact, it can be easily checked (cf. the proof of (A.1.11)) that the function

$$\rho(\zeta) := \begin{cases} (2\pi|\zeta_1 - \zeta|)^{-1}, & \frac{|\zeta_1 - \zeta_3|}{e^{2\pi}} \leq |\zeta_1 - \zeta| \leq (c_1 + 1)e^{2\pi}|\zeta_1 - \zeta_3|, \\ 0, & \text{otherwise,} \end{cases}$$

is admissible in the L-definition of the module of $\Gamma(\zeta_1, \zeta_3, \zeta_2, \mathbb{C})$. Therefore,

$$m(\zeta_1, \zeta_3, \zeta_2, \mathbb{C}) \leq \int_\mathbb{C} \rho^2 dm = \frac{1}{2\pi} \log(c_1 + 1) + 2 =: c_5.$$

By virtue of (1.10) we obtain

$$c_5 \geq m(\zeta_1, \zeta_3, \zeta_2, \mathbb{C}) \geq K^{-1} m(w_1, w_3, w_2, \mathbb{C}) \geq \frac{1}{2\pi K} \log \left| \frac{w_1 - w_2}{w_1 - w_3} \right|,$$

whence the estimate $|w_1 - w_2| \leq c_6|w_1 - w_3|$, $c_6 = c_6(c_5, K)$, is satisfied. Thus, in any case,

$$|w_1 - w_2| \leq c_2|w_1 - w_3|, \quad c_2 := 1 + c_6.$$

Replacing the points $\zeta_1, \zeta_2, \zeta_3$ by w_1, w_2, w_3 we conclude that the condition $|w_1 - w_2| \leq c_2|w_1 - w_3|$ implies the estimate $|\zeta_1 - \zeta_2| \leq c_1|\zeta_1 - \zeta_3|$, $c_1 = c_1(c_2, K)$.

(ii). First of all, we note that any points ζ_1, ζ_2 and ζ_3 satisfying the condition $|\zeta_1 - \zeta_2| \leq c_1|\zeta_1 - \zeta_3|$ also satisfy the inequality

$$0 \leq m(\zeta_1, \zeta_2, \zeta_3, \mathbb{C}) - \frac{1}{2\pi} \log \left| \frac{\zeta_1 - \zeta_3}{\zeta_1 - \zeta_2} \right| \leq c_7, \quad c_7 = c_7(c_1) > 0. \quad (1.12)$$

Indeed, if $|\zeta_1 - \zeta_3| \leq |\zeta_1 - \zeta_2| \leq c_1|\zeta_1 - \zeta_3|$, then inequality (1.12) follows from the nonnegativity of the module of a curve family and the corresponding analogue of the estimate (1.11). If $|\zeta_1 - \zeta_2| < |\zeta_1 - \zeta_3|$, then the left-hand part of inequality (1.12) follows by virtue of Theorem A.1.2 and (A.1.8) from the fact that $\Gamma(\zeta_1, \zeta_2, \zeta_3, \mathbb{C})$ contains the family of curves separating the boundary components of the annulus $\{\zeta : |\zeta_1 - \zeta_2| < |\zeta_1 - \zeta| < |\zeta_1 - \zeta_3|\}$. To establish the validity of the right-hand part of the inequality, we construct the metric

$$\rho(\zeta) := \begin{cases} (2\pi|\zeta - \zeta_1|)^{-1}, & e^{-2\pi}|\zeta_1 - \zeta_2| \leq |\zeta_1 - \zeta| \leq e^{2\pi}|\zeta_1 - \zeta_3|, \\ 0, & \text{otherwise ,} \end{cases}$$

which is admissible in the L-definition of the module of the curve family $\Gamma(\zeta_1, \zeta_2, \zeta_3, \mathbb{C})$ (cf. the proof of (A.1.11)). Thus,

$$m(\zeta_1, \zeta_2, \zeta_3, \mathbb{C}) \leq \int_\mathbb{C} \rho^2 dm = \frac{1}{2\pi} \int_{e^{-2\pi}|\zeta_1 - \zeta_2|}^{e^{2\pi}|\zeta_1 - \zeta_3|} \frac{dr}{r} = 2 + \frac{1}{2\pi} \log \left| \frac{\zeta_1 - \zeta_3}{\zeta_1 - \zeta_2} \right|.$$

Applying inequality (1.12) to the points w_1, w_2 and w_3, we obtain

$$0 \leq m(w_1, w_2; w_3; \mathbb{C}) - \frac{1}{2\pi} \log \left| \frac{w_1 - w_3}{w_1 - w_2} \right| \leq c_8, \quad c_8 = c_8(c_2) > 0. \quad (1.13)$$

Hence the inequality (1.2.21) is a simple combination of the estimates (1.10), (1.12), and (1.13). □

From Theorem 1.2.7 immediately follows the so-called $\mathcal{D}$-*property of quasiconformal mappings* (see [24]). This property was first established by Lavrentiev [109] for a mapping that is near to conformal.

Corollary 1.1. *Let $F(\zeta)$ be as in Theorem 1.2.7. Then an arbitrary circle $C(\zeta_0, r) := \{\zeta : |\zeta - \zeta_0| = r\}$, $\zeta_0 \in \mathbb{C}$, $r > 0$, is transformed by the mapping F so that*

$$\frac{\max_{\zeta \in C(\zeta_0, r)} |F(\zeta) - F(\zeta_0)|}{\min_{\zeta \in C(\zeta_0, r)} |F(\zeta) - F(\zeta_0)|} \leq \mathcal{D},$$

where the constant $\mathcal{D}$ depends only on the coefficient of quasiconformality of the mapping F.

B.1.7

It is very important for the analysis of quasiconformal mappings that Green's formula and the generalized Cauchy integral representation are valid. We give a suitable formulation of the results we are interested in (more detailed information can be obtained from the monographs [112], [111]).

Let f be defined in a domain $G \subset \mathbb{C}$. We say that f has L^p-*derivatives*, $p \geq 1$, in G if:

(i) f is ACL in G;

(ii) the partial derivatives f_x and f_y are locally L^p-integrable in G.

The class of functions f having L^p-derivatives is commonly referred to as the *Sobolev space* $W^1_{p,\text{loc}}$. All quasiconformal mappings have L^p-derivatives with some $p > 2$. This follows from the analytic definition and property 6° in Section B.1.4.

Theorem 1.2. *Suppose $f(z)$ has an L^1-derivative in the domain $G \subset \mathbb{C}$. If $U \subset \overline{U} \subset G$ is a domain bounded by a rectifiable curve, then the Green formula*

$$\int_{\partial U} f(\zeta) \, d\zeta = 2i \int_U f_{\overline{\zeta}}(\zeta) \, dm(\zeta)$$

holds.

For a proof, see [112, p. 150].

Corollary 1.3. *Under the conditions of Theorem 1.2, for each $z \in U$, the following integral representation is valid (see [112, p. 155]):*

$$f(z) = \frac{1}{2\pi i} \int_{\partial U} \frac{f(\zeta)}{\zeta - z} d\zeta - \frac{1}{\pi} \int_U \frac{f_{\bar{\zeta}}(\zeta)}{\zeta - z} dm(\zeta), \qquad (1.14)$$

where the two-dimensional Lebesgue integral in (1.14) is evaluated in the sense of Cauchy's principal value.

B.2 Quasiconformal Curves and Arcs

B.2.1

A class of Jordan curves in the extended complex plane $\overline{\mathbb{C}}$ can be considered as a class of images of a circle under a homeomorphism of the Riemann sphere onto itself. If, in addition to this, we require that the homeomorphism be quasiconformal, we obtain a new class of curves.

The image of the unit circle under a K-quasiconformal mapping of $\overline{\mathbb{C}}$ onto $\overline{\mathbb{C}}$ is called a *K-quasiconformal curve* or a *K-quasicircle*. A curve that for some $K \geq 1$ is a K-quasicircle is said to be a *quasiconformal curve* or a *quasicircle*.

A domain $G \subset \overline{\mathbb{C}}$ is called a *quasiconformal domain* or a *quasidisk* if it is bounded by a quasicircle.

The two-dimensional Lebesgue measure of a quasicircle is equal to zero. This immediately follows from the definition and (1.9). However, it is not necessary for a bounded quasicircle to be rectifiable. We refer the reader to the monograph [112, p. 104] for corresponding examples. A more interesting example was constructed by Belinskii (see [24, p. 42]). His quasiconformal curve having the "almost-smoothness" property is not even locally rectifiable.

The possibility to define a quasicircle with the help of geometrical terminology is a truly remarkable circumstance. In the statement presented below some such definitions are collected. The proof of the equivalence of these definitions can be found in the monographs of Ahlfors as well as those of Lehto and Virtanen that we have already cited. Beforehand, the reader will be acquainted with three definitions. The relation between these definitions and the notion of quasiconformality will be presented in Theorem 2.1 below.

A set $E \subset \overline{\mathbb{C}}$ is called *linearly locally connected* if there is a constant $c \geq 1$ such that for any $z \in \mathbb{C}$, $r > 0$, and the corresponding disk $D(z, r) = \{\zeta : |\zeta - z| < r\}$, the following conditions are satisfied:

(i) Any two points of the set $E \cap \overline{D(z, r)}$ can be joined by an arc in $E \cap \overline{D(z, cr)}$.

(ii) Any two points of the set $E \setminus D(z,r)$ can be joined by an arc in $E \setminus D(z, r/c)$.

The two pictures in Figure 2.1 illustrate the situation where (i) and (ii) come into play.

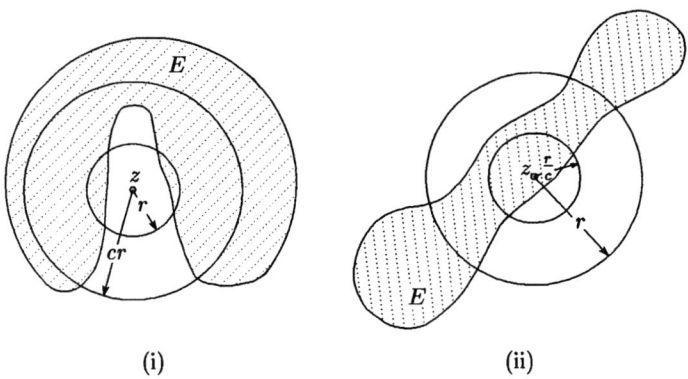

Fig. 2.1

Let G be a simply connected proper subdomain of $\mathbb{C}$. We shall call G a *uniform domain* if there are constants a and b such that an arbitrary pair of points $z_1, z_2 \in G$ can be joined by an arc $\alpha \subset G$ with the following properties:

(i) The Euclidean length of α satisfies the inequality
$$|\alpha| \leq a\,|z_1 - z_2|.$$

(ii) For all $z \in \alpha$,
$$\min(|\alpha_1|, |\alpha_2|) \leq b\,\mathrm{dist}\,(z, \partial G),$$
where α_1 and α_2 are the components of the set $\alpha \setminus \{z\}$ (see Figure 2.2).

The requirement that a quasiconformal mapping be sense-preserving ensures certain formal simplifications. On the other hand, most of the cited results also hold for an *antiquasiconformal* mapping, i.e., a quasiconformal mapping followed by the reflection $z \mapsto \bar{z}$.

Let L be a curve that is the common boundary of two (disjoint) domains G_1 and G_2. An antiquasiconformal mapping $y : \overline{\mathbb{C}} \to \overline{\mathbb{C}}$ with the properties $y(y(z)) = z$, $y(G_1) = G_2$, $y(G_2) = G_1$ that keeps the points of L invariant is called a *quasiconformal reflection* with respect to L.

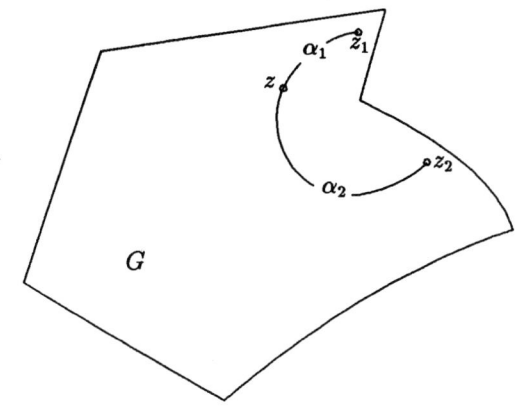

Fig. 2.2

Theorem 2.1. *Suppose $L \subset \overline{\mathbb{C}}$ is a Jordan curve. The following statements are equivalent:*

1° *L is a quasicircle.*

2° *For four arbitrary points z_1, z_2, z_3, z_4, that are situated on L in index-decreasing order (in accordance with the given orientation), the cross-ratio*
$$|z_1, z_4, z_2, z_3| := \left| \frac{(z_1 - z_2)(z_4 - z_3)}{(z_1 - z_3)(z_4 - z_2)} \right|$$
is uniformly bounded.

3° *There exists a quasiconformal reflection $y(z)$ with respect to L.*

4° *$L = \partial G$, where G is a uniform domain.*

5° *$L = \partial G$, where G is a linearly locally connected domain.*

6° *Every pair of points $z_1, z_2 \in L$ satisfies the relation*
$$\frac{\operatorname{diam} L(z_1, z_2)}{|z_1 - z_2|} = O(1). \tag{2.1}$$
Here $L(z_1, z_2)$ is the one of the two arcs of which $L \setminus \{z_1, z_2\}$ consists having smaller diameter.

7° *If G_j, $j = 1, 2$, denote the connected components of the set $\overline{\mathbb{C}} \setminus L$ and $\varphi_j(z)$ is a conformal mapping of G_j onto the unit disk $\mathbb{D}$ (or* onto its exterior Δ), then φ_j can be extended to a quasiconformal homeomorphism of the Riemann sphere onto itself.

For a proof, see [111, pp. 38–48].

There is a relation between the constants in the statements 1°–7° and the coefficient K of quasiconformality of the curve L. We do not dwell on this and refer the reader to [111].

B.2.2

A subarc of a K-quasiconformal curve is called a *K-quasiconformal* (or simply *quasiconformal*) *arc*.

Proof of Lemma 1.2.6. The proof of the equivalence of 1° and 6° in Theorem 2.1, which is known as *Ahlfors's theorem* (see [112, p. 100]), is substantially based on the fact that $\overline{\mathbb{C}} \setminus L$ consists of two components, and cannot automatically be carried over to the case of an arc L, although the final result (Lemma 1.2.6) repeats practically word by word the Ahlfors theorem.

(ii) $\Rightarrow$ (i) is evident by virtue of the Ahlfors theorem.

(i) $\Rightarrow$ (ii). The condition (i) means that the domain $\Omega := \overline{\mathbb{C}} \setminus L$ is a John domain; i.e., for every rectilinear crosscut $[a, b]$ of Ω,

$$\operatorname{diam} H \leq c\,|a - b|, \quad c = \text{const}, \qquad (2.2)$$

holds for the bounded component H of $\Omega \setminus [a, b]$ (cf. [147, p. 96] or [129]).

For the prime end $Z \in \tilde{L} := \tilde{\Omega} \setminus \Omega$ we put $z := |Z|$,

$$\Gamma_Z := \{\zeta : \zeta \in \Omega, \ \arg \Phi(\zeta) = \arg \Phi(Z)\},$$

where Φ is the corresponding conformal mapping $\Phi : \Omega \to \Delta$ normalized at ∞ by (1.2.1).

The result [147, p. 100, Corollary 5.3] yields

$$d(\zeta, L) \geq c_1 |\zeta - z|, \qquad \zeta \in \Gamma_Z, \qquad (2.3)$$

where $c_1 > 0$ depends only on the constant c in (2.2).

Further, we claim that

$$|\Gamma_Z(\zeta_1, \zeta_2)| \leq c_2 |\zeta_1 - \zeta_2|, \qquad \zeta_1, \zeta_2 \in \Gamma_Z, \qquad (2.4)$$

where c_2 also depends only on the c in (2.2), and $\Gamma_Z(\zeta_1, \zeta_2)$ denotes the subarc of Γ_Z between the points ζ_1 and ζ_2.

In fact, let $\tau := \Phi(Z)$, $w_j := \Phi(\zeta_j)$, $j = 1, 2$, $|w_2| > |w_1|$.

If $|w_1 - w_2| \leq (|w_2| - 1)/32$, then by Lemma 1.2.3 for $w \in [w_1, w_2]$ and $\zeta := \Psi(w) := \Phi^{-1}(w)$,

$$|\zeta - \zeta_2| \leq \frac{1}{2} d(\zeta_2, L),$$

which implies that

$$|\Gamma_Z(\zeta_1,\zeta_2)| = \int_{[w_1,w_2]} |\Psi'(w)|\,|dw| \leq 8\,d(\zeta_2,L) \int_{[w_1,w_2]} \frac{|dw|}{|w|-1}$$

$$\leq 16\,\frac{d(\zeta_2,L)}{|w_2|-1}|w_1-w_2| \leq 16^2\,|\zeta_1-\zeta_2|.$$

If $|w_1-w_2| > (|w_2|-1)/32$, then we can use the estimate (see [147, p. 97, Theorem 5.2])

$$|\Psi'(w)| \leq c_3|\Psi'(w_2)|\left(\frac{|w|-1}{|w_2|-1}\right)^{\alpha-1}, \qquad w \in [w_1,w_2],$$

where $0 < \alpha = \alpha(L) \leq 1$, $c_3 = c_3(L) > 0$, and Lemma 1.2.3 to obtain

$$|\Gamma_Z(\zeta_1,\zeta_2)| = \int_{[w_1,w_2]} |\Psi'(w)|\,|dw|$$

$$\leq 4\,c_3\,\frac{d(\zeta_2,L)}{(|w_2|-1)^\alpha} \int_{[w_1,w_2]} (|w|-1)^{\alpha-1}\,|dw|$$

$$\leq \frac{4\,c_3}{\alpha}\,d(\zeta_2,L) \leq c_4\,|\zeta_2-\zeta_1|.$$

Hence, in both cases we get (2.4).

By z_1 and z_2 we denote the endpoints of the arc L and put

$$\tau_j := \Phi(z_j),\quad L^j := \Gamma_{z_j}, \qquad j = 1,2$$

(see Figure 2.3).

Now we introduce the family Γ_1 of all crosscuts of Δ that separate in Δ the points τ_1 and τ_2 from ∞. For the module of this family we have, according to Theorems A.1.1, A.1.2 and (A.1.12),

$$m(\Gamma_1) \geq \frac{1}{2\pi}\log\frac{1}{|\tau_1-\tau_2|}. \qquad (2.5)$$

Next, we consider the following metric:

$$\rho(z) =: \begin{cases} (\operatorname{diam} L)^{-1}, & |z-z_1| \leq 2\operatorname{diam} L, \\ 0, & \text{otherwise}. \end{cases}$$

By virtue of our assumption (i), for any $\gamma \in \Gamma' := \Psi(\Gamma_1)$,

$$|\gamma| \geq c_5 \operatorname{diam} L, \quad c_5 = c_5(L).$$

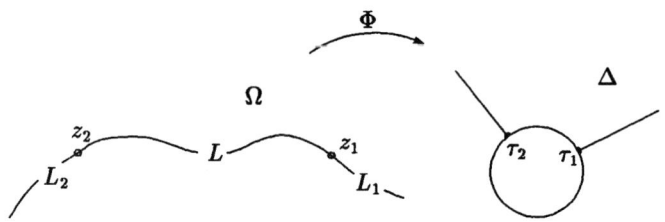

Fig. 2.3

Consequently, for any such arc γ,

$$\int_\gamma \rho(z)\,|dz| \geq c_6 := \min\{1, c_5\},$$

whence, by the definition of a module, we find that

$$m(\Gamma_1') \leq \int_{\mathbb{C}} \rho^2\,dm \left(\inf_{\gamma \in \Gamma_1'} \int_\gamma \rho\,|dz|\right)^{-2} \leq \frac{4\pi}{c_6^2}. \qquad (2.6)$$

Combining (2.5) and (2.6) we obtain

$$|\tau_1 - \tau_2| \geq c_7 = c_7(L) > 0. \qquad (2.7)$$

The following inequalities are also valid:

$$d(\zeta, L^2) \geq c_8\,|\zeta - z_1|, \qquad \zeta \in L^1, \qquad (2.8)$$

$$d(\zeta, L^1) \geq c_8\,|\zeta - z_2|, \qquad \zeta \in L^2, \qquad (2.9)$$

with some $c_8 = c_8(L) > 0$.

Indeed, suppose that $\zeta \in L^1$ is an arbitrary finite point. By $\zeta' \in L^2 \cup \{z_2\}$ we denote an arbitrary point of the arc $L^2 \cup \{z_2\}$ nearest to ζ. Next, we put $w := \Phi(\zeta), w' := \Phi(\zeta')$.

By virtue of (2.3), only the case where $|\zeta - \zeta'| < d(\zeta, L)$ is nontrivial. We consider the family Γ_2 of all curves $\gamma \subset \Delta$, that separate the point w and w' from $\mathbb{T}$ and ∞. Reasoning similarly to the proof of the estimates (2.5), (2.6) and taking into account (2.7), we obtain

$$\frac{1}{2\pi}\log\frac{d(\zeta, L)}{|\zeta - \zeta'|} \leq m(\Gamma_2) = m(\Phi(\Gamma_2)) \leq c_9 = c_9(L).$$

This gives us (2.8).

To prove estimate (2.9) we merely need to exchange the indices 1 and 2 in the discussion above.

With the help of the inequalities (1.2.19), (2.3), (2.4), (2.8), and (2.9) we easily come to the following conclusion: The relation (2.1) is satisfied for any pair of points on
$$L^3 := L^1 \cup L \cup L^2.$$
Thus, by Theorem 2.1, L^3 is quasiconformal and (ii) is fulfilled.

(ii) $\Rightarrow$ (iii). Note that (iii) means that there exists a quasiconformal mapping $Q : \overline{\mathbb{C}} \to \overline{\mathbb{C}}$ and a real number $0 < \theta_0 < \pi$ such that $Q^{-1}(L) = \{e^{i\theta} : |\theta| \leq \theta_0\}$.

If we set
$$Q(z) = F\left(i\frac{z-1}{z+1}\right),$$
condition (iii) is trivially fulfilled.

(iii) $\Rightarrow$ (ii). Let Q is given. Without loss of generality we can assume that $Q(\infty) = \infty$. Consider the curve
$$S := Q^{-1}(L) \cup \{re^{i\theta_0} : r \geq 1\} \cup \{re^{-i\theta_0} : r \geq 1\}.$$
It is quasiconformal because of Theorem B.2.1. Therefore, there exists a quasiconformal mapping $R : \overline{\mathbb{C}} \to \overline{\mathbb{C}}$ and a real number $0 < \eta_0 < \pi$ such that
$$R(\mathbb{T}) = S, \ R(-1) = \infty, \ R(\{e^{i\theta} : |\theta| \leq \eta_0\}) = Q^{-1}(L).$$
Setting
$$F(z) := Q \circ R \circ M\left(z \tan \frac{\eta_0}{2}\right),$$
where
$$M(z) := \frac{i+z}{i-z},$$
we get (ii).

A careful analysis of the proof shows that the constants K and c in Lemma 1.2.6 depend only on each other. $\square$

B.2.3

Now let L be a K-quasiconformal curve, and let $\Phi : \overline{\mathbb{C}} \to \overline{\mathbb{C}}$ be a K_1-quasiconformal extension of the conformal mapping of Ω onto Δ normalized by (1.2.1) (cf. Theorem 2.1). Note that K_1 can be taken equal to K^2 (see [3, Chapter IV]). We introduce the *level lines* of Φ by the formulas
$$L_u := \{z \in \Omega : |\Phi(z)| = 1 + u\}, \quad u > 0,$$
$$L_u^* := \{z \in G := \text{int } L : |\Phi(z)| = 1 - u\}, \quad 0 < u < 1.$$

Let $z, \zeta \in L$, $z \neq \zeta$, and let $L(z, \zeta)$ be the subarc of L joining z and ζ that is smaller in diameter. According to Theorem 2.1,

$$\frac{\operatorname{diam} L(z, \zeta)}{|z - \zeta|} \leq c = c(K). \tag{2.10}$$

Next, we note the following fact. Let $z \in \Omega$, γ a rectifiable crosscut of Ω separating z from ∞, and $\zeta \in \gamma$. Then the length of γ satisfies

$$\frac{|\gamma|}{|z - \zeta|} \geq \varepsilon = \varepsilon(K). \tag{2.11}$$

Indeed, let z_1 and $z_2 \in L$ be the endpoints of γ. By (2.10), we have

$$|z - \zeta| \leq \operatorname{diam} L(z_1, z_2) + |\gamma| \leq c|z_1 - z_2| + |\gamma| \leq (c+1)|\gamma|,$$

which is the assertion in (2.11).

Let $w = \Phi_u(z)$, $0 < u < 1$, be a conformal mapping of ext L_u^* on Δ with the normalization $\Phi_u(\infty) = \infty$, $\Phi_u'(\infty) > 0$, and let $\Psi_u := \Phi_u^{-1}$.

In what follows we shall often use the fact that if L is K-quasiconformal, then L_u^* is K^2-quasiconformal. Consequently, the mapping $\Phi_u(z)$ can be extended to a K^4-quasiconformal mapping of the plane onto itself, a fact that makes it possible to use Theorem 1.2.7.

Lemma 2.2. *For $\zeta \in L_u$ and $0 < u < \frac{1}{2}$,*

$$1 + u \leq |\Phi_u(\zeta)| \leq 1 + cu, \quad c = c(K) > 0. \tag{2.12}$$

Proof. The left-hand side of (2.12) is a simple consequence of Schwarz's lemma [75, p. 25]. In fact, the function $F(w) := 1/\Phi_u(\Psi(1/w))$ maps the unit disk $\mathbb{D}$ onto itself, with $F(0) = 0$. By Schwarz's lemma, $|F(w)| \leq |w|$, and if we set $w = 1/\Phi(\zeta)$, we obtain

$$|\Phi_u(\zeta)| \geq |\Phi(\zeta)| = 1 + u.$$

We now establish the right-hand side of (2.12). Below, $c_1, c_2, \ldots$ denote constants depending only on K. Let $\zeta \in L_u$, $t := \Phi_u(\zeta) \in S_u := \Phi_u(L_u)$. As $t_1 \in l_u := \Phi_u(L)$ we select any point such that $|t - t_1| = d(t, l_u)$.

If we apply Theorem 1.2.7, it is easy to verify the relations

$$|t| - 1 =: \delta(t) \asymp |t - t_1| \asymp |t_1| - 1 \asymp d(t_1, S_u),$$

where $a \asymp b$ means that both $a \leq c_1 b$ and $b \leq c_2 a$.

Let us note some properties of $\delta(t)$. Let t and $w \in S_u$. Then

$$\delta(t) \asymp \delta(w) \quad \text{if } |t - w| \leq c_1 \delta(t), \tag{2.13}$$

$$\delta(w) \leq c_2 \delta^\varepsilon(t) |t - w|^{1-\varepsilon} \quad \text{if } |t - w| \geq c_1 \delta(t), \tag{2.14}$$

B.2 Quasiconformal Curves and Arcs

where $\varepsilon > 0$ is sufficiently small.

In fact, (2.13) is a simple corollary of Theorem 1.2.7. To establish (2.14) we consider the family Γ_1 of all crosscuts of $\mathcal{D}_u := \operatorname{ext} l_u$ separating t from w and ∞, and the similar family Γ_2 of all crosscuts of $\mathcal{D}_u$ separating w from t and ∞.

Let $\Gamma'_j := \Phi \circ \Psi_u(\Gamma_j)$, $j = 1, 2$ (see Figure 2.4). Evidently,

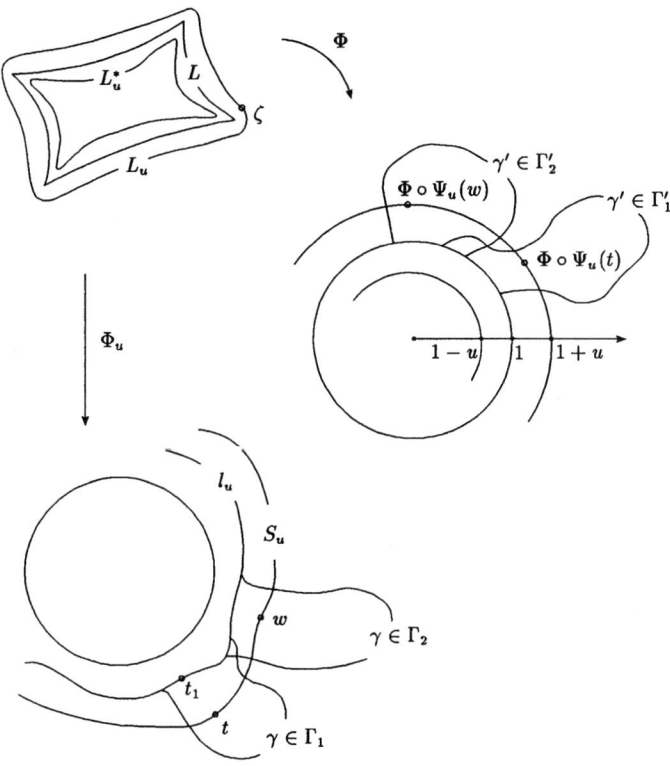

Fig. 2.4

$$m(\Gamma_1) = m(\Gamma'_1) = m(\Gamma'_2) = m(\Gamma_2). \tag{2.15}$$

By (A.1.14) we get

$$m(\Gamma_2) \leq \frac{1}{\pi} \log \frac{|\Phi \circ \Psi_u(t) - \Phi \circ \Psi_u(w)|}{u} + c_3,$$

which by virtue of Theorem 1.2.7 implies that

$$m(\Gamma_2) \le \frac{1}{\varepsilon_1 \pi} \log \frac{c_4 |t-w|}{d(w, l_u)} \le \frac{1}{\varepsilon_1 \pi} \log \frac{c_5 |t-w|}{\delta(w)}. \qquad (2.16)$$

On the other hand, if Γ_3 is the family of all crosscuts of $\mathcal{D}_u$ separating t and t_1 from w and ∞, then $\Gamma_3 \subset \Gamma_1$, and according to the comparison principle (cf. Theorem 1.1.2) and [25, Theorem 6] (see also [17, p. 100, Theorem 4.2]) we have

$$m(\Gamma_1) \ge m(\Gamma_3) \ge \frac{1}{2\pi} \log \frac{c_6 |t-w|}{\delta(t)}. \qquad (2.17)$$

We obtain (2.14) by combining (2.15)–(2.17).

Now consider the family Γ_4 of all crosscuts of $\mathcal{D}_u$ separating t from ∞. By (A.1.9) the module of $\Gamma_4' := \Phi \circ \Psi_u(\Gamma_4)$ can be estimated as follows:

$$m(\Gamma_4) = m(\Gamma_4') \ge \frac{1}{\pi} \log \frac{1}{u}. \qquad (2.18)$$

To obtain an upper bound for $m(\Gamma_4)$ we are going to construct a nonnegative Borel measurable function $\rho(v)$, $v \in \mathbb{C}$, satisfying the condition

$$L_\rho(\Gamma_4) := \inf_{\gamma \in \Gamma_4} \int_\gamma \rho(v) |dv| \ge 1 \qquad (2.19)$$

and in some sense being close to extremal.

We may suppose without loss of generality that $\operatorname{Im} t = 0$, $\operatorname{Re} t > 1$, and that u is sufficiently small. We set

$$\rho_1(v) := \begin{cases} (\pi |v-t| - c_1 \delta^\varepsilon(t) |v-t|^{1-\varepsilon})^{-1}, & \text{if } \operatorname{Re} v > t \\ & \text{and } c_2 \delta(t) \le |v-t| \le c_3, \\ 0, & \text{in all other cases}, \end{cases}$$

where ε is the constant in (2.14) and the c_j, $j = 1, 2, 3$, are chosen so large that when $\gamma \in \Gamma_4$ and $\gamma \subset \{v : v \in \overline{\mathcal{D}}_u, c_2 \delta(t) \le |v-t| \le c_3\}$, we have $\int_\gamma \rho_1(v) |dv| \ge 1$ (the existence of constants c_j with the specified properties is guaranteed by (2.14)). We also set

$$\rho_2(v) := \begin{cases} \dfrac{c_4}{\delta(t)}, & \text{if } |t-v| \le (c_2+1)\delta(t), \\ c_5, & \text{if } c_3 - 1 \le |t-v| \le c_6, \\ 0, & \text{in all other cases}, \end{cases}$$

where the constants c_j, $j = 4, 5, 6$, are chosen large enough so that

$$\int_\gamma \rho_2(v) |dv| \ge 1$$

for every arc $\gamma \in \Gamma_4$ with

$$\gamma \cap (\mathbb{C} \setminus \{v : c_2\delta(t) < |v-t| < c_3\}) \neq \emptyset.$$

The existence of such constants follows from the fact that the arc l_u is K^5-quasiconformal (where K is the coefficient of quasiconformality of L), and consequently, we can use (2.11).

Consider the function $\rho(v) := \max\{\rho_1(v), \rho_2(v)\}$. It satisfies (2.19), and by the definition of the module of a family of curves we thus have

$$m(\Gamma_4) \leq \int_\mathbb{C} \rho^2(v)\, dm(v) \leq \int_\mathbb{C} \rho_1^2(v)\, dm(v) + \int_\mathbb{C} \rho_2^2(v)\, dm(v)$$

$$\leq c_7 + \pi \int_{c_2\delta(t)}^{c_3} \frac{r\, dr}{(\pi r - c_1 \delta^\varepsilon(t) r^{1-\varepsilon})^2}$$

$$= c_7 + \int_{c_2\delta(t)}^{c_3} \frac{dr}{\pi r} + \frac{c_1 \delta^\varepsilon(t)}{\pi} \int_{c_2\delta(t)}^{c_3} \frac{(2\pi - c_1\delta^\varepsilon(t)/r^\varepsilon)\, dr}{(\pi - c_1\delta^\varepsilon(t)/r^\varepsilon)^2 r^{1+\varepsilon}}$$

$$\leq \frac{1}{\pi} \log \frac{c_8}{\delta(t)} + c_9 \delta^\varepsilon(t) \int_{c_2\delta(t)}^{c_3} \frac{dr}{r^{1+\varepsilon}} \leq \frac{1}{\pi} \log \frac{c_{10}}{\delta(t)}.$$

Combining (2.18) and the last inequality we obtain the right-hand side of (2.12). □

C
Constructive Theory of Functions of a Complex Variable

In 1951 Mergelyan proved his approximation theorem (see [123]), which gives an answer to the question: What conditions have to be imposed on a function f defined on a set $E \subset \mathbb{C}$ such that f can be approximated by polynomials arbitrarily closely and what conditions must be imposed on the set E so that an investigation of the uniform approximation of functions by polynomials on this set can be carried out? The set E has to be compact with connected complement, and the function f has to be continuous on E and analytic at interior points of E.

A typical problem of a constructive description of function classes defined on a compact set E (for example, the class of functions with a given majorization of their modulus of continuity) is the following: We have to determine the rate of polynomial approximation for these classes (*direct theorems*) and to prove that this rate guarantees a function being in the corresponding class (*inverse theorems*). If direct and inverse theorems are in full accordance, it is common to say that a class has a *constructive characterization*.

One of the main goals of this appendix is the solution of the following question arising from the proof of direct theorems: How does one construct suitable polynomials for the approximation of functions defined on different sets in $\mathbb{C}$? A significant role in the solution of this problem is played by the construction of approximating polynomials in the form of a convolution with Jackson kernels or their generalizations. To demonstrate the idea of transferring polynomial convolution to a complex domain we give the following typical example.

Let $E = \overline{G}$, where G is a bounded Jordan domain with rectifiable boundary $L = \partial G$. For $z \in G$, a function $f(z)$ that is analytic in a neighborhood of $\overline{G}$ can be represented by the Cauchy integral

$$f(z) = \frac{1}{2\pi i} \int_L \frac{f(\zeta)}{\zeta - z} d\zeta.$$

If for any $n \in \mathbb{N}$ it is possible to construct a polynomial

$$K_n(\zeta, z) = \sum_{k=0}^n a_k(\zeta) z^k \qquad (0.1)$$

with coefficients depending on ζ such that for $z \in G$,

$$\int_L \left| \frac{1}{\zeta - z} - \sum_{k=0}^n a_k(\zeta) z^k \right| |d\zeta| \to 0 \qquad \text{as } n \to \infty,$$

then, observing the fact that

$$p_n(f, z) := \frac{1}{2\pi i} \int_L f(\zeta) K_n(\zeta, z) d\zeta = \sum_{k=0}^n \left(\frac{1}{2\pi i} \int_L f(\zeta) a_k(\zeta) d\zeta \right) z^k$$

is also a polynomial of degree at most n, we have for $z \in G$,

$$|f(z) - p_n(f, z)| \to 0 \qquad \text{as } n \to \infty. \qquad (0.2)$$

Besides, if the function $f(z)$ satisfies some additional properties and the coefficients $a_k(\zeta)$, $k = 0, \ldots, n$, in (0.1) are suitably chosen, the convergence in (0.2) may appear as convergence in uniform norm in $\overline{G}$.

Many methods for the approximation of functions by polynomials (for different functions and different metrics) on sets in $\mathbb{C}$ use explicitly or implicitly the idea of the construction of a "good" preliminary approximation to the Cauchy kernel $1/(\zeta - z)$.

In 1967 Dzjadyk and Shvai (see [51]) proposed a general form of kernels that are polynomials with respect to z whose coefficients are meromorphic functions with respect to ζ:

$$K_{m,n}(\zeta, z) = \frac{1 - (1 - (\zeta - z)\tilde{\pi}_n(\zeta, z))^m}{\zeta - z},$$

where $m \in \mathbb{N}$ is a parameter and $\tilde{\pi}_n(\zeta, z)$ is some elementary kernel.

Later, it was shown by Dzjadyk [49] that the most suitable form for the elementary kernel is the following: $\tilde{\pi}_n(\zeta, z) = \pi_n(\zeta, z)$, where $\pi_n(\zeta, z)$ is introduced below by the formula (2.1) or (2.2).

C.1 Jackson Type Kernels

C.1.1

In 1911 Jackson (see [89]) introduced the following polynomial trigonometric kernels (*Jackson kernel*, for the proof of direct approximation theorems for 2π-periodic continuous functions:

$$J_n(t) := \frac{3}{2n(2n^2+1)} \left(\frac{\sin \frac{nt}{2}}{\sin \frac{t}{2}} \right)^4, \qquad n \in \mathbb{N}. \tag{1.1}$$

Note that an increase in the even exponent in (1.1) strengthens substantially the approximation properties of corresponding convolutions. Thus, later on, the generalized Jackson kernels were actively applied. We define them as follows.

The function

$$J_{nk}(t) := \frac{1}{b_{nk}} \left(\frac{\sin \frac{nt}{2}}{\sin \frac{t}{2}} \right)^{2(k+1)}, \qquad n, k \in \mathbb{N}, \tag{1.2}$$

where

$$b_{nk} = \frac{1}{2\pi} \int_{-\pi}^{\pi} \left(\frac{\sin \frac{nt}{2}}{\sin \frac{t}{2}} \right)^{2(k+1)} dt,$$

is called a *Jackson type kernel*.

The kernels $J_{nk}(t)$ possess the following properties.
1°.

$$\frac{1}{2\pi} \int_{-\pi}^{\pi} J_{nk}(t) dt = 1. \tag{1.3}$$

To prove it we apply the definition directly.

2°. For any fixed $k \in \mathbb{N}$ and arbitrary $n \in \mathbb{N}$ the kernel $J_{nk}(t)$ is an even nonnegative trigonometric polynomial of degree $(k+1)(n-1)$.

Indeed, since for $n \geq 2$,

$$\left(\frac{\sin \frac{nt}{2}}{\sin \frac{t}{2}} \right)^2 = \operatorname{Re} \left(\frac{e^{int/2} - e^{-int/2}}{e^{it/2} - e^{-it/2}} \right)^2 = n \left(1 + \sum_{\nu=1}^{n-1} u_{\nu n} \cos \nu t \right), \tag{1.4}$$

the definition (1.2) yields

$$J_{nk}(t) = \sum_{\nu=0}^{(k+1)(n-1)} v_{\nu n} \cos \nu t,$$

where $u_{\nu n}$ and $v_{\nu n}$ are positive numbers.

384   C. Constructive Theory of a Complex Variable

3°.
$$\left(\frac{2}{\pi}\right)^{2(k+1)} n^{2k+1} \leq b_{nk} \leq \frac{4}{3} n^{2k+1}. \tag{1.5}$$

For the proof we use the well-known inequalities

$$|\sin nt| \leq n|\sin t|, \qquad -\infty < t < \infty, \tag{1.6}$$

$$\sin t \geq \frac{2}{\pi} t, \qquad 0 \leq t \leq \frac{\pi}{2}, \tag{1.7}$$

$$\sin t \leq t, \qquad t \geq 0. \tag{1.8}$$

Applying these estimates we obtain

$$b_{nk} = \frac{1}{\pi} \int_0^\pi \left(\frac{\sin \frac{nt}{2}}{\sin \frac{t}{2}}\right)^{2(k+1)} dt$$

$$\leq \frac{1}{\pi} \int_0^{\pi/n} \frac{n^{2(k+1)}(\sin \frac{t}{2})^{2(k+1)}}{(\sin \frac{t}{2})^{2(k+1)}} dt + \frac{1}{\pi} \int_{\pi/n}^\pi \left(\frac{\pi}{t}\right)^{2(k+1)} dt$$

$$\leq n^{2k+1} + \pi^{2k+1} \int_{\pi/n}^\infty t^{-2(k+1)} dt$$

$$= n^{2k+1} \left(1 + \frac{1}{2k+1}\right) \leq \frac{4}{3} n^{2k+1}.$$

Analogously, making use of the inequalities (1.6)–(1.8) we obtain

$$b_{nk} \geq \frac{1}{\pi} \int_0^{\pi/n} \frac{(\frac{2}{\pi} \frac{nt}{2})^{2(k+1)}}{(\frac{t}{2})^{2(k+1)}} dt = \left(\frac{2}{\pi}\right)^{2(k+1)} n^{2k+1}.$$

4°. For any $\delta > 0$,

$$\int_\delta^\pi J_{nk}(t) dt \leq \frac{c_1}{(n\delta)^{2k+1}}, \qquad c_1 = c_1(k). \tag{1.9}$$

For the proof of (1.9) we use the inequalities (1.5) and (1.7) to write

$$\int_\delta^\pi J_{nk}(t) dt = \frac{1}{b_{nk}} \int_\delta^\pi \left(\frac{\sin \frac{nt}{2}}{\sin \frac{t}{2}}\right)^{2(k+1)} dt$$

$$\leq \frac{1}{b_{nk}} \int_\delta^\infty \left(\frac{\pi}{t}\right)^{2(k+1)} dt \leq \frac{\pi^{4(k+1)}}{2^{2(k+1)}(2k+1)(n\delta)^{2k+1}}.$$

5°. For any $j \in \mathbb{N}$, $j \leq 2k$,

$$\int_{-\pi}^{\pi} J_{nk}(t) |t|^j \, dt \leq c_2 \, n^{-j}, \quad c_2 = c_2(k), \tag{1.10}$$

whence we may conclude that for $j \in \mathbb{N}$, $j \leq 2k$,

$$\int_{-\pi}^{\pi} J_{nk}(t) (1 + n|t|)^j \, dt \leq c_3 = c_3(k). \tag{1.11}$$

In fact, (1.5) and (1.6) imply

$$\int_{-\pi}^{\pi} J_{nk}(t) |t|^j \, dt = 2 \int_0^{\pi} t^j J_{nk}(t) \, dt$$

$$\leq \frac{2}{b_{nk}} \left(\int_0^{\pi/n} t^j \, n^{2(k+1)} \, dt + \int_{\pi/n}^{\pi} t^j \left(\frac{\pi}{t}\right)^{2(k+1)} dt \right)$$

$$\leq c_2 \, n^{-j}.$$

C.1.2

The expansion of the kernel $J_{nk}(t)$ into a Fourier series with respect to the orthogonal function system $\{e^{i\nu t}\}_{\nu=-\infty}^{+\infty}$ has the form

$$J_{nk}(t) = \sum_{-(k+1)(n-1)}^{(k+1)(n-1)} l_{|\nu|} e^{i\nu t}, \tag{1.12}$$

where the l_ν are the Fourier coefficients and, by (1.3), $l_0 = 1$. Using (1.3), (1.8), and (1.10) it is easy to verify that for $\nu > 0$,

$$0 < 1 - l_\nu \leq c_4 \frac{\nu^2}{n^2}, \quad c_4 = c_4(k). \tag{1.13}$$

Indeed,

$$1 - l_\nu = \frac{1}{\pi} \int_0^{\pi} J_{nk}(t) \left(1 - \frac{e^{i\nu t} + e^{-i\nu t}}{2}\right) dt = \frac{1}{\pi} \int_0^{\pi} J_{nk}(t) (1 - \cos \nu t) \, dt$$

$$= \frac{2}{\pi} \int_0^{\pi} J_{nk}(t) \sin^2 \frac{\nu t}{2} \, dt \leq \frac{\nu^2}{2\pi} \int_0^{\pi} J_{nk}(t) \, t^2 \, dt \leq \frac{c_2 \, \nu^2}{2\pi \, n^2}.$$

C.2 Polynomial Kernels Approximating the Cauchy Kernel

C.2.1

In this section we consider polynomial kernels constructed with the help of the generalized Faber polynomials and trigonometric kernels of Jackson type. The reader can find more details concerning this theory in [50, Chapter IX, §7] or [17, Chapter 3, §3].

Let us consider an arbitrary continuum $E \subset \mathbb{C}$ with connected complement Ω and the sequence of standard generalized Faber polynomials $\Pi_n(z)$, $n \in \mathbb{N}$, associated with E (see Section 1.1.3). Let $\tilde{\Omega}$ denote the compactification of Ω by prime ends in the sense of Carathéodory (see Appendix A or [146, pp. 271–278]). Denote by $|\mathfrak{z}|$ the *impression* of a prime end $\mathfrak{z} \in \tilde{\Omega} \setminus \Omega$ (see [146, p. 276]). If $\mathfrak{z} = \zeta \in \Omega$, we set $|\mathfrak{z}| := \zeta$ and, for convenience, call it also the impression of the point ζ.

For any prime end or point $\mathfrak{z} \in \tilde{\Omega} \setminus \{\infty\}$, $n \in \mathbb{N}$, and $t \leq |\pi|$ we set

$$\tilde{\zeta}_t = \tilde{\zeta}_{1/n,t} := \Psi\left(\Phi(\mathfrak{z})\left(1 + \frac{1}{n}\right)e^{-it}\right), \quad \tilde{\zeta} := \tilde{\zeta}_0,$$

where Φ is the conformal mapping of Ω onto the exterior of the unit disk Δ normalized by the conditions (1.2.1) and $\Psi := \Phi^{-1}$.

For the construction of a suitable polynomial kernel approximating the Cauchy kernel $1/(\zeta - z)$ we use the following idea: First, the Cauchy kernel is replaced by a function $1/(\tilde{\zeta} - z)$ "near" to it. After that, we construct the polynomials in z (their coefficients depend on ζ) by a convolution of the function $1/(\tilde{\zeta}_t - z)$ with the trigonometric kernel $J_{nk}(t)$. A further improvement of the kernel's approximation properties is guaranteed by a special analytic construction.

Lemma 2.1. *Let $l \in \mathbb{N}$. For any integer $n > 1$, for all $z \in E$ and $\mathfrak{z} \in \tilde{\Omega} \setminus \{\infty\}$ the integral*

$$\pi_{ln}(\mathfrak{z}, z) := \frac{1}{2\pi} \int_{-\pi}^{\pi} \frac{J_{nl}(t)}{\tilde{\zeta}_t - z} \, dt \tag{2.1}$$

is a polynomial with respect to z of degree $(l+1)(n-1) - 1$. In addition,

$$\pi_{ln}(\mathfrak{z}, z) = \sum_{\nu=1}^{(l+1)(n-1)} \frac{l_\nu \, \Pi_\nu(z)}{(1 + 1/n)^\nu (\Phi(\mathfrak{z}))^\nu}. \tag{2.2}$$

Here, l_ν are the coefficients of the Fourier series (1.12).

C.2 Polynomial Kernels Approximating the Cauchy Kernel

Proof. According to (1.3.2) for $z \in E$ and $w = \Phi(\mathfrak{z})$, $\mathfrak{z} \in \tilde{\Omega} \setminus \{\infty\}$, the function $(\tilde{\zeta}_t - z)^{-1} = \left(\Psi\left((1+1/n)\, w\, e^{-it}\right) - z\right)^{-1}$ can be expanded into the series

$$\frac{1}{\Psi\left((1+1/n)we^{-it}\right) - z} = \sum_{k=1}^{\infty} \frac{\Pi_k(z)}{(1+1/n)^k\, e^{-ikt}\, w^k}. \tag{2.3}$$

After term-by-term integration in (2.1) with application of (2.3) and (1.12), we obtain (2.2). □

Let each prime end $\mathfrak{z} \in \tilde{\Omega} \setminus \Omega$ have a one-point impression $\zeta := |\mathfrak{z}|$, and let k, m, and r be fixed natural numbers, $l := \left[\frac{r}{2}\right] + k + 2$. We consider the polynomial kernel $K_{r,m,k,n}(\mathfrak{z}, z)$ associated with the continuum E to be the function

$$K_{r,m,k,n}(\mathfrak{z}, z) := \frac{1 - (1 - (\zeta - z)\pi_{ln}(\mathfrak{z}, z))^{km}}{\zeta - z}, \tag{2.4}$$

which is defined for $z \in E$ and $\zeta = |\mathfrak{z}|$, $\mathfrak{z} \in \tilde{\Omega} \setminus \{\infty\}$. The function $K_{r,m,k,n}(\zeta, z)$ is called the *Dzjadyk kernel*.

For all $n \in \mathbb{N}$, $n > 1$, this function will be a polynomial with respect to z of degree $(l+1)(n-1)km - 1$ and with coefficients depending on ζ.

C.2.2

To study the approximation properties of polynomial kernels we need the following result, which is essentially due to Belyi (cf. [25, Lemma 4] or [17, p. 65]). We consider the level lines

$$L_u := \{\zeta : |\Phi(\zeta)| = 1 + u\}, \qquad u > 0,$$

of the conformal mapping Φ and set $L_0 := L := \partial E$.

Lemma 2.2. *Let E be a continuum with connected complement. Let $\mathfrak{z} \in \tilde{\Omega} \setminus \{\infty\}$, $|\Phi(\mathfrak{z})| - 1 =: u \geq 0$, $|t| \leq \pi$. Then*

$$\frac{|\tilde{\zeta} - \tilde{\zeta}_t|}{d(\tilde{\zeta}, L_u)} \leq e^{8\pi}(1 + n\,|t|)^4. \tag{2.5}$$

Proof. We can assume that $|\tilde{\zeta} - \tilde{\zeta}_t| > d(\tilde{\zeta}, L_u)$. Consider the family Γ_1 of all crosscuts of $\Omega_u := \{\zeta : |\Phi(\zeta)| > 1 + u\}$ that separate the points $\tilde{\zeta}$ and $\tilde{\zeta}_t$, as well as the family Γ_2 of all curves that separate in the annulus

$$\left\{z : d(\tilde{\zeta}, L_u) < |z - \tilde{\zeta}| < |\tilde{\zeta} - \tilde{\zeta}_t|\right\}$$

its boundary components (see Figure 3.1). Since $\Gamma_2 > \Gamma_1$, we obtain by

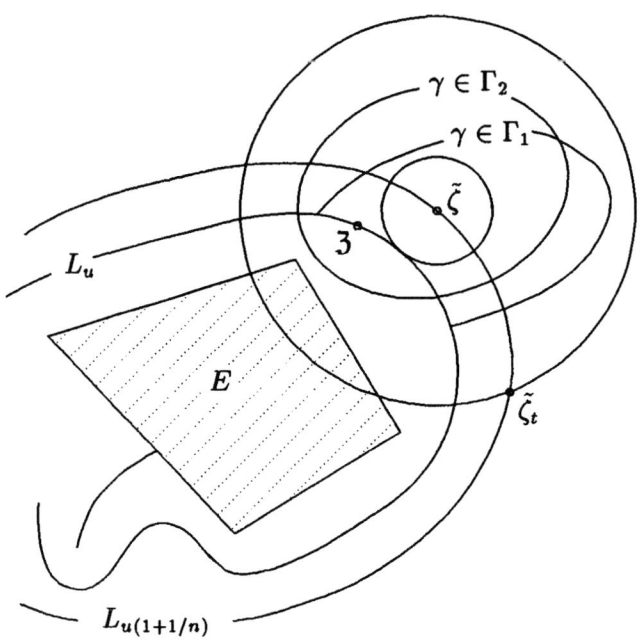

Fig. 3.1

Theorem A.1.2 and (A.1.8),

$$m(\Gamma_1) \geq m(\Gamma_2) = \frac{1}{2\pi} \log \frac{|\tilde{\zeta} - \tilde{\zeta}_t|}{d(\tilde{\zeta}, L_u)}.$$

On the other hand, by virtue of Theorem A.1.1 and (A.1.15),

$$m(\Gamma_1) = m(\Phi(\Gamma_1)) \leq 4 + \frac{2}{\pi} \log(1 + n|t|).$$

Comparing the last two inequalities we obtain (2.5). □

Corollary 2.3. *Taking in (2.5) $\tilde{\zeta}_{-t}$ instead of $\tilde{\zeta}$ and then changing $-t$ to t, we have*

$$\frac{|\tilde{\zeta} - \tilde{\zeta}_t|}{d(\tilde{\zeta}_t, L_u)} \leq e^{8\pi}(1 + n|t|)^4. \tag{2.6}$$

C.2 Polynomial Kernels Approximating the Cauchy Kernel

The following theorem describes the rate of approximation of the Cauchy kernels and their derivatives by the polynomial kernels (cf. [13]).

Theorem 2.4. *Let $E \subset \mathbb{C}$ be an arbitrary continuum with connected complement Ω such that each prime end $\mathfrak{z} \in \tilde{\Omega} \setminus \Omega$ has a one-point impression. Then, for $k \geq 2r$ and arbitrary $p = 0, \ldots, r$, $z \in E$, and $\mathfrak{z} \in \tilde{\Omega} \setminus \{\infty\}$, the polynomial kernel $K_{r,m,k,n}(\mathfrak{z}, z)$ satisfies the inequalities*

$$\left| \frac{\partial^p}{\partial z^p} \left(\frac{1}{\zeta - z} - K_{r,m,k,n}(\zeta, z) \right) \right|$$

$$\leq c_1 \frac{|\tilde{\zeta} - \zeta|^{km}}{|\zeta - z|^{p+1} |\tilde{\zeta} - z|^{km}} \left(1 + \left| \frac{\zeta - z}{\tilde{\zeta} - z} \right| \right)^{rkm},$$

and

$$\left| \frac{\partial^p}{\partial z^p} K_{r,m,k,n}(\mathfrak{z}, z) \right| \leq c_2 |\tilde{\zeta} - z|^{-p-1} \left(1 + \left| \frac{\zeta - z}{\tilde{\zeta} - z} \right| \right)^{km-1},$$

where $\zeta := |\mathfrak{z}|$ and the constants c_1 and c_2 are independent of z, $\mathfrak{z}$, and n.

Proof. First, we consider the polynomials $\pi_{ln}(\mathfrak{z}, z)$. For $z \in E$ and $\mathfrak{z} \in \tilde{\Omega} \setminus \{\infty\}$ the following relation holds:

$$\frac{\partial^p}{\partial z^p} \left(\frac{1}{\zeta - z} - \pi_{ln}(\mathfrak{z}, z) \right) = \frac{p!}{2\pi} \int_{-\pi}^{\pi} J_{nl}(t) \left(\frac{1}{(\zeta - z)^{p+1}} - \frac{1}{(\tilde{\zeta}_t - z)^{p+1}} \right) dt$$

$$= \frac{p!}{2\pi} \int_{-\pi}^{\pi} J_{nl}(t) \frac{(\tilde{\zeta}_t - \zeta) \sum_{s=0}^{p} (\tilde{\zeta}_t - z)^{p-s}(\zeta - z)^s}{(\zeta - z)^{p+1}(\tilde{\zeta}_t - z)^{p+1}} dt.$$

Next, since by (2.5) and (2.6),

$$\left| \frac{\tilde{\zeta}_t - \zeta}{\tilde{\zeta} - \zeta} \right| \leq 1 + \frac{|\tilde{\zeta}_t - \tilde{\zeta}|}{d(\tilde{\zeta}, L_u)} \leq 2e^{8\pi}(1 + n|t|)^4,$$

$$\left| \frac{\tilde{\zeta}_t - z}{\tilde{\zeta} - z} \right| \leq 1 + \frac{|\tilde{\zeta}_t - \tilde{\zeta}|}{d(\tilde{\zeta}, E)} \leq 2e^{8\pi}(1 + n|t|)^4,$$

$$\left| \frac{\tilde{\zeta} - z}{\tilde{\zeta}_t - z} \right| \leq 1 + \frac{|\tilde{\zeta} - \tilde{\zeta}_t|}{d(\tilde{\zeta}_t, E)} \leq 2e^{8\pi}(1 + n|t|)^4,$$

we obtain by (1.11) that

$$\left|\frac{\partial^p}{\partial z^p}\left(\frac{1}{\zeta-z}-\pi_{ln}(\mathfrak{z},z)\right)\right|$$

$$\preceq \frac{|\tilde\zeta-\zeta|(|\tilde\zeta-z|+|\zeta-z|)^p}{|\zeta-z|^{p+1}|\tilde\zeta-z|^{p+1}}\int_{-\pi}^{\pi}J_{nl}(t)(1+n|t|)^{4(p+1)}\,dt$$

$$\preceq \frac{|\tilde\zeta-\zeta|(|\tilde\zeta-z|+|\zeta-z|)^p}{|\zeta-z|^{p+1}|\tilde\zeta-z|^{p+1}} \preceq \frac{|\tilde\zeta-\zeta|}{|\zeta-z|^{p+1}|\tilde\zeta-z|}\left(1+\left|\frac{\zeta-z}{\tilde\zeta-z}\right|\right)^r,$$

if $4(r+1) \le 2l = 2\left[\frac{r}{2}\right] + 2k + 4$. But the last conditions is fulfilled since we have chosen $k \ge 2r$.

Analogously,

$$\left|\frac{\partial^p}{\partial z^p}\pi_{ln}(\mathfrak{z},z)\right| = \frac{p!}{2\pi}\left|\int_{-\pi}^{\pi}J_{nl}(t)\frac{dt}{(\tilde\zeta_t-z)^{p+1}}\right|$$

$$\preceq \frac{1}{|\tilde\zeta-z|^{p+1}}\int_{-\pi}^{\pi}J_{nl}(t)(1+n|t|)^{4(p+1)}\,dt$$

$$\preceq |\tilde\zeta-z|^{-p-1}.$$

The desired estimates of $K_{r,m,k,n}(\mathfrak{z},z)$ can be obtain as follows

$$\left|\frac{\partial^p}{\partial z^p}\left(\frac{1}{\zeta-z}-K_{r,m,k,n}(\mathfrak{z},z)\right)\right|$$

$$= \left|\frac{\partial^p}{\partial z^p}\left\{(\zeta-z)^{km-1}\left(\frac{1}{\zeta-z}-\pi_{ln}(\mathfrak{z},z)\right)^{km}\right\}\right|$$

$$\le \sum_{j=0}^{p}\binom{p}{j}\left|\frac{\partial^{p-j}}{\partial z^{p-j}}(\zeta-z)^{km-1}\right|\left|\frac{\partial^j}{\partial z^j}\left(\frac{1}{\zeta-z}-\pi_{ln}(\mathfrak{z},z)\right)^{km}\right|$$

$$\preceq \frac{|\tilde\zeta-\zeta|^{km}}{|\zeta-z|^{p+1}|\tilde\zeta-z|^{km}}\left(1+\left|\frac{\zeta-z}{\tilde\zeta-z}\right|\right)^{rkm},$$

and
$$\left|\frac{\partial^p}{\partial z^p}K_{r,m,k,n}(\mathfrak{z},z)\right|$$
$$=\left|\frac{\partial^p}{\partial z^p}\sum_{j=1}^{km}(-1)^j\binom{km}{j}(\zeta-z)^{j-1}\pi_{ln}(\mathfrak{z},z)^j\right|$$
$$\preceq\sum_{j=1}^{km}\sum_{s=0}^{p}\left|\frac{\partial^{p-s}}{\partial z^{p-s}}(\zeta-z)^{j-1}\right|\left|\frac{\partial^s}{\partial z^s}\pi_{ln}(\mathfrak{z},z)^j\right|$$
$$\preceq\frac{1}{|\tilde{\zeta}-z|^{p+1}}\left(1+\left|\frac{\zeta-z}{\tilde{\zeta}-z}\right|\right)^{km-1}.$$

□

C.2.3

Let $E = L$ be a rectifiable arc. Each point $\zeta \in L$ (except the endpoints) is an impression of two prime ends $\mathfrak{z}_1$ and $\mathfrak{z}_2$, which are defined as follows. We extend Φ to a homeomorphism between $\tilde{\Omega}$ and $\overline{\Delta}$ (see the Carathéodory theorem in [146, p. 272]). Each of the endpoints z_1 and z_2 of L corresponds to precisely one prime end. Let $w_j := \Phi(z_j) = e^{i\theta_j}$, $j = 1, 2$, $\theta_1 < \theta_2 < \theta_1 + 2\pi$,
$$C_1 := \{e^{i\theta} : \theta_1 < \theta < \theta_2\}, \quad C_2 := \mathbb{T}\setminus\overline{C_1}.$$
For $\zeta \in L$ we denote by $\mathfrak{z}_j$ the prime end with the properties
$$|\mathfrak{z}_j| = \zeta, \quad \Phi(\mathfrak{z}_j) \in \overline{C_j}.$$
For $\zeta, z \in L$, set
$$K_{r,m,k,n}^{(j)}(\zeta,z) := K_{r,m,k,n}(\mathfrak{z}_j,z).$$

Lemma 2.5. *With a suitable orientation of L,*
$$\sum_{j=1}^{2}\frac{(-1)^{j+1}}{2\pi i}\int_L K_{r,m,k,n}^{(j)}(\zeta,z)d\zeta \tag{2.7}$$
$$=\lim_{u\to+0}\frac{1}{2\pi i}\int_{L_u}K_{r,m,k,n}(\zeta,z)d\zeta = 1-\frac{a}{n^{km}}, \quad 0 < a \le c_1(k,m).$$

Proof. The conformal mapping $\Psi := \Phi^{-1}$ has a Laurent expansion in Δ of the form
$$\Psi(w) = cw + \alpha_0 + \frac{\alpha_1}{w} + \cdots, \quad c = \operatorname{cap} L.$$

Representing $\pi_{ln}(\zeta, z)$ by formula (2.2), we have

$$I_u := \frac{1}{2\pi i} \int_{L_u} K_{r,m,k,n}(\zeta, z) \, d\zeta$$

$$= \sum_{s=1}^{km} (-1)^{s-1} \binom{km}{s} \frac{1}{2\pi i} \int_{L_u} (\zeta - z)^{s-1} \pi_{ln}(\zeta, z)^s \, d\zeta$$

$$= \sum_{s=1}^{km} (-1)^{s-1} \binom{km}{s} \frac{1}{2\pi i} \int_L (\zeta - z)^{s-1}$$

$$\times \left(\sum_{\nu=1}^{(l+1)(n-1)} \frac{l_\nu \Pi_\nu(z)}{((1+1/n)\Phi(\zeta))^\nu} \right)^s d\zeta.$$

Since L is rectifiable, $\Psi'(w)$ has angular boundary values $\Psi'(e^{i\theta})$ for almost all $0 \leq \theta < 2\pi$, which form an integrable function on $|w|=1$ (see [147, p. 134]). Hence, the change of variable $\zeta = \Psi(w)$ in the last relation and passage to the limit yield the first equality in (2.7).

Further, the same change of variable implies

$$\lim_{u \to +0} I_u = \lim_{u \to +0} \sum_{s=1}^{km} (-1)^{s-1} \binom{km}{s} \frac{1}{2\pi i} \int_{|w|=1+u} (\Psi(w) - z)^{s-1}$$

$$\times \left(\sum_{\nu=1}^{(l+1)(n-1)} \frac{l_\nu \Pi_\nu(z)}{(1+1/n)^\nu w^\nu} \right)^s \Psi'(w) \, dw$$

$$= \lim_{u \to +0} \sum_{s=1}^{km} (-1)^{s-1} \binom{km}{s} \frac{1}{2\pi i} \int_{|w|=1+u} (cw + \alpha_0 - z + \alpha_1 w^{-1} + \cdots)^{s-1}$$

$$\times \left(\frac{l_1 \Pi_1(z)}{(1+1/n)w} + \cdots + \frac{l_{(l+1)(n-1)} \Pi_{(l+1)(n-1)}(z)}{(1+1/n)^{(l+1)(n-1)} w^{(l+1)(n-1)}} \right)^s$$

$$\times (c - \alpha_1 w^{-2} - \cdots) \, dw$$

$$= \sum_{s=1}^{km} (-1)^{s-1} \binom{km}{s} \left(\frac{l_1}{1+1/n} \right)^s = 1 - \left(1 - \frac{l_1}{1+1/n} \right)^{km}$$

$$= 1 - \left((1 - l_1) + \frac{l_1}{n+1} \right)^{km}.$$

Hence, taking into account (1.13) we obtain (2.7). □

C.2.4

Let $E \subset \mathbb{C}$ be a compact set and let $A(E)$ be the class of functions that are continuous on E and analytic in E°. For $f \in A(E)$ let us introduce the function
$$\omega_E(f, \delta) := \sup_{\substack{|z-\zeta|<\delta \\ z,\zeta \in E}} |f(z) - f(\zeta)|, \qquad \delta > 0,$$
which is called *modulus of continuity* for the function f.

The first result, which demonstrates that direct theorems in terms of a local estimate of the polynomial approximation rate by means of
$$\rho_{1/n}(z) := d(z, L_{1/n}), \quad z \in \mathbb{C}, \quad n \in \mathbb{N}$$
(in the form established after a series of Dzjadyk's works) can be proved for domains with nonrectifiable boundary, is represented by the following *Belyi theorem*.

Theorem 2.6. *Let G be a bounded quasidisk. If $f(z) \in A(\overline{G})$, then for each $n \in \mathbb{N}$, there is an algebraic polynomial $p_n \in \mathbb{P}_n$ such that, for all $z \in L = \partial G$,*
$$|f(z) - p_n(z)| \leq c\,\omega_{\overline{G}}(f, \rho_{1/n}(z)),$$
where $c > 0$ is independent of n, z, and f.

For a proof, see [25, Theorem 3] or [17, p. 119].

C.3 Inverse Theorems

C.3.1

To prove the key Theorem 3.5 of this section we follow the scheme given in the monographs [160] and [17, Chapter 6].

In what follows, E is a continuum with connected complement Ω.

We begin with an auxiliary fact characterizing, in some sense, the distortion of the distance by a conformal mapping.

Lemma 3.1. *For an arbitrary fixed point $z_0 \in E$ the relation*
$$\frac{1}{c_1}\left(1 + \frac{|z - z_0|}{d}\right) \leq |\Phi(z)| \leq c_1\left(1 + \frac{|z - z_0|}{d}\right), \qquad z \in \Omega, \qquad (3.1)$$
holds, where $d := \operatorname{diam} E$, $c_1 = 8e^{4\pi}$.

Proof. The desired inequality (3.1) will be obtained by estimating the module of the family Γ, which consists of all curves $\gamma \subset \Omega$ separating the continuum E from the points z and ∞ in the complex plane. If $z_0 \in E$ is an

394   C. Constructive Theory of a Complex Variable

arbitrary fixed point, $z \in \Omega \setminus \{\infty\}$, then the estimate

$$\frac{1}{2\pi} \log \frac{1}{2}\left(1 + \left|\frac{z-z_0}{d}\right|\right) \leq m(\Gamma) \leq \frac{1}{2\pi} \log 2\left(1 + \left|\frac{z-z_0}{d}\right|\right) + 2 \quad (3.2)$$

is valid.

Indeed, for $|z - z_0| \leq d$, the left-hand part of inequality (3.2) follows from the nonnegativity of the module of a curve family.

Now let $|z - z_0| > d$. Denote by Γ_1 the family of curves separating in the annulus $\{\zeta : d < |\zeta - z_0| < |z - z_0|\}$ its boundary components. By Theorem A.1.2 and (A.1.8),

$$m(\Gamma) \geq m(\Gamma_1) = \frac{1}{2\pi} \log \left|\frac{z-z_0}{d}\right|,$$

whence the left-hand part of estimate (3.2) follows.

To justify the validity of the right-hand part, we consider the metric

$$\rho(\zeta) = \begin{cases} (2\pi|\zeta - z_0|)^{-1}, & de^{-2\pi}/2 \leq |\zeta - z_0| \leq e^{2\pi}(|z - z_0| + d), \\ 0, & \text{otherwise}. \end{cases}$$

This metric is admissible in the L-definition of $m(\Gamma)$, i.e.,

$$\int_\gamma \rho(\zeta)\,|d\zeta| \geq 1, \qquad \gamma \in \Gamma$$

(compare with the proof of (A.1.11)). Consequently,

$$m(\Gamma) \leq \int_C \rho^2\,dm = \frac{1}{2\pi} \log 2\left(1 + \frac{|z-z_0|}{d}\right) + 2.$$

The same estimate as in (3.2) is also valid for the particular case where $E = \overline{\mathbb{D}}$, $z_0 = 0$, and the point z has been replaced by the point $\tau = \Phi(z)$:

$$\frac{1}{2\pi} \log \frac{1}{2}(1 + |\tau|) \leq m(\Phi(\Gamma)) \leq \frac{1}{2\pi} \log 2(1 + |\tau|) + 2. \quad (3.3)$$

If one takes into account the equality $m(\Gamma) = m(\Phi(\Gamma))$ due to the conformal invariance of the module (cf. Theorem A.1.1), then (3.1) is obtained by a simple comparison of the relations (3.2) and (3.3). □

Theorem 3.2. *Let E be a continuum with connected complement and suppose that for some $M_0 > 0$, $\rho > 0$, $z_0 \in L := \partial E$, and $k \in \mathbb{N}$ the polynomial $p_n \in \mathbb{P}_n$, $n \in \mathbb{N}$, satisfies the inequality*

$$|p_n(z)| \leq M_0\left(1 + \left|\frac{z-z_0}{\rho}\right|^k\right), \qquad z \in L. \quad (3.4)$$

Then, for all $z \in D(z_0, \rho) \cap \operatorname{int} L_{1/n}$ the estimate

$$|p_n(z)| \leq c_1 M_0 \qquad (3.5)$$

holds with some $c_1 = c_1(k) > 0$.

Proof. According to the Bernstein–Walsh lemma (cf. Lemma 1.1.11) only the case $0 < \rho < (\operatorname{diam} E)/2$ is nontrivial. Denote by E_0 the connected component of the intersection $\overline{D(z_0, \rho)} \cap L$ containing the point z_0. Let Φ_0 be the corresponding Riemann mapping function for the domain $\Omega_0 := \overline{\mathbb{C}} \setminus E_0$ with standard normalization at ∞.

By Lemma 3.1, we have

$$\frac{1}{c_2}\left(1 + \left|\frac{z - z_0}{d}\right|^k\right) \leq |\Phi_0(z)|^k \leq c_2\left(1 + \left|\frac{z - z_0}{d}\right|^k\right), \qquad z \in \Omega_0, \quad (3.6)$$

with some $c_2 = c_2(k) \geq 1$.

Next, we consider the function

$$f_1(z) := \frac{p_n(z)}{M_0\,\Phi_0(z)^k}, \qquad z \in E^\circ.$$

By virtue of (3.4) and (3.6) we conclude that

$$\lim_{\substack{z \in E^\circ \\ z \to L}} |f_1(z)| \leq c_2.$$

According to the generalized maximum principle (see Theorem 1.1.9) and (3.6) for all $z \in E^\circ \cap D(z_0, \rho)$, the inequality

$$|p_n(z)| \leq c_2\,M_0\,|\Phi_0(z)|^k \leq 2\,c_2^2\,M_0$$

is valid. Carrying out the analogous reasoning for the function

$$f_2(z) := \frac{p_n(z)}{\Phi(z)^n\,\Phi_0(z)^k\,M_0}$$

in Ω we obtain inequality (3.5) at points $z \in D(z_0, \rho) \cap \Omega$. $\square$

With the appropriate reasoning for the proof already given above, let us formulate the following statement, due to Tamrazov [174].

Lemma 3.3. *If for the function $f \in A(E)$, defined on a continuum E having connected complement, and some $M_0 > 0$, $z_0 \in \partial E$, $k \in \mathbb{N}$, and $\rho > 0$ the inequality*

$$|f(z)| \leq M_0\left(1 + \left|\frac{z - z_0}{\rho}\right|^k\right), \qquad z \in \partial E,$$

is satisfied, then

$$|f(z)| \leq c\,M_0, \qquad z \in E \cap \overline{D(z_0, \rho)},$$

with some $c = c(k) > 0$.

Theorem 3.4 (see [174] and [152]), given below, is an important consequence of Lemma 3.3. It allows us to estimate the modulus of continuity of the function on the entire set with the help of the modulus of continuity taken on the boundary of this set.

A positive, nondecreasing function $\omega(\delta)$, $\delta > 0$, that satisfies the conditions

$$\lim_{\delta \to +0} \omega(\delta) = 0,$$

$$\omega(t\delta) \leq c_1 t \omega(\delta), \qquad t \geq 1, \delta > 0,$$

with some constant $c_1 > 0$, is called a *function of modulus of continuity type*.

Theorem 3.4. *Let E be a continuum with connected complement and let the function $f \in A(E)$ satisfy the condition*

$$|f(z) - f(\zeta)| \leq \omega(|z - \zeta|), \qquad z, \zeta \in \partial E,$$

where $\omega(\delta)$ is a function of modulus of continuity type. Then

$$\omega_E(f, \delta) \leq c \omega(\delta), \qquad \delta > 0,$$

with some $c > 0$ independent of δ and E.

Proof. We begin with the remark that if for some $\delta > 0$ the points $z, \zeta \in E$ satisfy

$$|z - \zeta| \leq \delta, \quad |f(z) - f(\zeta)| = \omega_E(f, \delta), \tag{3.7}$$

then there is a pair of points with (3.7) such that at least one of them belongs to ∂E.

In fact, if $z \in E^\circ$ and $\zeta \in E^\circ$, we assume without loss of generality that

$$d := d(z, \partial E) \leq d(\zeta, \partial E).$$

Consider the function

$$g(\xi) := f(z + \xi d) - f(\zeta + \xi d), \qquad |\xi| \leq 1.$$

By the Cauchy formula,

$$\omega_E(f, \delta) = |g(0)| \leq \frac{1}{2\pi} \int_0^{2\pi} \left| f\left(z + e^{i\theta} d\right) - f\left(\zeta + e^{i\theta} d\right) \right| d\theta \leq \omega_E(f, \delta).$$

Therefore,

$$\left| f\left(z + e^{i\theta} d\right) - f\left(\zeta + e^{i\theta} d\right) \right| = \omega_E(f, \delta), \qquad 0 \leq \theta < 2\pi.$$

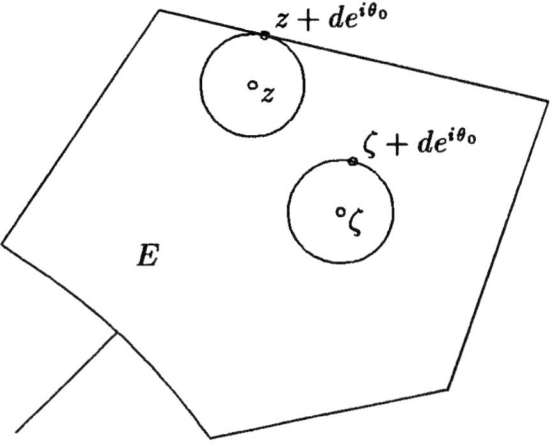

Fig. 3.2

With an appropriate choice of $\theta = \theta_0$ the point $z + e^{i\theta_0} d$ belongs to ∂E, and $z + e^{i\theta_0} d$ as well as $\zeta + e^{i\theta_0} d$ can be taken as a new pair of points satisfying (3.7) (cf. Figure 3.2).

Further, for each point $z_0 \in \partial E$ and $\delta > 0$, the auxiliary function $h(z) := f(z) - f(z_0)$ satisfies for $z \in \partial E$ the inequality

$$|h(z)| = |f(z) - f(z_0)| \leq \omega(|z - z_0|)$$
$$\leq \omega(\delta + |z - z_0|) \leq c_1 \omega(\delta) \left(1 + \left|\frac{z - z_0}{\delta}\right|\right),$$

where c_1 is the constant from the definition of a function of modulus of continuity type. Thus, according to Lemma 3.3,

$$|f(z) - f(z_0)| \leq c_2 \omega(\delta), \qquad z \in E \cap \overline{D(z_0, \delta)},$$

where $c_2 = c_2(c_1) > 0$. □

Theorem 3.5. *Suppose that on the boundary $L := \partial E$ of a continuum E having connected complement the polynomial $p_n \in \mathbb{P}_n$, $n \in \mathbb{N}$, satisfies the inequality*

$$|p_n(z)| \leq \omega(\rho_{1/n}(z)) \, \rho_{1/n}(z)^k, \qquad z \in L,$$

where $k \in \mathbb{N}_0$ and $\omega(t)$ is a function of modulus of continuity type. Then, for any natural number $s \in \mathbb{N}$, the inequality

$$|p_n^{(s)}(\zeta)| \leq c_1 \omega(\rho_{1/n}(z)) \rho_{1/n}(z)^{k-s}$$

holds for any

$$z \in L, \quad \zeta \in \overline{D\left(z, \rho_{1/n}(z)/2\right)},$$

where the constant c_1 is independent of z, ζ, and n.

Proof. For $z, \zeta \in L$ the polynomial p_n satisfies

$$|p_n(\zeta)| \leq \omega(\rho_{1/n}(\zeta)) \rho_{1/n}(\zeta)^k$$
$$\leq \omega\left(\rho_{1/n}(z) + |z-\zeta|\right) \left(\rho_{1/n}(z) + |z-\zeta|\right)^k$$
$$\leq c_2 \omega(\rho_{1/n}(z)) \rho_{1/n}(z)^k \left(1 + \frac{|z-\zeta|}{\rho_{1/n}(z)}\right)^{k+1}$$

with some $c_2 > 0$ independent of z, ζ, and n. Consequently, by virtue of Theorem 3.2, for some $c_3 = c_3(c_2, k) > 0$,

$$|p_n(\zeta)| \leq c_3 \omega(\rho_{1/n}(z)) \rho_{1/n}(z)^k, \quad \zeta \in \overline{D(z, \rho_{1/n}(z))}.$$

Using the Cauchy formula we find for $\zeta \in \overline{D(z, \rho_{1/n}(z)/2)}$ the desired estimate:

$$|p_n^{(s)}(\zeta)| = \frac{s!}{2\pi} \left| \int_{|\xi-z|=\rho_{1/n}(z)} \frac{p_n(\xi)}{(\xi-\zeta)^{s+1}} d\xi \right|$$
$$\leq c_3 \, 2^{s+1} \, \omega(\rho_{1/n}(z)) \rho_{1/n}(z)^{k-s}.$$

□

C.3.2

The result, given below in Theorem 3.6, was established first for continua with piecewise smooth boundary and $\omega(\delta) = \delta^\alpha$ (see [47]). The requirement that the modulus of continuity $\omega(\delta)$ be a Hölder function was removed in the paper [48]. Lebedev and Tamrazov [110] obtained the most general result (from the point of view of geometric properties of the continuum E) concerning the solution of the inverse problem (for details, see [175]).

Theorem 3.6. *Let E be a continuum with connected complement. If for the function $f \in A(E)$ there is a sequence of polynomials $\{p_n\}_{n=1}^\infty$, $p_n \in \mathbb{P}_n$, such that*

$$|f(z) - p_n(z)| \leq \omega(\rho_{1/n}(z)), \quad z \in \partial E, \, n \in \mathbb{N},$$

where $\omega(\delta)$ is a function of modulus of continuity type, then for $0 < \delta \leq d := \operatorname{diam} E$,

$$\omega_E(f, \delta) \leq c\delta \int_\delta^{2d} \frac{\omega(u)}{u^2}\, du, \tag{3.8}$$

where $c > 0$ is a constant independent of δ.

Proof. Let $z_0 \in \partial E$ be an arbitrary point, and let $\delta > 0$ be an arbitrary sufficiently small number.

For any monotonically increasing sequence $\{n_\nu\}_{\nu=1}^\infty$ of natural numbers the difference $f(z) - f(z_0)$, where $z \in E \cap \overline{D(z_0, \delta)}$, can be represented in the form

$$f(z) - f(z_0)$$
$$= p_{n_1}(z) + \sum_{\nu=1}^{N-1} \left(p_{n_{\nu+1}}(z) - p_{n_\nu}(z)\right) + f(z) - p_{n_N}(z)$$
$$- p_{n_1}(z_0) - \sum_{\nu=1}^{N-1} \left(p_{n_{\nu+1}}(z_0) - p_{n_\nu}(z_0)\right) - (f(z_0) - p_{n_N}(z_0))$$
$$= (p_{n_1}(z) - p_{n_1}(z_0)) + \sum_{\nu=1}^{N-1} (u_\nu(z) - u_\nu(z_0))$$
$$+ (f(z) - p_{n_N}(z)) - (f(z_0) - p_{n_N}(z_0)),$$

where

$$u_\nu(z) := p_{n_{\nu+1}}(z) - p_{n_\nu}(z),$$

and the number $N = N(z_0, \delta)$ can be chosen so that the condition

$$d_{N+1}(z_0) \leq 2\delta < d_N(z_0),$$

where

$$d_\nu(z) := \rho_{1/n_\nu}(z), \qquad \nu \in \mathbb{N},$$

is satisfied.

By integration along the line segment joining the points z_0 and z, we obtain

$$|f(z) - f(z_0)| \leq \int_{z_0}^z |p'_{n_1}(\zeta)|\, |d\zeta| + \sum_{\nu=1}^{N-1} \int_{z_0}^z |u'_\nu(\zeta)|\, |d\zeta|$$
$$+ |f(z) - p_{n_N}(z)| + |f(z_0) - p_{n_N}(z_0)|. \tag{3.9}$$

To estimate the right-hand part of this inequality we will use the notation $a \preceq b$ for the inequality $a \leq cb$ with $c > 0$ independent of z_0, z, and δ.

The estimate of the first term is elementary:

$$\int_{z_0}^{z} |p'_{n_1}(\zeta)|\,|d\zeta| \leq |z - z_0|\,\|p'_{n_1}\|_{\overline{D(z_0,\delta)}} \preceq \delta \preceq \omega(\delta). \tag{3.10}$$

The estimate of the other terms is more complicated. To this end, we have to choose a special sequence $\{n_\nu\}_{\nu=1}^{\infty}$. In our reasoning the choice of the sequence $\{n_\nu\}$ will depend on the point z_0. Let n_1 be fixed arbitrarily (for example, $n_1 := 16$). Let us note that for $z \in \partial E$, $n \geq n_1$,

$$\rho_{1/n}(z) \leq 2\,\rho_{1/(n+1)}(z). \tag{3.11}$$

Indeed, let

$$\rho_{1/(n+1)}(z) = |z - \zeta|, \quad \zeta \in L_{1/(n+1)},$$

and let $w = (1 + 1/(n+1))\,e^{i\tau} := \Phi(\zeta)$, $w' := (1 + 1/n)\,e^{i\tau}$, $\zeta' := \Psi(w')$. Since by Lemma 1.2.3,

$$|\zeta - \zeta'| = |\Psi(w) - \Psi(w')| \leq 16\,\frac{\rho_{1/(n+1)}(z)}{1/(n+1)}\,\frac{1}{n(n+1)} \leq \rho_{1/(n+1)}(z),$$

we have

$$\rho_{1/n}(z) \leq |z - \zeta| + |\zeta - \zeta'| \leq 2\,\rho_{1/(n+1)}(z),$$

which proves (3.11).

If n_ν is already constructed, then we denote by $n_{\nu+1}$ the maximal number among the natural numbers satisfying the inequality

$$\rho_{1/n}(z_0) \geq \frac{1}{4}\rho_{1/n_\nu}(z_0).$$

Evidently, such a choice of the sequence $\{n_\nu\}$ implies that the inequality

$$2 \leq \frac{d_\nu(z_0)}{d_{\nu+1}(z_0)} \leq 4$$

is true. The last two terms on the right-hand side of (3.9) can be estimated as follows. First,

$$|f(z_0) - p_{n_N}(z_0)| \leq \omega\,(d_N(z_0)) \leq \omega(4d_{N+1}(z_0)) \leq \omega(8\delta) \preceq \omega(\delta). \tag{3.12}$$

Since for all points $\zeta \in L$,

$$|f(\zeta) - p_{n_N}(\zeta)| \leq \omega\,(d_N(\zeta)) \leq \omega(d_N(z_0) + |z_0 - \zeta|)$$

$$\leq \omega(8\delta + |z_0 - \zeta|) \preceq \omega(\delta)\left(1 + \left|\frac{z_0 - \zeta}{\delta}\right|\right),$$

an application of Lemma 3.3 to the function $f - p_{n_N}$ yields

$$|f(z) - p_{n_N}(z)| \preceq \omega(\delta). \tag{3.13}$$

Further, since for $\zeta \in L$,

$$|u_\nu(\zeta)| \leq |f(\zeta) - p_{n_{\nu+1}}(\zeta)| + |f(\zeta) - p_{n_\nu}(\zeta)| \leq 2\omega(d_\nu(\zeta)) \preceq \omega(d_{\nu+1}(\zeta)),$$

it follows from Theorem 3.5 that for all $\zeta \in \overline{D(z_0, \delta)}$ and $\nu \leq N - 1$,

$$|u'_\nu(\zeta)| \preceq \frac{\omega(d_{\nu+1}(z_0))}{d_{\nu+1}(z_0)} \leq 8 \int_{d_{\nu+1}(z_0)}^{d_\nu(z_0)} \frac{\omega(t)}{t^2} dt.$$

Thus,

$$\sum_{\nu=1}^{N-1} \int_{z_0}^{z} |u'_\nu(\zeta)| \, |d\zeta| \preceq \delta \sum_{\nu=1}^{N-1} \int_{d_{\nu+1}(z_0)}^{d_\nu(z_0)} \frac{\omega(t)}{t^2} dt = \delta \int_{d_N(z_0)}^{d_1(z_0)} \frac{\omega(t)}{t^2} dt. \tag{3.14}$$

Combining the estimates (3.10)–(3.14) and using (3.9), we obtain

$$|f(z) - f(z_0)| \preceq \delta + \omega(\delta) + \delta \int_\delta^{2d} \frac{\omega(t)}{t^2} dt \preceq \delta \int_\delta^{2d} \frac{\omega(t)}{t^2} dt.$$

Since z_0 and z can be arbitrary (cf. the first part of the proof of Theorem 3.4), we have established the estimate (3.8). □

C.3.3

The following result may be proved in much the same way as Theorem 3.6.

Theorem 3.7. *Let E be a continuum with connected complement and let $f \in A(E)$ be such that*

$$E_n(f, E) = O(n^{-s}) \quad \text{as } n \to \infty, \tag{3.15}$$

with some $s > 5$. Then f has continuous derivatives on E up to the order $\left[\frac{s}{2}\right] - 2$.

Proof. We represent the function f in the form

$$f(z) = p_{16}(z) + \sum_{k=16}^{\infty} v_k(z), \quad z \in E,$$

where

$$v_k(z) := p_{k+1}(z) - p_k(z), \quad k \geq 16.$$

Since by Corollary A.2.5,
$$\rho_{1/n}(z) \geq \frac{\text{cap } E}{2n^2}, \quad z \in \partial E, \quad n \in \mathbb{N},$$
we have according to our assumption (3.15) and (3.11) for $z \in \partial E$,
$$v_k(z) = |p_{k+1}(z) - f(z)| + |f(z) - p_k(z)|$$
$$\preceq k^{-s} \leq \frac{2}{\text{cap } E} \rho_{1/k}(z) k^{-s+2} \leq \frac{4}{\text{cap } E} \rho_{1/(k+1)}(z) k^{-s+2}.$$

Applying Theorem 3.5 we obtain
$$\|v_k'\|_E \preceq k^{-s+2}. \tag{3.16}$$

The same reasoning yields
$$\|v_k''\|_E \preceq k^{-s+4}. \tag{3.17}$$

The estimate (3.16) implies that the series
$$g(z) = p_{16}'(z) + \sum_{k=16}^{\infty} v_k'(z)$$
converges uniformly on E to some function $g \in A(E)$ that for $n \geq 16$ satisfies
$$E_n(g, E) \leq \sum_{k=n+1}^{\infty} \|v_k'\|_E \preceq n^{-s+2}. \tag{3.18}$$

To show that g coincides with f' (which in particular means the existence of $f' \in A(E)$) we represent the difference
$$V := \frac{f(z) - f(z_0)}{z - z_0} - g(z_0), \quad z, z_0 \in E, \quad z \neq z_0,$$
in the form
$$V = \frac{1}{z - z_0} \int_{z_0}^{z} \int_{z_0}^{\zeta} p_{16}''(\xi) \, d\xi \, d\zeta + \frac{1}{z - z_0} \sum_{k=16}^{n-1} \int_{z_0}^{z} \int_{z_0}^{\zeta} v_k''(\xi) \, d\xi \, d\zeta$$
$$- \sum_{k=n}^{\infty} v_k'(z_0) + \frac{f(z) - p_n(z)}{z - z_0} - \frac{f(z_0) - p_n(z_0)}{z - z_0}.$$

Putting $\delta := |z - z_0|$ (which can be taken sufficiently small) we choose $n > 16$ such that
$$\frac{\text{cap } E}{2(n+1)^2} \leq \delta < \frac{\text{cap } E}{2n^2}.$$

By the Bernstein–Walsh lemma (cf. Lemma 1.1.11), inequalities (3.16), (3.17), and our assumption (1.4)

$$|V| \preceq \delta \, \|p''_{16}\|_E + \delta \sum_{k=1}^{n-1} k^{-s+4} + \sum_{k=n}^{\infty} k^{-s+2} + n^{-s+2}.$$

Thus,
$$\lim_{z \to z_0} V = 0,$$
whence $f' = g$ on E.

Taking into account (3.18) and repeating our reasoning we can prove the existence of $f^{(r)}$ for all $r \leq [s/2] - 2$. $\square$

C.4 Polynomial Approximation in Domains with Smooth Boundary

C.4.1

Let $G \subset \mathbb{C}$ be a finite Jordan domain. As before, consider the domain $\Omega := \overline{\mathbb{C}} \setminus \overline{G}$, the Riemann mapping function $\Phi : \Omega \to \Delta$, normalized as in (1.2.1), and its level lines

$$L_u := \{\zeta : |\Phi(\zeta)| = 1 + u\}, \qquad u > 0.$$

Throughout this section we assume that $L := \partial G$ is a Dini-smooth curve. In this case $\Phi'(z)$ has a continuous extension to $\overline{\Omega}$ with $\Phi'(z) \neq 0$ (see [147, p. 48]). Therefore, for any points z and $\zeta \in L$ the smallest (in the sense of the length) component $L(z, \zeta)$ of $L \setminus \{z, \zeta\}$ satisfies

$$|L(z,\zeta)| \leq c\,|z - \zeta|, \quad c = c(G) > 0. \tag{4.1}$$

By Theorem B.2.1, L is quasiconformal. Hence, the function

$$\rho_u(z) := \text{dist}\,(z, L_u)\,, \qquad z \in \mathbb{C},$$

has the following obvious property:

$$\rho_u(z) \asymp |z - z_u| \asymp u\,, \qquad z \in L, u > 0, \tag{4.2}$$

where $z_u := \Psi((1+u)\Phi(z))$. The first relation in (4.2) follows from Theorem 1.2.7; the second one is a direct consequence of (1.2.28).

The geometric property (4.1) of G implies that the modulus of continuity $\omega(\delta) := \omega_L(f, \delta)$ along L for an arbitrary function $f \in A(\overline{G})$ satisfies the inequality

$$\omega(t\,\delta) \leq c_1\, t\, \omega(\delta)\,, qquad t \geq 1, \quad \delta > 0, \tag{4.3}$$

with some constant $c_1 = c_1(c) > 0$.

Let $\kappa(\delta)$ be a positive and nondecreasing function for $\delta > 0$ satisfying

$$\lim_{\delta \to +0} \kappa(\delta) = 0 \tag{4.4}$$

and

$$\kappa(t\delta) \leq c_2 t^\alpha \kappa(\delta), \quad t \geq 1, \quad \delta > 0, \tag{4.5}$$

with some constants $c_2 \geq 1$ and $0 < \alpha < 1$.

Theorem 4.1. *Let G be bounded by a Dini-smooth curve L, $\kappa(\delta)$ as above, and let $f \in A(\overline{G})$. In order that*

$$c_1 \kappa\left(\frac{1}{n}\right) \leq E_n(f, \overline{G}) \leq c_2 \kappa\left(\frac{1}{n}\right), \quad n \in \mathbb{N}, \tag{4.6}$$

holds it is necessary and sufficient that

$$c_3 \kappa(\delta) \leq \omega_L(f, \delta) \leq c_4 \kappa(\delta), \quad 0 < \delta < 1, \tag{4.7}$$

where the pairs of constants c_1, c_2 and c_3, c_4 are mutually dependent and depend on G and the function $\kappa(\delta)$ only.

Proof. Assume that (4.6) holds. The right-hand part of (4.7) follows from (4.2), (4.5), and Theorem 3.6:

$$\omega(\delta) \preceq \delta \int_\delta^{2d} \frac{\kappa(u)}{u^2} du \preceq \delta^{1-\alpha} \kappa(\delta) \int_\delta^{2d} \frac{du}{u^{2-\alpha}} \preceq \kappa(\delta).$$

Next, for $0 < \delta < 1$, choosing the integer n such that $(n+1)^{-1} \leq \delta < n^{-1}$, we have by (4.2), (4.3), (4.6), and Theorems 2.6 and 3.4,

$$\kappa(\delta) \leq \kappa\left(\frac{1}{n}\right) \preceq E_n(f, \overline{G}) \preceq \omega\left(\frac{1}{n}\right) \preceq \omega\left(\frac{1}{n+1}\right) \leq \omega(\delta).$$

Hence, the left-hand inequality in (4.7) also holds.

Now let (4.7) be satisfied. The right-hand part of (4.6) follows from (4.2) and Theorems 2.6 and 3.4 used already above. Let us verify the correctness of the left-hand part of this estimate.

Let $p_n^* \in \mathbb{P}_n$, $n \in \mathbb{N}$, be such that

$$\|f - p_n^*\|_{\overline{G}} = E_n(f, \overline{G}).$$

We claim that for $z_0 \in L$, $z \in D(z_0, \rho_0/2) := \{\zeta : |\zeta - z_0| < \rho_0/2\}$, where $\rho_0 := \rho_{1/n}(z_0) \asymp 1/n$,

$$|p_n^{*'}(z)| \preceq n \kappa\left(\frac{1}{n}\right), \tag{4.8}$$

C.4 Polynomial Approximation in Domains with Smooth Boundary

and, consequently,

$$|p_n^*(z) - p_n^*(z_0)| = \left| \int_{[z_0,z]} p_n^{*'}(\zeta)\, d\zeta \right| \preceq \kappa\left(\frac{1}{n}\right) n\,|z - z_0|. \qquad (4.9)$$

Indeed, choose an integer s such that $2^s \leq n < 2^{s+1}$. The polynomial $p_n^*(z)$ can be rewritten in the form

$$p_n^*(z) = \sum_{j=0}^{s+1} v_j(z),$$

where

$$v_j(z) := \begin{cases} p_1^*(z) & j = 0, \\ p_{2^j}^*(z) - p_{2^{j-1}}^*(z) & 1 \leq j \leq s, \\ p_n^*(z) - p_{2^s}^*(z) & j = s+1. \end{cases}$$

For the polynomials $v_j(z)$ we have according to the right-hand side of (4.6)

$$\|v_j\|_{\overline{G}} \preceq \kappa(2^{1-j}) \preceq \kappa(2^{-j}), \qquad 1 \leq j \leq s+1.$$

Therefore, we obtain from Lemma 1.1.11 that

$$\|v_j\|_{D(z_0, \rho_j)} \preceq \kappa\left(2^{-j}\right),$$

where $\rho_j := \rho_{2^{-j}}(z_0) \asymp 2^{-j}$. Consequently, for $z \in D(z_0, \rho_0/2)$ we have

$$|v_j'(z)| \leq \frac{1}{2\pi} \int_{|\zeta - z_0| = \rho_j} \frac{|v_j(\zeta)|}{|\zeta - z|^2}\, |d\zeta| \preceq \kappa(2^{-j})\, 2^j.$$

Hence, if $z \in D(z_0, \rho_0/2)$, we get from (4.5) that

$$|p_n^{*'}(z)| \leq \sum_{j=0}^{s+1} |v_j'(z)| \preceq \sum_{j=1}^{s+1} \kappa(2^{-j})\, 2^j$$

$$\preceq \kappa(2^{-(s+1)})\, 2^{\alpha(s+1)} \sum_{j=1}^{s+1} 2^{j(1-\alpha)} \preceq \kappa\left(\frac{1}{n}\right) n,$$

which is the assertion in (4.8).

For an arbitrary point $z_0 \in L$ define

$$u(f, z_0, \delta) := \sup_{\substack{z \in \overline{G}, \\ |z - z_0| \leq \delta}} |f(z) - f(z_0)|.$$

By (4.9) we have for $0 < \delta \leq \rho_0/2$,

$$u(f, z_0, \delta) \leq u(f - p_n^*, z_0, \delta) + u(p_n^*, z_0, \delta) \leq 2E_n(f, \overline{G}) + c_1\kappa\left(\frac{1}{n}\right) n\,\delta.$$

Let $k \in \mathbb{N}$, $k > 1$, be such that $\delta := \rho_{1/(kn)}(z_0) < \rho_0/2$ (a more concrete choice of the number $k = k(\kappa, G)$ will be specified below).

According to (4.2) and (4.5),

$$c_1\kappa\left(\frac{1}{n}\right) n\,\delta \leq c_2\kappa\left(\frac{1}{kn}\right) k^{\alpha-1}.$$

Therefore,

$$2E_n(f, \overline{G}) \geq u(f, z_0, \delta) - c_2\kappa\left(\frac{1}{kn}\right) k^{\alpha-1}. \tag{4.10}$$

Choosing in (4.10) the point z_0 such that

$$u(f, z_0, \delta) \geq \omega(f, \delta) \geq \omega\left(f, \frac{c_3}{kn}\right) \geq c_4\omega\left(f, \frac{1}{kn}\right),$$

we successively obtain

$$E_n(f, \overline{G}) \geq c_4\omega\left(f, \frac{1}{kn}\right) - c_2\kappa\left(\frac{1}{kn}\right) k^{\alpha-1}$$

$$\geq \kappa\left(\frac{1}{kn}\right)(c_5 - c_2 k^{\alpha-1}) \geq \frac{c_5}{2}\kappa\left(\frac{1}{kn}\right) \succeq \kappa\left(\frac{1}{n}\right)$$

as soon as $k = \left[(2c_2/c_5)^{1/(1-\alpha)}\right] + 1$. $\square$

D
Miscellaneous Topics

In this appendix we discuss two topics that help us to understand more clearly some constructions in the proof of the discrepancy theorems.

The concept of a regularized distance to an arbitrary compact set $E \subset \mathbb{R}^n$ is described by E.M. Stein [166, pp. 170–171]. It is based on the decomposition of open sets into cubes and the partition of unity due to Whitney. For our purposes it is enough to assume that E is a continuum in the complex plane with connected complement Ω. In this case the notion of a regularized distance can be explained by making use of the properties of a conformal Riemann mapping of Ω onto the exterior of the unit disk.

The second topic concerns the interpretation of the Green function for a finite system of real intervals as the real part of some conformal mapping. Here we partially follow H. Widom [182, pp. 224–226].

D.1 The Regularized Distance

Let $E \subset \mathbb{C}$ be a continuum with connected complement $\Omega := \overline{\mathbb{C}} \setminus E$, and let $d(z) := d(z, E)$ denote the distance of z from E. This function is in general not smoother on $\Omega \setminus \{\infty\}$ than what the obvious Lipschitz-condition inequality

$$|d(z) - d(\zeta)| \leq |z - \zeta|, \quad z, \zeta \in \mathbb{C},$$

indicates.

408   D. Miscellaneous Topics

For several applications it is desirable to replace $d(z)$ by a regularized distance $\rho(z)$ that is infinitely differentiable for $z \in \Omega \setminus \{\infty\}$. In addition, this regularized distance has to have essentially the same behavior as $d(z)$.

Let $\Phi : \Omega \to \Delta$ be the Riemann conformal mapping with standard normalization (1.2.1) at infinity. The function

$$\rho(z) := \frac{|\Phi(z)| - 1}{2|\Phi'(z)|}, \qquad z \in \Omega \setminus \{\infty\}, \tag{1.1}$$

is called a *regularized distance* of z from E.

Applying Lemma 1.2.3 we have

$$\frac{1}{8} d(z) \leq \rho(z) \leq 2 d(z), \qquad z \in \Omega \setminus \{\infty\}. \tag{1.2}$$

To study the smoothness properties of $\rho(z)$ we note the following fact. Let the function $f(z)$, $z = x + iy$, be analytic and nonvanishing in $\Omega \setminus \{\infty\}$. A simple calculation yields that for any $z \in \Omega \setminus \{\infty\}$

$$|f|'_x = |f| (\log |f|)'_x = |f| \operatorname{Re} (\log f)'_z = |f| \operatorname{Re} \frac{f'_z}{f}, \tag{1.3}$$

$$|f|'_y = |f| (\log |f|)'_y = |f| \operatorname{Re} (i \log f)'_z = -|f| \operatorname{Im} \frac{f'_z}{f}. \tag{1.4}$$

Hence we conclude that

$$\left| |f|'_\xi \right| \leq |f'_z|, \qquad \xi = x \text{ or } y. \tag{1.5}$$

Differentiating (1.3) and (1.4) once more, we obtain for $j + k = 2$, $j, k \geq 0$,

$$\left| \frac{\partial^2 |f|}{\partial x^j \partial y^k} \right| \leq |f''_{zz}| + 2 \frac{|f'_z|^2}{|f|}. \tag{1.6}$$

Formulas (1.3) and (1.4) imply that $\rho(z) \in C^\infty(\Omega \setminus \{\infty\})$.

Next, we claim that for $z = x + iy \in \Omega \setminus \{\infty\}$, $j, k = 0, 1, 2$, $1 \leq j + k \leq 2$,

$$\left| \frac{\partial^{j+k}}{\partial x^j \partial y^k} \rho(z) \right| \leq c_1 \rho(z)^{1-j-k} \tag{1.7}$$

with an absolute constant $c_1 > 0$.

Indeed, inequality (1.7) follows from (1.5), (1.6), and (1.2.7) by differentiation of the formula (1.1) with respect to $\xi_j := x$ or y, $j = 1, 2$. Namely,

$$\frac{\partial \rho}{\partial \xi_1} = \frac{1}{2|\Phi'_z|^2} \left(|\Phi|'_{\xi_1} |\Phi'_z| - (|\Phi| - 1)|\Phi'_z|'_{\xi_1} \right)$$

and

$$\frac{\partial^2 \rho}{\partial \xi_1 \partial \xi_2} = \frac{1}{2|\Phi'_z|^4} \Big\{ \left(|\Phi|''_{\xi_1 \xi_2} |\Phi'_z| + |\Phi|'_{\xi_1} |\Phi'_z|'_{\xi_2} \right.$$
$$- |\Phi|'_{\xi_2} |\Phi'_z|'_{\xi_1} - (|\Phi| - 1)|\Phi'_z|''_{\xi_1 \xi_2} \big) |\Phi'_z|^2$$
$$- 2 \left(|\Phi|'_{\xi_1} |\Phi'_z| - (|\Phi| - 1) |\Phi'_z|'_{\xi_1} \right) |\Phi'_z| |\Phi'_z|'_{\xi_2} \Big\}.$$

If we know that for $k = 2, 3$,

$$|\Phi^{(k)}| \leq c_2(|\Phi(z)| - 1)\rho(z)^{-k}, \qquad z \in \Omega \setminus \{\infty\}, \tag{1.8}$$

with some absolute constant $c_2 > 0$, then (1.7) is proved.

In order to prove (1.8) we put $d := d(z)/32$ and note that by (1.2.8) for ζ with $|\zeta - z| = d$ we have

$$|\Phi(\zeta) - \Phi(z)| \leq \frac{1}{2}(|\Phi(\zeta)| - 1).$$

Therefore, for such ζ according to (1.2.7),

$$|\Phi'(\zeta)| \leq 4\frac{|\Phi(\zeta)| - 1}{d(\zeta)} \leq 10\frac{|\Phi(z)| - 1}{d(z)}.$$

Next, we apply the Cauchy formula and (1.2.7) to obtain that for $k = 2, 3$,

$$|\Phi^{(k)}(z)| = \frac{(k-1)!}{2\pi}\left|\int_{|\zeta - z| = d}\frac{\Phi'(\zeta)}{(\zeta - z)^k}d\zeta\right| \leq 10\,(k-1)!\,32^{k-1}\frac{|\Phi(z)| - 1}{d(z)^k}.$$

This completes the proof of (1.8) and, consequently, of (1.7).

D.2 Green's Function for a System of Intervals

Let $E = \bigcup_{\nu=1}^{k} I_\nu$ consist of a finite number of mutually disjoint intervals

$$I_\nu = [\alpha_\nu, \beta_\nu], \qquad \nu = 1, \ldots, k,$$

on the real axis, ordered so that

$$-\infty < \alpha_1 < \beta_1 < \alpha_2 < \cdots < \alpha_k < \beta_k < +\infty.$$

Our aim is to derive an interpretation of the Green function $G(z)$ of $\Omega := \overline{\mathbb{C}} \setminus E$ with pole at infinity in terms of some special conformal mappings F and T. Since $G(z) = G(\bar{z})$, it is natural to restrict our consideration to the upper half-plane $H_+ := \{z : \text{Im}\, z > 0\}$.

We introduce

$$q(z) := \prod_{\nu=1}^{k}(z - \alpha_\nu)(z - \beta_\nu)$$

and denote by $q(z)^{1/2}$ the branch of the square root of $q(z)$ that is asymptotically z^k near infinity. Note that $q(x)^{1/2}$ is real on $\mathbb{R} \setminus E$ and that the limiting values of $q(x)^{1/2}$ on E are purely imaginary.

Let
$$p(z) = \sum_{j=0}^{k-1} a_j z^j$$
be a polynomial with real coefficients. Consider the function
$$\int_{\alpha_1}^{z} p(\zeta) q(\zeta)^{-1/2} d\zeta, \qquad (2.1)$$
where the integration is performed along an arc in $\Omega\setminus\{\infty\}$ joining the points α_1 and z. The period of this integral around each I_j is purely imaginary, so that the real part of the integral (2.1) is a single-valued harmonic function, which near infinity has the form
$$a_{k-1} \log|z| + O(1).$$

The limiting values, from below and above I_j, $j = 2,\ldots,k$, of this harmonic function are
$$\sum_{\nu=1}^{j-1} \int_{\beta_\nu}^{\alpha_{\nu+1}} p(x) q(x)^{-1/2} dx.$$

Thus, if these expressions are all zero and $a_{k-1} = 1$, then the real part of (2.1) is the Green function $G(z)$, i.e., we have
$$G(z) = \operatorname{Re} F(z), \quad F(z) := \int_{\alpha_1}^{z} p(\zeta) q(\zeta)^{-1/2} d\zeta + ic, \qquad (2.2)$$
where $c \in \mathbb{R}$ is an arbitrary constant and the coefficients of $p(\zeta)$ are determined by the system of equations
$$\sum_{l=0}^{k-1} a_l \sum_{\nu=1}^{j-1} \int_{\beta_\nu}^{\alpha_{\nu+1}} x^l q(x)^{-1/2} dx = 0, \quad j = 2,\ldots,k,$$
$$a_{k-1} = 1.$$

To show that this system does have a solution it suffices to prove that the corresponding homogeneous system has only the trivial solution. But for the polynomial $p(z)$ corresponding to a solution of the homogeneous system the real part of (2.1) is harmonic in Ω (including $z = \infty$) and zero on E. Thus, the real part of the integral (2.1) is identically zero, and so the same is true for $p(z)$ itself.

The zeros $z_1,\ldots,z_{k-1}$ of the polynomial $p(z)$ are zeros of $F'(z)$. It is clear from the behavior of $G(z)$ on the real axis that there is exactly one z_j in each interval (β_j, α_{j+1}), $j = 1,\ldots,k-1$.

D.2 Green's Function for a System of Intervals

Hence, the restriction of the function F to the upper half-plane is the Schwarz–Christoffel transformation that maps H_+ conformally onto the domain V bounded by

$$\partial V = \{x + ic : x \geq 0\} \cup \{x + i(c-b) : x \geq 0\}$$
$$\cup \, [i(c-b), ic] \cup \bigcup_{j=1}^{k-1} \{x + iv_j : 0 \leq x \leq u_j\},$$

where u_j, $c - v_j$, and b are some positive real numbers, with the following boundary correspondence:

$$F(\infty) = \infty, \quad F(\alpha_1) = ic, \quad F(\beta_k) = i(c-b),$$
$$F(\beta_j) = F(\alpha_{j+1}) = iv_j, \quad F(z_j) = u_j + iv_j$$

(see [5, pp. 236–238] or [119, p. 325]).

Note that according to the properties of $q(z)$ the function $F'(z)$ is asymptotically

$$\frac{1}{z} + o\left(\frac{1}{z}\right)$$

near infinity. Therefore, for $R > 0$ large enough,

$$F(-R) - F(R) = \int_{\{z \in H_+ : |z| = R\}} F'(z)\, dz = i\pi + o(1),$$

and

$$b = \operatorname{Im}(F(-R) - F(R)) = \pi + o(1).$$

Since R is arbitrary, we have $b = \pi$.

Setting $c = b = \pi$ we obtain a conformal mapping F given by formula (2.2), which maps H_+ onto the polygonal domain V bounded by

$$\partial V = \{x + i\pi : x \geq 0\} \cup \{x : x \geq 0\}$$
$$\cup \, [0, i\pi] \cup \bigcup_{j=1}^{k-1} \{x + iv_j : 0 \leq x \leq u_j\}$$

(see Figure 4.1).

Next, we introduce the notion of Green lines. For $0 \leq v \leq \pi$, $v \neq v_j$, where $j = 1, \ldots, k-1$, the unbounded arc

$$\{z \in \overline{H_+} : \operatorname{Im} F(z) = v\}$$

is called a *Green line*.

Further, the arcs

$$\{z \in \overline{H_+} : \operatorname{Im} F(z) = v_j\} \cup [z_j, \alpha_{j+1}], \qquad j = 1, \ldots, k-1,$$

412 D. Miscellaneous Topics

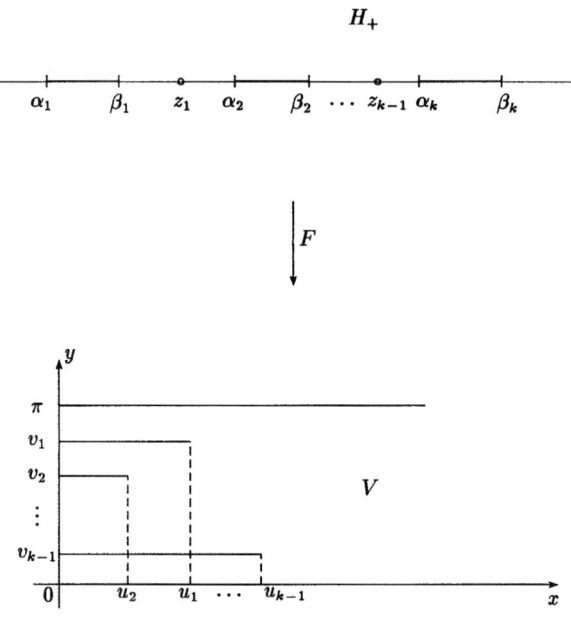

Fig. 4.1

and
$$\left\{z \in \overline{H_+} : \operatorname{Im} F(z) = v_j\right\} \cup [\beta_j, z_j], \qquad j = 1, \ldots, k-1,$$
are also called *Green lines*.

Finally, we take over the name *Green line* for each arc $\gamma \subset \mathbb{C} \setminus H_+$ such that $\{z : \bar{z} \in \gamma\}$ is a Green line.

Next, we introduce the function $T(z)$, $z \in \overline{\mathbb{C}} \setminus [\alpha_1, \beta_k]$, which is equal to $e^{F(z)}$ for $z \in H_+$, $\overline{T(\bar{z})}$ for $z \in \mathbb{C} \setminus \overline{H_+}$, and is extended continuously to $[-\infty, \alpha_1] \cup (\beta_k, +\infty]$. As a result we have constructed a conformal mapping T of $\overline{\mathbb{C}} \setminus [\alpha_1, \beta_k]$ onto

$$\Delta_E := \Delta \setminus \bigcup_{j=1}^{k-1} \left(\{re^{iv_j} : 1 < r \le e^{u_j}\} \cup \{re^{-iv_j} : 1 < r \le e^{u_j}\}\right)$$

with the normalization

$$T(\infty) = \infty, \quad T'(\infty) > 0$$

(see Figure 4.2).

D.2 Green's Function for a System of Intervals

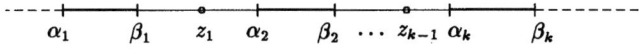

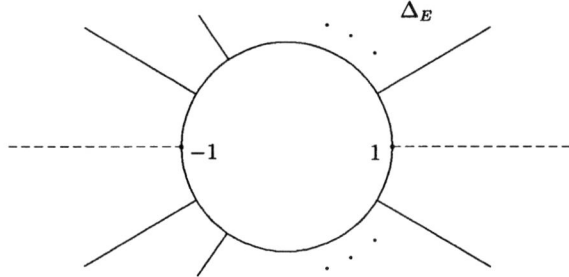

Fig. 4.2

It is clear that

$$G(z) = \log |T(z)|, \qquad z \in \mathbb{C} \setminus [\alpha_1, \beta_k]. \tag{2.3}$$

The pair (r, θ) such that $T(z) = re^{i\theta}$ are called the *Green coordinates* of the point $z \in \mathbb{C} \setminus [\alpha_1, \beta_k]$.

Bibliography

[1] F.G. Abdullaev (1986): On Orthogonal Polynomials in Domains with Quasiconformal Boundary. Dissertation. Donetsk (Russian).

[2] L.V. Ahlfors (1954): *On quasiconformal mappings.* J. Analyse Math., **3**: 1–58 and 207–208.

[3] L.V. Ahlfors (1966): Lectures on Quasiconformal Mappings. Princeton, N.J.: Van Nostrand.

[4] L.V. Ahlfors (1973): Conformal Invariants. New York: Mc Graw-Hill Book Comp.

[5] L.V. Ahlfors (1979): Complex Analysis. New York: Mc Graw-Hill Book Comp.

[6] L.V. Ahlfors, A. Beurling (1950): *Conformal invariants and function-theoretic null-sets.* Acta Math., **83**: 101–129.

[7] F. Amoroso, M. Mignotte (1996): *On the distribution of the roots of polynomials.* Ann. Inst. Fourier (Grenoble), **46**: 1275–1291.

[8] J.M. Anderson, W.H.J. Fuchs (1988): *Remarks on "almost best" approximation in the complex plane.* Complex Analysis (ed. J. Hersch, A. Huber): 17–25.

[9] V.V. Andrievskii (1980): *Some properties of continua with a piecewise quasiconformal boundary.* Ukrainian Math. J., **32**: 435–440 (Russian).

[10] V.V. Andrievskii (1981): *Direct theorems of approximation theory on quasiconformal arcs.* Math. USSR–Izv., **16**: 221–238.

[11] V.V. Andrievskii (1984): *Convergence of Bieberbach polynomials in domains with quasiconformal boundary.* Ukrainian Math J., **32**: 435–440 (Russian).

[12] V.V. Andrievskii (1985): *On the analogue of a de la Vallée Poussin's problem for arcs in the complex plane.* Anal. Math., **11**: 303–309.

[13] V.V. Andrievskii (1986): *The approximation characterization of classes of functions on continua of the complex plane.* Math. USSA-Sb., **53**: 69–87.

[14] V.V. Andrievskii (1990): *Constructive characterization of harmonic functions in domains with a quasiconformal boundary.* Math. USSR–Izv., **34**: 441–454.

[15] V.V. Andrievskii (1993): *Uniform polynomial approximation to analytic functions on a quasidisk.* J. Approx. Theory, **73**: 136–148.

[16] V.V. Andrievskii (1996): *Polynomial approximation to a piecewise analytic function on touching domains.* Complex Variables, **31**: 325–344.

[17] V.V. Andrievskii, V.I. Belyi, V.K. Dzjadyk (1995): Conformal Invariants in Constructive Theory of Functions of Complex Variable. Atlanta, Georgia: World Federation Publisher.

[18] V.V. Andrievskii, H.-P. Blatt (1997): *Erdős–Turán type theorems on piecewise smooth curves and arcs.* J. Approx. Theory, **88**: 109–134.

[19] V.V. Andrievskii, H.-P. Blatt (1997): *A discrepancy theorem on quasiconformal curves.* Constr. Approx., **13**: 363–379.

[20] V.V. Andrievskii, H.-P. Blatt (1999): *Erdős–Turán Type Theorems on Quasiconformal Curves and Arcs.* J. Approx. Theory, **97**: 334–365.

[21] V.V. Andrievskii, H.-P. Blatt, H.N. Mhaskar (2001): *A local discrepancy theorem.* Indag. Math., N.S., **12**(1): 23–39.

[22] V.V. Andrievskii, D. Gaier (1992): *Uniform convergence of Bieberbach polynomials in domains with piecewise quasianalytic boundary.* Mitt. Math. Sem. Giessen, **211**: 49–60.

[23] V.V. Andrievskii, St. Ruscheweyh (1994): *Maximal polynomial subordination to univalent functions in the unit disc.* Constr. Approx., **10**: 131–144.

[24] P.P. Belinskii (1974): General Properties of Quasiconformal Mappings. Novosibirsk: Nauka (Russian).

[25] V.I. Belyi (1977): *Conformal mappings and the approximation of analytic functions in domains with a quasiconformal boundary.* Math. USSR–Sb., **31**: 289–317.

[26] V.I. Belyi, V.M. Mikljukov (1974): *Some properties of conformal and quasiconformal mappings and direct theorems of the constructive theory of functions.* Izv. Akad. Nauk SSSR Ser. Mat., **38**: 1343–1361 (Russian).

[27] S.N. Bernstein (1952): *On the distribution of zeros of the polynomials tending to a continuous function positive on a given interval.* In: Complete Works, Vol. I. Akad. Nauk USSR, 443–451.

[28] H.-P. Blatt (1992): *On the distribution of simple zeros of polynomials.* J. Approx. Theory, **69**: 250–268.

[29] H.-P. Blatt (1997): *A discrepancy lemma for oscillating polynomials and sign changes of the error function of best approximants.* Ann. Num. Math., **4**: 55–66.

[30] H.-P. Blatt (1995): *The distribution of sign changes of the error function in best L^p-approximation.* In: Computational Methods and Function Theory, R.M. Ali, St. Rusheweyh, and E.B. Saff (eds.): 75–87.

[31] H.-P. Blatt, R. Grothmann (1991): *Erdős–Turán theorems on a system of Jordan curves and arcs.* Constr. Approx., **7**: 19–47.

[32] H.-P. Blatt, R. Grothmann, R.K. Kovacheva (2001): *On sign changes in weighted polynomial L^1-approximation.* Acta. Math. Hungar. (to appear).

[33] H.-P. Blatt, G.G. Lorentz (1988): *On a Theorem of Kadec.* In: Proceedings of the Conference on Constructive Theory of Functions, Varna, 1987, ed. Sendov, Petrushev, Ivanov, Maleev. Bulgarian Academy of Sciences: 56–64.

[34] H.-P. Blatt, H.N. Mhaskar (1993): *A general discrepancy theorem.* Ark. Mat., **31**: 219–246.

[35] H.-P. Blatt, H.N. Mhaskar (1995): *A discrepancy theorem concerning polynomials of best approximation in $L_w^p[-1,1]$.* Monatsh. Math., **120**: 91–103.

[36] H.-P. Blatt, H.N. Mhaskar (1998): *Some discrepancy theorems in the theory of weighted polynomial approximation.* J. Math. Anal. Appl., **219**: 312–330.

[37] H.-P. Blatt, E.B. Saff (1986): *Behavior of zeros of polynomials of near best approximation.* J. Approx. Theory, **46**: 323–344.

[38] H.-P. Blatt, E.B. Saff, M. Simkani (1988): *Jentzsch–Szegő type theorems for the zeros of best approximants.* J. London Math. Soc., **38**: 307–316.

[39] H.-P. Blatt, E.B. Saff, V. Totik (1989): *The distribution of extreme points in best complex polynomial approximation.* Constr. Approx., **5**: 357–370.

[40] A. Bloch, G. Pólya (1931): *On the roots of certain algebraic questions.* Proc. London Math. Soc., **33**: 101–114.

[41] P. Borwein (1984): *The relationship between the zeros of best approximation and differentiability.* Proc. Amer. Math. Soc., **92**: 528–532.

[42] P. Borwein, T. Erdélyi (1995): Polynomials and polynomial inequalities. New York: Springer-Verlag.

[43] L. Carleson (1964): *Mergelyan's theorem on uniform polynomial approximation.* Math. Scand., **15**: 167–175.

[44] E.W. Cheney (1966): Introduction to Approximation Theory. New York: Mc Graw-Hill.

[45] E.W. Cheney, D.E. Wulbert (1969): *The existence and unicity of best approximations*. Math. Scand., **24**: 113–140.

[46] R. Courant, D. Hilbert (1953): Methods of Mathematical Physics, Vol. I. New York: Interscience Publishers, Inc.

[47] V.K. Dzjadyk (1959): *On a problem of S.M. Nikolskii in a complex region.* Izv. Akad. Nauk SSSR Ser. Mat., **23**: 697–736 (Russian).

[48] V.K. Dzjadyk (1963): *Inverse theorems in the theory of approximation in complex domains.* Ukrain. Mat. Zh., **15**: 365–375 (Russian).

[49] V.K. Dzjadyk (1972): *The application of generalized Faber polynomials to the approximation of integrals of Cauchy type and functions of the classes A^r in domains with smooth and piecewise smooth boundary.* Ukrain. Mat. Zh., **24**: 3–19 (Russian).

[50] V.K. Dzjadyk (1977): Introduction to the Theory of Uniform Approximation of Functions by Polynomials. Moskow: Nauka (Russian).

[51] V.K. Dzjadyk, A.I. Shvai (1967): *A constructive characteristic of functions of Hölder classes in regions with piecewise smooth boundary and arbitrary positive exterior angles.* Dokl. Akad. Nauk Ukrain. SSR, Ser A, no 2: 1012–1015 (Ukrainian).

[52] P.L. Duren (1983): Univalent Functions. New York: Springer.

[53] P. Erdős, P. Turán (1938): *On Interpolation II. On the distribution of the fundamental points of Lagrange and Hermite interpolation.* Ann. of Math., **39**: 703–724.

[54] P. Erdős, P. Turán (1940): *On Interpolation III. Interpolatory Theory of Polynomials.* Ann. of Math., **41**: 510–553.

[55] P. Erdős, P. Turán (1940): *On the uniformly-dense distribution of certain sequences of points.* Ann. of Math., **41**: 162–173.

[56] P. Erdős, P. Turán (1948): *On a problem in the theory of uniform distribution. I and II.* Indag. Math., **10**: 370–378, 406–413.

[57] P. Erdős, P. Turán (1950): *On the distribution of roots of polynomials.* Ann. of Math., **51**: 105–119.

[58] R.P. Feinerman, D.J. Newman (1974): Polynomial Approximation. Baltimore: The Williams and Wilkins Company.

[59] M. Fekete, J.L. Walsh (1955): *On the asymptotic behaviour of polynomials with extremal properties, and of their zeros.* J. Anal. Math., **4**: 49–87.

[60] G. Freud (1969): Orthogonale Polynome. Basel & Stuttgart: Birkhäuser Verlag.

[61] W.H.J. Fuchs (1980): *On Chebyshev approximation on sets with several components*. Proceedings of the NATO Advanced Study Institute, Durham University, 1979. New York: Academic Press, 399–408.

[62] D. Gaier (1987): Lectures on Complex Approximation. Boston: Birkhäuser.

[63] D. Gaier (1987): *On a polynomial lemma of Andrievskii*. Arch. Math., **49**: 119–123.

[64] D. Gaier (1988): *On the convergence of Bieberbach polynomials in regions with corners*. Constr. Approx., **4**: 289–305.

[65] D. Gaier (1992): *On the convergence of Bieberbach polynomials in regions with piecewise analytic boundary*. Arch. Math., **58**: 462–470.

[66] D. Gaier (1997): *Complex approximation on touching domains*. Complex Variables, **34**: 325–342.

[67] T. Ganelius (1953): *Sequences of analytic functions and their zeros*. Ark. Mat. **3**: 1–50.

[68] T. Ganelius (1957): *Some applications of a lemma on Fourier series*. Académie Serbe des Sciences, Publications de l'Institut Mathématique (Belgrade), **11**: 9–18.

[69] Ya. L. Geronimus (1961): Orthogonal Polynomials. New York: Consultants Bureau.

[70] M. Götz (1997): *Discrepancy of signed measures on quasiconformal curves in terms of bounds for Green potentials*. Preprint: 97–001. Katholische Universität Eichstätt.

[71] M. Götz (1997): Diskrepanzabschätzungen für Maße und quantitative Verteilungsaussagen für energieextremale Punktsysteme. Dissertation, Katholische Universität Eichstätt.

[72] M. Götz (2000): *A discrepancy theorem in $\mathbb{R}^d$, $d \geq 3$*. Analysis, **20**: 303–323.

[73] M. Götz (2000): *On the distribution of weighted extremal points on a surface in $\mathbb{R}^d$, $d \geq 3$*. Potential Anal., **13**: 345–359.

[74] M. Götz, E.B. SAFF (2000): *Potential and Discrepancy estimates for weighted extremal points*. Constr. Approx., **16**: 541–557.

[75] G.M. Goluzin (1969): Geometric Theory of Functions of a Complex Variable, vol. 26 of Translations of Mathematical Monographs. Providence, R.I.: American Mathematical Society.

[76] R. Grothmann (1988): *On the zeros of sequences of polynomials*. J. Approx. Theory, **61**: 351–359.

[77] R. Grothmann (1989): *Ostrowski gaps, overconvergence and zeros of polynomials.* Approximation Theory VI: Vol 1 (ed. C.K. Chui, L.L. Schumaker, and J.D. Ward), Academic Press: 303–306.

[78] R. Grothmann (1992): *Interpolation Points and Zeros of Polynomials in Approximation Theory.* Habilitationsschrift. Katholische Universität Eichstätt.

[79] R. Grothmann (1996): *Distribution of interpolation points.* Ark. Mat., **34**: 103–117.

[80] R. Grothmann, E.B. Saff (1988): *On the behavior of zeros and poles of best uniform polynomial and rational approximants.* Nonlinear numerical methods and rational approximation (ed. A. Cuyt), Reidel, Dordrecht: 57–75.

[81] H. Grötzsch (1928): *Über die Verzerrung bei schlichten nichtkonformen Abbildungen und über eine damit zusammenhängende Erweiterung des Picardschen Satzes.* Sitzungsber. Sächs. Akad. Wiss. Leipzig Math.-Natur. Kl., **80**: 503–507.

[82] K. Haliste (1967): *Estimates of harmonic measure.* Ark. Mat., **6**: 1–31.

[83] W.K. Hayman, P.B. Kennedy (1976): Subharmonic Functions, Vol I. London: Academic Press.

[84] J. Hersch (1955): *Longueurs extrémales et théorie des fonctions.* Comment. Math. Helv., **29**: 301–337.

[85] J. Hüsing (1995): Diskrepanzsätze bei einseitigen Schranken für logarithmische Potentiale. Diplomarbeit, Katholische Universität Eichstätt.

[86] J. Hüsing (1998): Abschätzungen für die Diskrepanz von signierten Maßen durch deren Energienorm. Dissertation, Katholische Universität Eichstätt.

[87] J. Hüsing (2001): *Estimates on the discrepancy of a signed measure using its energy norm.* J. Approx. Theory (to appear).

[88] K.G. Ivanov, E.B. Saff, V. Totik (1991): *On the behavior of zeros of polynomials of best and near-best approximations.* Canad. J. Math., **43**: 1010–1021.

[89] D. Jackson (1911): Über die Genauigkeit der Annäherungen stetiger Funktionen durch ganze rationale Funktionen gegebenen Grades und trigonometrischen Summen gegebener Ordnung. Dissertation, Universität Göttingen.

[90] J.A. Jenkins (1965): Univalent Functions and Conformal Mappings, 2nd edn.. Berlin: Springer-Verlag.

[91] R. Jentzsch (1918): *Untersuchungen zur Theorie der Folgen analytischer Funktionen.* Acta Math., **41**: 219.

[92] M.I. Kadec (1960): *On the distribution of points of maximum deviation in the approximation of continuous functions by polynomials.* Uspekhi Mat. Nauk, **15**: 199–202; Amer. Math. Soc. Transl., **26**: 231–234.

[93] M.V. Keldysh (1939): *Sur l'approximation en moyenne quadratique des fonctions analytiques.* Mat. Sb., **5**: 391–401.

[94] W. Kleiner (1964): *Une condition de Dini–Lipschitz dans la théorie du potentiel.* Ann. Polon. Math., **XIV**: 117–130.

[95] W. Kleiner (1964): *Sur l'approximation de la représentation conforme par la méthode des points extrémaux de M. Leja.* Annales Pol. Math., **XIV**: 131–140.

[96] W. Kleiner (1964): *On the equilibrium of signed measures*, Colloq. Math., **11**: 257–278.

[97] J. Korevaar (1974): *Equilibrium distributions of electrons on roundish plane conductors I, II.* Indag. Math., **36**: 423–456.

[98] J. Korevaar, T. Geveci (1971): *Fields due to electrons on an analytic curve.* SIAM J. Math. Anal., **2**: 445–453.

[99] J. Korevaar, R.A. Kortram (1983): *Equilibrium distributions of electrons on smooth plane conductors.* Indag. Math., **45**: 203–209.

[100] T. Kővari (1971): *On the distribution of Fekete points* II. Mathematika (London), **18**: 40–49.

[101] T. Kővari, Ch. Pommerenke (1967): *On Faber polynomials and Faber expansions.* Math. Z., **99**: 193–206.

[102] T. Kővari, Ch. Pommerenke (1968): *On the distribution of Fekete points.* Mathematika (London), **15**: 70–75.

[103] M.F. Krawtchouk (1933): *Sur la distribution des racines des polynomes orthogonaux.* C. R. Acad. Sci., Paris, **196**: 739–741.

[104] A. Króo (1993): *On certain orthogonal polynomials, Nikolski- and Turán-type inequalities, and interpolatory properties of best approximants.* J. Approx. Theory, **73**: 162–179.

[105] A. Króo, F. Peherstorfer (1987): *Interpolatory properties of best L_1-approximation.* Math. Z., **196**: 249–257.

[106] A. Króo, E.B. Saff (1988): *The density of extreme points in complex polynomial approximation.* Proc. Amer. Math. Soc., **103**: 203–209.

[107] A. Króo, J. Swetits (1992): *On density of interpolation points, a Kadec type theorem and Saff's principle of contamination in L_p-approximation.* Constr. Approx., **8**: 87–103.

[108] N.S. Landkof (1972): Foundations of Modern Potential Theory. Berlin: Springer-Verlag.

[109] M.A. Lavrentiev (1935): *Sur une classe de représentations continues*. Mat. Sb., **42**: 407–423.

[110] N.A. Lebedev, P.M. Tamrazov (1970): *Inverse approximation theorems on regular compacta of the complex plane*. Izv. Akad. Nauk SSSR Ser. Mat., **34**: 1340–1390 (Russian).

[111] O. Lehto (1987): Univalent Functions and Teichmüller Spaces. New York: Springer-Verlag.

[112] O. Lehto, K.I. Virtanen (1973): Quasiconformal Mappings in the Plane, 2nd ed.. Berlin: Springer-Verlag.

[113] F.D. Lesley (1979): *On interior and exterior conformal mappings of the disk*. J. London Math. Soc., **20**: 67–78.

[114] F.D. Lesley (1985): *Conformal mappings of domains satisfying a wedge condition*. Proc. Amer. Math. Soc., **93**: 483–488.

[115] G.G. Lorentz (1966): Approximation of Functions. New York: Holt, Rinehart and Winston.

[116] G.G. Lorentz (1984): *Distribution of alternation points in uniform polynomial approximation*. Proc. Amer. Math. Soc., **92**: 401–403.

[117] K. Löwner (1919): *Über Extremumsätze bei der konformen Abbildung des Äußeren des Einheitskreises*. Math. Z., **3**: 65–77.

[118] D.S. Lubinsky, E.B. Saff (1987): Strong Asymptotics for Extremal Polynomials Associated with Weights on $\mathbb{R}$, Lecture Notes in Math., Vol. 1305. Berlin: Springer Verlag.

[119] A.I. Markushevich (1977): Theory of Functions of a Complex Variable III. New York: Chelsea Publishing Company.

[120] A. Máté, P. Nevai, V. Totik (1985): *Asymptotics for the ratio of leading coefficients of orthonormal polynomials on the unit circle*. Constr. Approx., **1**: 63 – 69.

[121] G. Meinardus (1967): Approximation of Functions. Theory and Numerical Methods. Berlin: Springer-Verlag.

[122] K. Menke (1976): *Über die Verteilung von gewissen Punktsystemen mit Extremaleigenschaften*. J. Reine Angew. Math., **283/284**: 421–435.

[123] S.N. Mergelyan (1951): *Some questions in constructive function theory*. Tr. Mat. Inst. Steklova, Akad. Nauk SSSR, **37**: 3–91 (Russian).

[124] H.N. Mhaskar (1996): Introduction to the theory of weighted polynomial approximation. Singapore: World Scientific.

[125] H.N. Mhaskar, E.B. Saff (1984): *Extremal problems for polynomials with exponential weights*. Trans. Amer. Math. Soc., **285**: 203–234.

[126] H.N. Mhaskar, E.B. Saff (1985): *Where does the sup norm of a weighted polynomial live?* Constr. Approx., **1**: 71–91.

[127] H.N. Mhaskar, E.B. Saff (1991): *On the distribution of zeros of extremal polynomials.* J. Approx. Theory, **65**: 279–300.

[128] M. Mignotte (1992): *Remarque sur une question relative à des fonctions conjugées*, C. R. Acad. Sci. Paris Sér. I, **t. 315**: 907–911.

[129] R. Näkki, J. Väisälä (1991): *John discs.* Expo. Math., **9**: 3–43.

[130] P. Nevai (1979): *Orthogonal Polynomials.* Mem. Amer. Math. Soc. 213, Providence, Rhode Island, USA.

[131] P. Nevai, V. Totik (1989): *Orthogonal polynomials and their zeros.* Acta Sci. Math, **53**: 99–114.

[132] R. Nevanlinna (1974): Eindeutige Analytische Funktionen. Berlin: Springer-Verlag.

[133] M.A. Newman (1961): Elements of the Topology of Plane Sets of Points. Cambridge University Press.

[134] D.J. Newman, H.S. Shapiro (1963): *Some theorems on Chebyshev approximation.* Duke Math. J., **30**: 673–681.

[135] M. Ohtsuka (1970): Dirichlet Problem, Extremal Length and Prime Ends. New York: Van Nostrand.

[136] A. Ostrowski (1922): *Über vollständige Gebiete gleichmässiger Konvergenz von Folgen analytischer Funktionen.* Abh. Math. Sem. Univ. Hamburg 1, **327**.

[137] N. Papamichael, E.B. Saff, J. Gong (1991): *Asymptotic behavior of zeros of Bieberbach polynomials.* J. Comput. Appl. Math., **34**: 325–342.

[138] A. Pfluger (1951): *Quasikonforme Abbildungen und logarithmische Kapazität.* Ann. Inst. Fourier (Grenoble), **2**: 69–80.

[139] R.R. Phelps (1966): *Čebyšev subspaces of finite dimension in L_1.* Proc. Amer. Math. Soc., **17**: 646–652.

[140] A. Pinkus (1985): n-Widths in Approximation Theory. Berlin: Springer-Verlag.

[141] A. Pinkus (1989): On L^1-approximation. Cambridge: Cambridge University Press.

[142] Ch. Pommerenke (1964): *Über die Faberschen Polynome schlichter Funktionen.* Math. Z., **85**: 197–208.

[143] Ch. Pommerenke (1965): *Polynome und konforme Abbildung.* Monatsh. Math., **69**: 58–61.

[144] Ch. Pommerenke (1967): *Über die Verteilung der Fekete-Punkte.* Math. Ann., **168**: 111–127.

[145] Ch. Pommerenke (1968): *Über die Verteilung der Fekete Punkte II.* Math. Ann., **179**: 212–218.

[146] Ch. Pommerenke (1975): Univalent Functions. Göttingen: Vandenhoeck and Ruprecht.

[147] Ch. Pommerenke (1992): Boundary behaviour of conformal maps. Berlin: Springer-Verlag.

[148] E.A. Rakhmanov (1983): *On the asymptotics of the ratio of orthogonal polynomials.* II, Math. USSR-Sb, **46**: 105–117

[149] S. Rickman (1966): *Characterization of quasiconformal arcs.* Ann. Acad. Sci. Fenn. Ser. A I Math., **395**: 1–30.

[150] T.H. Rivlin (1974): The Chebyshev Polynomials. New York: John Wiley & Sons.

[151] P.C. Rosenbloom (1955): *Distribution of zeros of polynomials.* Lectures on Functions of a Complex Variable. Ann. Arbor: 265–286.

[152] L.A. Rubel, A.L. Shields, B.A. Taylor (1975): *Mergelyan sets and the modulus of continuity of analytic functions.* J. Approx. Theory, **15**: 23–40.

[153] E.B. Saff, B. Shekhtman (1990): *Interpolatory properties of best L_2-approximants.* Indag. Math. (new series) **1**: 489–498.

[154] E.B. Saff, V. Totik (1989): *Polynomial approximation of piecewise analytic functions.* J. London Math. Soc., **39**: 487–498.

[155] E.B. Saff, V. Totik (1989): *Behavior of polynomials of best uniform approximation.* Trans. Amer. Math. Soc., **316**: 567–593.

[156] E.B. Saff, V. Totik (1997): Logarithmic Potentials with External Fields. Berlin: Springer-Verlag.

[157] E. Schmidt (1932): *Über algebraische Gleichungen vom Pólya–Bloch–Typus,* Sitzungsber. Preuss. Akad. Wiss. Phys.-Math. Kl., **XXII**: 321.

[158] I. Schur (1933): *Untersuchungen über algebraische Gleichungen I. Bemerkungen zu einem Satz von E. Schmidt.* Sitzungsber. Preuss. Akad. Wiss. Phys.-Math. Kl., **X**: 403–428.

[159] H.S. Shapiro (1971): Topics in Approximation Theory. Lecture Notes in Mathematics, Vol. 187. Berlin: Springer-Verlag.

[160] I.A. Shevchuk (1992): Approximation by Polynomials and Traces of Functions Continuous on a Segment. Kiev: Naukova Dumka (Russian).

[161] J. Siciak (1967): *Degree of convergence of some sequences in the conformal mapping theory.* Colloq. Math., **XVI**: 49–59.

[162] I.B. Simonenko (1978): *Convergence of Bieberbach polynomials in the case of a Lipschitz domain.* Math USSR-Izv., **13**: 166–174.

[163] I. Singer (1970): Best approximation in normed linear spaces by elements of linear subspaces. Berlin: Springer Verlag.

[164] P. Sjögren (1972): *Estimates of mass distributions from their potentials and energies.* Ark. Mat., **10**: 59–77.

[165] V.I. Smirnov, N.A. Lebedev (1968): Functions of a Complex Variable. Constructive Theory. Cambridge: Massachusetts Institute of Technology.

[166] E.M. Stein (1970): Singular Integrals and Differentiability Properties of Functions. Princeton, NJ: Princeton University Press.

[167] P.K. Suetin (1971): *Polynomials orthogonal over a region and Bieberbach polynomials.* Proc. Steklov Inst. Math., **100**. Providence, RI 1974. American Math. Society.

[168] P.K. Suetin (1998): Series of Faber Polynomials. Amsterdam: Gordon and Breach Science Publishers.

[169] G. Szegő (1922): *Über die Nullstellen der Polynome einer Folge, die in einem einfach zusammenhängenden Gebiet gleichmässig konvergiert.* Nachr. Akad. Wiss. Göttingen, Math.-Phys. Kl.,**II**: 137–143.

[170] G. Szegő (1922): *Über die Nullstellen von Polynomen, die in einem Kreis gleichmässig konvergieren.* Sitzungsber. Berliner Math. Ges., **21**: 59–64.

[171] G. Szegő (1924): *Über eine Eigenschaft der Exponentialreihe.* Sitzungsber. Berliner Math. Ges., **23**: 50–64.

[172] G. Szegő (1934): *Bemerkungen zu einem Satz von E. Schmidt über algebraische Gleichungen*, Sitzungsber. Berl. Akad.: 86–98

[173] G. Szegő (1975): Orthogonal Polynomials. Amer. Math. Soc., Colloquium Publications. Providence, Rhode Island.

[174] P.M. Tamrazov (1973): *Contour and solid structural properties of holomorphic functions of a complex variable.* Russian Math. Surveys, **28**: 141–173.

[175] P.M. Tamrazov (1975): Smoothnesses and Polynomial Approximations. Kiev: Naukova Dumka (Russian).

[176] A.F. Timan (1963): Theory of Approximation of Functions of a Real Variable. New York: Pergamon Press.

[177] V. Totik (1993): *Distribution of simple zeros of polynomials.* Acta Math., **170**: 1–28.

[178] M. Tsuji (1950): Potential Theory in Modern Function Theory. New York: Chelsea Publ. Co.

[179] Wu Xue-Mou (1963): *On Bieberbach polynomials.* Acta Math. Sinica, **13**: 145–151.

[180] J.L. Walsh (1959): *The analogue for maximally convergent polynomials of Jentzsch's theorem.* Duke Math. J., **26**: 605–616.

[181] J.L. Walsh (1969): Interpolation and Approximation by Rational Functions in the Complex Plane, 5th edition. Providence: American Mathematical Society.

[182] H. Widom (1969): *Extremal polynomials associated with a system of curves in the complex plane.* Adv. in Math., **3**: 127–232.

[183] W. Wirtinger (1927): *Zur formalen Theorie der Funktionen von mehreren komplexen Veränderlichen.* Math. Ann., **97**: 357–375.

[184] A. Zygmund (1977): Trigonometric series. Cambridge: Cambridge University Press.

Notation

$[a]$	integral part of a p. 160
$a(\delta)$	outer one-sided bound of U^σ p. 130
$a(\delta) = a(\delta, \sigma, \varepsilon, \Gamma)$	supremum norm of U^σ on $B_\varepsilon(\Gamma) \cap L_\delta$ (local bound) p. 115
$A(E)$	set of functions $f \in C(E)$ analytic on E° p. 45
$A(\rho)$	ρ-area of the metric ρ pp. 24, 342
$a_{\inf}(\delta)$	outer bound for the infimum of U^σ p. 130
$a_{\inf}^+(\delta) = a_{\inf}(\delta)$	outer bound for the infimum of U^σ p. 150
$a_{\inf}^-(\delta)$	inner bound for the infimum of U^σ p. 150
a_{kl}	Grunsky coefficients p. 37
$A_n(f)$	set of extreme points of $f - p_n^*(f)$ p. 242
$A_{\sigma,\tau}(J)$	(σ, τ)-neighborhood of J p. 58
$A_\sigma(J)$	σ-neighborhood of J p. 69
$\widetilde{A}_\sigma(I)$	neighboring set of the interval I p. 90
$a_{\sup}(\delta)$	outer bound for the supremum of U^σ p. 130

428   Notation

$a_{\sup}^+(\delta) = a_{\sup}(\delta)$	outer bound for the supremum of U^σ p. 150
$a_{\sup}^-(\delta)$	inner bound for the supremum of U^σ p. 150
$B_\varepsilon^*(\Gamma)$	neighboring set of $\Phi(\Gamma)$ in Δ p. 115
$B_\varepsilon(\Gamma)$	neighborhood of Γ p. 115
$\mathbb{C}$	complex plane p. 1
cap E	capacity of E p. 11
$C(E)$	set of continuous functions on E p. 45
$C^1(\mathbb{C})$	set of real-valued functions on $\mathbb{C}$ with compact support and continuous derivative p. 46
$C^\infty(G)$	set of real-valued functions on $\mathbb{C}$ with compact support and continuous derivatives of all orders p. 47
C_ρ	circle with center 0 and radius ρ, $\rho > 0$ p. 49
$\mathbb{D}$	unit disk p. 2
$d(A,B)$	distance between A and $B = \operatorname{dist}(A,B)$ p. 2
$D(a,r)$	disk with center a and radius r pp. 2, 47
Δ	exterior of the unit circle p. 2
Δu	Laplacian of u pp. 3, 6
Δ_1, Δ_2	subdomains in Δ associated with a Jordan arc p. 29
$\Delta_\Gamma \widetilde{G}$	net change on Γ of the conjugate $\widetilde{G}$ of Green's function G p. 79
$\Delta_n = \Delta_n(E)$	nth discriminant of E p. 17
$\delta_\nu(\rho)$	inner bound of the logarithmic potential $U^{\mu_L - \nu}$ pp. 130, 155
$\delta_{p,E}$	$= \varepsilon_{p,E} - U^{\mu_E - \nu_p}(z_0)$ with fixed $z_0 \in \operatorname{int} L' \subset E$ where L' is a curve p. 85
$\delta_{p,L}$	$= \varepsilon_{p,L} - U^{\mu_L - \nu_p}(z_0)$ with fixed $z_0 \in \operatorname{int} L$ where L is a curve p. 57
diam A	diameter of A p. 28
dm	Lebesgue measure in the plane pp. 6, 24

$D[\sigma]$	discrepancy of σ p. 96
D_f	dilatation quotient p. 362
$D_\Gamma[\sigma]$	local discrepancy of σ on Γ p. 108
$\overline{E}$	closure of E p. 2
$\varepsilon^*_{p,L}(\sigma)$	one-sided bound for $U^{\mu_L - \nu_p}$ p. 74
$\varepsilon_n(w)$	characteristic number with respect to the weight function $w(x)$ p. 255
$\varepsilon_{p,E}$	global upper bound for $U^{\mu_E - \nu_p}$ p. 79
$\varepsilon_{p,L}$	global upper bound for $U^{\mu_L - \nu_p}$ p. 57
$\varepsilon_\sigma(\delta)$	supremum norm of U^σ on L_δ p. 96
ext L	outer domain of the Jordan curve L p. 2
E°	interior of E p. 2
E'_σ	interior of L'_σ p. 80
$E_{n,p,Q}$	weighted L^p-norm of $T_{n,p,Q}$ with respect to w_Q p. 275
$E_n(f, E)$	minimal uniform deviation of f on E with respect to $\mathbb{P}_n$ pp. 45, 188
E_r	$:= E \cup \{z \in \Omega \;:\; G(z) < \log(1+r)\},\ r > 0$ pp. 15, 45, 78
E_σ	interior of the level line L_σ pp. 69, 114
$E_{\sigma,\tau}$	domain bounded by L^+_σ and L^-_τ p. 58
$f \asymp g$	ordering symbol pp. 48, 376
$f \preceq g$	ordering symbol p. 48
$f \succeq g$	ordering symbol p. 48
$\mathcal{F}_n(E)$	nth Fekete point set of E p. 226
$f_z, f_{\bar{z}}$	formal derivatives of f p. 361
$\overline{\gamma}$	closed Jordan arc p. 2
$\Gamma(Z, \zeta_1, \zeta_2)$	family of curves and arcs seperating in Ω the prime end Z and the point ζ_1 from ζ_2 and ∞ p. 323

430   Notation

$\Gamma_\infty(K,G)$	family of curves and crosscuts of G seperating K from ∞ p. 357
$\gamma_z(r)$	crosscut used in the integrated versions of the composition laws p. 25
$|\text{grad } u(z)|$	Euclidean length of the gradient of $u(z)$ p. 4
$G(z,a) = G_\Omega(z,a)$	Green's function for Ω with pole at a p. 15
$G(z) = G(z,\infty)$	Green's function for Ω with pole at ∞ p. 14
int L	inner domain of the Jordan curve L p. 2
$I[\sigma]$	energy of a positive Borel measure σ p. 10
$J_{nk}(t)$	Jackson type kernel p. 383
J_σ^+	neighboring set of J p. 58
J_τ^-	neighboring set of J p. 58
$K(z)$	averaging and smoothing kernel p. 46
$K(z,\zeta)$	Bergman kernel function p. 216
$K_{r,m,k,n}(\zeta,z)$	Dzjadyk kernel pp. 330, 387
$\lambda(\Gamma) := 1/m(\Gamma)$	extremal length of the curve family Γ p. 342
λ_n^*	normalized extremal measure on $[-1,1]$ p. 275
$L(\zeta,\xi)$	subarc of L joining the points ζ and ξ of L pp. 70, 376
L_u^*	level line $|\Phi(z)| = 1 - u$, $0 < u < 1$ p. 375
$L^2(G)$	set of functions f analytic in G such that $|f|^2$ is Lebesgue-integrable on G p. 215
$L_1^2(G)$	set of functions g analytic in G such that $g(z_0) = 0$ and $|g'|^2$ is Lebesgue-integrable on G p. 337
$L_r'^-$	level line of the normalized conformal mapping of int L' onto $\mathbb{D}$ p. 85
L_σ'	level line of the conformal mapping Φ_b p. 80
L_r	level line of Green's function G pp. 15, 68, 78, 114, 130
$L_\rho(\Gamma)$	minimal ρ-length of the curve family Γ pp. 24, 342

Notation 431

$L_r^+ = L_r$	level line of Green's function G pp. 58, 127, 150
L_r^-	level line of $\varphi(z)$ pp. 58, 127, 150
L_u	level line $|\Phi(z)| = 1 + u$, $u > 0$ p. 375
$\mathcal{M}(E)$	positive unit Borel measures supported on E p. 10
mes(E)	Lebesgue measure of $E \in \mathbb{R}$ p. 46
$m(\Gamma)$	module of the curve family Γ pp. 24, 342
$M(Q)$	(conformal) module of the quadrilateral $Q = Q(z_1, z_2, z_3, z_4)$ p. 363
μ_E	equilibrium measure of E pp. 11, 21
$\mu_\infty(K, G)$	reduced module of the family $\Gamma_\infty(K, G)$ p. 357
$\mu_n \xrightarrow{*} \mu$	μ_n is weak*-convergent to μ p. 14
$\mathbb{M}_n$	set of monic polynomials of degree n p. 17
$\mathbb{N}$	natural numbers p. 16
$\nu(X)$	normalized counting measure of X pp. 40, 249
$\widehat{\nu}_n$	balayage measure of ν_n p. 313
ν_p	zero-counting measure of the polynomial p p. 40
$\mathbb{N}_0$	natural numbers and 0 p. 16
Ω	exterior of E, $\Omega = \overline{\mathbb{C}} \setminus E$ pp. 10, 21, 188
Ω_1, Ω_2	subdomains of Ω defined for a Jordan arc p. 30
$\mathcal{P}$	admissible functions or metrics p. 342
$\partial/\partial \mathbf{n} = \partial/\partial \mathbf{n}_z$	normal derivative p. 6
$\overline{\partial} = \frac{\partial}{\partial \overline{z}}$	formal derivative with respect to $\overline{z}$ p. 361
$\partial = \frac{\partial}{\partial z}$	formal derivative with respect to z p. 361
∂E	boundary of E p. 2
$\partial_\alpha f(z)$	directional derivative of f at z p. 362
Φ	conformal mapping of Ω onto $\mathbb{D}$ p. 20
$\varphi \perp \mathbb{P}_n$	sign function φ orthogonal to $\mathbb{P}_n$ p. 266

432   Notation

$\varphi(z)$	conformal mapping of int L onto $\mathbb{D}$ p. 57
Φ_1, Φ_2	quasiconformal extension of the restriction of Φ to Ω_1, respectively Ω_2 p. 30
$\Phi_k(z)$	Faber polynomials, $k = 0, 1, \ldots$ p. 37
Φ_u	normalized conformal mapping of ext L_u^* onto Δ p. 376
$\Pi_n(z)$	Faber standard generalized polynomial of order $n-1$ p. 37
ψ	inverse mapping of φ p. 57
Ψ_u	inverse mapping of Φ_u p. 376
$p_n^* = p_n^*(f)$	best uniform polynomial approximant of f p. 188
$\mathcal{P}_A$	set of metrics with $A(\rho) \leq 1$ p. 342
$\mathcal{P}_L(\Gamma)$	set of metrics with minimal ρ-length ≥ 1 p. 342
$\mathbb{P}_n$	set of polynomials of degree $\leq n$ p. 16
$P_n(x; a, b)$	Pollaczek polynomials p. 304
q.e.	quasi-everywhere p. 11
$Q(z_1, z_2, z_3, z_4)$	quadrilateral with successive points z_1, z_2, z_3, z_4 p. 363
$q_n(z)$	nth Fekete polynomial p. 17
$\mathbb{R}$	real numbers p. 2
$\rho_u(z)$	distance of z to L_u, $u > 0$ p. 403
$S(I)$	strip orthogonal to the interval I p. 86
$\mathrm{supp}(\sigma)$	support of the measure σ p. 10
$s_a = s_a(Q)$	minimal Euclidian length of the curve family Γ_a of all crosscuts that join the a-sides of Q p. 363
$\mathbb{T}$	unit circle p. 2
T_n	Chebyshev polynomial p. 17
$T_{n,p}$	monic polynomial of minimal L_τ^p-norm p. 261
$T_{n,p,Q}$	monic L^p-extremal polynomials on $\mathbb{R}$ p. 275

$t_n(E)$	Chebyshev numbers p. 17
$U^{\mu_E - \nu_p} = U^{\mu_E} - U^{\nu_p}$	difference of the logarithmic potentials of μ_E and ν_p p. 40
U^σ	logarithmic potential of σ pp. 10, 40
$U_{n,\tau}$	Chebyshev polynomial of the second kind of degree n with respect to τ p. 268
$V(E)$	minimal logarithmic energy with respect to $\mathcal{M}(E)$ p. 10
$W^1_{p,\text{loc}}$	Sobolev space p. 368
$w_Q(x)$	weight function on $\mathbb{R}$ p. 274
$\tilde{\zeta}$	p. 386
ζ_L	projection of ζ on L pp. 58, 69, 127
$\tilde{\zeta}_t$	p. 386
$\|\cdot\|_A$	supremum norm on A pp. 16, 96
$\|\cdot\|_\infty = \|\cdot\|_{\infty,\mathbb{R}}$	L^∞-norm on $\mathbb{R}$ p. 275
$\|\cdot\|_{p,A}$	L^p-norm on $A \subset \mathbb{R}$, $1 \leq p \leq \infty$ p. 275
$\|\cdot\|_p = \|\cdot\|_{p,\mathbb{R}}$	L^p-norm on $\mathbb{R}$ p. 275

Index

(+)-point 266
(−)-point 266
(σ, τ)-neighborhood 58
K-quasicircle 369
ρ-area 341
σ-neighborhood 69, 90

ACL 364
adjacent 322, 355
Ahlfors's theorem 27, 372
alternant 248
alternating sequence 257
analytic completion 3
approximant
    best L^p_τ- − 45, 254
    best uniform − 45
approximation
    best uniform − 45
    near-best − 193
arc 2
    K-quasiconformal − ... 27, 372
    Dini-smooth − 32
    locally rectifiable − 341

    piecewise Dini-smooth − ... 32
    quasiconformal − 27, 372
    separating − 348
    smooth − 2
averaging kernel 47

Belyi theorem 393
Bergman kernel function 216
Bernstein–Walsh lemma 16
Beurling's criterion 345
Bieberbach polynomial . 216, 220
boundary 2

capacity 11
Carathéodory extension
    theorem 20
Cauchy–Riemann equations ... 3
Chebyshev constant 18
Chebyshev numbers 17
chordal metric 1
Christoffel–Darboux formula
    on the unit circle 313
Christoffel–Darboux kernel . 298

Christoffel number 301
closure 2
coefficient of
    quasiconformality 28
comparison principle 25, 343
composition laws 25, 344
    integrated version of – 25
Condition A 274
conformal invariance 24
conformally equivalent 363
conformal radius 215
constructive characterization 381
continuum 15
convergence
    weak* – 14
    weak – 14
crosscut 2
curve 2
    K-quasiconformal – ... 27, 369
    Dini-smooth – 32
    locally rectifiable – 341
    piecewise Dini-smooth – ... 32
    quasiconformal – 27, 369
    separating – 348
    smooth – 2
curve family 24, 341

de la Vallée Poussin property 322
derivative
    L^p- – 368
    complex – 361
    directional – 362
    formal – 361
    Radon–Nikodym – ... 254, 298
dilatation
    maximal – 364
    quotient 362
direct theorem 381
discrepancy 96
    local – 95
discriminant 17
distance 2
    regularized – 408
distortion theorem 22

domain 2
    quasiconformal – 27, 369
    regular – 15
    simply connected – 2
    uniform – 370
Dzjadyk kernel 227, 330, 387

endpoints 2
energy 10
energy integral 178
equilibrium distribution 11
exterior 2
extremal length 342
extremal metric 342
extreme point 242

family
    of arcs 24
    of curves 24
Fekete points 17
Fekete point set 17, 226
function
    absolutely continuous – on
        lines 364
    admissible
        in the L-definition 342
    admissible – 24, 341
    Bergman kernel – 216
    harmonic – 3
    lower semicontinuous – 6
    of modulus of continuity
        type 396
    subharmonic – 6
    superharmonic – 6
    upper semicontinuous – 6
fundamental theorem of
    Frostman 11

generalized maximum
    principle 14
Green's formula 6
Green coordinates 413
Green function

level line of –.............. 15
Green line 90, 411, 412
Grunsky coefficients 37

harmonic conjugate........... 3
harmonic measure 8
Helly's selection theorem..... 14
homeomorphism............. 20
    sense-preserving –........ 362
Hurwitz's theorem........... 19

impression.................. 386
inequality
    Grunsky –.................. 38
    Nikol'skiĭ-type –.......... 255
    Remez –................... 46
    Rengel's –................ 364
infinite-finite-range inequality
    for the uniform norm..... 278
    for weighted L^p-norms.... 279
interior....................... 2
inverse theorem............. 381

Jackson kernel......... 155, 383
Jackson type kernel......... 383
Jacobian 362
Jordan arc................... 2
    closed – 2
    open –..................... 2
Jordan curve.................. 2
Jordan domain................ 2
Joukowski map 13, 30

Koebe's one-quarter theorem 22
Koebe's distortion theorem .. 22

Laplace equation............. 3
Laplacian..................... 3
Laplacian operator........... 3
level line............... 15, 375
line segment 2
local discrepancy 108
logarithmic potential 10

discretizing a –........... 281

mapping
    K-quasiconformal – . 364, 365
    antiquasiconformal –..... 370
    quasiconformal –......... 364
        $\mathcal{D}$-property of a –...... 368
        analytic definition of a – 365
        geometric definition of a
        –..................... 364
    regular K-quasiconformal
    –....................... 362
    regular quasiconformal – . 362
maximum principle........... 3
mean-value property.......... 3
measure
    a-value counting –........ 208
    balayage –........... 214, 313
    discrepancy of a –......... 96
    equilibrium –.............. 11
    extremal – 277
    local discrepancy of a –... 108
    normalized counting –..... 40
    support of a –............. 10
    zero counting –............ 40
metric.................. 24, 341
Mhaskar–Rakhmanov–Saff
    number.................... 276
minimum property 298
module 24, 342, 363
    A-definition of a –........ 342
    L-definition of a –........ 342
    quasi-invariance of a –..... 28
    reduced – 357
modulus..................... 174
modulus of continuity 393
MRS number............... 276

neighborhood................. 2
net change................... 79
norm
    L^p- – on $A \in \mathbb{R}$........... 275
    L^p_τ- –................. 45, 254
    Chebyshev –.............. 16

supremum –................ 16
uniform –................. 16

Poisson
   integral..................... 4
   kernel...................... 4
polynomial
   L^p-extremal –............. 275
   Bieberbach –.............. 216
   Chebyshev –............... 17
   Chebyshev – of second
     kind.................... 268
   Faber –.................... 37
   Faber standard generalized
     –......................... 37
   Fekete –................... 17
   monic..................... 16
   Pollaczek –............... 304
potential
   inner bound of a –....... 150
   outer bound of a –... 130, 150

q.e............................11
quadrilateral............... 362
   a-sides of a –............. 363
   conjugate –............... 363
quasi-everywhere............ 11
quasicircle.................. 369
quasiconformal reflection... 370
quasidisk.............. 27, 369

regular boundary............ 15
Riemann mapping........... 21

Riemann mapping theorem.. 20
Riemann sphere.............. 1
Robin constant
   modified –................ 277
Rouché's theorem............ 19

Schwarz's formula............. 4
set
   admissible –............... 12
   linearly locally connected
     –......................... 369
sign change................. 266
sign function............... 266
   equivalent –.............. 266
   orthogonal –............. 266
Sobolev space.............. 368
Szegő class................. 298

transfinite diameter.......... 17
Tsuji points................ 318

Vandermonde determinant... 17
Vandermonde expression.... 146
Vitali's theorem............. 20
VP-property................ 322

weight function
   Freud-type –.............. 296
   generalized Jacobi –...... 262
   generalized Pollaczek –... 305
   GP –..................... 305
   on a domain.............. 208